Mexican Mathematicians in the World

Trends and Recent Contributions

Contemporary Mathematics

775

Aportaciones Matemáticas
Sociedad Matemática Mexicana

Mexican Mathematicians in the World

Trends and Recent Contributions

IV Meeting
Reunión de Matemáticos Mexicanos en el Mundo
June 10–15, 2018
Casa Matemática Oaxaca
Oaxaca, Mexico

Fernando Galaz-García
Cecilia González-Tokman
Juan Carlos Pardo Millán
Editors

American Mathematical Society

2020 *Mathematics Subject Classification.* Primary 53Cxx, 83Cxx, 46Lxx, 37Axx, 55Pxx, 35Cxx, 47Axx, 17Bxx, 11Fxx, 22Exx.

For additional information and updates on this book, visit
www.ams.org/bookpages/conm-775

Library of Congress Cataloging-in-Publication Data

Names: Galaz-García, Fernando, editor. | Reunión de Matemáticos Mexicanos en el Mundo (4th : 2018 : Oaxaca, Mexico)
Title: Mexican mathematicians in the world : trends and recent contributions / Fernando Galaz-García, Cecilia González-Tokman, Juan Carlos Pardo Millán, editors.
Description: Providence : American Mathematical Society, [2021] | Series: Contemporary mathematics, 0271-4132 ; volume 775 | Aportaciones matematicas / Sociedad Matemática Mexicana | This volume contains the proceedings of the IV Meeting of Mexican Mathematicians Abroad (IV Reunión de Matemáticos Mexicanos en el Mundo), held at Casa Matemática Oaxaca (CMO) in Oaxaca, Mexico, on June 10-15, 2018–Preface. | Includes bibliographical references.
Identifiers: LCCN 2021019776 | ISBN 9781470465360 (paperback) | 9781470467289 (ebook)
Subjects: LCSH: Mathematics–Mexico–Congresses. | Mathematicians–Mexico–Congresses. | Women mathematicians–Mexico–Congresses. | AMS: Differential geometry – Global differential geometry. | Relativity and gravitational theory – General relativity. | Functional analysis – Selfadjoint operator algebras. | Dynamical systems and ergodic theory. | Algebraic topology – Homotopy theory. | Partial differential equations. | Operator theory – General theory of linear operators. | Nonassociative rings and algebras – Lie algebras and Lie superalgebras. | Number theory – Discontinuous groups and automorphic forms. | Topological groups, Lie groups.
Classification: LCC QA27.M49 M49 2021 | DDC 510.972–dc23
LC record available at https://lccn.loc.gov/2021019776

---

10 9 8 7 6 5 4 3 2 1 26 25 24 23 22 21

Contents

Preface

This volume contains the proceedings of the *IV Meeting of Mexican Mathematicians Abroad* (*IV Reunión de Matemáticos Mexicanos en el Mundo*), held at Casa Matemática Oaxaca (CMO) in Oaxaca, Mexico, on June 10–15, 2018. This was the fourth in a series of ongoing biannual meetings connecting Mexican mathematicians residing abroad with their colleagues in Mexico. These events aim, on the one hand, at showcasing the research of Mexican mathematicians based abroad and, on the other hand, at strengthening academic connections among participants, Mexican institutions, and the broader mathematical community.

These conferences have been well received by the Mexican mathematical community, having had a visible positive impact on many levels. This can be readily seen when considering the number of new collaborations, international grant proposals, organization and participation in specialized events and, perhaps most importantly, the (re)incorporation of various speakers to Mexican academic institutions.

On this occasion, the meeting consisted of five days of activities, including 18 scientific talks, three poster sessions, three panels and discussion fora, and a public outreach talk, all supplemented with plenty of interaction time during meals and coffee breaks. This arrangement allowed for intense interactions and networking among the 53 workshop participants, who came from more than 10 countries and 10 Mexican states.

As in previous meetings, there were also opportunities for dialogue and analysis of issues relevant to the wider mathematical community and to colleagues working abroad. There was a round-table discussion on mobility, exchange, and opportunities for returning to Mexico (*Movilidad, Intercambio, Reincorporación*), featuring established members of the Mexican mathematical and scientific community: Alejandro Adem (University of British Columbia University), Daniel Juan (Centro de Ciencias Matemáticas, UNAM), Víctor Rivero (Centro de Investigación en Matemáticas), Gelasio Salazar (Universidad Autónoma de SLP and former President of the Mexican Mathematical Society), and Julia Tagüeña (CONACYT). In addition to this, there was an open forum led by Humberto Gutiérrez (Universidad de Guadalajara) and Silvia Reyes (Universidad Tecnológica de la Mixteca), heads of their departments of mathematics, and early-career researchers who recently joined Mexican institutions: Ana Rechtman (Instituto de Matemáticas, UNAM), Felipe García Ramos (CONACYT - Universidad Autónoma de San Luis Potosí), Alfredo Nájera (CONACYT - UNAM, Oaxaca), and Gerardo Hernández (UNAM, Juriquilla). Opportunities for graduate studies were discussed in a third forum.

The scientific talks were recorded using BIRS video facilities and are publicly available at the workshop website

`https://www.birs.ca/events/2018/5-day-workshops/18w5142.`

The organizing committee was composed of a mix of mathematicians based in Mexico and abroad:

- Luz Angélica Caudillo (University of British Columbia Canada),
- Felipe García (Universidad Autónoma de San Luis Potosí, CONACYT Mexico),
- Cecilia González-Tokman (The University of Queensland, Australia),
- Xavier Gómez-Mont (CIMAT, Mexico),
- Rita Jiménez Rolland (IMATE Oaxaca, UNAM, Mexico),
- César A. Lozano (UNAM, Unidad Oaxaca, CONACYT, Mexico),
- César Maldonado (Instituto Potosino de Investigación Científica y Tecnológica, Mexico),
- Juan Pablo Maldonado (Czech Technical University, Czech Republic),
- Ana Rechtman Bulajich (Université de Strasbourg, France, and IMATE, UNAM, Mexico),
- José Antonio Seade Kuri (IMATE, UNAM, Mexico).

All the affiliations in the preceding list, as well as the affiliations in the lists of speakers and participants, are those at the time the meeting took place.

The conference was made possible thanks to the financial support of Casa Matemática Oaxaca, CONACYT, FORDECyT, Sociedad Matemática Mexicana, Instituto de Física de la UASLP, Instituto de Matemáticas de la UNAM, and the logistic support of Claudia Arias (CMO), Karla Soto (UNAM), and América Carrasco (SMM). We thank the conference speakers and the participants for their commitment to making the meeting a success, the authors and referees who contributed their time and work to these proceedings, and the editorial committees and staff of Aportaciones Matemáticas and Contemporary Mathematics for their support in preparing this volume.

Fernando Galaz-García
Durham University

Cecilia González-Tokman
The University of Queensland

Juan Carlos Pardo Millán
CIMAT

Organizers at the Fourth Meeting of Mcxican Mathematicians Abroad, CMO, June 10–15, 2018. (Photo by CMO)

Poster session at the Fourth Meeting of Mexican Mathematicians Abroad, CMO, June 10–15, 2018. (Photo by CMO)

Participants at the Fourth Meeting of Mexican Mathematicians Abroad, CMO, June 10–15, 2018. (Photo by CMO)

List of Talks and Speakers

- On the non-linear prescribed k-curvature problem on hypersurfaces of hyperbolic and anti-de sitter space
  Daniel Ballesteros Chávez, Durham University (UK)

- An example of the Deligne–Mumford compactification
  Araceli Bonifant, University of Rhode Island (USA)

- On the Mumford–Tate conjecture for Abelian varieties
  Victoria Cantoral Farfán, International Centre for Theoretical Physics (Italy)

- A non-commutative topological dimension
  Jorge Castillejos, Katholieke Universiteit Leuven (Belgium)

- Isoperimetric and Sobolev inequalities on magnetic graphs
  Javier Chávez Domínguez, University of Oklahoma (USA)

- Positive Ricci Curvature and Actions of High Dimensional Tori
  Diego Corro Tapia, Karlsruher Institut für Technology (Germany)

- Metastable convergence of ergodic averages: the continuous logic viewpoint
  Eduardo Dueñez, Spelman College (USA)

- Einstein metrics on 4-manifolds
  Ixchel Dzohara Gutiérrez Rodríguez, Universidad de Santiago de Compostela (Spain)

- Identifying unique observations in storm super-resolved microscopic images via a spatio-temporal model
  Alejandra Donaji Herrera Reyes, University of British Columbia (Canada)

- Dressing-up surfaces
  Alfredo Hubard, Université Marne-La Vallée (France)

- Singularly Perturbed Ordinary Differential Equations
  Hildeberto Jardón, Technical University Munich (Germany)

- Some connections between functional and volumetric inequalities in convex geometry
  Carlos Hugo Jiménez Gómez, PUC-Rio (Brasil)

- Modelling the effect of temperature on sleep quality
  Alicia Prieto Langarica,Youngstown University (USA)

- Determinism and Chaos (Public Outreach Talk)
  Ana Rechtman Bulajich, Université de Strasbourg (France) and UNAM (Mexico)

- The hybridizable discontinuous Galerkin method: An application to magnetic equilibrium in fusion reactors
  Tonatiuh Sánchez-Vizuet, Courant Institute of Mathematical Sciences (USA)

- Synthetic curvature and isometries
  Jaime Santos Rodríguez, Universidad Autónoma de Madrid (Spain)

- The minimum quadratic non-residue and related problems
  Enrique Treviño, Lake Forest College (USA)

- The set-intersection problem and its applications
  Paola Vera Licona, University of Connecticut Health Center (USA)

- Microlocal and semiclassical analysis
  Alejandro Uribe, University of Michigan, Ann Arbor (USA)

List of Participants

Alejandro Adem
University of British Columbia, Canada

Alma Rosario Arévalo Loyola
UNAM, Mexico

Daniel Ballesteros Chávez
Durham University, UK

Araceli Bonifant
University of Rhode Island, USA

Victoria Cantoral Farfán
ICTP, Italy

Herón Cárdenas Cruz
Universidad Autónoma de Chiapas, Mexico

Jorge Castillejos
Katholieke Universiteit Leuven, Belgium

Luz Angélica Caudillo Mata
University of British Columbia, Canada

Jonathan Chávez Casillas
University of Rhode Island, USA

Javier Chávez Domínguez
University of Oklahoma, USA

Diego Corro Tapia
Karlsruher Insitut für Technologie, Germany

Eduardo Dueñez
Spelman College, USA

Manuel Alejandro García Acosta
UNAM, Mexico

Felipe García-Ramos
Universidad Autónoma de San Luis Potosí, Mexico

Tania Raquel Garibay Valladolid
Universidad Michoacana de San Nicolás de Hidalgo, Mexico

Cecilia González-Tokman
The University of Queensland, Australia

Humberto Gutiérrez Pulido
Universidad de Guadalajara, Mexico

Ixchel Dzohara Gutiérrez Rodríguez
Universidade de Santiago de Compostela, Spain

Gerardo Hernández Dueñas
UNAM, Juriquilla, Mexico

Alejandra Donaji Herrera Reyes
University of British Columbia, Canada

Alfredo Hubard
Université Marne la Vallée, Paris Est, France

Delio Jaramillo Vélez
CIMAT, Mexico

Hildeberto Jardón Kojakhmetov
Technische Universität München, Germany

Carlos Hugo Jiménez Gómez
PUC-Rio, Brazil

Rita Jiménez Rolland
UNAM, Mexico

Daniel Juan Pineda
Centro de Ciencias Matemáticas, UNAM, Morelia, Mexico

Porfirio Leandro León Álvarez
UNAM, Mexico

Luis Lomelí
Pontificia Universidad Católica de Valparaíso, Chile

César Lozano Huerta
CONACYT-UNAM, Oaxaca, Mexico

César Maldonado
Instituto Potosino de Investigación Científica y Tecnológica, Mexico

Miguel Maldonado
Universidad Autónoma de Zacatecas, Mexico

Gari Yamel Peralta Alvarez
Universidad de Colima, Mexico

Miguel Angel Prado Godoy
Universidad de Guanajuato, Mexico

Alicia Prieto Langarica
Youngstown State University, USA

Andrea Quintanilla Carranza
Facultad de Ciencias, UNAM, Mexico

Ana Rechtman Bulajich
Université de Strasbourg, France and UNAM, Mexico

Silvia Reyes
Universidad Tecnológica de la Mixteca, Mexico

Mónica Andrea Reyes Quiroz
Universidad Autónoma del Estado de México, Mexico

Víctor Manuel Rivero Mercado
CIMAT, Mexico

Diego Rodríguez Guzmán
Universidad Autónoma de Aguascalientes, Mexico

Edgardo Roldán Pensado
Centro de Ciencias Matemáticas, UNAM, Morelia, Mexico

Gelasio Salazar
Universidad Autónoma de San Luis Potosí, Mexico

Fernando Saldaña García
CIMAT, Mexico

Tonatiuh Sánchez-Vizuet
New York University, USA

Jaime Santos Rodríguez
Universidad Autónoma de Madrid, Spain

Arelis Serrato Martínez
Facultad de Ciencias, UNAM, Mexico

Alina Sotolongo Aguiar
CIMAT, Mexico

Julia Tagüeña
CONACYT, Mexico

Enrique Treviño López
Lake Forest College, USA

Alejandro Uribe
University of Michigan, Ann Arbor, USA

Eduardo Iván Velázquez Richards
UNAM, Mexico

Paola Vera Licona
University of Connecticut Health Center, USA

Contemporary Mathematics
Volume **775**, 2021
https://doi.org/10.1090/conm/775/15586

A survey on extensions of Riemannian manifolds and Bartnik mass estimates

Armando J. Cabrera Pacheco and Carla Cederbaum

ABSTRACT. Mantoulidis and Schoen developed a novel technique to handcraft asymptotically flat extensions of Riemannian manifolds $(\Sigma \cong \mathbb{S}^2, g)$, with g satisfying $\lambda_1 := \lambda_1(-\Delta_g + K(g)) > 0$, where λ_1 is the first eigenvalue of the operator $-\Delta_g + K(g)$ and $K(g)$ is the Gaussian curvature of g, with control on the ADM mass of the extension. Remarkably, this procedure allowed them to compute the Bartnik mass in this so-called minimal case; the Bartnik mass is a notion of quasi-local mass in General Relativity which is very challenging to compute. In this survey, we describe the Mantoulidis–Schoen construction, its impact and influence in subsequent research related to Bartnik mass estimates when the minimality assumption is dropped, and its adaptation to other settings of interest in General Relativity.

1. Mathematical relativity

In the context of Mathematical Relativity, a Riemannian manifold (M^3, γ) with non-negative scalar curvature $R(\gamma)$ represents a "time-symmetric time-slice" of a spacetime satisfying the "dominant energy condition". From the point of view of the Cauchy problem in General Relativity, we can consider (M^3, γ) as a time-symmetric *initial data set* for the Einstein Equations, and the dominant energy condition $R(\gamma) \geq 0$ as a compatibility condition that will allow this manifold to evolve into a physically relevant spacetime containing (M^3, γ) as a submanifold. The initial value problem in General Relativity has been widely studied and we refer the interested reader to the books by Choquet-Bruhat [**18**] and Ringström [**53**].

In this survey, we are mostly interested in *asymptotically flat Riemannian manifolds.* Consider a gravitating system in which all the matter is contained inside a compact region K. Then, as we move away from K, the gravitational field will get weaker, and if we go far enough away, we will increasingly feel as if there was no gravitation exerted on us anymore. For a time-symmetric initial data set (M^3, γ) modeling this gravitating system, this means that as we move away from $K \subset M^3$, the metric γ approaches the Euclidean metric δ. More precisely, a Riemannian manifold (M^3, γ) is said to be *asymptotically flat* if $M^3 \setminus K$ is diffeomorphic to $\mathbb{R}^3 \setminus \overline{B_1(0)}$ via a map Φ which defines a set of *coordinates at infinity*, and in these

2020 *Mathematics Subject Classification.* Primary 53C21.

The authors are grateful to the Carl Zeiss Foundation for its generous support. Work of the second author was supported by the Institutional Strategy of the University of Tübingen (Deutsche Forschungsgemeinschaft, ZUK 63).

coordinates the metric takes the form $\gamma_{ij} = \delta_{ij} + \mathcal{O}_2(|x|^{-p})$ with $p > \frac{1}{2}$. In addition, we require the scalar curvature $R(\gamma)$ of γ to be integrable over M^3.

For example, consider the famous *Schwarzschild spacetime of mass* $m > 0$, which models the gravitational field surrounding a static, non-rotating, rotationally symmetric black hole or celestial body of mass m in vacuum. The spacetime manifold can be written as $\overline{M}^4 = \mathbb{R} \times M^3 = \mathbb{R} \times (2m, \infty) \times \mathbb{S}^2 \ni (t, r, x)$, with the Lorentzian (spacetime) metric defined by

$$\overline{\gamma}_m = -\left(1 - \frac{2m}{r}\right) dt^2 + \left(1 - \frac{2m}{r}\right)^{-1} dr^2 + r^2 g_*,$$

where g_* denotes the standard round metric on $\mathbb{S}^2$. The Riemannian manifold M^3 defined by $\{t = 0\}$ is called the *(spatial) Schwarzschild manifold of mass* m, and clearly, it consists of $M^3 = (2m, \infty) \times \mathbb{S}^2$ with the Riemannian metric

$$\gamma_m = \left(1 - \frac{2m}{r}\right)^{-1} dr^2 + r^2 g_*. \tag{1.1}$$

It can be checked that $R(\gamma_m) \equiv 0$, and evidently γ_m approaches the Euclidean metric δ as $r \longrightarrow \infty$.

Remarkably, by imposing the decay conditions on the metric described above, an asymptotically flat manifold (M^3, γ) behaves in a similar way to a Schwarzschild manifold and we can "detect" the total mass of the gravitating system it models — or its *total mass* for short — from its asymptotic behavior. This notion of total mass is called ADM mass (defined by Arnowitt, Deser, and Misner in [**5**]) and it is denoted by $m_{\text{ADM}}(M^3, \gamma)$; whenever there is no risk of confusion we will drop the reference to the underlying manifold M^3. As the reader will discover, for this survey we do not need to use the explicit definition of the ADM mass, rather we will only need to keep in mind that it is measured from the asymptotic behavior of (M^3, γ) and that a Schwarzschild manifold of mass m has, in fact, $m_{\text{ADM}}(\gamma_m) = m$. We avoid any digress towards technical definitions and convergence issues by referring the interested reader to [**5**], and to the works of Bartnik [**6**] and Chruściel [**20**].

With a notion of total mass at hand, the next natural step is to determine if, under some reasonable physical assumptions, it is non-negative. This is known as the Positive Mass Theorem and was proven independently by Schoen and Yau [**56**] using minimal surface theory and by Witten [**66**] using spinors.

THEOREM. *(Positive Mass Theorem) Let (M^3, γ) be a complete, asymptotically flat manifold with $R(\gamma) \geq 0$. Then $m_{\text{ADM}}(M^3, \gamma) \geq 0$, and equality holds if and only if (M^3, γ) is isometric to the Euclidean space $(\mathbb{R}^3, \delta)$.*

The important Riemannian Penrose Inequality states that the mass of an asymptotically flat Riemannian manifold is bounded below in terms of the area of 2-surfaces representing black holes. In the time-symmetric case (that is, for Riemannian manifolds), black holes are represented by compact minimal surfaces, i.e., compact surfaces with vanishing mean curvature, which are called *horizons*. A horizon is said to be *outer-minimizing* if it minimizes area among all surfaces that enclose it. As an example, consider the surface $\{r = 2m\}$ in a Schwarzschild manifold of mass $m > 0$ — but note that, as it is written, $\{r = 2m\}$ is in fact not part of the Schwarzschild manifold, and if we naively try to include it as an inner boundary, γ_m would appear to degenerate there. Nevertheless, one can perform a change of coordinates that allows a smooth extension of γ_m to the boundary (see Section 3.2).

THEOREM (Riemannian Penrose Inequality). *Let (M^3, γ) be an asymptotically flat manifold with $R(\gamma) \geq 0$ and such that its boundary ∂M is an outer-minimizing horizon. Then*

$$m_{\text{ADM}}(M^3, \gamma) \geq \sqrt{\frac{|\partial M|_\gamma}{16\pi}}, \tag{1.2}$$

where $|\partial M|_\gamma$ represents the area of the boundary of M with respect to γ. Moreover, equality holds if and only if (M^3, γ) is isometric to a Schwarzschild manifold of positive mass.

The Riemannian Penrose Inequality has been proven by Huisken and Ilmanen [**33**] using a weak version of inverse mean curvature flow (IMCF) when the horizon has a single connected component, and by Bray [**9**] using conformal flow and thereby allowing for multiple connected components. There is plenty of literature available about the Riemannian Penrose Inequality and the more general (spacetime) Penrose Inequality; we particularly recommend to consult Mars' survey article [**45**] which also covers other geometric inequalities from General Relativity that will appear in this survey.

An interesting question in Mathematical Relativity is whether it is possible to define a local notion of mass contained in a small region $\Omega \subset M^3$. For various conceptual as well as physical reasons, it turns out to be more convenient in many cases to work with localized notions of mass defined in terms of the geometry of $\Sigma := \partial\Omega$. Such notions of mass are referred to as *quasi-local mass* notions. We now define a highly important example of such a notion.

DEFINITION 1.1. Let Σ be a surface in a Riemannian manifold (M^3, γ) with induced metric g and mean curvature H. The *Hawking mass* of the triplet (Σ, g, H) is defined as

$$\mathfrak{m}_{\text{H}}(\Sigma, g, H) := \sqrt{\frac{|\Sigma|_g}{16\pi}} \left(1 - \frac{1}{16\pi} \int_\Sigma H^2 \, d\sigma \right),$$

where $|\Sigma|_g$ is the area of Σ with respect to the metric g and $d\sigma$ denotes the area form on Σ.

Note that, using the notion of Hawking mass, we can simply rewrite (1.2) as

$$m_{\text{ADM}}(M^3, \gamma) \geq \mathfrak{m}_{\text{H}}(\partial M, \gamma|_{\partial M}, H \equiv 0). \tag{1.3}$$

Another important notion of quasi-local mass, which will be one of the main subjects in this survey, is the *Bartnik mass*, defined by Bartnik in [**7**].

DEFINITION 1.2. A triple $(\Sigma \cong \mathbb{S}^2, g, H)$, where g is a Riemannian metric and $H \geq 0$ is a smooth function on Σ is called *Bartnik data*. Bartnik data $(\Sigma \cong \mathbb{S}^2, g, H)$ with $H \equiv 0$ are called *minimal Bartnik data*.

For given Bartnik data $(\Sigma \cong \mathbb{S}^2, g, H)$, we consider the set of admissible extensions $\mathcal{A}$, consisting of all asymptotically flat manifolds (M^3, γ) such that $R(\gamma) \geq 0$, the boundary ∂M is outer-minimizing, and $(\partial M, \gamma|_{\partial M}) \cong (\Sigma, g)$ with mean curvature H.

DEFINITION 1.3. The *Bartnik mass* of Bartnik data $(\Sigma \cong \mathbb{S}^2, g, H)$ is defined as

$$\mathfrak{m}_{\mathrm{B}}(\Sigma \cong \mathbb{S}^2, g, H) := \inf\{m_{\mathrm{ADM}}(M^3,\gamma) \,|\, (M^3,\gamma) \in \mathcal{A}\}. \tag{1.4}$$

Evidently, by definition, the Bartnik mass is extremely complicated to compute. However, notice that when $H = 0$, an admissible extension satisfies the conditions of the Riemannian Penrose Inequality (Theorem 1), and hence, the Hawking mass provides a *lower bound* for the Bartnik mass. The original derivation leading to the notion of Bartnik mass does in fact not readily extend to the case of minimal Bartnik data. We suppress this delicate debate here.

By Huisken and Ilmanen's proof of the Riemannian Penrose Inequality [**33**], we know that

$$\mathfrak{m}_{\mathrm{H}}(\Sigma \cong \mathbb{S}^2, g, H) \leq m_{\mathrm{ADM}}(M^3,\gamma) \tag{1.5}$$

for any admissible extension (M^3,γ) of given Bartnik data $(\Sigma \cong \mathbb{S}^2, g, H)$, again with equality if and only if the Bartnik data $(\Sigma \cong \mathbb{S}^2, g, H)$ suitably embed into a Schwarzschild manifold of positive mass. This allows to compute the Bartnik mass in the case of spherical symmetry — i.e., in the case of Bartnik data that isometrically embed as a centered round sphere into a Schwarzschild manifold of positive mass m; in that case, the Bartnik and Hawking mass both coincide with the mass parameter m.

From Definition 1.3, it is apparent that computing the Bartnik mass of given Bartnik data is a challenging task. However, also from its definition, we see that any extension of given Bartnik data will provide an *upper bound* for the Bartnik mass. This in part motivated Bartnik himself to study asymptotically flat "quasi-spherical" extensions of intrinsically round Riemannian 2-spheres (Σ, g) in [**8**]. Other extensions using similar techniques to those in [**8**] are the works of Smith and Weinstein [**61,62**], Shi and Tam [**58**], Smith [**60**] and Lin [**37**]. Building on the work of Lin in [**37**], Lin and Sormani constructed asymptotically flat extensions to estimate the Bartnik mass in [**38**].

In 2016, Mantoulidis and Schoen [**43**] first succeeded in computing the Bartnik mass of non-spherically symmetric Bartnik data. More concretely, they were able to compute the Bartnik mass of minimal Bartnik data $(\Sigma \cong \mathbb{S}^2, g, H \equiv 0)$ in the case when g satisfies $\lambda_1 := \lambda_1(-\Delta_g + K(g)) > 0$, where λ_1 denotes the first eigenvalue of the operator $-\Delta_g + K(g)$, and $K(g)$ is the Gaussian curvature of g. To do so, they constructed asymptotically flat extensions of minimal Bartnik data $(\Sigma \cong \mathbb{S}^2, g, H \equiv 0)$ in such a way that they can control their ADM masses, ultimately leading them to prove that $\mathfrak{m}_{\mathrm{B}}(\Sigma \cong \mathbb{S}^2, g, H \equiv 0) = \mathfrak{m}_{\mathrm{H}}(\Sigma \cong \mathbb{S}^2, g, H \equiv 0)$. Their work has inspired related constructions, among other results, that provide estimates for Bartnik data with $H \neq 0$. The Mantoulidis–Schoen construction together with its adaptations and modifications, specifically those useful to obtain Bartnik mass estimates, will be the main topic of this survey.

Very recently, building on work by Shi and Tam [**58**] and by Miao [**48**], Miao and Xie constructed extensions of vanishing scalar curvature for Bartnik data of positive Gaussian curvature in [**51**]. In particular, for the minimal case they recover that $\mathfrak{m}_{\mathrm{B}}(\Sigma \cong \mathbb{S}^2, g, H \equiv 0) = \mathfrak{m}_{\mathrm{H}}(\Sigma \cong \mathbb{S}^2, g, H \equiv 0)$. See [**51**] for details.

For the sake of completeness, let us note that it is common in the literature, for very good reasons, to request that admissible extensions for given Bartnik data satisfy that (M^3,γ) contains no other minimal surfaces (homologous to ∂M), except perhaps for ∂M — instead of requesting the outer-minimizing condition on the boundary that we used above, see e.g. [**9, 33**], and the work of Bray and Chruściel [**10**]. In this survey, it will not make any difference which condition is chosen since all the extensions considered here will satisfy both.

We further remark that there are in fact also other definitions of Bartnik mass in the literature. This raises natural questions about the relations or equivalences among them, which will not be discussed here. We refer the interested reader to the works of Jauregui [**35**] and McCormick [**46**], which specifically address these questions of equivalence of various definitions of Bartnik mass. For more information on Bartnik mass, see also the review [**28**] by Galloway, Miao, and Schoen.

The structure of this survey is as follows. In Section 2, we intuitively explain the Mantoulidis–Schoen construction and describe works related to Bartnik mass estimates in a historical order. In Section 3, we describe in detail key steps used to construct these types of extensions and how to obtain Bartnik mass estimates from them. In Section 4, we describe in more detail the specific Bartnik mass estimates obtained with this method. Finally, in Section 5, we discuss the current state of the art and other related problems.

Some useful considerations. In what follows, we will give a brief description of higher dimensional Riemannian manifolds in the context of Mathematical Relativity, of asymptotically hyperbolic Riemannian manifolds, and of charged Riemannian manifolds arising as time-symmetric initial data sets for the Einstein–Maxwell Equations. We recommend that the reader skips this part upon first reading and returns to it as it becomes necessary.

1.1. Higher dimensions. A Riemannian manifold (M^{n+1},γ) of dimension $n+1$ is said to be *asymptotically flat* if, as before, outside a compact set, it is diffeomorphic to $\mathbb{R}^{n+1}\setminus\overline{B_1(0)}$. In this coordinate chart at infinity, we ask that $\gamma_{ij}=\delta_{ij}+\mathcal{O}_2(|x|^{-p})$ for $p>\frac{n-2}{2}$. In addition, $R(\gamma)$ must be integrable. As in the 3-dimensional case, by imposing these decay conditions on the metric, the total mass $m_{\mathrm{ADM}}(M^{n+1},\gamma)$ is well-defined. The Positive Mass Theorem also holds in higher dimensions. It was proven by Schoen and Yau for $2\le n<7$ [**55**] (the case $n=7$ follows from the work of Smale [**59**]) and for spin manifolds in general dimension by Witten [**66**]. Recently, Schoen and Yau established the Positive Mass Theorem in all dimensions [**57**]. The Riemannian Penrose Inequality in higher dimensions, proven by Bray and Lee [**11**], states that for an asymptotically flat manifold (M^{n+1},γ) with $R(\gamma)\ge 0$ (for $2\le n<7$), such that its boundary ∂M is an outer-minimizing horizon, we have

$$m_{\mathrm{ADM}}(M^{n+1},\gamma)\ge\frac{1}{2}\left(\frac{|\partial M|_\gamma}{w_n}\right)^{\frac{n-1}{n}},\tag{1.6}$$

where ω_n denotes the volume of the standard sphere $\mathbb{S}^n$ and $|\partial M|_\gamma$ represents the area of the boundary of M with respect to γ as before. Equality holds if and only if (M^{n+1},γ) is isometric to a suitably defined (spatial) Schwarzschild manifold of dimension $n+1$ of positive mass.

1.2. Asymptotically hyperbolic manifolds. Roughly speaking, a Riemannian manifold (M^3,γ) is said to be *asymptotically hyperbolic*, if, outside a compact set, it is diffeomorphic to $\mathbb{H}^3\setminus\overline{B_1^{\mathbb{H}}(0)}$, and the metric γ approaches the metric

$\gamma_H = (1+r^2)^{-1}dr^2 + r^2 g_*$ of the hyperbolic space $\mathbb{H}^3$ as $r \longrightarrow \infty$. In this setting, the notion of total mass is more subtle and has been defined by X. Wang [**65**] and by Chruściel and Herzlich [**23**]; here, it will be denoted it by $m_{\mathrm{AH}}(M^3, \gamma)$. It vanishes on hyperbolic space, $m_{\mathrm{AH}}(\mathbb{H}^3, \gamma_H) = 0$. The dominant energy condition for a time-symmetric initial data set can also be derived in this context — after an appropriate rescaling of the so-called cosmological constant in the Einstein Equations — and amounts to $R(\gamma) \geq -6$. The asymptotically hyperbolic Positive Mass Theorem (with rigidity $m_{\mathrm{AH}}(M^3, \gamma) = 0$ if and only if (M^3, γ) is isometric to hyperbolic space) holds for spin manifolds in all dimensions as was shown under restrictions on the asymptotics by X. Wang [**65**] and without such restrictions by Chruściel and Herzlich [**23**], and under some restrictions on the geometry at infinity as was shown by Andersson, Cai, and Galloway [**4**], by Chruściel, Galloway, Nguyen, and Paetz [**22**], and by Chruściel and Delay [**21**]. Sakovich [**54**] has announced a proof of the 3-dimensional asymptotically hyperbolic Positive Mass Theorem in full generality (with extra asymptotic assumptions needed for asserting rigidity). The rigidity case has been settled under fully general asymptotic assumptions by Huang, Jang and Martin [**32**].

The conjectured asymptotically hyperbolic Riemannian Penrose Inequality states that if (M^3, γ) is an asymptotically hyperbolic manifold with boundary ∂M which consists of an outer-minimizing horizon, then

$$m_{\mathrm{AH}}(M^3, \gamma) \geq \sqrt{\frac{|\partial M|_\gamma}{16\pi}} \left(1 + \frac{|\partial M|_\gamma}{4\pi}\right), \tag{1.7}$$

where $|\partial M|_\gamma$ represents the area of the boundary of M with respect to γ as before.

Equality holds if and only if (M^{n+1}, γ) is isometric to an AdS-Schwarzschild manifold of positive mass: The *(spatial) AdS-Schwarzschild manifold of mass* $m > 0$ can be described as the manifold $(r_+, \infty) \times \mathbb{S}^2$, with the Riemannian metric

$$\gamma_{m,\mathrm{AdS}} = \left(1 + r^2 - \frac{2m}{r}\right)^{-1} dr^2 + r^2 g_*,$$

where r_+ is the largest positive zero of $1 + r^2 - \frac{2m}{r}$ and g_* again denotes the standard round metric on $\mathbb{S}^2$. This inequality has been proven in special cases by Dahl, Giquaud and Sakovich in [**24**], complemented by the work of de Lima and Girão in [**26**]; and by Ambrozio in [**2**].

The *hyperbolic Hawking mass* of a closed surface (Σ, g) inside M^3 with mean curvature H is defined as

$$\mathfrak{m}_{\mathrm{H}}^{\mathrm{AH}}(\Sigma, g, H) := \sqrt{\frac{|\Sigma|_g}{16\pi}} \left(1 - \frac{1}{16\pi} \int_\Sigma (H^2 - 4)\, d\sigma\right).$$

In [**15**], mimicking the definition of Bartnik mass in the asymptotically flat case, McCormick and the authors defined a notion of *asymptotically hyperbolic Bartnik mass.* Given Bartnik data $(\Sigma \cong \mathbb{S}^2, g, H)$, we can define the set of admissible extensions $\mathcal{A}$ as the set of asymptotically hyperbolic manifolds (M^3, γ) with $R(\gamma) \geq -6$ such that their boundary ∂M is outer-minimizing and isometric to (Σ, g) with mean curvature H (cf. [**15**]). The *asymptotically hyperbolic Bartnik mass* is given by

$$\mathfrak{m}_{\mathrm{B}}^{\mathrm{AH}}(\Sigma, g, H) := \inf\{m_{\mathrm{AH}}(M^3, \gamma) \,|\, (M^3, \gamma) \in \mathcal{A}\}.$$

Clearly, if the asymptotically hyperbolic Riemannian Penrose conjecture is proven to be true, the hyperbolic Hawking mass would provide a lower bound for the hyperbolic Bartnik mass for an outer-minimizing connected minimal boundary.

1.3. Initial data sets for the Einstein–Maxwell equations. In the context of solutions to the Einstein–Maxwell Equations, i.e., when the Einstein equations are generalized to incorporate electromagnetic effects, and assuming that the magnetic field vanishes, a time-symmetric initial data set satisfying the dominant energy condition corresponds to a triple (M^3, γ, E), where (M^3, γ) is an asymptotically flat Riemannian manifold and E is a divergence-free vector field on (M^3, γ), modeling the *electric field*, satisfying

$$R(\gamma) \geq 2|E|^2_\gamma.$$

An very important example is given by the *(spatial) Reissner–Nordström manifold* of charge Q and mass $m > |Q|$, is defined as the manifold $(r_+, \infty) \times \mathbb{S}^2$, with the Riemannian metric given by

$$\gamma_{m,Q} := \left(1 - \frac{2m}{r} + \frac{Q^2}{r^2}\right)^{-1} dr^2 + r^2 g_*,$$

where $r_+ := m + \sqrt{m^2 - Q^2}$ and g_* again denotes the standard round metric on $\mathbb{S}^2$. When $m > |Q|$ we call this a *sub-extremal* spatial Reissner–Nordström manifold. The corresponding electric field is given by

$$E_{m,Q} := \frac{Q}{r^2}\sqrt{1 - \frac{2m}{r} + \frac{Q^2}{r^2}}\,\partial_r.$$

The Riemannian Penrose Inequality in this context, called the *Riemannian Penrose Inequality with (electric) charge*, was proven by Jang [**34**] assuming the existence of a smooth solution to the inverse mean curvature flow (IMCF); with the weak formulation of the IMCF by Huisken and Ilmanen [**33**], its general proof (for a connected outer-minimizing horizon) is now complete (see the works of Disconzi and Khuri [**27**] and Mars [**45**]). The corresponding rigidity statement was proven by Disconzi and Khuri in [**27**]. The Riemannian Penrose inequality with (electric) charge can be stated as follows: Let (M^3, γ, E) be a triple such that (M^3, γ) is an asymptotically flat Riemannian manifold and E is a divergence-free vector field on M^3 such that $R(\gamma) \geq 2|E|^2_\gamma$. Moreover, suppose that ∂M consists of a connected, outer-minimizing minimal surface. Then

$$m_{\mathrm{ADM}}(M^3, \gamma) \geq \sqrt{\frac{|\partial M|_\gamma}{16\pi}}\left(1 + \frac{4\pi}{|\partial M|_\gamma} Q^2_\infty\right), \tag{1.8}$$

where

$$Q_\infty := \lim_{r \longrightarrow \infty} \frac{1}{4\pi} \int_{\mathbb{S}^2_r} \gamma(E, N_r)\, d\sigma_r$$

denotes the *total (electric) charge of* (M^3, γ, E), with $\mathbb{S}^2_r \subset \mathbb{R}^3$ the coordinate sphere of radius r, N_r the (Euclidean) outer unit normal to $\mathbb{S}^2_r$, and $d\sigma_r$ the (Euclidean) area form on $\mathbb{S}^2_r$. Equality holds if and only if (M^3, γ) is isometric to (an exterior region of) a sub-extremal Reissner–Nordström manifold of mass $m = m_{\mathrm{ADM}}(M^3, \gamma)$ and charge Q_∞, and E corresponds to E_{m,Q_∞} (see [**45**] and the review article by Dain and Gabach-Clement [**25**] for more detailed expositions of initial data sets for the Einstein–Maxwell Equations and related geometric inequalities).

Let (Σ, g) be a closed surface contained in (M^3, γ, E). The *total charge contained in* Σ is defined as $Q := \frac{1}{4\pi}\int_\Sigma \gamma(E, N)\, d\sigma$, where N is the outward unit normal to Σ in (M^3, γ) and $d\sigma$ is the induced area form of Σ. Let H denote the mean curvature of Σ in (M^3, γ), then the *charged Hawking mass* of (Σ, g, H, Q) is given by

$$\mathfrak{m}_{\mathrm{H}}^{\mathrm{CH}}(\Sigma, g, H, Q) := \sqrt{\frac{|\Sigma|_g}{16\pi}}\left(1 + \frac{4\pi Q^2}{|\Sigma|_g} - \frac{1}{16\pi}\int_\Sigma H^2\, d\sigma\right).$$

As in the other settings described above, it is possible to define a set of admissible extensions $\mathcal{A}$ for suitable charged Bartnik data, leading to an ad-hoc version of Bartnik mass in this scenario which was done by Alaee and the authors in [**1**]. For given *charged Bartnik data* $(\Sigma \cong \mathbb{S}^2, g, H, Q)$, where (Σ, g, H) is as above and $Q \in \mathbb{R}$, we say that a triple (M^3, γ, E) is an admissible extension of (Σ, g, H, Q) if (M^3, γ) is an asymptotically flat Riemannian manifold with outer-minimizing boundary ∂M isometric to (Σ, g) with mean curvature H in (M^3, γ) and E is a smooth vector field on M^3, interpreted as an electric field, such that $R(\gamma) \geq |E|_\gamma^2$ and $\operatorname{div}_\gamma E = 0$, and such that Q is the total charge in ∂M. Then, we define the *charged Bartnik mass* as

$$\mathfrak{m}_{\mathrm{B}}^{\mathrm{CH}}(\Sigma, g, H, Q) := \inf\{m_{\mathrm{ADM}}(M^3, \gamma) \,|\, (M^3, \gamma) \in \mathcal{A}\}.$$

2. The Mantoulidis–Schoen construction

Recall that, as discussed in Section 1, the Bartnik mass is remarkably difficult to compute. In [**43**], Mantoulidis and Schoen computed the Bartnik mass for minimal Bartnik data — $(\Sigma \cong \mathbb{S}^2, g, H)$ is *minimal Bartnik data* if $H \equiv 0$. To do this, given $(\Sigma \cong \mathbb{S}^2, g, H \equiv 0)$ satisfying $\lambda_1 := \lambda_1(-\Delta_g + K(g)) > 0$, where $K(g)$ is the Gaussian curvature and λ_1 denotes the first eigenvalue of $-\Delta_g + K(g)$, they constructed asymptotically flat extensions with non-negative scalar curvature, such that the ADM mass of the extensions can be made arbitrarily close to the optimal value of (1.2). The condition on λ_1 arises naturally in the theory of minimal surfaces, and its connection to Mathematical Relativity comes from the fact that outer-minimizing horizons in an asymptotically flat initial data set with non-negative scalar curvature are, in fact, stable minimal spheres. The construction can be described by the following four steps:

(1) Using the uniformization theorem, write $g = e^{2w} g_*$, where w is a smooth function on Σ. Connect g to g_* by a smooth path of metrics given by $g(t) = e^{2(1-t)w} g_*$, $t \in [0, 1]$.
(2) Modify $\{g(t)\}_{0 \leq t \leq 1}$ by a t-dependent family of diffeomorphisms on Σ in a suitable way, and use it to construct a *collar extension* of (Σ, g) with positive scalar curvature, that is, a Riemannian 3-manifold with topology $[0, 1] \times \Sigma$, such that $\Sigma_0 = \{0\} \times \Sigma$ is isometric to (Σ, g) and $\Sigma_1 = \{1\} \times \Sigma$ is round. This collar extension is constructed so that Σ_0 is minimal and outer-minimizing. Additionally, it is rotationally symmetric in a region near Σ_1, and $|\Sigma_1|_{g(1)}$ can be made arbitrarily close to $|\Sigma_0|_g$. See dotted black part in Figure 1.

(3) For any $m > \sqrt{\frac{|\Sigma|_g}{16\pi}}$, consider a Schwarzschild manifold of mass m and deform it in a rotationally symmetric way in a small region near the horizon so that it has positive (instead of vanishing) scalar curvature in this region. See dashed black part in Figure 1.
(4) Smoothly glue these two manifolds via a positive scalar curvature bridge. See dashed red part in Figure 1.

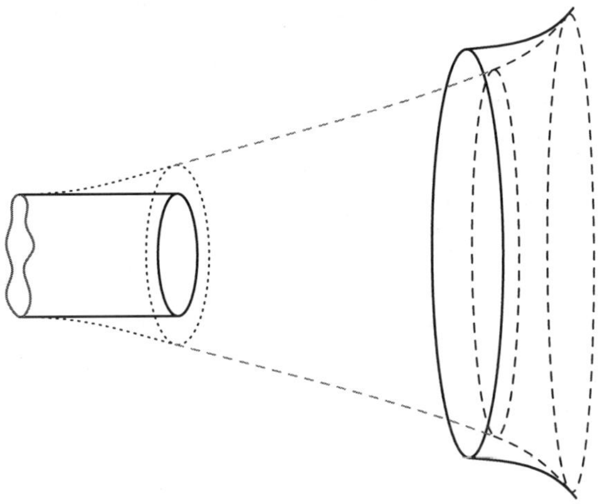

FIGURE 1. The Mantoulidis–Schoen construction [**43**]. The blue surface represents the given surface (Σ, g), dotted 3-manifold represents the collar extension, the black dashed 3-manifold represents the defomed Schwarzschild manifold of mass m, the red dashed 3-manifold represents the positive scalar bridge.

The resulting manifold is a smooth, asymptotically flat manifold with non-negative scalar curvature, such that its boundary is isometric to (Σ, g), minimal and outer-minimizing, and hence, an admissible extension of $(\Sigma, g, H \equiv 0)$. Since this manifold is exactly Schwarzschild of mass m outside a compact set, its ADM mass is equal to m, which can be made arbitrarily close to the optimal value in (1.2), proving that

$$m_B(\Sigma, g, H \equiv 0) = \sqrt{\frac{|\Sigma|_g}{16\pi}} = \mathfrak{m}_{\mathrm{H}}(\Sigma, g, H \equiv 0).$$

Besides providing an exact value for the Bartnik mass of suitable minimal Bartnik data, the Mantoulidis–Schoen construction can be thought as evidence for instability of the Riemannian Penrose Inequality (Theorem 1) in the sense that, given $(\Sigma \cong \mathbb{S}^2, g)$, the ADM mass of the extension can be made arbitrarily close to the optimal value of (1.2), while the geometry of the horizon could be far away from being round. For a more detailed summary of the results of Mantoulidis and Schoen [**43**] and interesting consequences thereof, see [**42**] by the same authors.

From this perspective and from considering the Riemannian Penrose Inequality in higher dimensions (1.6), it is natural to ask whether a construction similar to this one could be performed in higher dimensions. Miao and the first named author answered this question positively in [**17**], by extending the arguments developed in [**43**] to higher dimensions. More precisely, they showed that given $(\Sigma \cong \mathbb{S}^n, g, H \equiv 0)$ $(n \geq 3)$ with $R(g) > 0$ for $n = 3$ and for $n \geq 4$ with the additional assumption that (Σ, g) can be isometrically embedded into $(\mathbb{R}^{n+1}, \delta)$ as a star-shaped surface, then, for any $m > 0$ such that

$$m > \frac{1}{2}\left(\frac{|\Sigma|_g}{\omega_n}\right)^{\frac{n-1}{n}},$$

where ω_n denotes the volume of $\mathbb{S}^n$ and $|\Sigma|_g$ the volume of (Σ, g), there exists an asymptotically flat extension with non-negative scalar curvature and ADM mass equal to m, such that its boundary is isometric to $(\Sigma \cong \mathbb{S}^n, g)$, minimal and outer-minimizing. A key step in this construction is the existence of paths connecting the given metric g to a round metric (corresponding to Step (1) above). For $n = 3$, this is possible since the space of metrics with positive scalar curvature on $\mathbb{S}^3$ is path connected, which was proved by Marques using Ricci flow with surgeries in [**44**]. For $n \geq 4$, extrinsic geometric flows developed by Gerhardt [**30**] and Urbas [**63**] can be used.

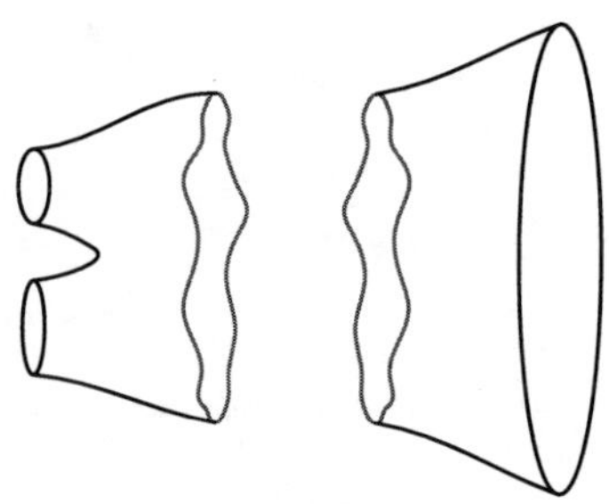

FIGURE 2. The Miao–Xie construction [**50**]. The left part depicts the manifold Ω, in which the CMC boundary component Σ is depicted in blue. The right part depicts the collar extension of Σ.

Motivated by the quasi-local mass problem in General Relativity, in [**50**], Miao and Xie investigated the effect of non-negative scalar curvature in compact bodies inside a Riemannian manifold. More precisely, they considered compact 3-dimensional Riemannian manifolds Ω with boundary $\partial\Omega = \Sigma_h \cup \Sigma$, where $\Sigma \cong \mathbb{S}^2$ has positive constant mean curvature H, and Σ_h is a minimal surface (possibly disconnected), and such that there are no other minimal surfaces in Ω. Inspired by [**43**], they constructed collar extensions of (Σ, g) where g is a non-round metric induced on Σ by Ω, as in Step (2) above, but with the property that the mean curvature of the boundary Σ is equal to H. Then, they attach this collar extension to Ω along Σ and apply a result by Miao [**48**] to obtain

$$\mathfrak{m}_{\mathrm{H}}(\Sigma_1) \geq \sqrt{\frac{|\Sigma_h|}{16\pi}}.$$

Assuming a smallness condition on H (only in terms of (Σ, g)), they were able to estimate $\mathfrak{m}_{\mathrm{H}}(\Sigma_1)$ in terms of $\mathfrak{m}_{\mathrm{H}}(\Sigma)$ and quantities that depend only on (Σ, g), which translates to relations between Σ_h and Σ, and information about Σ itself when $\Sigma_h = \emptyset$. This construction and its consequences provide insights about the interaction of the scalar curvature and geometry of the vicinity of a horizon in a Riemannian 3-manifold. We refer to the interested reader to [**50**] for the precise statements and a more detailed description. The collar extensions used by Miao and Xie in [**50**], which will be discussed in Section 4, have the remarkable property that the Hawking mass of $\Sigma_t = \{t\} \times \Sigma$ is controlled along the collar.

In [**16**], McCormick, Miao, and the authors developed analytic tools to smoothly glue compact, rotationally symmetric manifolds with boundary Ω_{rot} to (an exterior region of) a Schwarzschild manifold, inspired by the tools developed in [**43**]. The main difference is that in [**16**], the set of Schwarzschild manifolds for which this gluing technique works depends on the Hawking mass of the "outer" boundary of Ω_{rot}, instead of on its area. Since the collar extensions constructed by Miao and Xie in [**50**] give a very precise control on the Hawking mass growth along the collar, and since by Step (2) above they are rotationally symmetric close to the "outer" boundary, combined with the gluing construction in [**16**], they produce asymptotically flat extensions of a given surface (Σ, g) such that its mean curvature in the extension is equal to a prescribed, sufficiently small positive constant H. Furthermore, the resulting manifolds are admissible extensions in the sense of Section 1, so this result provided new estimates for *constant mean curvature (CMC)* Bartnik data (see Figure 3).

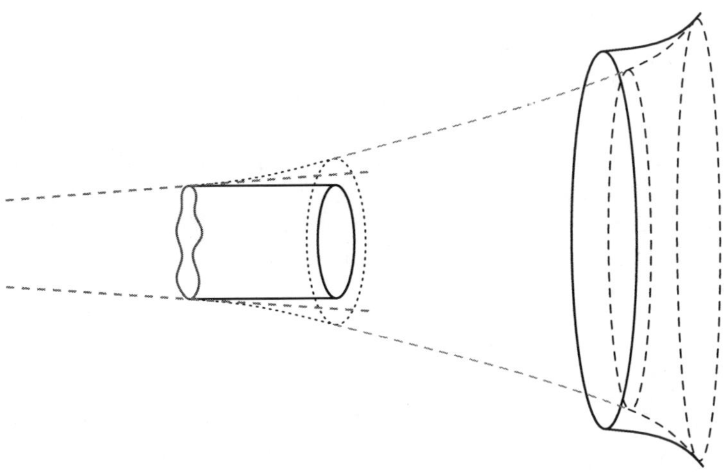

FIGURE 3. The modification of the Mantoulidis–Schoen construction to allow a CMC boundary in [**16**]. The given surface (Σ, g) is represented in blue and the given constant mean curvature is represented by the dashed green line. As before, the collar extension is represented by the dotted part and the deformed Schwarzschild manifold by the dashed black line. The positive scalar bridge is represented by the red dashed part.

After the success of the Mantoulidis–Schoen construction and its modifications [**16, 50**] in computing the Bartnik mass for minimal Bartnik data and in estimating the Bartnik mass of CMC Bartnik data in the context of asymptotically flat manifolds, McCormick and the authors investigated in [**15**] if similar constructions could be carried out in the context of asymptotically hyperbolic manifolds. After formulating an analog to the Bartnik mass in the asymptotically hyperbolic case, see Section 1.2, similar results for both the minimal and the CMC case are obtained, following the techniques in [**16, 43, 50**].

Later, Miao, Y. Wang, and Xie [**49**] refined the construction of the collar extensions by Miao and Xie [**50**], removing the smallness condition on H. In particular, this refinement allowed them to find conditions on the intrinsic metric of the boundary Σ of a domain Ω in a Riemannian manifold (M^3, γ) with non-negative scalar curvature that guarantees the positivity of the Hawking mass of Σ, previously known for stable CMC surfaces by work of Christodoulou and Yau [**19**]. In view of Bartnik mass estimates for CMC Bartnik data $(\Sigma \cong \mathbb{S}^2, g, H)$, the refined collar construction allowed Miao, Y. Wang, and Xie [**49**] to estimate the Bartnik mass without the smallness assumption on H needed in [**16**], instead simply requiring the Hawking mass of the given Bartnik data to be non-negative. They also improved the collar construction and obtained related Bartnik mass estimates in the asymptotically hyperbolic case.

Very recently, Alaee and the authors formulated an ad-hoc definition of a charged Bartnik mass in the context of the Einstein–Maxwell Equations [**1**], see Section 1.3, and computed its value for charged minimal Bartnik data. This construction and computation were applied and generalized to higher dimensions by Peñuela Díaz in [**52**], utilizing techniques and results of [**17**], and using the Ricci flow existence and convergence results for compact Riemannian manifolds with pointwise $\frac{1}{4}$-pinched sectional curvature developed by Brendle and Schoen in [**12, 13**].

Also very recently, the techniques of [**15**] were combined by Gehring in [**29**] with those developed in [**17**], to handle the minimal case in the asymptotically hyperbolic setting in higher dimensions $n \geq 3$.

The Mantoulidis–Schoen construction, its parts and modifications, have been used for various purposes besides constructing admissible extensions for the Bartnik mass. In this survey, we will only focus on the applications mentioned above. However, in Section 5 we provide a list of references to works where aspects of this construction were used.

3. Collar extensions, gluing procedures, and Bartnik mass estimates

We will now describe the Mantoulidis–Schoen construction and its consequent extensions to more general settings in some detail. For the sake of presentation, we provide the descriptions and computations in the most general form known to us, which is a result of the contributions made in the articles mentioned in Section 2. To see the precise statements, we invite the reader to consult those works individually.

This section is divided into three parts. The first one is devoted to describing the collar extensions in general, the second one to explaining the gluing techniques in different contexts, and the last one to staging the previous two parts in the context of Bartnik mass estimates. While the first two are relevant in higher dimensions $n \geq 2$ and may well have or are known to have applications in other settings, Bartnik mass estimates will only be addressed for $n = 2$ because of the predominantly physically motivated interest in quasi-local notions of mass.

3.1. Collar extensions.

Let $(\Sigma \cong \mathbb{S}^n, g)$ be an n-dimensional Riemannian manifold. A *collar extension* of Σ is an $(n+1)$-dimensional Riemannian manifold (M, γ), where $M = [0,1] \times \Sigma$, with a Riemannian metric of the form

$$\gamma := v(t, \cdot)^2 dt^2 + h(t), \tag{3.1}$$

where v is a positive smooth function on $[0,1] \times \Sigma$ and $\{h(t)\}_{0 \leq t \leq 1}$ is a smooth family of metrics on Σ, satisfying $h(0) = g$. Here, we chose the interval $[0,1]$ for convenience. Notice that the condition $h(0) = g$ prescribes the metric on what we call the *inner boundary*, making it isometric to (Σ, g).

Denote $\frac{d}{dt}$ by $'$. Our convention for the mean curvature is that spheres in Euclidean space have positive mean curvature with respect to the outward unit normal. The *outward* unit normal to Σ_t in a collar extension is the one pointing in the direction of ∂_t. Hence, a direct computation shows that the mean curvature for the level set $\Sigma_t = \{t\} \times \Sigma$, is given by

$$H := H(t, \cdot) = \frac{1}{2v(t, \cdot)} \mathrm{tr}_{h(t)} h'(t). \tag{3.2}$$

This is one of the very useful relations between the family of metrics $\{h(t)\}_{0 \le t \le 1}$ and the geometry of the collar extension (M, γ). For example, if we would like (M, γ) to have a minimal inner boundary, then we would need to arrange $\mathrm{tr}_{h(t)} h'(t)|_{t=0} = 0$.

We now proceed to compute the scalar curvature $R(\gamma)$ of the collar extension (3.1). Denoting the scalar curvature and the second fundamental form of Σ_t in M by $R(h)$ and $\mathrm{II} := \mathrm{II}(t, \cdot)$, respectively, and the outward unit normal to Σ_t by $N = N(t, \cdot)$, using the Gauss equation we know that

$$\begin{aligned} R(\gamma) &= 2\mathrm{Ric}_\gamma(N, N) + R(h) - H^2 + |\mathrm{II}|^2, \\ &= 2\mathrm{Ric}_\gamma(N, N) + R(h) - \frac{1}{4v^2} \left(\mathrm{tr}_h h'\right)^2 + \frac{1}{4v^2} |h'|_h^2. \end{aligned}$$

Direct computations give

$$\mathrm{Ric}_\gamma(N, N) = -2\frac{\Delta_h v}{v} - \frac{\mathrm{tr}_h h''}{v^2} + \frac{\partial_t v \mathrm{tr}_h h'}{v^3} + \frac{1}{2v^2} |h'|_h^2.$$

Therefore, we have

$$\begin{aligned} R(\gamma) &= \frac{2}{v}\left(-\Delta_h v + \frac{R(h)}{2} v\right) \\ &\quad + \frac{1}{v^2}\left(-\mathrm{tr}_h h'' + \frac{\partial_t v}{v} \mathrm{tr}_h h' - \frac{1}{4}\left(\mathrm{tr}_h h'\right)^2 + \frac{3}{4}|h'|_h^2\right). \end{aligned} \tag{3.3}$$

We are interested in the case where $h(t) = F(t)^2 g(t)$, where $F(t) \ge 1$ is the *radial profile* of the collar extension, which is a smooth function on $[0, 1]$ with $F(0) = 1$, and $\{g(t)\}_{0 \le t \le 1}$ is a smooth family of metrics satisfying

(i) $g(0) = g$ and $g(1)$ is round,
(ii) $g'(t) \equiv 0$ for $t \in [\theta, 1]$ for some $0 < \theta < 1$, and
(iii) $\mathrm{tr}_{g(t)} g'(t) \equiv 0$ for all $t \in [0, 1]$.

Depending on the situation, the smooth family of metrics $\{g(t)\}_{0 \le t \le 1}$ will be required to satisfy given special curvature conditions. If we denote the set of all metrics satisfying the respective given special curvature conditions by $\mathscr{M}$, the latter can then be expressed as

(iv) $g(t) \in \mathscr{M}$ for all $t \in [0, 1]$.

REMARK 3.1. Property (i) prescribes the metric at the boundaries of the collar extension, while (ii) implies that (M, γ) is rotationally symmetric for $t \in [\theta, 1]$. The property (iii) can be seen to be equivalent to requiring that the volume form $dV_{g(t)}$ is preserved along the path $\{g(t)\}_{0 \le t \le 1}$.

REMARK 3.2. It is possible to associate two scaling-invariant quantities to the path constructed above. These quantities can be regarded as a measurement of how round (Σ, g) is, and they are given by

$$\alpha := \frac{1}{4} \max_{[0,1]\times\Sigma} |g'(t)|^2_{g(t)}, \tag{3.4}$$

and

$$\beta := \min_{[0,1]\times\Sigma} r_o^n \frac{R(g(t))}{n(n-1)}, \tag{3.5}$$

where $r_o := \left(\frac{|\Sigma|_g}{\omega_n}\right)^{\frac{1}{n}}$ is the *volume radius* of (Σ, g). Note that by (iii) this implies that $|\Sigma|_{g(t)} \equiv \omega_n r_o^n$ for all $t \in [0,1]$. Note that if g is round then a constant path can be chosen, and for this path, $\alpha = 0$ and $\beta = 1$. The quantities α and β were defined and studied in detail by Miao and Xie [**50**] in dimension $n = 2$.

The existence of a path satisfying (i)-(iii) follows after we know that a volume-preserving smooth family of metrics $\{\widetilde{g}(t)\}_{0\leq t\leq 1} \subset \mathscr{M}$ satisfying (i) exists: Indeed, by reparametrizing the family $\{g(t)\}_{0\leq t\leq 1}$ through a composition with a bump function, one can achieve (ii) (which of course preserves (i) and (iv)). Subsequently, one can find a 1-parameter family of diffeomorphisms $\{\phi_t\}$ on Σ such that $g(t) := \phi_t^*(\widetilde{g}(t))$ satisfies (iii), by solving a 1-parameter family of Poisson problems on $(\Sigma, \widetilde{g}(t))$ (in [**43**], the method was inspired by what is called "Moser's trick" in symplectic geometry). Note that this does not affect the validity of (i), (ii), and (iv). See [**43**, Lemma 1.2] for details when $n = 2$ and [**17**, Lemma 4.1] for the extension to arbitrary dimensions.

With this choice of $\{h(t)\}_{0\leq t\leq 1}$ (i.e., $h(t) = F(t)^2 g(t)$ and $\{g(t)\}_{0\leq t\leq 1}$ satisfying (i)–(iv) above), we have

$$\begin{aligned}
\operatorname{tr}_h h' &= 2n\frac{F'}{F} + \operatorname{tr}_g g' = 2n\frac{F'}{F},\\
|h'|^2_h &= 4n\frac{(F')^2}{F^2} + 4\frac{F'}{F}\operatorname{tr}_g g' + |g'|^2_g = 4n\frac{(F')^2}{F^2} + |g'|^2_g, \text{ and}\\
\operatorname{tr}_h h'' &= 2n\left[\frac{(F')^2}{F^2} + \frac{F''}{F}\right] + 4\frac{F'}{F}\operatorname{tr}_g g' + \operatorname{tr}_g g''\\
&= 2n\left[\frac{(F')^2}{F^2} + \frac{F''}{F}\right] + |g'|^2_g,
\end{aligned}$$

since $\operatorname{tr}_g g' \equiv 0$ by (iii). Using (3.3), the scalar curvature of the collar extension is given by

$$\begin{aligned}
R(\gamma) &= \frac{2}{F^2 v}\left(-\Delta_g v + \frac{R(g)}{2} v\right)\\
&\quad + \frac{1}{v^2}\left(2n\frac{F'}{F}\frac{\partial_t v}{v} - \frac{n}{F^2}\left[(n-1)F'^2 + 2FF''\right] - \frac{1}{4}|g'|^2_g\right).
\end{aligned} \tag{3.6}$$

Furthermore, from (3.2), we find that the mean curvature of Σ_t is given by

$$H(t) = \frac{n}{v(t,\cdot)}\frac{F'(t)}{F(t)}. \tag{3.7}$$

For the remainder of this survey, we assume that $h(t) = F(t)^2 g(t)$ in (3.1) as described above.

This technique for constructing collar extensions of given data $(\Sigma \cong \mathbb{S}^n, g)$ corresponds to Step (2) in the Mantoulidis–Schoen construction. In what follows, we will explain how to glue the collar manifold to an exterior region of a rotationally symmetric model manifold. When this manifold is a 3-dimensional Schwarzschild manifold of positive mass, this corresponds to Steps (3) and (4) in the Mantoulidis–Schoen construction.

3.2. Smooth gluing procedures. A collar extension as in (3.1) with $h(t) = F(t)^2 g(t)$, where the smooth path of metrics $\{g(t)\}_{0\le t\le 1}$ satisfies (i)-(iv), has the special feature that it is rotationally symmetric for $t \in [\theta, 1]$, provided that $v(t,\cdot)$ is constant on Σ_t for $t \in [\theta, 1]$ (which will be the case for all choices of v considered here). Since Schwarzschild manifolds and all other model manifolds considered here are rotationally symmetric, Step (4) in the Mantoulidis–Schoen construction and its various generalizations can thus be phrased as a problem concerning how to smoothly glue two rotationally symmetric manifolds.

More precisely, consider two rotationally symmetric Riemannian manifolds $(M_1, \gamma_1) = ([a_1, b_1]\times\mathbb{S}^n, ds^2+f_1(s)^2 g_*)$ and $(M_2, \gamma_2) = ([a_2, b_2]\times\mathbb{S}^n, ds^2+f_2(s)^2 g_*)$, both with non-negative scalar curvature. Is there a way to smoothly glue these two manifolds in such a way that the non-negativity of the scalar curvature is preserved? Without losing generality, we may assume that $f_1(b_1) \le f_2(a_2)$ and $b_1 < a_2$. Then, a natural approach will be to attach them together via a piece of a cone (see Figure 4). Note that, in general, the resulting manifold would of course not be smooth. To smooth out this manifold, one could perform a mollification, however, this could disrupt the non-negativity of the scalar curvature (say, if the scalar curvature at $s = a_1$ or $s = a_2$ is equal to 0, as it is for a Schwarzschild manifold).

To make some room for the mollification to work, we impose the condition that M_1 and M_2 have strictly positive scalar curvature. In [**43**], Mantoulidis and Schoen proved an analytic tool to smoothly glue two rotationally symmetric manifolds with positive scalar curvature for $n = 2$ (under some extra conditions that we will state in a moment). This tool was later generalized to other settings, see below.

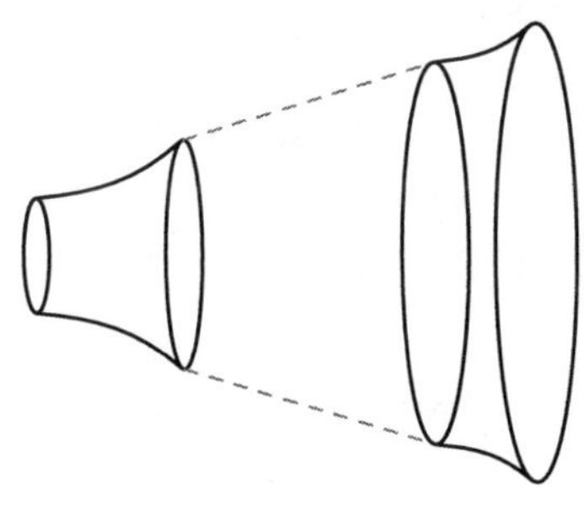

FIGURE 4. The black solid part represent two rotationally symmetric Riemannian manifolds. The red dashed part represents a piece of a cone attached between them. The joining points will not be smooth in general.

Observe that a metric of the form $\gamma = v_0^2 dt^2 + F_0(t)^2 g_*$ on $[a_0, b_0] \times \mathbb{S}^n$, where v_0 is a positive constant and F_0 is a smooth positive function on $[a_0, b_0]$ (which will be the case for the collar extensions described above on $[\theta, 1]$, see also Section 4) can be written in the form $\gamma = ds^2 + f(s)^2 g_*$ after the change of variables $s = v_0 t$, with $f(s) := F(\frac{s}{v_0})$. Hence, the following lemma can be applied to the discussed collar extensions.

LEMMA 3.3 (Gluing Lemma [**43**, Lemma 2.2]). *Let $f_i\colon [a_i, b_i] \longrightarrow \mathbb{R}$ be two smooth functions ($i = 1, 2$). Suppose that*

(a) *$f_i, f_i', f_i'' > 0$ for $i = 1, 2$,*
(b) *the scalar curvature of the metric $\gamma_i = ds^2 + f_i(s)^2 g_*$ on $[a_i, b_i] \times \mathbb{S}^2$ is positive for $i = 1, 2$, and*
(c) *$f_1(b_1) < f_2(a_2)$ and $f_1'(b_1) = f_2'(a_2)$,*

then, after translating the intervals so that $(a_2 - b_1)f_1'(b_1) = f_2(a_2) - f_1(b_1)$, there exists a smooth function $f\colon [a_1, b_2] \longrightarrow \mathbb{R}$ such that

(A) *$f, f' > 0$ on $[a_1, b_2]$,*
(B) *$f \equiv f_1$ on $\left[a_1, \frac{a_1+b_1}{2}\right]$, and $f \equiv f_2$ on $\left[\frac{a_2+b_2}{2}, b_2\right]$, and*
(C) *the metric $\gamma_f := ds^2 + f(s)^2 g_*$ has positive scalar curvature on $[a, b] \times \mathbb{S}^2$.*

The Schwarzschild manifold (1.1) can smoothly be extended to the boundary in such a way that we can write it as the manifold $[0, \infty) \times \mathbb{S}^2$ with metric given by

$$\gamma_m = ds^2 + u_m(s)^2 g_*, \tag{3.8}$$

where u_m is a smooth function such that

(a) $u_m(0) = 2m$,
(b) $u_m'(s) = \sqrt{1 - \frac{2m}{u_m(s)}}$, and
(c) $u_m''(s) = \frac{m}{u_m(s)^2}$.

However, recall that the scalar curvature of a Schwarzschild manifold is zero, thus one can not directly apply Lemma 3.3. It turns out that it is possible to perform a deformation in a small region in a Schwarzschild manifold in order to push up the scalar curvature a bit [**43**, Lemma 2.3]. After this deformation, Mantoulidis and Schoen adjust the parameters of the collar extension and the deformation so that the conditions in Lemma 3.3 are satisfied. Consequently, by considering collar extensions for which the area of $\Sigma_1 = \{1\} \times \Sigma$ is not much bigger than the area of $\Sigma_0 = \{0\} \times \Sigma$, they obtain the desired manifold [**43**, Proof of Theorem 2.1]. This procedure was extended to higher dimensions by Miao and the first named author in [**17**], with appropriate generalizations of all relevant notions, see also Section 1.1.

Following the general idea in [**43**], in the context of CMC Bartnik data and $n = 2$, McCormick, Miao, and the authors developed a (smooth) gluing tool by relaxing the conditions in Lemma 3.3 and expressing the conditions needed for the gluing in terms of the Hawking mass of the surface Σ_1 in the collar extension. This defines a set of Schwarzschild manifolds that can be glued to a given rotationally symmetric manifold. For more details, in particular for an explanation why the size of the Hawking mass of Σ_1 becomes relevant, see Section 4.2. We state this smooth gluing tool in the following proposition.

PROPOSITION 3.4 (CMC-Gluing to a Schwarzschild manifold [**16**, Prop. 2.1]). *Consider a metric $\gamma_f := ds^2 + f(s)^2 g_*$ on $[a, b] \times \mathbb{S}^2$, where f is a smooth, positive, and increasing function on $[a, b]$. Suppose that*

(a) *γ_f has positive scalar curvature,*
(b) *$\Sigma_b = \{b\} \times \mathbb{S}^2$ has positive mean curvature, and*
(c) *$\mathfrak{m}_{\mathrm{H}}(\Sigma_b) \geq 0$.*

Then, for any $m > \mathfrak{m}_{\rm H}(\Sigma_b)$, the manifold $([a,b]\times\mathbb{S}^2,\gamma_f)$ can be smoothly glued to (an exterior region of) a Schwarzschild manifold of mass m, such that the resulting asymptotically flat manifold is rotationally symmetric, has non-negative scalar curvature and its coordinate spheres have positive constant mean curvature.

Even though Proposition 3.4 is essentially a combination of Steps (3) and (4) of the original construction, it has the advantage of avoiding a parameter matching procedure to perform the gluing, hence, it is suitable for applications in other situations, since it gives quantitative conditions in terms of the Hawking mass of $\Sigma_b = \{b\}\times\mathbb{S}^2$ which allows to perform the gluing.

Later, McCormick and the authors [**15**] developed a gluing tool similar to Proposition 3.4 to glue rotationally symmetric manifolds to AdS-Schwarzschild manifolds, also for $n = 2$, (cf. Section 1.2), which will be presented next.

PROPOSITION 3.5 (Gluing to an AdS-Schwarzschild manifold [**15**, Prop. 3.3]). *Consider a metric $\gamma_f := ds^2 + f(s)^2 g_*$ on $[a,b]\times\mathbb{S}^2$, where f is a smooth, positive, and increasing function on $[a,b]$. Suppose that*

(a) *γ_g has scalar curvature $R(\gamma_f) > -6$,*
(b) *Σ_b has positive mean curvature, and*
(c) *$\mathfrak{m}_{\rm H}^{\rm AH}(\Sigma_b) \geq -f(b)^3$.*

Then, for any $m > \mathfrak{m}_{\rm H}^{\rm AH}(\Sigma_b)$, the manifold $([a,b]\times\mathbb{S}^2,\gamma_f)$ can be smoothly glued to (an exterior region of) an AdS-Schwarzschild manifold of mass m, such that the resulting asymptotically hyperbolic manifold has scalar curvature bigger than or equal to -6 and its coordinate spheres have positive constant mean curvature.

Following [**15, 17**], this proposition has very recently been extended to higher dimensions by Gehring in [**29**], with appropriate generalizations of all relevant notions, see also Sections 1.1 and 1.2.

In the context of asymptotically flat manifolds (M^3,γ) together with a vector field E on M^3 playing the role of an electric field, that is, to systems (M^3,γ,E) corresponding to time-symmetric initial data sets for the Einstein–Maxwell Equations (see Section 1.3), Alaee and the authors [**1**] developed the next proposition which provides a gluing tool to smoothly glue rotationally symmetric Riemannian manifolds with an electric field to a Reissner–Nordström manifold of positive mass and sufficiently small charge Q. The main difference between the following result and the previous ones relies on the extra work needed to also obtain a smooth extension of the electric field E satisfying the desired properties.

PROPOSITION 3.6 (Gluing to a Reissner–Nordström manifold [**1**, Prop. 4.1]). *Consider the metric $\gamma_f := ds^2 + f(s)^2 g_*$ on $[a,b]\times\mathbb{S}^2$, where f is a smooth, positive, and increasing function on $[a,b]$. Suppose that*

(a) *the scalar curvature of γ_f satisfies*
$$R(\gamma_f) > 2|E|^2_{\gamma_f},$$
where $E := Qf^{-2}\partial_s$ for some constant $Q\in\mathbb{R}$,
(b) *$f(b) > |Q|$,*
(c) *Σ_b has positive mean curvature, and*
(d) *$\mathfrak{m}_{\rm H}^{\rm CH}(\Sigma_b) \geq |Q|$.*

Then, for any $m > \mathfrak{m}_{\rm H}^{\rm CH}(\Sigma_b)$, the manifold $([a,b]\times\mathbb{S}^2,\gamma_f)$ can be smoothly glued to (an exterior region of) a sub-extremal Reissner–Nordström manifold of mass m

and charge Q. Moreover, the vector field E can be smoothly extended to a divergence free vector field $\widetilde{E}$, which eventually coincides with the electric field of the Reissner–Nordström manifold. In addition, the resulting asymptotically flat manifold has scalar curvature bigger than or equal to $2|\widetilde{E}|^2$ and its coordinate spheres have positive constant mean curvature.

Following [**1**, **17**], this proposition has very recently been extended to higher dimensions by Peñuela Díaz in [**52**], with appropriate generalizations of all relevant notions, see also Sections 1.1 and 1.3.

It can be shown using maximum principle methods from minimal and CMC surface theory that Riemannian manifolds (M^n, γ) foliated by positive constant mean curvature spheres automatically have outer-minimizing inner boundary. This allows us to use the above Lemmas and Propositions towards constructing admissible extensions for Bartnik data as introduced in Section 1.

3.3. Obtaining Bartnik mass estimates. Suppose that $(\Sigma \cong \mathbb{S}^2, g, H)$ are given 2-dimensional Bartnik data. It follows from the definition of Bartnik mass that in order to obtain an upper bound for $\mathfrak{m}_{\mathrm{B}}(\Sigma \cong \mathbb{S}^2, g, H)$, it suffices to construct any admissible extension of the data — in general, constructing admissible extensions is a very difficult problem. The Mantoulidis–Schoen construction has proven to be a very useful method of handcrafting admissible extensions for certain types of Bartnik data. We start by presenting the general strategy to obtain Bartnik mass estimates (using a Mantoulidis–Schoen type construction).

Let $(\Sigma \cong \mathbb{S}^2, g, \mathcal{D})$ be given 2-dimensional Bartnik data, where $g \in \mathscr{M}$ (some suitable set of metrics satisfying a curvature constraint) and $\mathcal{D}$ is a tuple formed by the remaining quantities to be prescribed at the boundary. For example, for minimal Bartnik data $\mathcal{D} = H \equiv 0$, while for minimal charged Bartnik data $\mathcal{D} = (H \equiv 0, Q)$. Let $\mathcal{A}$ denote the class of admissible extensions, which depends on the setting we are interested in, and let $\mathfrak{m}_{\mathrm{H}}^{\mathcal{A}}$ and $\mathfrak{m}_{\mathrm{B}}^{\mathcal{A}}$ denote the corresponding notions of Hawking and Bartnik mass, see Section 1.

The following procedure allows to obtain an upper bound for $\mathfrak{m}_{\mathrm{B}}^{\mathcal{A}}(\Sigma \cong \mathbb{S}^2, g, \mathcal{D})$:

(I) Prove the existence of an area-preserving smooth path of metrics in $\mathscr{M}$ that connects g to a round metric.

(II) Construct a suitable collar extension of $(\Sigma \cong \mathbb{S}^2, g, \mathcal{D})$ in such a way that $\mathcal{D}$ is prescribed appropriately at the inner boundary Σ_0; additionally arrange the inner boundary Σ_0 to be outer-minimizing. Estimate the corresponding Hawking mass (which depends on the type of extensions we are considering), $\mathfrak{m}_{\mathrm{H}}^{\mathcal{A}}$, at the outer boundary Σ_1 of the collar extension.

(III) Use a gluing tool to connect the collar extension to a model space of total mass $m > \mathfrak{m}_{\mathrm{H}}^{\mathcal{A}}(\Sigma_1)$. Since the resulting manifold is an admissible extension, we obtain

$$\mathfrak{m}_{\mathrm{B}}^{\mathcal{A}}(\Sigma \cong \mathbb{S}^2, g, \mathcal{D}) \leq m.$$

4. Recent results on Bartnik mass estimates

In this section, we present some recent results about Bartnik mass estimates that are obtained via Mantoulidis–Schoen type constructions. For the sake of presentation, we will deviate from the historical order and first discuss the case when the Bartnik data is minimal, i.e., treat extensions of $(\Sigma \cong \mathbb{S}^2, g, H \equiv 0)$ in the asymptotically flat, asymptotically hyperbolic, and electrically charged setting.

Then, we will present the results for CMC Bartnik data, i.e., for $(\Sigma \cong \mathbb{S}^2, g, H)$ with H a positive constant, in the asymptotically flat and hyperbolic settings. Finally, we briefly discuss some progress regarding estimates for non-CMC Bartnik data. For a historical recollection of these results see Section 2.

4.1. Minimal Bartnik data. One of the most remarkable features of the Mantoulidis–Schoen construction is that it provides a way to compute the Bartnik mass of minimal Bartnik data as was first done in [**43**].

Motivated by the fact that horizons in a time-symmetric initial data set are stable minimal surfaces, in [**43**], Mantoulidis and Schoen consider the set of metrics

$$\mathscr{M}^+ := \{g \text{ is a metric on } \Sigma \,|\, \lambda_1(-\Delta_g + K(g)) > 0\} \tag{4.1}$$

where λ_1 denotes the first eigenvalue of $-\Delta_g + K(g)$, and $K(g)$ denotes the Gaussian curvature of the 2-surface (Σ, g). Using the uniformization theorem, given any metric $g \in \mathscr{M}^+$, they obtain an area-preserving path of metrics $\{g(t)\}_{0\leq t\leq 1}$ in $\mathscr{M}^+$ connecting g to g_*, the round metric on $\mathbb{S}^2$ (see Section 2). As discussed in Section 3, this gives rise to a family of metrics satisfying (i)-(iv) on page 13, which we will again denote by $\{g(t)\}_{\{0\leq t\leq 1\}}$.

Set $v = Au(t, \cdot)$, where A is a positive constant and $u > 0$ is an eigenfunction corresponding to the eigenvalue $\lambda_1(t) > 0$. Here, the eigenfunctions u need to be chosen so that u is smooth on $[0,1] \times \mathbb{S}^2$ (see [**43**, Lemma A.1]). In addition, they need to be L^2-normalized with respect to $d\sigma_{g(t)}$. Then (3.6) becomes

$$\begin{aligned} R(\gamma) = {} & \frac{2}{F(t)^2}\lambda_1(t) \\ & + \frac{1}{A^2u^2}\left(4\frac{F'(t)}{F(t)}\frac{\partial_t u}{u} - \frac{2}{F(t)^2}\left[F'(t)^2 + 2F(t)F''(t)\right] - \alpha\right). \end{aligned} \tag{4.2}$$

It is clear that, by compactness, choosing A sufficiently large in terms of u, F (and their derivatives), and α, the last term of Equation (4.2) can be made arbitrarily small. Hence, we can choose A so that $R(\gamma) > 0$, since $\lambda_1(t) > 0$ by assumption. However, by making A larger, the length of the collar extension increases as well. In order to specify the data near the boundary in such a way that the boundary is outer-minimizing, set $F(t) := (1 + \varepsilon t^2)^{\frac{1}{2}}$, so the collar metric takes the form

$$\gamma = A_o^2 u(t, \cdot)^2 dt^2 + (1 + \varepsilon t^2) g(t), \tag{4.3}$$

where $A_o > 0$ is chosen sufficiently large so that $R(\gamma) > 0$. Using (3.7), we know that the mean curvature at $\Sigma_0 = \{0\} \times \Sigma$ is $H(0) \equiv 0$, and positive for Σ_t with $t > 0$. One can check directly that by taking ε small, the area $|\Sigma_1|_{g(1)}$ can be made arbitrarily close to $|\Sigma|_{g_o}$.

Using property (ii) of $\{g(t)\}_{0\leq t\leq 1}$ from page 13, and since $g(1)$ is a round metric, one can directly check that $u := u(t, \cdot)$ is a constant for $t \in [\theta, 1]$. Then one can perform the change of variables $s = A_o ut$ to write the metric γ in the form required by Proposition 3.4. Therefore, it follows that for any $m > \sqrt{\frac{|\Sigma|_{g_o}}{16\pi}}$, there is an admissible extension of $(\Sigma \cong \mathbb{S}^2, g_o, H_o \equiv 0)$ with ADM mass equal to m. Since in this case the Riemannian Penrose Inequality (1.2) establishes that $\sqrt{\frac{|\Sigma|_{g_o}}{16\pi}}$ is a lower bound for the Bartnik mass of the Bartnik data $(\Sigma \cong \mathbb{S}^2, g_o, H_o \equiv 0)$, we immediately obtain the following theorem.

THEOREM 4.1 (Bartnik mass of minimal Bartnik data (c.f. [**43**, Theorem 2.1])). *Let $(\Sigma \cong \mathbb{S}^2, g_o, H_o \equiv 0)$ be minimal Bartnik data satisfying $\lambda_1 > 0$, where λ_1 is the first eigenvalue of $-\Delta_{g_o} + K(g_o)$, with $K(g_o)$ denoting the Gaussian curvature of (Σ, g_o). Then*

$$\mathfrak{m}_{\mathrm{B}}(\Sigma \cong \mathbb{S}^2, g_o, H_o \equiv 0) = \sqrt{\frac{|\Sigma|_{g_o}}{16\pi}}.$$

In the asymptotically hyperbolic setting (see Section 1.2), we are interested in constructing asymptotically hyperbolic extensions with scalar curvature bounded below by -6. In [**15**], McCormick and the authors considered the set $\mathscr{M}^+$ above and the set

$$\mathscr{K}^- := \{g \text{ is a metric on } \Sigma \,|\, K(g) > -3\}.$$

The existence of the smooth area-preserving path in $\mathscr{K}^-$ connecting g to a round metric is given by Ricci flow in dimension two studied by Hamilton [**31**]. Again, denote by $\{g(t)\}_{0\le t\le 1}$ the modified family of metrics satisfying (i)-(iv) from page 13.

Using the same type of collar extensions given by (4.3), the condition $R(\gamma) \ge -6$ is equivalent via (4.2) to

$$A^2u^2(\lambda_1(t) + 3F(t)^2) + 2\frac{F'(t)}{F(t)}\frac{\partial_t u}{u} - \frac{1}{F(t)^2}\left[F'(t)^2 + 2F(t)F''(t)\right] - \frac{1}{2}\alpha \ge 0.$$

Since $F(t) \ge 1$ for all $t \in [0,1]$, it is clear that by requiring $\{g(t)\}_{0\le t\le 1}$ to lie in $\mathscr{M}^+$ or $\mathscr{K}^-$, it suffices to make $R(\gamma) > -6$ (by taking A sufficiently large). Similarly as in the asymptotically flat case, these collar extensions together with Proposition 3.5 produce admissible extensions, ultimately leading to the following estimate.

THEOREM 4.2 (Asymptotically hyperbolic Bartnik mass estimate for minimal Bartnik data [**15**, Theorem 1.1]). *Let $(\Sigma \cong \mathbb{S}^2, g_o, H_o \equiv 0)$ be minimal Bartnik data satisfying either $\lambda_1 > 0$, where λ_1 is the first eigenvalue of the operator $-\Delta_{g_o} + K(g_o) > 0$, or $K(g_o) > -3$, recalling that $K(g_o)$ denotes the Gaussian curvature of (Σ, g_o). Then*

$$\mathfrak{m}_{\mathrm{B}}^{\mathrm{AH}}(\Sigma \cong \mathbb{S}^2, g_o, H_o \equiv 0) \le \sqrt{\frac{|\Sigma|_{g_o}}{16\pi}}\left(1 + \frac{|\Sigma|_{g_o}}{4\pi}\right).$$

If, in addition, we suppose that the asymptotically hyperbolic Riemannian Penrose Inequality (1.7) holds, we would obtain the exact value

$$\mathfrak{m}_{\mathrm{B}}^{\mathrm{AH}}(\Sigma \cong \mathbb{S}^2, g_o, H_o \equiv 0) = \sqrt{\frac{|\Sigma|_{g_o}}{16\pi}}\left(1 + \frac{|\Sigma|_{g_o}}{4\pi}\right).$$

For the Einstein–Maxwell setting, an electric charge Q is part of the given Bartnik data, see Section 1.3. Given charged Bartnik data $(\Sigma \cong \mathbb{S}^2, g, H \equiv 0, Q)$ where Q is a real number and $g \in \mathscr{M}^+$ defined as in (4.1), Alaee and the authors [**1**] consider the path of metrics $\{g(t)\}_{0\le t\le 1}$ given by the uniformization theorem and satisfying (i)-(iv) from page 13 as before. Additionally, we define the useful quantity

$$\kappa := \inf_{\Sigma\times[0,1]} \lambda_1(t), \tag{4.4}$$

where $\lambda_1(t) := \lambda_1(-\Delta_{g(t)} + K(g(t)))$ denotes the first eigenvalue of the operator $-\Delta_{g(t)} + K(g(t))$ which smoothly depends on $t \in [0,1]$.

In this case, the main difference to the uncharged case treated in [**43**] is that in addition to the collar extension (4.3), it is necessary to construct a suitable divergence free electric vector field E along the collar, such that $R(\gamma) \geq 2|E|^2_\gamma$. Using (4.2) and taking A sufficiently big as before, we can arrange the collar metric to satisfy $R(\gamma) > 2|E|^2_\gamma$ and simultaneously make $\mathfrak{m}_{\mathrm{H}}^{\mathrm{CH}}(\Sigma_1)$ arbitrarily close to the optimal value in (1.8) for a suitable choice of E, which together with Proposition 3.6 gives the following result.

THEOREM 4.3 (Charged Bartnik mass estimate for minimal charged Bartnik data (c.f. [**1**, Theorem 5.1])). *Let $(\Sigma \cong \mathbb{S}^2, g_o, H_o \equiv 0, Q_o)$ be charged Bartnik data satisfying $\lambda_1 > 0$, where λ_1 is the first eigenvalue of $-\Delta_{g_o} + K(g_o)$ with $K(g_o)$ denoting the Gaussian curvature of (Σ, g_o), $Q_o < r_o^2$, and $\kappa > \frac{Q_o^2}{r_o^4}$, where κ is given in* (4.4). *Then,*

$$\mathfrak{m}_{\mathrm{B}}^{\mathrm{CH}}(\Sigma \cong \mathbb{S}^2, g_o, H_o \equiv 0, Q_o) = \sqrt{\frac{|\Sigma|_{g_o}}{16\pi}} + \sqrt{\frac{\pi}{|\Sigma|_{g_o}}} Q_o^2.$$

As we have seen, for minimal Bartnik data, Mantoulidis–Schoen type extensions can be arranged to arbitrarily approach the optimal value in the corresponding Riemannian Penrose Inequality (or rather Conjecture in the asymptotically hyperbolic case). This is partly due to the fact that the size of the constant A does not play a role in enforcing zero mean curvature at the inner boundary (see (3.7)), which we will see below to be rather different in the case of CMC Bartnik data.

4.2. CMC Bartnik data. In this part, we restrict our attention to CMC Bartnik data $(\Sigma \cong \mathbb{S}^2, g, H)$, that is, we assume that $H \equiv H_o$ is a positive constant. In addition, we assume that $K(g) > 0$. We remind the reader that outer-minimizing Bartnik data $(\Sigma \cong \mathbb{S}^2, g, H)$ satisfy the generalized Penrose inequality (1.5), namely

$$\mathfrak{m}_{\mathrm{H}}(\Sigma \cong \mathbb{S}^2, g, H) \leq \mathfrak{m}_{\mathrm{B}}(\Sigma \cong \mathbb{S}^2, g, H).$$

Now, given CMC Bartnik data $(\Sigma \cong \mathbb{S}^2, g_o, H_o)$, the set of metrics considered by McCormick, Miao, and the authors in [**16**] is

$$\mathscr{K}^+ := \{g \text{ is a metric on } \Sigma \,|\, K(g) > 0\}.$$

As before, the uniformization theorem is used to show the existence of the desired area-preserving path, and we denote by $\{g(t)\}_{0\leq t\leq 1}$ the modified path satisfying (i)-(iv) from page 13. For the collar metric, set $v := A$, where A is a positive constant, so that

$$\gamma = A^2 dt^2 + F(t)^2 g(t).$$

Then (3.6) takes the form

$$R(\gamma) = \frac{2}{F(t)^2} K(g(t)) + \frac{1}{A^2}\left(-\frac{2}{F(t)^2}\left[F'(t)^2 + 2F(t)F''(t)\right] - \alpha\right). \tag{4.5}$$

As before, by making A large in terms of F (and its derivatives) and α, the last term can be made arbitrarily small. In contrast to the minimal case (where the size of A played no role in prescribing the mean curvature at the boundary), the size of A will now affect the size of the mean curvature at the boundary, since by (3.7), we have

$$H(0) = \frac{2}{A} F'(0).$$

Thus, by making A large, we necessarily restrict the freedom of H_o. The goal is then to produce collar extensions in which A is chosen in some optimal sense, and such that they appropriately propagate the Hawking mass of the given data along the collar (instead of the area which needed to be propagated appropriately in the minimal case). Using the collar extensions developed by Miao and Xie [**50**], the Hawking mass at the end of the collar can be estimated in terms of the Hawking mass of (Σ, g_o, H_o), α and β, see Remark 3.2. The idea in [**50**] is to use a piece of the neck of a suitable Schwarzschild manifold as the collar extension.

More precisely, let $m \in (-\infty, \frac{r_o}{2}]$ where $r_o := \sqrt{\frac{|\Sigma|_{g_o}}{4\pi}}$ denotes the area radius of the Bartnik data. The metric of part of the neck part of the (potentially negative mass) Schwarzschild manifold corresponding to $[r_o, \infty) \times \mathbb{S}^2$ can be written as before as

$$\gamma_m = ds^2 + u_{m,r_o}(s)^2\, g_*,$$

where u_{m,r_o} is a smooth function on $[0,\infty)$ satisfying

(a) $u_{m,r_o}(0) = r_o$,
(b) $u'_{m,r_o}(s) = \sqrt{1 - \frac{2m}{u_{m,r_o}(s)}}$, and
(c) $u''_{m,r_o}(s) = \frac{m}{u_{m,r_o}(s)^2}$.

Therefore, for the collar extension

$$\gamma_c := A dt^2 + \frac{u_{m,r_o}(Akt)^2}{r_o^2} g(t), \tag{4.6}$$

with $k = \frac{H_o r_o}{2}\left(1 - \frac{2m}{r_o}\right)^{-1/2}$, we have by (4.5) that

$$R(\gamma_c) \geq 2u_{m,r_o}^{-2}\left(\beta - k^2 - \frac{1}{2}\alpha A^{-2} u_m^2(At)\right). \tag{4.7}$$

Assuming the smallness condition

$$\frac{1}{4}H_o^2 r_o^2 < \frac{\beta}{1+\alpha} \tag{4.8}$$

and picking m so that

$$\frac{1}{4}H_o^2 r_o^2 < \frac{\beta}{1+\alpha}\left(1 - \frac{2m}{r_o}\right),$$

they assert that there exists A_o such that

(a) $R(\gamma_c) > 0$,
(b) Σ_0 is isometric to (Σ, g_o) with mean curvature H_o,
(c) Σ_t has positive mean curvature for all $t \in [0,1]$, and
(d) the Hawking mass of Σ_1 can be estimated by

$$\mathfrak{m}_{\mathrm{H}}(\Sigma_1) \leq \begin{cases} \frac{1}{4}H_o r_o A_o(1-k^2) + \mathfrak{m}_{\mathrm{H}}(\Sigma, g_o, H_o), & \text{if } m < 0, \\ \frac{1}{2}A_o k(1-k^2) + \mathfrak{m}_{\mathrm{H}}(\Sigma, g_o, H_o), & \text{if } m \geq 0. \end{cases}$$

All the computations above are explained in [**50**].

Recalling that for $t \in [\theta, 1]$ the collar extension is rotationally symmetric, one readily sees that Proposition 3.4 is applicable for any $m_* > \mathfrak{m}_{\mathrm{H}}(\Sigma_1)$. It is possible to perform an optimality analysis on the choice of m and we refer the interested reader to [**50**] for details. In particular, by setting $m = 0$, we obtain the following estimate.

THEOREM 4.4 (Bartnik mass estimate for CMC Bartnik data [**16**, Theorem 1.1]). *Let* $(\Sigma \cong \mathbb{S}^2, g_o, H_o)$ *be CMC Bartnik data with Gaussian curvature* $K(g_o) > 0$. *Suppose that* $\frac{1}{4}H_o^2 r_o^2 < \frac{\beta}{1+\alpha}$, *then*

$$\mathfrak{m}_{\mathrm{B}}(\Sigma, g_o, H_o) \leq \left[1 + \left(\frac{\alpha \frac{H_o^2 r_o^2}{4}}{\beta - (1+\alpha)\frac{H_o^2 r_o^2}{4}}\right)^{1/2}\right] \mathfrak{m}_{\mathrm{H}}(\Sigma, g_o, H_o).$$

In [**49**], Miao, Y. Wang, and Xie refined the construction of collar extensions for CMC Bartnik data by proving the existence of an optimal constant A_o in (4.6), which makes the right hand side of (4.7) equal to 0, for a suitable choice of m. By using a limiting argument on m, they were able to obtain more general estimates for $\mathfrak{m}_{\mathrm{H}}(\Sigma_1)$ than those in [**50**], which in particular hold without the smallness assumption (4.8). They obtained the following estimate.

THEOREM 4.5 (Bartnik mass estimate for CMC Bartnik data (c.f. [**49**, Theorem 1.3])). *Let* $(\Sigma \cong \mathbb{S}^2, g_o, H_o)$ *be CMC Bartnik data with Gaussian curvature* $K(g_o) > 0$ *and satisfying* $\frac{H_o r_o}{2} \leq 1$. *Then*

$$\mathfrak{m}_{\mathrm{B}}(\Sigma \cong \mathbb{S}^2, g_o, H_o) \leq C r_o (1 + \zeta_{g_o} \frac{1}{2} H_o r_o) \zeta_{g_o} \frac{1}{2} H_o r_o + \mathfrak{m}_{\mathrm{H}}(\Sigma, g_o, H_o),$$

where C *is an absolute constant and*

$$\zeta_{g_o} := \inf \left(\frac{\alpha}{2\beta}\right)^{1/2},$$

where the infimum is taken over the set of paths $\{g(t)\}_{0 \leq t \leq 1}$ *in* $\mathscr{K}^+$ *connecting* g_o *to a round metric and satisfying* $\mathrm{tr}_{g(t)} g'(t) \equiv 0$ *for* $t \in [0,1]$.

REMARK 4.6. When the value of H_o is small, the estimate in Theorem 4.5 improves that in Theorem 4.4; for a detailed comparison of the estimates, see [**49**].

REMARK 4.7. In [**38**], Lin and Sormani obtained interesting estimates for CMC Bartnik data $(\Sigma \cong \mathbb{S}^2, g_o, H_o)$ with normalized area $|\Sigma_o|_{g_o} = 4\pi$. Even though their method also relies on the construction of asymptotically flat extensions, their extensions and the character of their estimates are different in nature to the ones discussed here. For a precise statement we refer the reader to [**38**].

In the asymptotically hyperbolic case, by using the collar extensions constructed by Miao and Xie [**50**] and radial profile extensions that mimic a piece of an AdS-Schwarzschild manifold neck, the following estimate is obtained in [**15**].

THEOREM 4.8 (Asymptotically hyperbolic Bartnik mass estimate for CMC Bartnik data [**15**, Theorem 1.3]). *Let* $(\Sigma \cong \mathbb{S}^2, g_o, H_o)$ *be CMC Bartnik data with Gaussian curvature* $K(g_o) > -3$. *Assume further that their Hawking mass satisfies*

$$\mathfrak{m}_{\mathrm{H}}^{\mathrm{AH}}(\Sigma, g_o, H_o) > -\left(\frac{|\Sigma|_{g_o}}{4\pi}\right)^{\frac{3}{2}}.$$

Then its asymptotically hyperbolic Bartnik mass satisfies

$$\mathfrak{m}_{\mathrm{B}}^{\mathrm{AH}}(\Sigma, g_o, H_o) \leq \mathfrak{m}_{\mathrm{H}}^{\mathrm{AH}}(\Sigma, g_o, H_o) + \mathcal{E}_1(H_o, \alpha),$$

where $\mathcal{E}_1(H_o, \alpha) \longrightarrow 0$ *as* H_o *or* α *tend to* 0, *with* α *given by* (3.4).

Using a similar approach as in the asymptotically flat case, Miao, Y. Wang, and Xie [**49**] also refined the construction of the collar extensions in this setting and correspondingly obtained an estimate for the asymptotically hyperbolic Bartnik mass of CMC Bartnik data under the condition that the Bartnik data bound a Riemannian domain with a negative lower bound on the scalar curvature. Even though the estimate in [**49**] holds for asymptotically hyperbolic extensions with scalar curvature bounded below by -6κ, for a parameter $\kappa > 0$, for the sake of exposition we state it for $\kappa = 1$.

THEOREM 4.9 (Asymptotically hyperbolic Bartnik mass estimate for CMC Bartnik data (c.f. [**49**, Theorem 4.1])). *Let $(\Sigma \cong \mathbb{S}^2, g_o, H_o)$ be CMC Bartnik data with $K(g_o) > -3$ bounding a compact domain Ω with scalar curvature greater than or equal to -6. Then its asymptotically hyperbolic Bartnik mass satisfies*

$$\mathfrak{m}_{\mathrm{B}}^{\mathrm{AH}}(\Sigma, g_o, H_o) \leq \mathfrak{m}_{\mathrm{H}}^{\mathrm{AH}}(\Sigma, g_o, H_o) + \mathcal{E}_2(H_o, \xi_o),$$

where $\mathcal{E}_2(H_o, \xi_o) \longrightarrow 0$ as H_o tends to 0. Here, ξ_o is a constant that depends only on g_o.

To see the exact form of $\mathcal{E}_1$ and $\mathcal{E}_2$, as well as the definition of ξ_o, we invite the reader to consult [**15**] and [**49**], respectively.

4.3. Non-CMC Bartnik data. Suppose that $(\Sigma \cong \mathbb{S}^2, g, H)$ are given Bartnik data for which $H \geq 0$ is not necessarily constant. What can we say about estimates of their Bartnik mass? Being able to construct explicit asymptotically flat extensions with the techniques discussed here seems to be a challenging problem, and it is not clear whether it is possible at all. However, it is possible to use the Bartnik mass estimates for CMC Bartnik data described above to obtain estimates for non-CMC Bartnik data, as was pursued by McCormick [**46**].

As was mentioned in Section 1, there are variants of the definition of Bartnik mass in the literature. These differences are mainly related to the boundary conditions of the extensions and to what is commonly called a "non-degeneracy" condition imposed on the set of admissible extensions (see [**35, 46**] and the references cited therein). It is thus of high interest to reconcile these definitions, which is a non-trivial problem. This reconciliation has been done for at least two cases, independently by Jauregui [**35**] and McCormick [**46**]. In one case, studied by McCormick [**46**], a convexity condition is imposed on Bartnik data $(\Sigma \cong \mathbb{S}^2, g, H > 0)$, which is given by (c.f. [**47**])

$$2K(g) - 2H\Delta_g\left(\frac{1}{H}\right) - \frac{1}{2}H^2 > 0. \tag{4.9}$$

The other case, considered by Jauregui [**35**], is when the "non-degeneracy" condition consists of assuming the boundary of the asymptotically flat extensions to be *strictly* outer-minimizing — note that in this survey we only assume the extensions to be (weakly) outer-minimizing. More precisely, given Bartnik data $(\Sigma \cong \mathbb{S}^2, g, H > 0)$, we say that they are *locally extendable* if it is possible to construct a collar extension of the Bartnik data, with non-negative scalar curvature as in Section 3.

Under the corresponding assumptions, Jauregui and McCormick show that different definitions of Bartnik mass coincide. McCormick also obtains estimates for the Bartnik mass of non-CMC Bartnik data [**46**]. Combining their results, we have the following estimate.

THEOREM 4.10 (Bartnik mass estimate for non-CMC data (cf. [**35**, **46**])). *Let* $(\Sigma \cong \mathbb{S}^2, g, H)$ *be Bartnik data with* $H > 0$. *Assume either that the Bartnik data* $(\Sigma \cong \mathbb{S}^2, g, H)$ *are locally extendable or that the convexity condition* (4.9) *holds. Then*

$$\mathfrak{m}_{\mathrm{B}}(\Sigma, g, H) \leq \mathfrak{m}_{\mathrm{B}}(\Sigma, g, \min_{\Sigma} H). \tag{4.10}$$

REMARK 4.11. In Theorem 4.10, it is implicitly understood that if the Bartnik data $(\Sigma \cong \mathbb{S}^2, g, H)$ are assumed to be locally extendable, then the extra condition that ∂M be strictly outer-minimizing in the definition of admissible extensions and thus of the Bartnik mass (1.4) is required. If the convexity condition (4.9) is assumed, then the estimate works for the definition of Bartnik mass in (1.4) used in this survey.

REMARK 4.12. In [**67**], using a different methodology to the ones discussed here, Wiygul estimated the Bartnik mass for almost CMC Bartnik data C^2-close to the standard sphere. His techniques allowed him to study the asymptotic behavior of the Bartnik mass for small spheres. See [**67**] for details.

5. Conclusions and open problems

The Mantoulidis–Schoen construction [**43**] is a novel way of handcrafting extensions of 2-dimensional Riemannian manifolds $(\Sigma \cong \mathbb{S}^2, g)$, suited to produce admissible extensions in the context of Bartnik mass. Moreover, it is optimal in the sense that it eventually leads to a lower bound on the Bartnik mass of minimal Bartnik data $(\Sigma \cong \mathbb{S}^2, g, H \equiv 0)$, which, together with the Riemannian Penrose Inequality, establishes that

$$\mathfrak{m}_{\mathrm{B}}(\Sigma, g, H \equiv 0) = \sqrt{\frac{|\Sigma|_g}{16\pi}},$$

provided that $\lambda_1(-\Delta_g + K(g)) > 0$. In particular, it also suggests that the Riemannian Penrose Inequality is unstable, in the sense that a manifold can almost achieve equality in (1.2), while being far away from being a Schwarzschild manifold (since the minimal Bartnik data, i.e., the horizon, can be arranged to be highly non-round).

We have seen that this procedure extends to higher dimensions [**17**], and remains optimal in the context of the higher dimensional Riemannian Penrose Inequality, thus suggesting the instability of the higher dimensional Riemannian Penrose Inequality in the same sense as above. It is interesting to know that this technique led to similar instability phenomena in the the asymptotically hyperbolic [**29**] and the electrically charged [**52**] cases.

Currently, the authors are using the gluing tools developed in [**16**] to produce minimizing sequences for the Bartnik mass in the minimal case [**14**]. This will shed light on subtle aspects of the stability of the Riemannian Penrose Inequality and will likely have implications on our understanding of the role of the outer-minimizing condition in the definition of Bartnik mass. Similar considerations should apply to higher dimensions.

We have also seen that the Mantoulidis–Schoen construction can be adjusted to compute the Bartnik mass and to give Bartnik mass estimates for minimal Bartnik data in the electrically charged and asymptotically hyperbolic settings, respectively.

Modifications of the Mantoulidis–Schoen construction allow to obtain estimates for the Bartnik mass of CMC Bartnik data [**16**]. However, the estimates obtained there are not optimal in any sense. In the light of the gluing tools developed in [**16**], is plausible that by picking a different type of collar extensions, the Bartnik mass estimates for CMC Bartnik data could be improved. It is evident from the work of Miao and Xie [**50**] and the recent work of Miao, Y. Wang, and Xie [**49**], that constructing (and improving) collar extensions is a very challenging problem.

As discussed briefly in this survey, two recent works by Jauregui [**35**] and McCormick [**46**] also address questions related to the outer-minimizing condition in the definition of Bartnik mass. Remarkably, in both works, ideas à la Mantoulidis–Schoen are used.

We would like to point out that there have been other applications of the ideas by Mantoulidis and Schoen and their modifications different from those discussed in this survey: the works of Anderson and Jauregui [**3**], Li and Mantoulidis [**36**], Mantoulidis and Miao [**39, 40**], Mantoulidis, Miao and Tam [**41**], and McCormick and Miao [**47**]. Unfortunately, discussing those in detail would go far beyond the scope of this survey.

All the considerations in this survey, and indeed all the relevant works known to the authors using, generalizing, or applying the Mantoulidis–Schoen construction, focus on time-symmetric initial data sets for (or time-slices of) the Einstein Equations — possibly in higher dimensions, or assuming different asymptotics related to a cosmological constant, or incorporating (electric) charge. This leads to two natural questions:

Going beyond time-symmetry. It would be natural to study whether a similar construction of initial data sets satisfying the dominant energy condition can be performed without assuming time-symmetry. Ideally, such a construction would give us a better understanding of the (conjectured) Penrose Inequality for initial data sets beyond the Riemannian (i.e., time-symmetric) context. Furthermore, it could allow to compute admissible extensions for a non-time-symmetric version of Bartnik mass [**7**] and thus potentially allow to compute or estimate this mass in special cases. In the 3-dimensional, asymptotically flat, non-charged setting, this would entail constructing Riemannian manifolds (M^3, γ) carrying a symmetric $(0,2)$-tensor field K to be thought of as the second fundamental form of (M, γ) in the ambient spacetime. The dominant energy condition would then read

$$R(\gamma) - |K|^2_\gamma + (\mathrm{tr}_\gamma K)^2 \geq |\mathrm{div}_\gamma K - d(\mathrm{tr}_\gamma K)|_\gamma, \tag{5.1}$$

and the definition of asymptotic flatness would extend to K by requesting that $K_{ij} = \mathcal{O}_1(r^{-p-1})$ as $r \longrightarrow \infty$ while exchanging integrability of $R(\gamma)$ for integrability of both sides of (5.1). The corresponding asymptotic model space would be the so-called Kerr initial data sets, arising as special time-slices of the important (sub-extremal) Kerr spacetimes modelling rotating black holes or celestial bodies of prescribed mass m and angular momentum a with $m \geq |a|$, in vacuum, see e.g. Wald [**64**]. These are not spherically, but still axially symmetric. Minimal Bartnik data would need to be replaced by "MOTS" Bartnik data, for which in particular $H = \mathrm{tr}_\Sigma K$ would be requested to hold on $\Sigma \cong \mathbb{S}^2$. Constructing admissible extensions for MOTS Bartnik data would thus require

(a) to come up with a new idea of how to construct suitable axisymmetric "collars", including a construction for K, near MOTS, replacing the condition $R(\gamma) > 0$ by the (strict) version of (5.1), and
(b) to be able to smoothly glue such a collar to a suitable time-slice of a sub-extremal Kerr spacetime with suitably chosen mass m, while ensuring (5.1) along the gluing bridge and that the inner boundary of the constructed initial data set is outer-minimizing.

Studying the time-evolution of the constructed initial data. From a physics perspective, it is natural to ask what spacetimes arising by time-evolution under the Einstein Equations from initial data sets such as those described in this survey will look like. This question silently assumes that one chooses what is a called a *matter model* describing the matter present in the gravitating system one wishes to model mathematically. Mantoulidis and Schoen [**43**] as well as Alaee and the authors [**1**] give examples of matter models which are compatible with time-evolution of the respective constructed time-symmetric initial data sets. It would be desirable to know more generally which matter models are compatible with time-evolution under the Einstein Equations for the various generalizations of the Mantoulidis–Schoen construction described in this survey.

Acknowledgments

We would like to thank the anonymous referee, Jeffrey L. Jauregui, Christos Mantoulidis, Stephen McCormick, Pengzi Miao, and Richard M. Schoen for helpful comments and interesting conversations regarding this survey. We thank Axel Fehrenbach for creating the pictures for this survey.

References

[1] Aghil Alaee, Armando J. Cabrera Pacheco, and Carla Cederbaum, *Asymptotically flat extensions with charge*, Adv. Theor. Math. Phys. **23** (2019), no. 8, 1951–1980, DOI 10.4310/atmp.2019.v23.n8.a1. MR4101657

[2] Lucas C. Ambrozio, *On perturbations of the Schwarzschild anti-de Sitter spaces of positive mass*, Comm. Math. Phys. **337** (2015), no. 2, 767–783, DOI 10.1007/s00220-015-2360-6. MR3339162

[3] Michael T. Anderson and Jeffrey L. Jauregui, *Embeddings, immersions and the Bartnik quasi-local mass conjectures*, Ann. Henri Poincaré **20** (2019), no. 5, 1651–1698, DOI 10.1007/s00023-019-00786-3. MR3942233

[4] Lars Andersson, Mingliang Cai, and Gregory J. Galloway, *Rigidity and positivity of mass for asymptotically hyperbolic manifolds*, Ann. Henri Poincaré **9** (2008), no. 1, 1–33, DOI 10.1007/s00023-007-0348-2. MR2389888

[5] R. Arnowitt, S. Deser, and C. W. Misner, *Coordinate invariance and energy expressions in general relativity*, Phys. Rev. (2) **122** (1961), 997–1006. MR127946

[6] Robert Bartnik, *The mass of an asymptotically flat manifold*, Comm. Pure Appl. Math. **39** (1986), no. 5, 661–693, DOI 10.1002/cpa.3160390505. MR849427

[7] Robert Bartnik, *New definition of quasilocal mass*, Phys. Rev. Lett. **62** (1989), no. 20, 2346–2348, DOI 10.1103/PhysRevLett.62.2346. MR996396

[8] Robert Bartnik, *Quasi-spherical metrics and prescribed scalar curvature*, J. Differential Geom. **37** (1993), no. 1, 31–71. MR1198599

[9] Hubert L. Bray, *Proof of the Riemannian Penrose inequality using the positive mass theorem*, J. Differential Geom. **59** (2001), no. 2, 177–267. MR1908823

[10] Hubert L. Bray and Piotr T. Chruściel, *The Penrose inequality*, The Einstein equations and the large scale behavior of gravitational fields, Birkhäuser, Basel, 2004, pp. 39–70. MR2098913

[11] Hubert L. Bray and Dan A. Lee, *On the Riemannian Penrose inequality in dimensions less than eight*, Duke Math. J. **148** (2009), no. 1, 81–106, DOI 10.1215/00127094-2009-020. MR2515101

[12] Simon Brendle and Richard Schoen, *Manifolds with* 1/4*-pinched curvature are space forms*, J. Amer. Math. Soc. **22** (2009), no. 1, 287–307, DOI 10.1090/S0894-0347-08-00613-9. MR2449060

[13] Simon Brendle and Richard Schoen, *Curvature, sphere theorems, and the Ricci flow*, Bull. Amer. Math. Soc. (N.S.) **48** (2011), no. 1, 1–32, DOI 10.1090/S0273-0979-2010-01312-4. MR2738904

[14] Armando J. Cabrera Pacheco and Carla Cederbaum, *Minimizing sequences for Bartnik's quasi-local mass*, work in progress.

[15] Armando J. Cabrera Pacheco, Carla Cederbaum, and Stephen McCormick, *Asymptotically hyperbolic extensions and an analogue of the Bartnik mass*, J. Geom. Phys. **132** (2018), 338–357, DOI 10.1016/j.geomphys.2018.06.010. MR3836784

[16] Armando J. Cabrera Pacheco, Carla Cederbaum, Stephen McCormick, and Pengzi Miao, *Asymptotically flat extensions of CMC Bartnik data*, Classical Quantum Gravity **34** (2017), no. 10, 105001, 15, DOI 10.1088/1361-6382/aa6921. MR3649699

[17] Armando J. Cabrera Pacheco and Pengzi Miao, *Higher dimensional black hole initial data with prescribed boundary metric*, Math. Res. Lett. **25** (2018), no. 3, 937–956, DOI 10.4310/MRL.2018.v25.n3.a10. MR3847341

[18] Yvonne Choquet-Bruhat, *General relativity and the Einstein equations*, Oxford Mathematical Monographs, Oxford University Press, Oxford, 2009. MR2473363

[19] D. Christodoulou and S.-T. Yau, *Some remarks on the quasi-local mass*, Mathematics and general relativity (Santa Cruz, CA, 1986), Contemp. Math., vol. 71, Amer. Math. Soc., Providence, RI, 1988, pp. 9–14, DOI 10.1090/conm/071/954405. MR954405

[20] Piotr T. Chruściel, *On the invariant mass conjecture in general relativity*, Comm. Math. Phys. **120** (1988), no. 2, 233–248. MR973533

[21] Piotr T. Chruściel and Erwann Delay, *The hyperbolic positive energy theorem*, `arXiv:1901.05263v1`, 2019.

[22] Piotr T. Chruściel, Gregory J. Galloway, Luc Nguyen, and Tim-Torben Paetz, *On the mass aspect function and positive energy theorems for asymptotically hyperbolic manifolds*, Classical Quantum Gravity **35** (2018), no. 11, 115015, 38, DOI 10.1088/1361-6382/aabed1. MR3801943

[23] Piotr T. Chruściel and Marc Herzlich, *The mass of asymptotically hyperbolic Riemannian manifolds*, Pacific J. Math. **212** (2003), no. 2, 231–264, DOI 10.2140/pjm.2003.212.231. MR2038048

[24] Mattias Dahl, Romain Gicquaud, and Anna Sakovich, *Penrose type inequalities for asymptotically hyperbolic graphs*, Ann. Henri Poincaré **14** (2013), no. 5, 1135–1168, DOI 10.1007/s00023-012-0218-4. MR3070749

[25] Sergio Dain and María E. Gabach-Clement, *Geometrical inequalities bounding angular momentum and charges in General Relativity*, Living Reviews in Relativity **21** (2018), no. 5, 74pp.

[26] Levi Lopes de Lima and Frederico Girão, *An Alexandrov-Fenchel-type inequality in hyperbolic space with an application to a Penrose inequality*, Ann. Henri Poincaré **17** (2016), no. 4, 979–1002, DOI 10.1007/s00023-015-0414-0. MR3472630

[27] Marcelo M. Disconzi and Marcus A. Khuri, *On the Penrose inequality for charged black holes*, Classical Quantum Gravity **29** (2012), no. 24, 245019, 18, DOI 10.1088/0264-9381/29/24/245019. MR3002957

[28] Gregory J. Galloway, Pengzi Miao, and Richard Schoen, *Initial data and the Einstein constraint equations*, General relativity and gravitation, Cambridge Univ. Press, Cambridge, 2015, pp. 412–448. MR3644040

[29] Penelope Gehring, *Construction of higher dimensional asymptotically hyperbolic initial data sets with minimal boundary*, Master thesis, Tübingen University, 2019.

[30] Claus Gerhardt, *Flow of nonconvex hypersurfaces into spheres*, J. Differential Geom. **32** (1990), no. 1, 299–314. MR1064876

[31] Richard S. Hamilton, *The Ricci flow on surfaces*, Mathematics and general relativity (Santa Cruz, CA, 1986), Contemp. Math., vol. 71, Amer. Math. Soc., Providence, RI, 1988, pp. 237–262, DOI 10.1090/conm/071/954419. MR954419

[32] Lan-Hsuan Huang, Hyun Chul Jang, and Daniel Martin, *Mass rigidity for hyperbolic manifolds*, Comm. Math. Phys. **376** (2020), no. 3, 2329–2349, DOI 10.1007/s00220-019-03623-0. MR4104551

[33] Gerhard Huisken and Tom Ilmanen, *The inverse mean curvature flow and the Riemannian Penrose inequality*, J. Differential Geom. **59** (2001), no. 3, 353–437. MR1916951

[34] Shahar Hod, *Cosmic censorship, area theorem, and self-energy of particles*, Phys. Rev. D (3) **66** (2002), no. 2, 024016, 4, DOI 10.1103/PhysRevD.66.024016. MR1936620

[35] Jeffrey L. Jauregui, *Smoothing the Bartnik boundary conditions and other results on Bartnik's quasi-local mass*, J. Geom. Phys. **136** (2019), 228–243, DOI 10.1016/j.geomphys.2018.11.005. MR3885243

[36] Chao Li and Christos Mantoulidis, *Positive scalar curvature with skeleton singularities*, Math. Ann. **374** (2019), no. 1-2, 99–131, DOI 10.1007/s00208-018-1753-1. MR3961306

[37] Chen-Yun Lin, *Parabolic constructions of asymptotically flat 3-metrics of prescribed scalar curvature*, Calc. Var. Partial Differential Equations **49** (2014), no. 3-4, 1309–1335, DOI 10.1007/s00526-013-0623-7. MR3168634

[38] Chen-Yun Lin and Christina Sormani, *Bartnik's mass and Hamilton's modified Ricci flow*, Ann. Henri Poincaré **17** (2016), no. 10, 2783–2800, DOI 10.1007/s00023-016-0483-8. MR3546987

[39] Christos Mantoulidis and Pengzi Miao, *Mean curvature deficit and quasi-local mass*, Nonlinear analysis in geometry and applied mathematics, Harv. Univ. Cent. Math. Sci. Appl. Ser. Math., vol. 1, Int. Press, Somerville, MA, 2017, pp. 99–107. MR3729087

[40] Christos Mantoulidis and Pengzi Miao, *Total mean curvature, scalar curvature, and a variational analog of Brown-York mass*, Comm. Math. Phys. **352** (2017), no. 2, 703–718, DOI 10.1007/s00220-016-2767-8. MR3627410

[41] Christos Mantoulidis, Pengzi Miao, and Luen-Fai Tam, *Capacity, quasi-local mass, and singular fill-ins*, J. Reine Angew. Math. **768** (2020), 55–92, DOI 10.1515/crelle-2019-0040. MR4168687

[42] Christos Mantoulidis and Richard M. Schoen, *The curvature on a black hole boundary*, CQG+- geometry, 2015.

[43] Christos Mantoulidis and Richard M. Schoen, *On the Bartnik mass of apparent horizons*, Class. Quantum Grav. **32** (2016), no. 20, 205002, 16pp.

[44] Fernando Codá Marques, *Deforming three-manifolds with positive scalar curvature*, Ann. of Math. (2) **176** (2012), no. 2, 815–863, DOI 10.4007/annals.2012.176.2.3. MR2950765

[45] Marc Mars, *Present status of the Penrose inequality*, Classical Quantum Gravity **26** (2009), no. 19, 193001, 59, DOI 10.1088/0264-9381/26/19/193001. MR2545137

[46] Stephen McCormick, *Gluing Bartnik extensions, continuity of the Bartnik mass, and the equivalence of definitions*, Pacific J. Math. **304** (2020), no. 2, 629–653, DOI 10.2140/pjm.2020.304.629. MR4062784

[47] Stephen McCormick and Pengzi Miao, *On a Penrose-like inequality in dimensions less than eight*, Int. Math. Res. Not. IMRN **7** (2019), 2069–2084, DOI 10.1093/imrn/rnx181. MR3938317

[48] Pengzi Miao, *On a localized Riemannian Penrose inequality*, Comm. Math. Phys. **292** (2009), no. 1, 271–284, DOI 10.1007/s00220-009-0834-0. MR2540078

[49] Pengzi Miao, Yaohua Wang, and Naqing Xie, *On Hawking mass and Bartnik mass of CMC surfaces*, Math. Res. Lett. **27** (2020), no. 3, 855–885, DOI 10.4310/MRL.2020.v27.n3.a12. MR4216572

[50] Pengzi Miao and Naqing Xie, *On compact 3-manifolds with nonnegative scalar curvature with a CMC boundary component*, Trans. Amer. Math. Soc. **370** (2018), no. 8, 5887–5906, DOI 10.1090/tran/7500. MR3803150

[51] Pengzi Miao and Naqing Xie, *Bartnik mass via vacuum extensions*, Internat. J. Math. **30** (2019), no. 13, 1940006, 11, DOI 10.1142/S0129167X19400068. MR4044002

[52] Alejandro Peñuela Díaz, *On the stability of the higher dimensional Riemannian Penrose inequality with charge*, Master thesis, Tübingen University, 2019.

[53] Hans Ringström, *The Cauchy problem in general relativity*, ESI Lectures in Mathematics and Physics, European Mathematical Society (EMS), Zürich, 2009, DOI 10.4171/053. MR2527641

[54] Anna Sakovich, *The Jang equation and the positive mass theorem in the asymptotically hyperbolic setting*, Comm. Math. Phys. **386** (2021), no. 2, 903–973, DOI 10.1007/s00220-021-04083-1. MR4294283

[55] Richard M. Schoen and Shing Tung Yau, *Complete manifolds with nonnegative scalar curvature and the positive action conjecture in general relativity*, Proc. Nat. Acad. Sci. U.S.A. **76** (1979), no. 3, 1024–1025, DOI 10.1073/pnas.76.3.1024. MR524327

[56] Richard Schoen and Shing Tung Yau, *On the proof of the positive mass conjecture in general relativity*, Comm. Math. Phys. **65** (1979), no. 1, 45–76. MR526976

[57] Richard Schoen and Shing Tung Yau, *Positive scalar curvature and minimal hypersurface singularities*, arXiv:1704.05490v1, 2017.

[58] Yuguang Shi and Luen-Fai Tam, *Positive mass theorem and the boundary behaviors of compact manifolds with nonnegative scalar curvature*, J. Differential Geom. **62** (2002), no. 1, 79–125. MR1987378

[59] Nathan Smale, *Generic regularity of homologically area minimizing hypersurfaces in eight-dimensional manifolds*, Comm. Anal. Geom. **1** (1993), no. 2, 217–228, DOI 10.4310/CAG.1993.v1.n2.a2. MR1243523

[60] Brian Smith, *Black hole initial data with a horizon of prescribed intrinsic and extrinsic geometry*, Complex analysis and dynamical systems IV. Part 2, Contemp. Math., vol. 554, Amer. Math. Soc., Providence, RI, 2011, pp. 229–244, DOI 10.1090/conm/554/10971. MR2884405

[61] Brian Smith and Gilbert Weinstein, *On the connectedness of the space of initial data for the Einstein equations*, Electron. Res. Announc. Amer. Math. Soc. **6** (2000), 52–63, DOI 10.1090/S1079-6762-00-00081-0. MR1777856

[62] Brian Smith and Gilbert Weinstein, *Quasiconvex foliations and asymptotically flat metrics of non-negative scalar curvature*, Comm. Anal. Geom. **12** (2004), no. 3, 511–551. MR2128602

[63] John I. E. Urbas, *On the expansion of starshaped hypersurfaces by symmetric functions of their principal curvatures*, Math. Z. **205** (1990), no. 3, 355–372, DOI 10.1007/BF02571249. MR1082861

[64] Robert M. Wald, *General relativity*, University of Chicago Press, Chicago, IL, 1984, DOI 10.7208/chicago/9780226870373.001.0001. MR757180

[65] Xiaodong Wang, *The mass of asymptotically hyperbolic manifolds*, J. Differential Geom. **57** (2001), no. 2, 273–299. MR1879228

[66] Edward Witten, *A new proof of the positive energy theorem*, Comm. Math. Phys. **80** (1981), no. 3, 381–402. MR626707

[67] David Wiygul, *The Bartnik-Bray outer mass of small metric spheres in time-symmetric 3-slices*, Comm. Math. Phys. **358** (2018), no. 1, 269–293, DOI 10.1007/s00220-017-3005-8. MR3772038

Department of Mathematics, Universität Tübingen, Tübingen 72076, Germany
Email address: cabrera@math.uni-tuebingen.de

Department of Mathematics, Universität Tübingen, Tübingen 72076, Germany
Email address: cederbaum@math.uni-tuebingen.de

Contemporary Mathematics
Volume **775**, 2021
https://doi.org/10.1090/conm/775/15587

Four-dimensional homogeneous Kähler Ricci solitons

Esteban Calviño-Louzao, Eduardo García-Río, Ixchel Gutiérrez-Rodríguez, and Ramón Vázquez-Lorenzo

ABSTRACT. We show that four-dimensional homogeneous Kähler Ricci solitons are Einstein, rigid or an algebraic Ricci soliton on the only 3-symmetric space.

1. Introduction

A Riemannian manifold (M, g) is a *Ricci soliton* if there is a vector field X on M so that

$$\mathcal{L}_X g + \rho = \lambda g, \tag{1.1}$$

where $\mathcal{L}$ is the Lie derivative, ρ denotes the Ricci tensor and $\lambda \in \mathbb{R}$. Ricci solitons are self-similar solutions of the Ricci flow $\frac{\partial}{\partial t} g(t) = -2\rho(g(t))$, i.e., they are fixed points of the flow up to diffeomorphisms and rescaling. While Ricci flat metrics are the genuine fixed points of the flow, Einstein metrics remain constant up to scaling under the flow. Indeed, if $g(0)$ is an Einstein metric satisfying $\rho(0) = \lambda g(0)$, then $g(0)$ evolves with the flow as $g(t) = (1 - 2\lambda t)g(0)$. A Ricci soliton (M, g, X) is called *shrinking*, *steady* or *expading* if $\lambda > 0$, $\lambda = 0$ or $\lambda > 0$, respectively. We refer to [**6, 7**] for more information and references on Ricci solitons.

If the vector field $X = \nabla f$ is a gradient, then (1.1) becomes $\operatorname{Hes}_f + \rho = \lambda g$ for some potential function f and (M, g, f) is called a *gradient Ricci soliton*. A gradient Ricci soliton is called *rigid* if (M, g) splits as a product $\mathbb{R}^k \times N$, where N is Einstein with $\rho = \lambda g$ and the potential function $f(\,\cdot\,) = \frac{\lambda}{4}\|\pi_{\mathbb{R}^k}(\,\cdot\,)\|^2$ is determined by the projection on the Euclidean factor [**16**]. While homogeneous gradient Ricci solitons are rigid [**15**], there are irreducible homogeneous manifolds admitting generic Ricci solitons. Many of them are constructed by using the notion of algebraic Ricci soliton.

Based on the fact that Ricci solitons are self-similar solutions of the Ricci flow, Lauret [**12**] considered Lie groups and searched for fixed points of the flow up to automorphisms of the Lie group instead of diffeomorphisms. Let G be a Lie group with Lie algebra $\mathfrak{g}$. A left-invariant metric $\langle\,\cdot\,,\,\cdot\,\rangle$ on G is called an *algebraic Ricci soliton* if

$$\mathfrak{D} = \operatorname{Ric} - \lambda \operatorname{id} \tag{1.2}$$

2020 *Mathematics Subject Classification.* Primary 53C30, 53C44, 53C25.

Supported by projects MTM2016-75897-P and ED431C 2019/10 with FEDER funds (Spain).

is a derivation of the Lie algebra, where Ric denotes the Ricci operator $\langle \mathrm{Ric}\, X, Y\rangle = \rho(X,Y)$. Let $\mathfrak{D}$ be a derivation given by (1.2) and let φ_t denote the one-parameter family of automorphisms of G determined by $d\varphi_t|_e = \exp \frac{t}{2}\mathfrak{D}$. Then the vector field X given by $X(p) = \frac{d}{dt}\varphi_t(p)|_{t=0}$ satisfies (1.1), thus defining a Ricci soliton on G (see [**10, 12, 13**] for more information on homogeneous Ricci solitons).

Our first purpose on this note is to prove the following result which gives an explicit description of four-dimensional homogeneous Kähler manifolds.

THEOREM 1.1. *A complete and simply connected four-dimensional Riemannian Kähler manifold is homogeneous if and only if it is symmetric or isometric to the* 3*-symmetric space.*

A consequence of the previous result provides an explicit description of four-dimensional homogeneous Kähler Ricci solitons as follows.

COROLLARY 1.2. *A four-dimensional complete and simply connected Riemannian homogeneous Kähler Ricci soliton is either Einstein, rigid or the* 3*-symmetric space, which is itself an algebraic soliton.*

We fix some notation on Kähler surfaces, homogeneous spaces and generalized symmetric spaces in Section 2. By using previous work of Bérard-Bergery [**2**] and Ovando [**14**] we analyze the homogeneous Kähler surfaces in Section 3 to proof Theorem 1.1. Finally the proof of Corollary 1.2 is given in Section 4.

2. Preliminaries

2.1. Four-dimensional geometry. We work at the purely algebraic level. Let $\{e_1, e_2, e_3, e_4\}$ be an orthonormal basis of an inner product vector space of Riemannian signature $(V, \langle\cdot,\cdot\rangle)$ and let $\{e^1, e^2, e^3, e^4\}$ be the associated dual basis. Further, let $e^{ij} = e^i \wedge e^j$ and consider the induced metric on $\Lambda^2(V)$ given by $\langle\langle x\wedge y, z\wedge w\rangle\rangle = \langle x,z\rangle\langle y,w\rangle - \langle y,z\rangle\langle x,w\rangle$. In dimension four the Hodge star operator $\star$ is an endomorphism of the space of 2-forms which satisfies $\star^2 = \mathrm{id}_{\Lambda^2}$. Hence it induces a splitting $\Lambda^2 = \Lambda^+ \oplus \Lambda^-$, where $\Lambda^\pm = \{\alpha \in \Lambda^2 : \star\alpha = \pm\alpha\}$ denote the spaces of self-dual and anti-self-dual 2-forms. Furthermore,

$$E_1^\pm = (e^{12} \pm e^{34})/\sqrt{2}, \quad E_2^\pm = (e^{13} \mp e^{24})/\sqrt{2}, \quad E_3^\pm = (e^{14} \pm e^{23})/\sqrt{2}$$

is an orthonormal basis of $\Lambda^\pm = \mathrm{span}\{E_1^\pm, E_2^\pm, E_3^\pm\}$.

Since Λ^2 splits under the action of the Hodge star operator so does the curvature operator acting on the space of 2-forms as

$$R = \frac{\tau}{12}\,\mathrm{id}_{\Lambda^2} + \rho_0 + W^+ + W^-$$

where $W^\pm = \frac{1}{2}(W \pm \star W)$ denote the self-dual and the anti-self-dual Weyl curvature operators, τ is the scalar curvature and ρ_0 denotes the trace-free Ricci tensor.

2.2. Almost Hermitian four-manifolds. An *almost Hermitian manifold* is a Riemannian manifold (M, g) equipped with an orthogonal almost complex structure (i.e., a $(1,1)$-tensor field J satisfying $J^2 = -\mathrm{id}$ and $J^*g = g$). Any almost Hermitian structure (g, J) gives rise to a non-degenerate 2-form $\Omega(X,Y) = g(JX, Y)$ inducing an orientation which agrees with the one defined by the almost complex structure. Hence, orienting M by the almost complex structure, the Kähler form Ω is self-dual. Hence Ω defines a section of Λ^+ and conversely, any section of

Λ^+ (resp., Λ^-) defines an almost Hermitian structure on M (resp., on the manifold M with the opposite orientation).

For every almost Hermitian four-dimensional manifold $\Lambda^+ = \mathbb{R}\Omega \oplus \langle \mathbb{R}\Omega \rangle^\perp$ and thus the self-dual part of the Weyl curvature operator can be written as

$$W^+ = \left(\begin{array}{c|c} \frac{\kappa}{6} & W_2^+ \\ \hline {}^t(W_2^+) & W_3^+ - \frac{\kappa}{12}\,\mathrm{id}_{\Lambda^2} \end{array} \right)$$

where κ is the conformal scalar curvature, W_2^+ corresponds to the part of W^+ that interchanges the two factors of the splitting of Λ^+ and W_3^+ is a trace-free, self-adjoint endomorphism of $\langle \mathbb{R}\Omega \rangle^\perp$. Moreover, the traceless part of the Ricci tensor (which is not necessarily J-invariant) decomposes into its invariant and anti-invariant components $\rho_0 = \rho_0^i + \rho_0^a$, where $\rho_0^i(X,Y) = \frac{1}{2}(\rho_0(X,Y) + \rho_0(JX,JY))$ and $\rho_0^a(X,Y) = \frac{1}{2}(\rho_0(X,Y) - \rho_0(JX,JY))$ are J-invariant and J-anti-invariant, respectively.

(M,g,J) is said to be *almost Kähler* if the 2-form Ω is closed and (M,g,J) is said to be *Kähler* if, in addition, the almost complex structure is integrable (or, equivalently, the Kähler form Ω is parallel). Curvature identities for Kähler manifolds show that $\rho_0^a = 0$ and $W_2^+ = W_3^+ = 0$ in the Kähler setting. Apostolov, Armstrong and Drăghici showed in [**1**] that there is only one non-Kähler, almost Kähler four-dimensional manifold for which both the Ricci and the Weyl curvatures have the same algebraic symmetries as they have for a Kähler metric (see Theorem 2.2).

2.3. Homogeneous spaces. A connected Riemannian manifold (M,g) is said to be *homogeneous* if the group of isometries acts transitively on M. Further a Kähler manifold (M,g,J) is homogeneous if there is a group of J-preserving isometries which acts transitively on M. Sekigawa showed in [**17**] that any three-dimensional complete and simply connected homogeneous Riemannian manifold is a symmetric space or it is isometric to a Lie group with left-invatiant metric. Bérard-Bergery extended this result to the four-dimensional setting so that one has

THEOREM 2.1. [**2**] *Let (M,g) be a four-dimensional complete and simply connected homogeneous Riemannian manifold. Then (M,g) is either symmetric or it is isometric to a Lie group with a left-invariant metric.*

In particular, M is one of the groups $\widetilde{SL(2,\mathbb{R})} \times \mathbb{R}$, $SU(2) \times \mathbb{R}$ or it is a solvable Lie group. Four-dimensional solvable Lie algebras are obtained as extensions of the three-dimensional unimodular Lie algebras: the abelian Lie algebra $\mathfrak{r}^3$, the Heisenberg algebra $\mathfrak{h}_3$, the Poincaré algebra $\mathfrak{e}(1,1)$ and the Euclidean algebra $\mathfrak{e}(2)$. It is an important observation that four-dimensional Lie groups which admit a left-invariant non-degenerate closed 2-form are necessarily solvable [**8**]. In particular any four-dimensional Kähler Lie group is solvable thus excluding the cases $\widetilde{SL(2,\mathbb{R})} \times \mathbb{R}$ and $SU(2) \times \mathbb{R}$.

2.4. 3-symmetric spaces. Let (M,g,J) be an almost Hermitian manifold. Then $\Theta = -\frac{1}{2}\,\mathrm{id} + \frac{\sqrt{3}}{2}J$ satisfies $\Theta^3 = \mathrm{id}$ and it determines a family of local cubic diffeomorphisms, i.e., a differentiable function $p \in M \mapsto \vartheta_p$ where ϑ_p is the diffeomorphism defined in a suitable small neighborhood of p so that $d\vartheta_p = \Theta$ and

p is the only fixed point of ϑ_p. Conversely given a family $p \mapsto \vartheta_p$ of local cubic diffeomorphisms, there exists an almost complex structure $\mathfrak{J}$ on M determined by $d\vartheta_p = -\frac{1}{2}\operatorname{id} + \frac{\sqrt{3}}{2}\mathfrak{J}_p$ (see [**9**]). A Riemannian *locally* 3*-symmetric space* is a manifold (M, g) equipped with a family of local cubic diffeomorphisms which are $\mathfrak{J}$-holomorphic isometries. Whenever the local cubic diffeomorphisms are globally defined one says that M is a 3*-symmetric space.*

There is only one Riemannian 3-symmetric space in dimension four, which is realized as the following metric on $\mathbb{R}^4$ with coordinates (t, x, y, z) (see [**11**]):

$$\begin{aligned} g &= \left((\sqrt{1+t^2+x^2} - t)dy^2 + (\sqrt{1+t^2+x^2} + t)dz^2 - 2x\,dydz\right) \\ &\quad + \beta\left((1+x^2)dt^2 + (1+t^2)dx^2 - 2tx\,dtdx\right)/(1+t^2+x^2), \end{aligned}$$

where β is a non-zero constant. Viewed as a homogeneous space G/K, the four-dimensional 3-symmetric space corresponds to the matrix groups (see [**11**])

$$G = \begin{pmatrix} a & b & u \\ c & d & v \\ 0 & 0 & 1 \end{pmatrix}, \text{ and } K = \begin{pmatrix} \cos t & -\sin t & 0 \\ \sin t & \cos t & 0 \\ 0 & 0 & 1 \end{pmatrix}, \text{ where } \det\begin{pmatrix} a & b \\ c & d \end{pmatrix} = 1.$$

A straightforward calculation shows that the self-dual and anti-self-dual Weyl curvature operators satisfy $W^\pm = \pm\operatorname{diag}[\frac{1}{2\beta}, -\frac{1}{4\beta}, -\frac{1}{4\beta}]$ and thus $W^\pm$, have a distinguished eigenvalue. Hence any four-dimensional 3-symmetric space is naturally equipped with an almost complex structure J_+ and an opposite almost complex structure J_- determined by the one-dimensional eigenspaces $\ker(W^\pm \mp \frac{1}{2\beta}\operatorname{id}_{\Lambda^\pm})$. Moreover $(\mathbb{R}^4, g, J_+, J_-)$ is almost Kähler and opposite Kähler (i.e., $d\Omega_+ = 0$ and $\nabla J_- = 0$) and the curvature of the almost Kähler structure has the algebraic symmetries of the Kähler case (see, for example [**5**]). Now the previously mentioned rigidity result of Apostolov, Armstrong and Drăghici can be stated as follows.

THEOREM 2.2. [**1**] *Any strictly almost Kähler four-dimensional manifold whose curvature satisfies $\rho_0^a = 0$, and $W_2^+ = W_3^+ = 0$ is locally isometric to the unique four-dimensional 3-symmetric space.*

3. Homogeneous Kähler surfaces. The proof of Theorem 1.1

By Theorem 2.1 and the results in [**8**], a complete and simply connected Riemannian homogeneous Kähler four-manifold (M, g, J) is either symmetric or isometric to a solvable Lie group with a left-invariant Kähler structure. Since Ovando classified left-invariant indefinite Kähler structures on four-dimensional Lie groups, in what remains of this section we will examine the different possibilities in [**14**]. The proof of Theorem 1.1 follows after a case by case analysis of Ovando's classification. Following the notation in [**14**], the Kähler structures corresponding to the Lie algebras $\mathfrak{rh}_3$, $\mathfrak{r}'_2$, $\mathfrak{r}_{4,-1,-1}$ and $\mathfrak{d}_{4,1}$ are of neutral signature (cf. [**14**, Corollary 3.12]). Therefore, we omit those cases in what follows and consider separately the other posibilities.

3.1. The Lie algebra $\mathfrak{rr}_{3,0}$. The Lie algebra $\mathfrak{rr}_{3,0}$ is described, with respect to a suitable basis $\{e_1, e_2, e_3, e_4\}$, by the non-zero bracket (up to the usual symmetries) given by $[e_1, e_2] = e_2$. $\mathfrak{rr}_{3,0}$ admits a left-invariant Kähler structure determined by the complex structure $Je_1 = e_2$, $Je_3 = e_4$ and Kähler form $\Omega = ae^{12} + be^{34}$, where $ab \neq 0$. Then the associated metric is given by

$$g = a(e^1 \circ e^1 + e^2 \circ e^2) + b(e^3 \circ e^3 + e^4 \circ e^4).$$

Hence, the metric is either Riemannian or of neutral signature depending on the sign of ab. The Ricci operator takes the form $\mathrm{Ric} = -\frac{1}{a}\,\mathrm{diag}[1,1,0,0]$ with respect to the orthogonal basis $\{e_1, e_2, e_3, e_4\}$ and it is parallel. Hence the underlying manifold is isometric to the product of the Euclidean plane and a surface with constant Gauss curvature $G = \mathbb{R}^2 \times N(-\frac{1}{a})$, and therefore symmetric.

3.2. The Lie algebra $\mathfrak{rr}'_{3,0}$. The Lie algebra $\mathfrak{rr}'_{3,0}$ is described, with respect to a suitable basis $\{e_1, e_2, e_3, e_4\}$, by the non-zero brackets (up to the usual symmetries) given by $[e_1, e_2] = -e_3$, and $[e_1, e_3] = e_2$. $\mathfrak{rr}'_{3,0}$ admits a left-invariant Kähler structure determined by the complex structure $Je_1 = e_4$, $Je_2 = e_3$ and Kähler form $\Omega = ae^{14} + be^{23}$, where $ab \neq 0$. Then the associated metric is given by

$$g = a(e^1 \circ e^1 + e^4 \circ e^4) + b(e^2 \circ e^2 + e^3 \circ e^3).$$

Hence, the metric is Riemannian or of neutral signature. Moreover, the curvature tensor vanishes and the homogeneous space is flat.

3.3. The Lie algebra $\mathfrak{r}_2\mathfrak{r}_2$. The Lie algebra $\mathfrak{r}_2\mathfrak{r}_2$ is described, with respect to a suitable basis $\{e_1, e_2, e_3, e_4\}$, by the non-zero brackets (up to the usual symmetries) given by $[e_1, e_2] = e_2$, and $[e_3, e_4] = e_4$. $\mathfrak{r}_2\mathfrak{r}_2$ admits a left-invariant Kähler structure determined by the complex structure $Je_1 = e_2$, $Je_3 = e_4$ and Kähler form $\Omega = ae^{12} + be^{34}$, where $ab \neq 0$. Then the associated metric is given by

$$g = a(e^1 \circ e^1 + e^2 \circ e^2) + b(e^3 \circ e^3 + e^4 \circ e^4).$$

A straightforward calculation shows that the Ricci operator, when expressed on the orthogonal basis $\{e_1, e_2, e_3, e_4\}$, takes the form $\mathrm{Ric} = -\,\mathrm{diag}[\frac{1}{a}, \frac{1}{a}, \frac{1}{b}, \frac{1}{b}]$. Moreover, the Ricci tensor is parallel and hence the homogeneous space is the product of two surfaces of constant Gauss curvature $G = N_1(-\frac{1}{a}) \times N_2(-\frac{1}{b})$, and thus symmetric.

3.4. The Lie algebra $\mathfrak{r}'_{4,0,\lambda}$. The Lie algebra $\mathfrak{r}'_{4,0,\lambda}$ is described, with respect to a suitable basis $\{e_1, e_2, e_3, e_4\}$, by the non-zero brackets

$$[e_1, e_4] = -e_1, \quad [e_2, e_4] = \lambda e_3, \quad [e_3, e_4] = -\lambda e_2.$$

In this case, there exist two possible Kähler structures.

3.4.1. *Case 1.* $\mathfrak{r}'_{4,0,\lambda}$ admits a left-invariant Kähler structure determined by the complex structure $J_1e_4 = e_1$, $J_1e_2 = e_3$ and Kähler form $\Omega_1 = ae^{14} + be^{23}$, where $ab \neq 0$. Then the associated metric is given by

$$g_1 = -a(e^1 \circ e^1 + e^4 \circ e^4) + b(e^2 \circ e^2 + e^3 \circ e^3).$$

Hence, the metric is Riemannian or of neutral signature. A straightforward calculation shows that the Ricci operator, when expressed in the orthogonal basis $\{e_1, e_2, e_3, e_4\}$, takes the form $\mathrm{Ric} = \frac{1}{a}\,\mathrm{diag}[1,0,0,1]$ and it is parallel. Hence, the homogeneous space is the product of the Euclidean plane and a surface with constant Gauss curvature $G = \mathbb{R}^2 \times N(\frac{1}{a})$ and thus symmetric.

3.4.2. *Case 2.* $\mathfrak{r}'_{4,0,\lambda}$ admits a left-invariant Kähler structure determined by the complex structure $J_2e_4 = e_1$, $J_2e_2 = -e_3$ and Kähler form $\Omega_2 = ae^{14} + be^{23}$, where $ab \neq 0$. Then the associated metric is given by

$$g_2 = -a(e^1 \circ e^1 + e^4 \circ e^4) - b(e^2 \circ e^2 + e^3 \circ e^3).$$

Proceeding exactly as in 3.4.1, the homogeneous space is the product of the Euclidean plane and a surface with constant Gauss curvature $G = \mathbb{R}^2 \times N(\frac{1}{a})$, and thus symmetric.

3.5. The Lie algebra $\mathfrak{d}_{4,2}$. The Lie algebra $\mathfrak{d}_{4,2}$ is described, with respect to a suitable basis $\{e_1, e_2, e_3, e_4\}$, by the non-zero brackets

$$[e_1, e_2] = e_3, \quad [e_1, e_4] = -2e_1, \quad [e_2, e_4] = e_2, \quad [e_3, e_4] = -e_3.$$

In this case, there exist two Kähler structures as follows. An important fact is that none of the Kähler structures on $\mathfrak{d}_{4,2}$ is symmetric.

3.5.1. *Case 1.* $\mathfrak{d}_{4,2}$ admits a left-invariant Kähler structure determined by the complex structure $J_1 e_4 = -e_2$, $J_1 e_1 = e_3$ and Kähler form $\Omega_1 = a(e^{14} + e^{23}) + be^{24}$, where $a \neq 0$. Then the associated metric is given by

$$g_1 = b(e^2 \circ e^2 + e^4 \circ e^4) + 2a(e^1 \circ e^2 + e^3 \circ e^4).$$

Hence, the metric is of neutral signature and we therefore omit this case.

3.5.2. *Case 2.* $\mathfrak{d}_{4,2}$ admits a left-invariant Kähler structure determined by the complex structure $J_2 e_4 = -2e_1$, $J_2 e_2 = e_3$ and Kähler form $\Omega_2 = ae^{14} + be^{23}$, where $ab \neq 0$. Then the associated metric is given by

$$g_2 = \frac{a}{2} e^1 \circ e^1 + 2ae^4 \circ e^4 + b(e^2 \circ e^2 + e^3 \circ e^3).$$

Hence, the metric is Riemannian or of neutral signature.

A straightforward calculation shows that the Ricci operator, when expressed in the orthogonal basis $\{e_1, e_2, e_3, e_4\}$, takes the form $\mathrm{Ric} = -\frac{3}{a}\,\mathrm{diag}[1, 0, 0, 1]$. Furthermore the self-dual and anti-self-dual Weyl curvature operators are given by

$$W^+ = \mathrm{sign}(ab) \begin{pmatrix} \frac{1}{2a} & 0 & 0 \\ 0 & \frac{1}{2a} & 0 \\ 0 & 0 & -\frac{1}{a} \end{pmatrix}, \quad W^- = \mathrm{sign}(ab) \begin{pmatrix} -\frac{1}{2a} & 0 & 0 \\ 0 & -\frac{1}{2a} & 0 \\ 0 & 0 & \frac{1}{a} \end{pmatrix},$$

where $\mathrm{sign}(ab)$ denotes the sign of ab, which determines the signature of the metric. Then $\ker(W^\pm \pm \mathrm{sign}(ab)\frac{1}{a}\,\mathrm{id}_{\Lambda^2})$ define one-dimensional subspaces in $\Lambda^\pm$. The corresponding 2-forms $\Omega_\pm = \sqrt{2}E_3^\pm$ define an almost Hermitian structure (g, J_+) and an opposite almost Hermitian structure (g, J_-) where $J_\pm e_4 = -2e_1$, $J_\pm e_2 = \pm e_3$.

Now, an easy calculation shows that the opposite almost Hermitian structure (g, J_-) is almost Kähler (i.e., $dE_3^- = 0$) and the Ricci tensor is $J_\pm$-invariant ($J_+ = J_2$). Moreover, since the anti-self-dual Weyl curvature operator W^- satisfies the identities $W_2^- = W_3^- = 0$, Theorem 2.2 shows that the underlying structure is isometric to the 3-symmetric space in the Riemannian setting.

3.6. The Lie algebra $\mathfrak{d}_{4,\frac{1}{2}}$. The Lie algebra $\mathfrak{d}_{4,\frac{1}{2}}$ is described, with respect to a suitable basis $\{e_1, e_2, e_3, e_4\}$, by the non-zero brackets

$$[e_1, e_2] = e_3, \quad [e_1, e_4] = -\frac{1}{2} e_1, \quad [e_2, e_4] = -\frac{1}{2} e_2, \quad [e_3, e_4] = -e_3.$$

In this case, there exist two Kähler structures (isometric to the complex hyperbolic plane) as follows.

3.6.1. *Case 1.* $\mathfrak{d}_{4,\frac{1}{2}}$ admits a left-invariant Kähler structure determined by the complex structure $J_1 e_4 = e_3$, $J_1 e_1 = e_2$ and Kähler form $\Omega_1 = a(e^{12} - e^{34})$, where $a \neq 0$. Then the associated metric is given by

$$g_1 = a(e^1 \circ e^1 + e^2 \circ e^2 + e^3 \circ e^3 + e^4 \circ e^4).$$

A straightforward calculation shows that the Ricci operator is a multiple of the identity $\mathrm{Ric} = -\frac{3}{2a}\,\mathrm{id}$ and thus Einstein. Furthermore $W^- = 0$ and $W^+ =$

$\operatorname{diag}[-\frac{1}{a}, \frac{1}{2a}, \frac{1}{2a}]$ and thus it corresponds to the complex hyperbolic space of constant holomorphic sectional curvature $H = -\frac{1}{a}$.

3.6.2. *Case 2.* $\mathfrak{d}_{4,\frac{1}{2}}$ admits a left-invariant Kähler structure determined by the complex structure $J_2 e_4 = e_3$, $J_2 e_1 = -e_2$ and Kähler form $\Omega_2 = a(e^{12} - e^{34})$ where $a \neq 0$. Then the associated metric, given by

$$g_2 = -a(e^1 \circ e^1 + e^2 \circ e^2 - e^3 \circ e^3 - e^4 \circ e^4)$$

is of neutral signature. Hence we avoid this case, which corresponds to a neutral signature Kähler manifold of constant holomorphic sectional curvature $H = -\frac{1}{a}$.

3.7. The Lie algebra $\mathfrak{d}'_{4,\lambda}$. The Lie algebra $\mathfrak{d}'_{4,\lambda}$ is described, with respect to a suitable basis $\{e_1, e_2, e_3, e_4\}$, by the non-zero brackets ($\lambda \geq 0$)

$$[e_1, e_2] = e_3, \quad [e_1, e_4] = e_2 - \frac{\lambda}{2} e_1, \quad [e_2, e_4] = -(e_1 + \frac{\lambda}{2} e_2), \quad [e_3, e_4] = -\lambda e_3.$$

In this case, there exist four Kähler structures whose geometry reduces to the following two cases. Each of them is isometric to a suitable complex hyperbolic plane.

3.7.1. *Case 1.* $\mathfrak{d}'_{4,\lambda}$ admits a left-invariant Kähler structure determined by the complex structure $J_1 e_4 = e_3$, $J_1 e_1 = e_2$ and Kähler form $\Omega_1 = a(e^{12} - \lambda e^{34})$, where $a \neq 0$ and $\lambda > 0$. Then the associated metric is given by

$$g_1 = a\,(e^1 \circ e^1 + e^2 \circ e^2 + \lambda\, e^3 \circ e^3 + \lambda e^4 \circ e^4).$$

Hence, the metric is Riemannian and a straightforward calculation shows that the Ricci operator takes the form $\operatorname{Ric} = -\frac{3\lambda}{2a} \operatorname{id}$. Moreover $W^- = 0$ and $W^+ = \frac{\lambda}{2a} \operatorname{diag}[-2, 1, 1]$ depending on the signature of the metric. Hence it corresponds to the Riemannian or the neutral signature complex hyperbolic plane of constant holomorphic sectional curvature $H = -\frac{\lambda}{a}$, and thus symmetric.

Proceeding in a completely analogous way the opposite Kähler structure J_1' given by $J_1' e_4 = e_3$, $J_1' e_1 = -e_2$ is isometric to the complex hyperbolic space.

3.7.2. *Case 2.* $\mathfrak{d}'_{4,\lambda}$ admits a left-invariant Kähler structure determined by the complex structure $J_2 e_4 = -e_3$, $J_2 e_1 = e_2$ and Kähler form $\Omega_2 = a(e^{12} - \lambda e^{34})$, where $a \neq 0$ and $\lambda > 0$. Then the associated metric is given by

$$g_2 = -a\,(e^1 \circ e^1 + e^2 \circ e^2 - \lambda\, e^3 \circ e^3 - \lambda e^4 \circ e^4).$$

Hence the metric is of neutral signature and it is of constant holomorphic sectional curvature $H = \frac{\lambda}{a}$.

Proceeding in a completely analogous way the opposite Kähler structure J_2' given by $J_2' e_4 = -e_3$, $J_2' e_1 = -e_2$ is isometric to the complex hyperbolic space.

4. The proof of Corollary 1.2

Let (M, g, J) be a four-dimensional homogeneous Kähler manifold. We distinguish the two cases in Theorem 1.1. If (M, g, J) is the 3-symmetric space corresponding to 3.5.2, then a straightforward calculation shows that $\mathfrak{D} = \operatorname{Ric} + \frac{3}{a} \operatorname{id}$ is a derivation of the Lie algebra $\mathfrak{d}_{4,2}$ and thus it defines an algebraic Ricci soliton (see also **[13]** for a classification of four-dimensional solvsolitons).

If (M, g, J) is symmetric then it is either Einstein or the product of two oriented surfaces of constant Gauss curvatures $N_1(c_1) \times N_2(c_2)$. It was shown in **[10]** that homogeneous Ricci solitons admitting a transitive semi-simple group of isometries are necessarily Einstein. Hence the only posible non-Einsteinian cases occur when

M is isometric to $\mathbb{R}^2 \times N(c)$. Then it is a rigid gradient Ricci soliton with potential function $f = \frac{c}{2}\|\pi_{\mathbb{R}^2}\|^2$ where $\pi_{\mathbb{R}^2}$ is the projection on the Euclidean factor, which completes the proof of Corollary 1.2.

References

[1] Vestislav Apostolov, John Armstrong, and Tedi Drăghici, *Local rigidity of certain classes of almost Kähler 4-manifolds*, Ann. Global Anal. Geom. **21** (2002), no. 2, 151–176, DOI 10.1023/A:1014779405043. MR1894944

[2] Lionel Bérard-Bergery, *Les espaces homogènes riemanniens de dimension 4* (French), Riemannian geometry in dimension 4 (Paris, 1978/1979), Textes Math., vol. 3, CEDIC, Paris, 1981, pp. 40–60. MR769130

[3] Andrzej Borowiec, Mauro Francaviglia, and Igor Volovich, *Anti-Kählerian manifolds*, Differential Geom. Appl. **12** (2000), no. 3, 281–289, DOI 10.1016/S0926-2245(00)00017-6. MR1764334

[4] Esteban Calviño-Louzao, Eduardo García-Río, María Elena Vázquez-Abal, and Ramón Vázquez-Lorenzo, *Local rigidity and nonrigidity of symplectic pairs*, Ann. Global Anal. Geom. **41** (2012), no. 2, 241–252, DOI 10.1007/s10455-011-9279-8. MR2876697

[5] Esteban Calviño-Louzao, Eduardo García-Río, María Elena Vázquez-Abal, and Ramón Vázquez-Lorenzo, *Geometric properties of generalized symmetric spaces*, Proc. Roy. Soc. Edinburgh Sect. A **145** (2015), no. 1, 47–71, DOI 10.1017/S0308210513001480. MR3304575

[6] Huai-Dong Cao, *Recent progress on Ricci solitons*, Recent advances in geometric analysis, Adv. Lect. Math. (ALM), vol. 11, Int. Press, Somerville, MA, 2010, pp. 1–38. MR2648937

[7] Bennett Chow, Sun-Chin Chu, David Glickenstein, Christine Guenther, James Isenberg, Tom Ivey, Dan Knopf, Peng Lu, Feng Luo, and Lei Ni, *The Ricci flow: techniques and applications. Part I*, Mathematical Surveys and Monographs, vol. 135, American Mathematical Society, Providence, RI, 2007. Geometric aspects, DOI 10.1090/surv/135. MR2302600

[8] Bon Yao Chu, *Symplectic homogeneous spaces*, Trans. Amer. Math. Soc. **197** (1974), 145–159, DOI 10.2307/1996932. MR342642

[9] Alfred Gray, *Riemannian manifolds with geodesic symmetries of order* 3, J. Differential Geometry **7** (1972), 343–369. MR331281

[10] Michael Jablonski, *Homogeneous Ricci solitons are algebraic*, Geom. Topol. **18** (2014), no. 4, 2477–2486, DOI 10.2140/gt.2014.18.2477. MR3268781

[11] Oldřich Kowalski, *Generalized symmetric spaces*, Lecture Notes in Mathematics, vol. 805, Springer-Verlag, Berlin-New York, 1980. MR579184

[12] Jorge Lauret, *Ricci soliton homogeneous nilmanifolds*, Math. Ann. **319** (2001), no. 4, 715–733, DOI 10.1007/PL00004456. MR1825405

[13] Jorge Lauret, *Ricci soliton solvmanifolds*, J. Reine Angew. Math. **650** (2011), 1–21, DOI 10.1515/CRELLE.2011.001. MR2770554

[14] Gabriela P. Ovando, *Invariant pseudo-Kähler metrics in dimension four*, J. Lie Theory **16** (2006), no. 2, 371–391. MR2197598

[15] Peter Petersen and William Wylie, *On gradient Ricci solitons with symmetry*, Proc. Amer. Math. Soc. **137** (2009), no. 6, 2085–2092, DOI 10.1090/S0002-9939-09-09723-8. MR2480290

[16] Peter Petersen and William Wylie, *Rigidity of gradient Ricci solitons*, Pacific J. Math. **241** (2009), no. 2, 329–345, DOI 10.2140/pjm.2009.241.329. MR2507581

[17] Kouei Sekigawa, *On some 3-dimensional curvature homogeneous spaces*, Tensor (N.S.) **31** (1977), no. 1, 87–97. MR464115

Consellería de Cultura, Educación e Ordenación Universitaria, Edificio Administrativo San Caetano, 15781 Santiago de Compostela, Spain
Email address: estebcl@edu.xunta.es

Faculty of Mathematics, University of Santiago de Compostela, 15782 Santiago de Compostela, Spain
Email address: eduardo.garcia.rio@usc.es

Department of Mathematics, University of Vigo, Esc. Sup. de Enx. Informática, Campus de Ourense, 32004 Ourense, Spain
Email address: ixchel.dzohara.gutierrez.rodriguez@uvigo.es

Department of Mathematics, IES de Ribadeo Dionisio Gamallo, 27700 Ribadeo, Spain
Email address: ravazlor@edu.xunta.es

Contemporary Mathematics
Volume **775**, 2021
https://doi.org/10.1090/conm/775/15588

C*-algebras and their nuclear dimension

Jorge Castillejos

ABSTRACT. We review the notion of nuclear dimension for C*-algebras introduced by Winter and Zacharias. We explain why it is a non-commutative version of topological dimension. After presenting several examples, we give a brief overview of the state of the art.

Introduction

Typically, *C*-algebras* are defined as complex Banach algebras with an involution that satisfy the C*-identity. Precisely, a complex algebra $\mathcal{A}$ with an involution is a C*-algebra if it has a complete norm that makes the operations continuous and satisfies the *C*-identity*,

$$\|a^*a\| = \|a\|^2, \qquad a \in \mathcal{A}.$$

Thanks to the Gelfand-Naimark theorem, we can also define C*-algebras as subalgebras of the algebra of bounded operators on a Hilbert space [**GN43**]. This immediately leads us to the theory of *operator algebras*.

Operator algebras are subalgebras of the algebra of bounded operators on a Hilbert space. In particular, C*-algebras are those that are closed and self-adjoint. Motivated by the fact that unital commutative C*-algebras are always isomorphic to the algebra of complex-valued continuous functions on a compact Hausdorff space, we view general C*-algebras as non-commutative topological spaces. This area of mathematics originated from the groundbreaking work of von Neumann on the mathematical formulation of quantum mechanics [**vN55**], and nowadays has interesting connections with other areas of mathematics such as group theory, dynamical systems, geometry, logic, and quantum information.

Many endeavours have been directed towards understanding the structure of C*-algebras. In particular, the well-behaved class of *nuclear* C*-algebras has been deeply studied in the past [**Tak64, Lan73, CE78, Kir77**]. Roughly speaking, a C*-algebra is nuclear if it can be approximated by matrix algebras and nuclearity can also be considered as the C*-version of *amenability* for groups.

By viewing C*-algebras as non-commutative topological spaces, Winter introduced several notions of non-commutative topological dimension for nuclear C*-algebras using Čech's *covering dimension* as a model [**Win03, Win09**]. These

2020 *Mathematics Subject Classification.* Primary 46L05, 46L35, 46L85, 43A07.
The author was supported by long term structural funding - Methusalem grant of the Flemish Government.

notions were later refined by Winter and Kirchberg [**KW04**], and by Winter and Zacharias [**WZ10**]. Roughly speaking, the covering dimension of a topological space is obtained by colouring finite open covers of the space and finding the minimum number of colours that can be used in such a way that open sets with the same colour do not intersect each other; the dimension is equal to the minimum number of colours minus one. For instance, the covering dimension of the interval $[0,1]$ is equal to 1 because we need at least 2 colours.

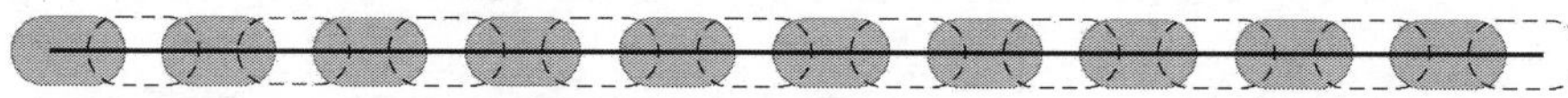

By Gelfand duality, unital commutative C*-algebras are always equal to the algebra of complex-valued continuous functions on some compact Hausdorff space X. The covering dimension of the underlying space X is reflected by certain finite dimensional approximations of the algebra of continuous functions. These approximations are used to model several versions of non-commutative covering dimension. After several iterations, the definitive notion appeared as *nuclear dimension*. This revolutionary notion became fundamental in the understanding of the fine structure of nuclear C*-algebras and played a major role in the *classification programme* of simple separable unital nuclear C*-algebras. We refer to [**Win18**] for an overview.

The classification programme of nuclear C*-algebras was initiated by Elliott in his seminal work [**Ell76**] where he classified all unital *approximately finite dimensional* C*-algebras via *ordered K-theory*. He managed to extend his classification theorem to larger classes of C*-algebras [**Ell93b**]. Based on this, he was prompted to conjecture that we can classify simple separable nuclear and unital C*-algebras with K-theory and *traces* [**Ell93a**]. For technical reasons, we must also assume that the C*-algebras satisfy the *Universal Coefficient Theorem* (UCT). This technical condition can roughly be considered as being "very weak homotopy equivalent" to a commutative C*-algebra. It is unknown if there is a separable nuclear C*-algebra that does not satisfy the UCT.

To the surprise of many, this conjecture was verified for large classes of C*-algebras [**DE02, EGLP96, EG96, EE93, EGL07, Lin03, Lin04, Lin11, Win14, Win16**], including the spectacular Kirchberg-Phillips classification of Kirchberg algebras that satisfy the UCT [**Phi00**]. However, counterexamples to this conjecture were found by Rørdam and Toms [**Rør03, Tom08**]. It became clearer then that some regularity condition must be added to the conjecture [**ET08**].

The missing ingredient in Elliott's conjecture was precisely finite nuclear dimension. Indeed, the aforementioned counterexamples to the conjecture have infinite nuclear dimension! After several decades of work, one of the major achievements in C*-algebra theory was completed: the classification of simple separable unital C*-algebras with finite nuclear dimension that satisfy the UCT [**GLN18, GLN19, EGLN, TWW17**].

In this document, we will review Winter and Zacharias' notion of nuclear dimension. We will make a detailed analysis of the commutative case as a motivation for the general definition. Then, we provide several examples of C*-algebras and discuss their nuclear dimension.

Let us finish this introduction by explaining the structure of this document. In Section 1 we review basic notions of C*-algebras. In Section 2 we discuss covering dimension and its non-commutative version, nuclear dimension. After this, we introduce several examples of C*-algebras and their nuclear dimension: approximately finite dimensional C*-algebras (Section 3), group C*-algebras (Section 4), C*-algebras associated to dynamical systems (Section 5) and Cuntz algebras (Section 6). In Section 7, we explain a trichotomy for nuclear dimension in the simple separable case. We finish this survey by proving that the nuclear dimension of $C(X)$ is equal to the covering dimension of the compact space X.

1. Basics of C*-algebras

Let us briefly introduce C*-algebras. A more comprehensive introduction can be found at [**Mur90**]. Given a complex Hilbert space $\mathcal{H}$, a linear operator $T : \mathcal{H} \to \mathcal{H}$ is *bounded* if there exists a positive constant M such that

$$\|T(h)\| \leq M\|h\|, \qquad h \in \mathcal{H}. \tag{1.1}$$

It is well known that a linear operator is continuous if and only if it is bounded. Let us denote by $B(\mathcal{H})$ to the set of bounded operators on $\mathcal{H}$. This set can be endowed with the structure of a complete normed algebra. Indeed, the sum and scalar multiplication are defined pointwise, the product is given by composition and the operator norm is defined as

$$\|T\| := \sup_{\|h\|=1} \|T(h)\|. \tag{1.2}$$

It is an standard fact that $B(\mathcal{H})$ is complete with respect to this norm. But $B(\mathcal{H})$ has another interesting operation, a natural involution that is given by the adjoint. Given an operator $T \in B(\mathcal{H})$, there exists a unique operator $T^* \in B(\mathcal{H})$ that satisfies

$$\langle T(h), g\rangle = \langle h, T^*(g)\rangle, \qquad g, h \in \mathcal{H}. \tag{1.3}$$

The operator T^* is called the *adjoint* of T and the map $T \mapsto T^*$ defines an *involution*, i.e. a conjugate linear map such that $(T^*)^* = T$ and $(ST)^* = T^*S^*$.

Another fundamental property of $B(\mathcal{H})$ is the C*-*identity*,

$$\|T\|^2 = \|T^*T\|, \qquad T \in B(\mathcal{H}). \tag{1.4}$$

The importance of this identity relies on the fact that it forces the norm to be determined by the algebraic structure of $B(\mathcal{H})$. Indeed, the norm of T^*T depends solely on its *spectrum* $\sigma(T^*T)$, the subset of all scalars $\lambda \in \mathbb{C}$ such that $T^*T - \lambda 1_{B(\mathcal{H})}$ is not an invertible operator.

Definition 1.1. A subalgebra $\mathcal{A}$ of $B(\mathcal{H})$ is a C*-*algebra* if it is closed with respect to the operator norm and closed under taking adjoints.

As already mentioned before, this is not the typical definition, but thanks to the Gelfand-Neimark theorem [**Mur90**, Section 3.4.1], our definition agrees with the typical one. The C*-identity has strong consequences in the structure of C*-algebras, for instance the norm on a C*-algebra is unique and any *-*homomorphism*[1] $\varphi : \mathcal{A} \to \mathcal{B}$ is contractive and hence continuous.

An easy example of a C*-algebra is $\mathbb{C}$, the complex numbers, where involution is simply given by complex conjugation. More generally, it follows from the definition that $B(\mathcal{H})$ is a C*-algebra. When the Hilbert space is of finite dimension n, the algebra of bounded operators is simply $M_n(\mathbb{C})$, the algebra of $n \times n$ matrices with

[1] A *-*homomorphism* $\varphi : \mathcal{A} \to \mathcal{B}$ is a linear map such that $\varphi(ab) = \varphi(a)\varphi(b)$ and $\varphi(a^*) = \varphi(a)^*$ for al $a, b \in \mathcal{A}$.

entries in $\mathbb{C}$, endowed with the operator norm, matrix multiplication as product and adjoint given by $(a_{ij})^*_{i,j} = (\overline{a_{ji}})_{i,j}$. In fact, any finite dimensional C*-algebra is always a direct sum of finitely many matrix algebras endowed with the operator norm. Notice that it is crucial to endow the operator norm. For instance, $\mathbb{C}^2$ endowed with the norm $\|(z_1, z_2)\|_1 = |z_1| + |z_2|$ is not a C*-algebra but it is if instead we endow it with the supremum norm $\|(z_1, z_2)\|_\infty = \max\{|z_1|, |z_2|\}$.

1.1. Commutative C*-algebras. Given a compact Hausdorff space X, we can always construct a C*-algebra by considering

$$C(X) := \{f : X \to \mathbb{C} \mid f \text{ is continuous}\}. \tag{1.5}$$

Operations are given by pointwise addition, pointwise multiplication, involution as $f^*(x) = \overline{f(x)}$ and norm given by

$$\|f\|_\infty = \sup_{x \in X} |f(x)|. \tag{1.6}$$

The algebra $C(X)$ acts on the Hilbert space $L^2(X, \mu)$ (where μ is a Borel measure) by multiplication.

A C*-algebra $\mathcal{A}$ is *unital* if it has a multiplicative unit, denoted by $1_\mathcal{A}$. Observe that the algebra $C(X)$ is unital, where the unit $1_{C(X)}$ is the constant function 1. However, not every C*-algebra is unital. For instance, if X is a locally compact Hausdorff space which is not compact, the corresponding algebra $C(X)$ is not normed [**Dal00**, Proposition 2.1.14] so we have to consider some subalgebra instead.

A continuous function $f : X \to \mathbb{C}$ *vanishes at infinity* if for all $\epsilon > 0$ there is a compact subset $K \subseteq X$ for which $|f(x)| \leq \epsilon$ if $x \in X \setminus K$. Set

$$C_0(X) := \{f \in C(X) \mid f \text{ vanishes at infinity}\}. \tag{1.7}$$

In this subalgebra, $\|\cdot\|_\infty$ is a norm and $C_0(X)$ is indeed a non-unital commutative C*-algebra since the constant function 1 does not vanishes at infinity. Observe that if X is compact, then $C(X) = C_0(X)$.

We now state Gelfand's theorem that characterises commutative C*-algebras. We refer to [**Mur90**, Theorem 2.1.10] for the details of its proof. Two C*-algebras $\mathcal{A}$ and $\mathcal{B}$ are *isomorphic*, $\mathcal{A} \cong \mathcal{B}$, if there is an *-isomorphism $\varphi : \mathcal{A} \to \mathcal{B}$.

THEOREM 1.2 (Gelfand). *If $\mathcal{A}$ is a commutative* C*-*algebra, then $\mathcal{A} \cong C_0(X)$ for some locally compact Hausdorff space X. In particular, $\mathcal{A}$ is unital if and only if X is compact.*

This actually defines a contravariant functor from the category of compact Hausdorff spaces with continuous maps to the category of unital C*-algebras and unital *-homomorphisms. This is the reason why C*-algebraist like to say that C*-algebras are non-commutative topological spaces. In fact, this analogy can be pushed a bit further as many topological properties are equivalent to C*-algebraic properties of the corresponding commutative algebra: unital - compact, non-unital - locally compact but not compact, closed ideal - open subset, unitisation - compactification, etc.

2. Nuclear dimension

This section is devoted to explaining an approach for defining a regularity condition that was deeply investigated by Winter, Kirchberg and Zacharias [**Win03**, **Win09**, **KW04**, **WZ10**]. Remarkably, this notion became a driving force in the classification programme of simple separable C*-algebras.

The idea behind this notion is simple: since we can view C*-algebras as noncommutative topological spaces, we can define a notion of dimension for C*-algebras using topological dimension as a model. The first step at implementing this approach is identifying the notion of topological dimension that is most convenient for our purposes. Let us focus on this first.

2.1. Topological dimension. Defining the dimension of topological spaces is a rather old problem and it goes back to work of Brouwer, Lebesgue and Poincaré. There exist different notions of dimension, for instance *small inductive dimension* (introduced independently by Urysohn and Menger [**Ury22**, **Men23**]) and *large inductive dimension* (introduced by Čech [**Čech31**]). We will focus mostly in another notion introduced by Čech in [**Čech33**], *covering dimension*, that was motivated by previous work of Lebesgue [**Leb11**]. This notion is more suitable for our purposes but it turns out that these three different theories agree on metrisable spaces. We refer to [**Pea75**] and [**Eng78**] for more comprehensive reviews of these notions.

Consider a finite open cover $\mathcal{U}$ of a topological space X and a natural number n. The *order* of $\mathcal{U}$ is at most n if any point of X is contained in at most $n+1$ open sets of $\mathcal{U}$.

DEFINITION 2.1. Let X be a topological space. The *covering dimension* of X is at most n, denoted by $\dim X \leq n$, if any finite open cover of X has an open refinement of order at most n. The covering dimension of X is the minimum n for which $\dim X \leq n$.

This notion agrees with our intuition that the topological dimension of $\mathbb{R}^n$ must be equal to n [**Pea75**, Theorem 3.2.7], even though the proof of this seemingly obvious fact is not easy! (see Figure 1). Another interesting fact about covering dimension is the following: A compact Hausdorff space has covering dimension equal to zero if and only if it is *totally disconnected* [**Pea75**, Proposition 3.1.3]. For instance, the Cantor set has covering dimension equal to zero.

When a topological space X is normal, a stronger form of covering dimension can be obtained: $\dim X \leq n$ if and only if for any finite open cover $\mathcal{V}$ there exists an open finite refinement $\mathcal{U}$ that can be decomposed into $n+1$ pairwise disjoint families $\mathcal{U}_0, \ldots, \mathcal{U}_n$ such that if U, V are elements of the same $\mathcal{U}_i$ then $U \cap V = \emptyset$ [**KW04**, Proposition 1.5]. This characterisation provides a nice interpretation of covering dimension in terms of colourings, *i.e.* we can colour the refinement $\mathcal{U}$ using $n+1$ colours, assigning the colour i to each element of $\mathcal{U}_i$. The key fact of this colouring is that any two open sets with the same colour do not intersect each other.

This is the exact form of topological dimension that will be used as a model for nuclear dimension. Roughly speaking, we will identify the coloured finite open covers produced by covering dimension with certain kind of finite dimensional approximations.

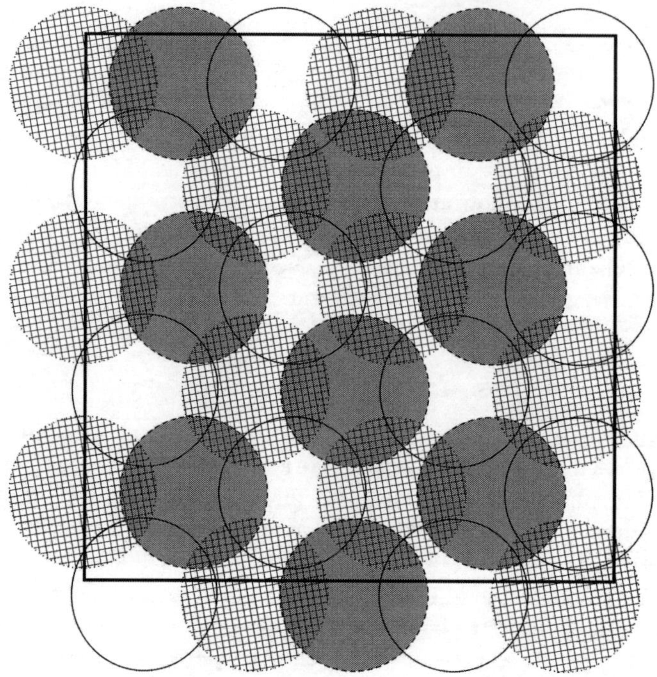

FIGURE 1. Covering dimension of the square is at most 2

2.2. Completely positive maps of order zero. Let us introduce all the ingredients we need for constructing the finite dimensional approximations induced by covering dimension.

An element in a C*-algebra is *positive* if it is self-adjoint and its spectrum is contained in the positive real numbers. The set of positive elements of $\mathcal{A}$ is denoted by $\mathcal{A}_+$. A linear map $\varphi : \mathcal{A} \to \mathcal{B}$ is *positive* if it maps positive elements into positive elements, *i.e.* $\varphi(\mathcal{A}_+) \subseteq \mathcal{B}_+$. This is a much broader class of maps between C*-algebras than *-homomorphisms. For instance, there are no non-zero *-homomorphisms between $M_2(\mathbb{C})$ and $\mathbb{C}$ but there are infinitely many positive maps from $M_2(\mathbb{C})$ to $\mathbb{C}$. An easy example can be the *unnormalised trace* $\mathrm{Tr} : M_2(\mathbb{C}) \to \mathbb{C}$ given by

$$\mathrm{Tr}\left(\begin{bmatrix} a_{11} & a_{12} \\ a_{21} & a_{22} \end{bmatrix}\right) = a_{11} + a_{22}. \tag{2.1}$$

We will denote by $M_n(\mathcal{A})$ to the C*-algebra of $n \times n$ matrices over $\mathcal{A}$. This algebra can also be seen as the tensor product $M_n(\mathbb{C}) \otimes \mathcal{A}$.

DEFINITION 2.2. A linear map $\varphi : \mathcal{A} \to \mathcal{B}$ is *completely positive* if, for all $n \in \mathbb{N}$, the map $\varphi^{(n)} : M_n(\mathcal{A}) \to M_n(\mathcal{B})$, given by

$$\varphi^{(n)}([a_{ij}]_{i,j}) = [\varphi(a_{ij})]_{i,j}, \tag{2.2}$$

if positive. For brevity, we will use c.p. for completely positive and c.p.c. for completely positive and contractive.

If φ is a *-homomorphism, then each $\varphi^{(n)}$ is a *-homomorphism as well. Hence the *-homomorphisms φ is a c.p. map. However, the converse is not true. For instance, let $p \in \mathcal{A}$ be a non trivial *projection* (i.e. $p = p^* = p^2$), then it is straightforward to verify that the map $a \mapsto pap$ is c.p.c. but it is not a *-homomorphism in general.

Thanks to a theorem due to Stinespring [**Sti55**, Theorem 1], we know the structure of c.p. maps: A bounded linear map $\varphi : \mathcal{A} \to B(\mathcal{H})$ is c.p. if and only if there exists a Hilbert space $\widetilde{\mathcal{H}}$, a *-homomorphism $\pi : \mathcal{A} \to B(\widetilde{\mathcal{H}})$, and an operator $V : H \to \widetilde{\mathcal{H}}$ such that

$$\varphi(a) = V^*\pi(a)V, \qquad a \in \mathcal{A}. \tag{2.3}$$

Completely positive maps are an essential ingredient in the approximation theory of C*-algebras and we refer to [**Pau02**] for a detailed introduction to c.p. maps. However, this class of maps does not preserve the structure enough for some of our purposes. The right class of maps was introduced by Winter when the domain is finite dimensional and by Winter and Zacharias in general [**Win03, WZ09**].

DEFINITION 2.3. A c.p. map $\varphi : \mathcal{A} \to \mathcal{B}$ is *order zero* if it preserves orthogonality, *i.e.* $\varphi(a)\varphi(b) = 0$ if $a, b \in \mathcal{A}_+$ satisfy $ab = 0$.

This class of maps lies between *-homomorphisms and c.p. maps. Indeed, c.p. maps of the form $a \mapsto pap$ are not order zero in general and order zero maps of the form $a \mapsto \lambda a$, with $\lambda \in (0,1)$, are not *-homomorphisms. As before, we have a theorem á la Stinespring that unravels the structure of order zero maps.

THEOREM 2.4 ([**WZ09**]). *Let $\varphi : \mathcal{A} \to \mathcal{B}$ be an order zero map. Then there exists a unital* C*-*algebra $\mathcal{M}$ containing $\varphi(A)$,*[2] *$h \in \mathcal{M}_+$ commuting with $\varphi(A)$, and a *-homomorphism $\pi : \mathcal{A} \to \mathcal{M}$ such that*

$$\varphi(a) = h\pi(a) = \pi(a)h, \qquad a \in \mathcal{A}. \tag{2.4}$$

If $\mathcal{A}$ is unital, then $h = \varphi(1_{\mathcal{A}})$.

This theorem enables us to obtain a canonical bijection between c.p.c. order zero maps $\mathcal{A} \to \mathcal{B}$ and *-homomorphisms $C_0((0,1],\mathcal{A}) \to \mathcal{B}$ [**WZ09**, Corollary 4.1]. The C*-algebra $C_0((0,1],\mathcal{A})$ is called the *cone over* $\mathcal{A}$ and it is given by

$$C_0((0,1],\mathcal{A}) := \{g : (0,1] \to \mathcal{A} \mid g \text{ is continuous and} \lim_{t\to 0} g(t) = 0\}. \tag{2.5}$$

This algebra is generated by the functions $f \otimes a : (0,1] \to \mathcal{A}$ given by

$$(f \otimes a)(t) = f(t)a, \tag{2.6}$$

where $f \in C_0(0,1]$, $a \in A$ and $t \in (0,1]$.

The bijection is precisely given in the following way: if $\varphi : \mathcal{A} \to \mathcal{B}$ is an order zero map, then the map $\rho_\varphi : C_0((0,1],\mathcal{A}) \to \mathcal{B}$ given on elementary tensors by

$$\rho_\varphi(f \otimes a) := f(h)\pi(a) \tag{2.7}$$

is a *-homomorphism. Conversely, if $\rho : C_0((0,1],\mathcal{A}) \to \mathcal{B}$ is a *-homomorphism, then the map $\varphi_\rho : \mathcal{A} \to \mathcal{B}$ given by

$$\varphi_\rho(a) := \rho(\mathrm{id}_{(0,1]} \otimes a) \tag{2.8}$$

is a c.p.c. order zero map. This is one of the key features of order zero maps and also explains why order zero maps are so useful. They avoid *topological obstructions* in contrast to the situation with *-homomorphisms, where due to topological

[2] A precise description of the algebra $\mathcal{M}$ can be found in [**WZ09**].

obstructions, non-trivial *-homomorphisms might not exist. Indeed, the C*-algebra $C_0((0,1],\mathcal{A})$ is homotopy equivalent to the zero C*-algebra. Hence, order zero maps do not carry any topological information.

2.3. Covering dimension of $C(X)$. Before defining nuclear dimension, we will discuss the commutative case as the motivating example for the proper definition. In order to simplify our equations, we will write $a \approx_\epsilon b$ instead of $\|a-b\| < \epsilon$.

Consider a second countable compact Hausdorff space X of covering dimension equal to n, $\epsilon > 0$ and let $\mathfrak{F}$ be a finite set of $C(X)$. By compactness of X, we can find a finite open cover $\mathcal{V}$ of X such that if $x, y \in V$, for some $V \in \mathcal{V}$, then $f(x) \approx_\epsilon f(y)$ for all $f \in \mathfrak{F}$.

Using the covering dimension of X, we can find a finite open refinement $\mathcal{U}$ of order $n+1$. So we can decompose it in the following way: $\mathcal{U} = \mathcal{U}_0 \sqcup \ldots \sqcup \mathcal{U}_n$ and $U \cap V = \emptyset$ if U and V are elements $\mathcal{U}_k$ for some $k = 0, \ldots, n$. Fix points $x_U \in U$ for each $U \in \mathcal{U}$ and set a c.p.c. map $\psi : C(X) \to \mathbb{C}^{\mathcal{U}}$ by

$$\psi(f) := (f(x_U))_{U \in \mathcal{U}}, \qquad f \in C(X). \tag{2.9}$$

Let $(h_U)_{U \in \mathcal{U}}$ be a *partition of unity* subordinated to the cover $\mathcal{U}$. Recall that this means that (h_U) is a family of continuous functions $X \to \mathbb{R}_+$ such that the support of h_U is contained in U and $\sum_{U \in \mathcal{U}} h_U(x) = 1$ for all $x \in X$. Set a c.p. map $\varphi : \mathbb{C}^{\mathcal{U}} \to C(X)$ by

$$\varphi((x_U)) := \sum_{U \in \mathcal{U}} x_U h_U, \qquad (x_U) \in \mathbb{C}^{\mathcal{U}}. \tag{2.10}$$

It is straightforward to obtain

$$\varphi \circ \psi(f)(y) = \sum_{U \in \mathcal{U}} f(x_U) h_U(y) = \sum_{y \in U} f(x_U) h_U(y) \approx_\epsilon f(y). \tag{2.11}$$

This is the standard proof of the fact that $C(X)$ is a *nuclear* C*-algebra (see [**BO08**, Proposition 2.4.2]). However, this approximation we have constructed has more properties. Let $\varphi_k : \mathbb{C}^{\mathcal{U}} \to C(X)$ be the c.p. map given by restricting φ to the family $\mathcal{U}_k$, i.e.

$$\varphi_k((x_U)_{U \in \mathcal{U}}) := \sum_{U \in \mathcal{U}_k} x_U h_U. \tag{2.12}$$

If we view $\mathbb{C}^{\mathcal{U}} = \bigoplus_{k=0}^n \mathbb{C}^{\mathcal{U}_k}$, then $\varphi_k = \varphi|_{\mathbb{C}^{\mathcal{U}_k}}$. Each φ_k is a c.p.c. order zero map and this follows from the fact that any two open sets in $\mathcal{U}_k$ are disjoint. This is precisely where the extra topological properties of the refinement $\mathcal{U}$ come into play. Therefore, the c.p. map φ is the sum of $n+1$ c.p.c. order zero maps,

$$\varphi = \sum_{k=0}^{n} \varphi_k. \tag{2.13}$$

The nuclear dimension of $C(X)$ will be defined as the number of order zero summands that the second map φ has minus one. The preceding discussion shows that the non-commutative covering dimension of $C(X)$ is at most n. The definition of nuclear dimension mimics the approximations we have described above for $C(X)$ and this is exactly why we refer to it as a non-commutative covering dimension.

2.4. The nuclear dimension of a C*-algebra. After discussing the commutative case, let us summarise the key properties of the completely positive approximations constructed in this case:

(1) we can approximate the identity map $\mathrm{id}_{C(X)}$ with a finite dimensional algebra $\mathbb{C}^r$ and maps $C(X) \xrightarrow{\psi} \mathbb{C}^r \xrightarrow{\varphi} C(X)$, i.e. for a finite subset of $C(X)$ and up to some small error, $\mathrm{id}_{C(X)} \approx \varphi\psi$,
(2) the map $\psi : C(X) \to \mathbb{C}^r$ is c.p.c.,
(3) there is a decomposition of $\mathbb{C}^r$ into $n+1$ finite dimensional subalgebras, $\mathbb{C}^r = \bigoplus_{k=0}^n \mathbb{C}^{m_i}$, so that each restriction $\varphi|_{\mathbb{C}^{m_i}}$ is a c.p.c. order zero map. Thus φ can be expressed as the sum of $n+1$ order zero maps.

This leads us to the actual definition of nuclear dimension introduced by Winter and Zacharias, where they replaced the commutative algebras with more general ones.

DEFINITION 2.5 ([**WZ10**]). A C*-algebra $\mathcal{A}$ has *nuclear dimension* at most n, $\dim_{\mathrm{nuc}} A \leq n$, if for any finite set $\mathfrak{F} \subseteq A$ and $\epsilon > 0$, there exist finite dimensional C*-algebras $F_0, F_1 \ldots, F_n$ and maps $\psi : \mathcal{A} \to \bigoplus_{k=0}^n F_k$, $\varphi : \bigoplus_{k=0}^n F_k \to \mathcal{A}$ such that

(i) $\|a - \varphi \circ \psi(a)\| < \epsilon$ for all $a \in \mathfrak{F}$,
(ii) ψ is c.p.c.,
(iii) $\varphi_k := \varphi|_{F_k}$ is a c.p.c. map of order zero for $i = 0, \ldots, n$.

The name 'nuclear dimension' comes from the fact that C*-algebras with approximations of the form $\mathcal{A} \xrightarrow{\psi} F \xrightarrow{\varphi} \mathcal{A}$, with F finite dimensional and ψ, φ are c.p.c. maps, are called *nuclear*, see [**BO08**, Chapter 2] for more information about nuclearity. It follows that C*-algebras with finite nuclear dimension are nuclear but the converse is false, for instance $C([0,1]^{\mathbb{N}})$.

An interesting variant of the previous definition is *decomposition rank*, denoted by $\mathrm{dr}\,(\mathcal{A})$. This precursor notion, introduced by Winter and Kirchberg in [**KW04**], is very similar to nuclear dimension but it asks the second map φ to be contractive. Observe that in the definition of nuclear dimension we only know that the norm of φ is at most $n+1$. This seemingly insubstantial change is rather subtle and deep since it imposes stronger structural properties (see Section 6). Note that we always have

$$\dim_{\mathrm{nuc}} \mathcal{A} \leq \mathrm{dr}\, \mathcal{A}. \tag{2.14}$$

We refer to [**Win18**] for a more detailed explanation.

After the discussion in the previous subsection, the following theorem should not be a great surprise.

THEOREM 2.6 ([**WZ10**, Proposition 2.4]). *Let X be a compact Hausdorff space. Then*

$$\dim_{\mathrm{nuc}} C(X) = \dim X. \tag{2.15}$$

We already showed that $\dim_{\mathrm{nuc}} C(X) \leq \dim X$ but the other direction is a bit more subtle. In Section 8, we will present a direct proof of this fact.

We finish this section by pointing out that the notion of nuclear dimension is a well-behaved dimension theory and it has good permanence properties. For instance,

$$\begin{aligned}\dim_{\mathrm{nuc}} \mathcal{A} \oplus \mathcal{B} &= \max\{\dim_{\mathrm{nuc}} \mathcal{A}, \dim_{\mathrm{nuc}} \mathcal{B}\},\\ \dim_{\mathrm{nuc}} \mathcal{A} \otimes \mathcal{B} &\leq (\dim_{\mathrm{nuc}} \mathcal{A} + 1)(\dim_{\mathrm{nuc}} \mathcal{B} + 1) - 1,\\ \max\{\dim_{\mathrm{nuc}} I, \dim_{\mathrm{nuc}} \mathcal{A}/I\} &\leq \dim_{\mathrm{nuc}} \mathcal{A} \leq \dim_{\mathrm{nuc}} I + \dim_{\mathrm{nuc}} \mathcal{A}/I + 1, \qquad (2.16)\end{aligned}$$

where in the last inequality, I is a closed two sided ideal of $\mathcal{A}$ [**WZ10**, Proposition 2.3].

In the following sections of this survey, we present more examples of C*-algebras and discuss recent developments regarding their nuclear dimension.

3. Zero dimensional objects

A natural question regarding nuclear dimension is to identify the C*-algebras with nuclear dimension exactly equal to zero. We already mentioned that a compact Hausdorff space has covering dimension zero if and only if it is totally disconnected. In conjunction with Theorem 2.6, this immediately provides us a characterisation of unital commutative C*-algebras with nuclear dimension equal to zero: $C(X)$ has nuclear dimension zero if and only if X is totally disconnected.

In the non-commutative setting, it is also straightforward to see that finite dimensional C*-algebras have nuclear dimension equal to zero, but we would also like to know if there exist infinite dimensional non-commutative C*-algebras with nuclear dimension equal to zero. The answer is precisely those C*-algebras that locally "look like" finite dimensional C*-algebras.

DEFINITION 3.1. A separable C*-algebra $\mathcal{A}$ is *approximately finite dimensional* (AF) if there exists an increasing sequence of finite dimensional C*-subalgebras $(F_n)_{n=1}^{\infty}$ such that

$$\mathcal{A} = \overline{\bigcup_{n=1}^{\infty} F_n}. \qquad (3.1)$$

Bratteli showed in [**Bra72**, Theorem 2.2] that a separable C*-algebra $\mathcal{A}$ is AF if and only if for any finite subset $\mathfrak{F} \subseteq \mathcal{A}$ and $\epsilon > 0$ there exists a finite dimensional C*-subalgebra $F \subseteq \mathcal{A}$ such that

$$\operatorname{dist}(a, F) := \inf_{x \in F} \|a - x\| < \epsilon, \qquad a \in \mathfrak{F}. \qquad (3.2)$$

This is a local characterisation of AF-algebras that is very useful when computing their nuclear dimension.

Before going further, let us consider some examples of an AF-algebra. The C*-algebra of compact operators on a separable Hilbert space $\mathbb{K}(\mathcal{H})$ is a (non-unital) AF-algebra. Indeed, suppose $(e_n)_{n=1}^{\infty}$ is an orthonormal basis of $\mathcal{H}$. The projection $e_i \otimes e_i : \mathcal{H} \to \mathbb{C}e_i \subseteq \mathcal{H}$ is given by

$$e_i \otimes e_i(h) := \langle h, e_i \rangle e_i, \qquad h \in \mathcal{H}. \qquad (3.3)$$

Let p_n be the projection $\sum_{i=1}^{n} e_i \otimes e_i$. It is not difficult to see that

$$\mathbb{K}(\mathcal{H}) = \overline{\bigcup_{n=1}^{\infty} p_n \mathbb{K}(\mathcal{H}) p_n} \quad \text{and} \quad p_n \mathbb{K}(\mathcal{H}) p_n \cong M_n(\mathbb{C}). \qquad (3.4)$$

This shows that $\mathbb{K}(\mathcal{H})$ is an AF-algebra.

Another example of an AF-algebra is the algebra of continuous functions on the Cantor set. In fact, any commutative C*-algebra $C_0(X)$ is AF if and only if X is totally disconnected; this shows that the commutative C*-algebras with nuclear dimension equal to zero that we already identified are indeed AF-algebras.

Winter showed that the zero dimensional objects for nuclear dimension are precisely the AF-algebras. This essentially relies on the local characterisation of AF-algebras that we discussed above.

THEOREM 3.2 ([**WZ10**]). *A separable* C*-*algebra* $\mathcal{A}$ *is approximately finite dimensional if and only if* $\dim_{\mathrm{nuc}} \mathcal{A} = 0$.

4. C*-algebras associated to groups

In this section, we will explain how to construct C*-algebras associated to discrete groups. Let us recall some basic definitions first.

An operator $U \in B(\mathcal{H})$ is a *unitary* if $U^*U = UU^* = 1_{\mathcal{H}}$. We denote by $\mathcal{U}(\mathcal{H})$ to the set of unitaries of $B(\mathcal{H})$. This set has the structure of a group. Indeed, by definition, the inverse of a unitary is precisely its adjoint, which is a unitary as well. Note that $\mathcal{U}(\mathcal{H})$ is also closed under multiplication, so it is a group.

The *group ring* $\mathbb{C}\Gamma$ is given by

$$\mathbb{C}\Gamma = \Big\{ \sum_{g \in \Gamma} a_g \cdot g \;\Big|\; a_g \in \mathbb{C} \text{ and } a_g = 0 \text{ except for finitely many} \Big\}. \tag{4.1}$$

A *unitary representation* of a discrete group Γ on a Hilbert space $\mathcal{H}$ is a group homomorphism $\pi : \Gamma \to \mathcal{U}(\mathcal{H})$. It follows that a unitary representation of Γ extends by linearity to a representation of $\mathbb{C}\Gamma$ into $B(\mathcal{H})$. Naturally, a unitary representation of Γ induces a seminorm on $\mathbb{C}\Gamma$ by simply taking

$$\|x\|_\pi := \|\pi(x)\|_{B(\mathcal{H})}, \qquad x \in \mathbb{C}\Gamma. \tag{4.2}$$

An important example of a unitary representation of a discrete group Γ is the *left regular representation.* Consider the Hilbert space

$$\ell^2(\Gamma) := \Big\{ a : \Gamma \to \mathbb{C} \;\Big|\; \sum_{g \in \Gamma} |a_g|^2 < \infty \Big\}, \tag{4.3}$$

and define $\lambda : \Gamma \to B(\ell^2(\Gamma))$ by

$$\lambda_g(a)(h) = a(g^{-1}h), \qquad g, h \in \Gamma,\; a \in \ell^2(\Gamma) \tag{4.4}$$

The seminorm $\|\cdot\|_\lambda$ is in fact a norm since the vector $\delta_e \in \ell^2$ separates points of $\lambda(\mathbb{C}\Gamma)$.

The *reduced group* C*-*algebra of* Γ, denoted by $\mathrm{C}^*_r(\Gamma)$, is given by completing $\mathbb{C}\Gamma$ with respect to the C*-norm $\|\cdot\|_\lambda$ induced by the left regular representation λ,

$$\mathrm{C}^*_r(\Gamma) := \overline{\mathbb{C}\Gamma}^{\|\cdot\|_\lambda}. \tag{4.5}$$

Another norm that we can consider is the *universal norm* given by

$$\|x\|_u = \sup_\pi \|x\|_\pi, \qquad x \in \mathbb{C}\Gamma, \tag{4.6}$$

where the supremum is taken over all unitary representations π of Γ. The *universal group* C*-*algebra of* Γ, denoted by $\mathrm{C}^*(\Gamma)$, is given by completing $\mathbb{C}\Gamma$ with respect to the universal norm, i.e.

$$\mathrm{C}^*(\Gamma) := \overline{\mathbb{C}\Gamma}^{\|\cdot\|_u}. \tag{4.7}$$

In particular, if the group Γ is finite then the reduced and universal group C*-algebras are finite dimensional, and in fact, they are equal to the group ring. This can be deduced from the fact that $\mathbb{C}\Gamma$ is finite dimensional and the uniqueness of the C*-norm.

Another interesting example is when Γ is commutative. In this case, again the reduced and universal group C*-algebras agree. But more is true: they are isomorphic to the C*-algebra of continuous functions on its *Pontryagin dual* $\widehat{\Gamma}$,

$$\mathrm{C}^*(\Gamma) \cong \mathrm{C}^*_r(\Gamma) \cong C(\widehat{\Gamma}). \tag{4.8}$$

This immediately tells us that, for instance, $\mathrm{C}^*(\mathbb{Z}) \cong \mathrm{C}^*_r(\mathbb{Z}) \cong C(\mathbb{T})$.

The previous examples show that, in some cases, reduced and universal group C*-algebras agree. An interesting and natural question is for which class these two group C*-algebras agree. The answer is related to the famous *Banach-Tarsky paradox* that states that the sphere $\mathbb{S}^2 \subseteq \mathbb{R}^3$ can be decomposed into finitely many pieces that can be reassembled as two copies of $\mathbb{S}^2$. A similar statement is true for spheres in higher dimensions but interestingly this cannot be done for $\mathbb{S}^1 \subseteq \mathbb{R}^2$. Despite the apparently geometric nature of the paradox, this is truly a measure theoretic phenomenon. This was identified by von Neumann, who introduced the concept of amenability [**von29**].

DEFINITION 4.1. A discrete group Γ is *amenable* if there exists a a *finitely additive measure* μ on Γ that satisfies $\mu(\Gamma) = 1$ and $\mu(gA) = \mu(A)$ for $g \in \Gamma$ and $A \subseteq \Gamma$.

The reason behind the Banach-Tarsky paradox for $\mathbb{S}^2$ lies in the fact that the group of isometries of $\mathbb{R}^3$ is not amenable. Meanwhile, this paradox does not hold for $\mathbb{S}^1$ precisely because the group of isometries of $\mathbb{R}^2$ is indeed amenable.

Examples of amenable discrete groups can be found in the finite groups, abelian groups, virtually solvable groups and finitely generated groups of sub-exponential growth. On the other hand, the free groups $\mathbb{F}_n$ are the canonical examples of non amenable groups. Moreover, any group that contains a copy of a free group is not amenable.

There exist a vast amount of characterisations of amenable groups (see [**Pat88**]). Let us present two of them that relate amenability with properties of group C*-algebras.

THEOREM 4.2 ([**BO08**, Theorem 2.6.8]). *Let Γ be a discrete group. The following are equivalent:*

(i) Γ *is amenable,*
(ii) $\mathrm{C}^*_r(\Gamma) \cong \mathrm{C}^*(\Gamma)$,
(iii) $\mathrm{C}^*_r(\Gamma)$ *is a nuclear* C*-*algebra.*

This theorem immediately shows that if Γ is not amenable, its reduced group C*-algebra $\mathrm{C}^*_r(\Gamma)$ is not a nuclear C*-algebra and hence $\dim_{\mathrm{nuc}} \mathrm{C}^*_r(\Gamma) = \infty$.

What can we say about the nuclear dimension of C*-algebras associated to amenable groups? First of all, we already know that if Γ is commutative, its group C*-algebra is isomorphic to $C(\widehat{\Gamma})$. By Theorem 2.6, $\dim_{\mathrm{nuc}} \mathrm{C}^*_r(\Gamma) = \dim \widehat{\Gamma}$. As a consequence, we obtain examples of amenable groups with infinite nuclear dimension, for instance $\mathbb{Z}^{\mathbb{Z}}$. Finding amenable groups with infinite nuclear dimension appears harder if we restrict to finitely generated groups. However, such examples do exist: the nuclear dimension of $C_r^*(\mathbb{Z} \wr \mathbb{Z})$ is infinite [**GK10, Win12**]. We finish our discussion of groups and their nuclear dimension by presenting a class of amenable groups with finite nuclear dimension.

THEOREM 4.3 ([**EGM19**]). *Let* Γ *be a finitely generated virtually nilpotent group. Then* $\dim_{\mathrm{nuc}} C_r^*(\Gamma) < \infty$.

In the previous theorem, the upper bound depends on the *Hirsch number* of Γ.

5. C*-algebras associated to dynamical systems

Let X be a compact Hausdorff space and consider a homeomorphism $\alpha : X \to X$. This homeomorphism induces an action of $\mathbb{Z}$ into $C(X)$, that we will also denote by α, in the following way:

$$\alpha_n(f) = f \circ \alpha^{-n}, \qquad f \in C(X), n \in \mathbb{Z}. \tag{5.1}$$

This action gives rise to a C*-algebra $C(X) \rtimes_\alpha \mathbb{Z}$ contained in $B(L^2(X) \otimes \ell^2(\mathbb{Z}))$ that is generated by a copy of $C(X)$ and a unitary $u \in B(L^2(X) \otimes \ell^2(\mathbb{Z}))$ that encodes the action, i.e.

$$ufu^* = f \circ \alpha^{-1}, \qquad f \in C(X). \tag{5.2}$$

One class of interesting examples of this construction are the *rotation algebras* $\mathcal{A}_\theta$. Let $\mathbb{T} \subseteq \mathbb{C}$ be the unit circle in the complex plane. For $\theta \in [0, 1)$, define $r_\theta : \mathbb{T} \to \mathbb{T}$ by $r_\theta(z) := e^{2\pi i\theta} z$. The corresponding crossed product is denoted by $\mathcal{A}_\theta := C(\mathbb{T}) \rtimes_{r_\theta} \mathbb{Z}$. These algebras are called rotation algebras and are also commonly referred as non-commutative tori. This name is motivated by the fact that $\mathcal{A}_0 \cong C(\mathbb{T}^2)$.

These algebras have received much attention since they are some of the most natural examples of non-commutative manifolds [**Rie90**]. For instance, $\mathcal{A}_\theta$ is isomorphic to $\mathcal{A}_\eta$ if and only if $\theta = \eta$ or $\theta = 1-\eta$ [**PV80, Rie81, Yin86**]. Furthermore, if θ is rational, $\mathcal{A}_\theta$ is a subalgebra of $M_k(C(\mathbb{T}^2))$ for some $k \in \mathbb{N}$ [**Rie83**]. On the other hand, if θ is irrational, $\mathcal{A}_\theta$ is a simple C*-algebra and locally "looks like" $C(\mathbb{T}) \otimes F$ where F is a finite dimensional C*-algebra [**EE93**].

This shows that the nuclear dimension of the rotation algebra $\mathcal{A}_\theta$ is one when θ is irrational, or two when θ is rational. Indeed, this essentially follows from the fact that $\dim_{\mathrm{nuc}} M_k(\mathbb{T}^2) = \dim \mathbb{T}^2 = 2$, and $\dim_{\mathrm{nuc}} C(\mathbb{T}) \otimes F = \dim \mathbb{T} = 1$ [**WZ10**, Example 6.1].

In general, understanding the internal structure of $C(X) \rtimes_\alpha \mathbb{Z}$ can be a very challenging task. However, it was recently shown in [**HW17**] that when the covering dimension of X is finite, the nuclear dimension of the C*-algebra induced by any homeomorphism is finite as well. Precisely

$$\dim_{\mathrm{nuc}} C(X) \rtimes_\alpha \mathbb{Z} \leq 2(\dim X)^2 + 6 \dim X + 4. \tag{5.3}$$

This is a surprising result because it does not impose conditions on the homeomorphism or on the space X (apart from having finite covering dimension).

The C*-algebra $C(X) \rtimes_\alpha \mathbb{Z}$ is an example of a more general construction. Given a C*-algebra $\mathcal{A} \subseteq B(\mathcal{H})$, a discrete group Γ and a group action $\alpha : \Gamma \curvearrowright A$, we can always form the *reduced crossed product* C*-algebra $\mathcal{A} \rtimes_{\alpha,r} \Gamma$ that is contained in $B(\mathcal{H} \otimes \ell^2(\Gamma))$. In general, determining the structure of crossed products can be a very difficult task but there exist recent spectacular results that enable us to estimate the nuclear dimension of certain crossed products [**Sza15, HWZ15, EN17, HSWW17, SWZ17, KS18**].

6. Cuntz algebras

Let $\mathcal{H}$ be a separable infinite dimensional Hilbert space. For any natural number n, we can decompose $\mathcal{H}$ as $\mathcal{H}_1 \oplus \cdots \oplus \mathcal{H}_n$ with $\mathcal{H}_i \cong \mathcal{H}$ for all i. Then, we can find elements $v_1, \ldots, v_n \in B(\mathcal{H})$ implementing an isomorphism between $\mathcal{H}$ and $\mathcal{H}_i$ via conjugation that satisfy

$$\sum_{i=1}^{n} v_i v_i^* = 1_{B(\mathcal{H})} \quad \text{and} \quad v_1^* v_1 = v_2^* v_2 = \ldots = v_n^* v_n = 1_{B(\mathcal{H})}. \tag{6.1}$$

The *Cuntz algebra* $\mathcal{O}_n$ is the C*-subalgebra of $B(\mathcal{H})$ generated by such elements $v_1, \ldots, v_n$. Similarly, we can find a sequence $(u_i)_{i\in\mathbb{N}}$ of elements in $B(\mathcal{H})$ satisfying

$$u_i^* u_i = 1_{B(\mathcal{H})}, \qquad (u_i u_i^*)(u_j u_j^*) = 0, \qquad i \neq j, \quad i, j \in \mathbb{N}. \tag{6.2}$$

The *Cuntz algebra* $\mathcal{O}_\infty$ is the C*-subalgebra of $B(\mathcal{H})$ generated by the sequence $(u_i)_{i\in\mathbb{N}}$.

These algebras were introduced by Cuntz in [**Cun77**] and they possess very interesting properties. For instance, each $\mathcal{O}_n$ is simple, nuclear, unital, and given any two non-zero positive elements $a, b \in \mathcal{O}_n$, there exists $x \in \mathcal{O}_n$ such that $a = x^*bx$. Simple C*-algebras that satisfy this property are called *purely infinite*. There exist spectacular results concerning Cuntz algebras (and purely infinite C*-algebras) that are beyond the scope of this survey. We refer to [**RS02**] for a comprehensive overview.

We briefly mentioned the notion of *decomposition rank* in Section 2.4. This stronger notion requires approximations $\mathcal{A} \xrightarrow{\psi} F \xrightarrow{\phi} \mathcal{A}$ similar to the approximations in Definition 2.5 but also requires the second map φ to be contractive, i.e. $\|\varphi\| \leq 1$. Finite decomposition rank imposes stronger structural properties on the algebra than finite nuclear dimension, for instance *strong quasidiagonality* [**KW04**, Proposition 5.1]. Cuntz algebras are not even *quasidiagonal*, and hence, they cannot have finite decomposition rank. However, these algebras do have finite nuclear dimension.

THEOREM 6.1 ([**WZ10**, Theorem 7.4], [**End15**, Theorem 4.1]). *For any $n \in \mathbb{N} \cup \{\infty\}$, $\dim_{\text{nuc}} \mathcal{O}_n = 1$.*

This theorem is a special case of a more general result. Cuntz algebras are particular examples of separable simple purely infinite and nuclear C*-algebras – a class of C*-algebras commonly referred to as the *Kirchberg algebras*. It was showed in [**BBS$^+$19, RSS15**] that the nuclear dimension of any Kirchberg algebra is exactly 1. More generally, for any non-zero separable nuclear C*-algebra $\mathcal{A}$, the nuclear dimension of $\mathcal{A} \otimes \mathcal{O}_\infty$ is finite [**Sza17**]. It was recently showed that $\dim_{\text{nuc}} \mathcal{A} \otimes \mathcal{O}_\infty = 1$ [**BGSW**].

7. General simple C*-algebras

Now, we focus our attention on the nuclear dimension of simple separable C*-algebras. It turns out that for this class of C*-algebras, there exists an interesting trichotomy. We emphasize that commutative C*-algebras apart from $\mathbb{C}$ are never simple.

THEOREM 7.1 ([**CET**$^+$, **CE**]). *The nuclear dimension of a simple separable* C*-*algebra is* $0, 1$ *or* ∞.

Simple C*-algebras with infinite nuclear dimension include C*-algebras that are not nuclear, as $C^*_r(\mathbb{F}_2)$, but also the exotic Villadsen algebras [**Vil98**, **Vil99**] and the famous examples constructed by Rørdam [**Rør03**] and Toms [**Tom08**]. On the other hand, unital simple (UCT) C*-algebras with finite nuclear dimension are those that can be classified using their K-theory and traces [**GLN18**, **EGLN**, **TWW17**] (see [**Win18**] for a survey on the subject).

The reason behind this trichotomy is that finite nuclear dimension imposes strong conditions in the algebra, and one of those conditions is so strong that it actually forces the nuclear dimension to be at most one. In order to explain this more clearly, let us introduce an exotic C*-algebra: the Jiang-Su algebra $\mathcal{Z}$ [**JS99**]. This algebra can be seen as an infinite-dimensional C*-version of the complex numbers and it has many interesting properties. This algebra, which locally looks like some subalgebra of $C([0,1], M_{pq}(\mathbb{C}))$ where p and q are relatively prime numbers, is simple and separable with $\dim_{\mathrm{nuc}} \mathcal{Z} = 1$. We refer to [**RW10**] for a detailed explanation of the Jiang-Su algebra $\mathcal{Z}$.

In a notable tour de force, Winter showed that finite nuclear dimension implies tensorial absorption of the Jiang-Su algebra, $\mathcal{A} \otimes \mathcal{Z} \cong \mathcal{A}$, when the algebra is separable simple unital and non elementary[3] [**Win12**]. This result was later extended to the non-unital case by Tikuisis [**Tik14**].

It was conjectured by Toms and Winter that the converse direction should hold as well and this was verified under certain extra conditions, see [**MS14**, **SWW15**, **BBS**$^+$**19**]. Recently, this direction was fully verified, first for unital C*-algebras in [**CET**$^+$], then extended to the non-unital case in [**CE**]. It was actually shown that if the C*-algebra is $\mathcal{Z}$-stable then its nuclear dimension is at most one.

By the results mentioned above, if the nuclear dimension of a simple separable C*-algebra $\mathcal{A}$ is finite, then it has to absorb the Jiang-Su algebra $\mathcal{Z}$, and so it must have nuclear dimension at most 1. Finally, if $\mathcal{A}$ is AF then its dimension is exactly zero; otherwise, it has nuclear dimension equal to one.

The following theorem summarises all that we have said so far.

THEOREM 7.2 ([**Win12**, **Tik14**, **CET**$^+$, **CE**]). *Let* $\mathcal{A}$ *be a simple separable nuclear and non elementary* C*-*algebra. The following are equivalent:*

(i) $\dim_{\mathrm{nuc}} \mathcal{A} < \infty$,
(ii) $\dim_{\mathrm{nuc}} \mathcal{A} \leq 1$,
(iii) $\mathcal{A} \otimes \mathcal{Z} \cong \mathcal{A}$.

8. Nuclear dimension of commutative algebras

As already explained in Section 2.3, the nuclear dimension of the C*-algebra $C(X)$ agrees with the covering dimension of X. In his early work on developing non-commutative covering dimension, Winter proved that the covering dimension of a second countable locally compact Hausdorff space X agrees with the *completely positive rank* of $C_0(X)$ [**Win03**]. This notion was a precursor concept of decomposition rank and nuclear dimension. Later on, when decomposition rank and nuclear dimension were introduced, a direct proof of this fact was not given; instead, its

[3] A C*-algebra $\mathcal{A}$ is elementary if it is isomorphic to $M_n(\mathbb{C})$ for some n, or the compacts $\mathbb{K}(\mathcal{H})$.

proof relied on the original proof concerning completely positive rank and some relations between this former concept and nuclear dimension. In this section, we present a direct proof of this fact in the compact and Hausdorff setting. The proof presented here originally appeared in [**Cas16**].

Let X be a compact Hausdorff space. We already showed in Section 2.3 how to prove that $\dim_{\mathrm{nuc}} C(X) \leq \dim X$. Before going into the proof of the opposite direction, let us briefly sketch it first.

We will consider any finite open cover $\mathcal{U}$ of X and a partition of unity $(h_r)_{r=1}^m$ subordinated to $\mathcal{U}$. For a sufficiently small ϵ and assuming $\dim_{\mathrm{nuc}} C(X) = n$, we can find a finite dimensional algebra of the form $F = \bigoplus_{k=0}^n F^{(k)}$, maps $\psi : C(X) \to F$ and $\varphi : F \to C(X)$, with $\varphi = \sum_{k=0}^n \varphi_k$ where each φ_k is order zero, such that they approximate the partition of unity $\{h_1, \ldots, h_m\}$. By the structure of order zero maps (Theorem 2.4), $F \cong \mathbb{C}^s$ for some $s \in \mathbb{N}$. We will set $f_i := (0, \ldots, 0, 1, 0, \ldots, 0) \in \mathbb{C}^r$ where 1 is in the i-th position. The family

$$\mathcal{W}_0 = \left\{ \varphi_k (f_i)^{-1} \left(\left(\frac{1}{m(n+1)} - \epsilon, \infty \right) \right) \,\middle|\, i = 1, \ldots, r; \, k = 0, \ldots, n \right\} \tag{8.1}$$

will be a cover of X of order at most n. We will finish the proof by showing that there exists a subcover of $\mathcal{W}_0$ that refines $\mathcal{U}$.

We now do the proof in detail. Suppose $\dim_{\mathrm{nuc}} C(X) = n$ and let $\mathcal{U} = \{U_1, \ldots, U_m\}$ be a finite open cover of X. Without loss of generality we can assume $m \geq 2$. Consider a partition of unity $(h_r)_{r=1}^m$ subordinated to $\mathcal{U}$. Find $\epsilon > 0$ small enough such that

$$\epsilon < \frac{1}{3m(n+1)}. \tag{8.2}$$

By the choice of ϵ, the following inequality holds:

$$\frac{5\epsilon}{3} < 2\epsilon < \frac{1}{m(n+1)} - \epsilon \leq \frac{1}{m(n+1)} - \frac{\epsilon}{m(n+1)^2}. \tag{8.3}$$

Set $\mathfrak{F} := \{1_{C(X)}, h_1, \ldots, h_m\}$ and using that $C(X)$ has nuclear dimension equal to n, we find a finite dimensional C*-algebra of the form $F = \bigoplus_{k=0}^n F^{(k)}$, maps $\psi : C(X) \to F$ and $\varphi : F \to C(X)$ such that

(i) ψ is c.p.c.,
(ii) $\varphi_k := \varphi|_{F^{(k)}}$ is a c.p.c. order zero map,
(iii) for every $x \in \mathfrak{F}$

$$\|x - \varphi\psi(x)\| < \frac{\epsilon}{m(n+1)}. \tag{8.4}$$

By Theorem 2.4, there exists a unital C*-algebra $\mathcal{M}$ containing $\varphi(F^{(k)})$ and a *-homomorphism $\pi : F^{(k)} \to \mathcal{M}$ commuting with $\varphi(1_{F^{(k)}})$ such that

$$\varphi(a) = \varphi(1_{F^{(k)}})\pi(a) = \pi(a)\varphi(1_{F^{(k)}}). \tag{8.5}$$

The algebra $\mathcal{M}$ given by Theorem 2.4 is commutative if the codomain is commutative. In fact, $\mathcal{M}$ can be assumed to be equal to $C(\beta U)$, where U is an open subset of X and βU is the Stone-Čech compactification of U.

The algebra $F^{(k)}$ is finite dimensional, so it is a finite direct sum of matrix algebras, say $F^{(k)} = \bigoplus_{i=1}^{m_k} M_{d_i}(\mathbb{C})$. Using that the kernel of an order zero map is an ideal of the domain, we can assume, since each $M_{d_i}(\mathbb{C})$ is simple, that $\varphi_k|_{M_{d_i}(\mathbb{C})}$ is injective. Let $a, b \in M_{d_i}(\mathbb{C})$, using the commutativity of $\mathcal{M}$, we have

$$\varphi(ab) = \varphi(1_{F^{(k)}})\pi(a)\pi(b) = \varphi(1_{F^{(k)}})\pi(b)\pi(a) = \varphi(ba). \tag{8.6}$$

Since $\varphi|_{M_{d_i}(\mathbb{C})}$ is injective, we conclude that $M_{d_i}(\mathbb{C})$ is commutative for $i = 1, \ldots, m_k$. This yields $d_i = 1$, and hence, $F^{(k)} = \mathbb{C}^{m_k}$ for some $m_k \in \mathbb{N}$.

In particular, for each $r \leq m$, we can write

$$\psi(h_r) = \left(\lambda_r^{(0)}, \ldots, \lambda_r^{(n)}\right) \in \bigoplus_{k=0}^{n} \mathbb{C}^{m_k} \tag{8.7}$$

where

$$\lambda_r^{(k)} = \left(\lambda_{1,r}^{(k)}, \ldots, \lambda_{m_k,r}^{(k)}\right) \in \mathbb{C}^{m_k}. \tag{8.8}$$

Set $e_i^{(k)} := (0, \ldots, 0, 1, 0, \ldots, 0) \in \mathbb{C}^{m_k}$ where the 1 is in the i-th position. Hence $1_{F^{(k)}} = \sum_{i=1}^{m_k} e_i^{(k)}$. Using this notation, we can write the image of h_r in the following way:

$$\varphi\psi(h_r) = \varphi\left(\sum_{i=1}^{m_k} \lambda_{i,r}^{(k)} e_i^{(k)}\right) = \sum_{k=0}^{n}\sum_{i=1}^{m_k} \lambda_{i,r}^{(k)} \varphi_k\left(e_i^{(k)}\right). \tag{8.9}$$

Similarly, since $1_{C(X)} = \sum_{r=1}^{m} h_r$,

$$\varphi\psi\left(1_{C(X)}\right) = \sum_{r=1}^{m}\sum_{k=0}^{n}\sum_{i=1}^{m_k} \lambda_{i,r}^{(k)} \varphi_k\left(e_i^{(k)}\right). \tag{8.10}$$

Observe that the functions $\varphi_k\left(e_1^{(k)}\right), \ldots, \varphi_k\left(e_{m_k}^{(k)}\right)$ are pairwise orthogonal since φ_k is an order zero map. This implies that if $\varphi_k\left(e_i^{(k)}\right)(x) > 0$ then $\varphi\left(e_j^{(k)}\right)(x) = 0$ for $j \neq i$. Hence, pointwise, the sum in (8.10) is at most the sum of $m(n+1)$ strictly positive summands.

For each $e_i^{(k)}$, define

$$\begin{aligned} W_i^{(k)} &:= \varphi_k\left(e_i^{(k)}\right)^{-1}\left(\left(\frac{1}{m(n+1)} - \epsilon, \infty\right)\right) \\ &= \left\{x \in X \;\middle|\; \varphi_k\left(e_i^{(k)}\right)(x) > \frac{1}{m(n+1)} - \epsilon\right\}. \end{aligned} \tag{8.11}$$

Then set

$$\mathcal{W}_0 := \left\{W_i^{(k)} \;\middle|\; 0 \leq k \leq n;\ 1 \leq i \leq m_k\right\}. \tag{8.12}$$

The order of $\mathcal{W}_0$ is at most n because φ is the sum of $n+1$ order zero maps. Indeed, for each $k = 0, 1, \ldots, n$, set $\mathcal{W}_0^{(0)} := \left\{W_i^{(0)} \mid i = 1, \ldots, m_0\right\}$ and

$$\mathcal{W}_0^{(k)} := \left\{W_i^{(k)} \mid i = 1, \ldots, m_k\right\} \setminus \bigcup_{j=0}^{k-1} \mathcal{W}_0^{(j)}, \qquad 1 \leq k \leq n. \tag{8.13}$$

It is immediate that $\mathcal{W}_0 = \bigsqcup_{k=0}^{n} \mathcal{W}_0^{(k)}$, and since the functions $\varphi_k\left(e_1^{(k)}\right), \ldots, \varphi_k\left(e_{m_k}^{(k)}\right)$ are pairwise orthogonal, we have that $W_i^{(k)} \cap W_j^{(k)} = \emptyset$ for $i \neq j$.

Momentarily fix $W_j^{(s)}$ for some j and s. We will show that if it is not contained in some element U_r of the original cover $\mathcal{U}$, then the corresponding coefficient $\lambda_{j,r}^{(s)}$ in equation (8.10) is "small". Precisely, suppose there exists $U_r \in \mathcal{U}$ such that $W_j^{(s)} \cap (X \setminus U_r) \neq \emptyset$, we will prove $\lambda_{j,r}^{(s)} < 5\epsilon/3$.

Let $x \in W_j^{(s)} \cap (X \setminus U_r)$. Using that the support of h_r is contained in U_r, we obtain $h_r(x) = 0$. By hypothesis

$$\|h_r - \varphi\psi(h_r)\| < \frac{\epsilon}{m(n+1)}, \tag{8.14}$$

then

$$\sum_{k=0}^{n}\sum_{i=1}^{m_k} \lambda_{i,r}^{(k)} \varphi_k\left(e_i^{(k)}\right)(x) \overset{(8.9)}{=} \varphi\psi(h_r)(x) < \frac{\epsilon}{m(n+1)}. \tag{8.15}$$

Observe that all summands are positive, hence

$$\lambda_{j,r}^{(s)} \varphi_s\left(e_j^{(s)}\right)(x) < \frac{\epsilon}{m(n+1)}. \tag{8.16}$$

Using that $x \in W_j^{(s)}$ and

$$\epsilon \overset{(8.2)}{<} \frac{1}{3m(n+1)} < \frac{2}{5m(n+1)}, \tag{8.17}$$

we obtain

$$\begin{aligned} \frac{\epsilon}{m(n+1)} &\overset{(8.16)}{>} \lambda_{j,r}^{(s)} \varphi_s\left(e_j^{(s)}\right)(x) \overset{(8.11)}{>} \lambda_{j,r}^{(s)}\left(\frac{1}{m(n+1)} - \epsilon\right) \\ &\overset{(8.17)}{>} \lambda_{j,r}^{(s)}\left(\frac{1}{m(n+1)} - \frac{2}{5m(n+1)}\right) = \frac{3\lambda_{j,r}^{(s)}}{5m(n+1)}. \end{aligned} \tag{8.18}$$

Thus

$$\lambda_{j,r}^{(s)} < \frac{5\epsilon}{3}. \tag{8.19}$$

Next step is proving that for any $x \in X$ there is at least one coefficient $\lambda_{j,r}^{(s)}$ which is "large enough", i.e. $\lambda_{j,r}^{(s)} \geq 5\epsilon/3$.

Let $x \in X$ and h_r such that $h_r(x) \geq \frac{1}{m}$. Such function exists by the *pigeonhole principle* and because $(h_1, \ldots, h_m)$ is a partition of unity. Then, by (8.14), we have

$$\sum_{k=0}^{n}\sum_{i=1}^{m_k} \lambda_{i,r}^{(k)} \varphi_k\left(e_i^{(k)}\right)(x) \overset{(8.9)}{=} \varphi\psi(h_r)(x) \geq \frac{1}{m} - \frac{\epsilon}{m(n+1)}. \tag{8.20}$$

Then, by the pigeonhole principle again, there exists $s \in \{0, \ldots, n\}$ such that

$$\sum_{i=1}^{m_s} \lambda_{i,r}^{(s)} \varphi_s\left(e_i^{(s)}\right)(x) = \varphi_s\psi(h_r)(x) \geq \frac{1}{m(n+1)} - \frac{\epsilon}{m(n+1)^2}. \tag{8.21}$$

The functions $\varphi_s\left(e_1^{(s)}\right), \ldots, \varphi_s\left(e_{m_s}^{(s)}\right)$ are pairwise orthogonal, thus all the summands in the previous inequality are equal to zero but one. Hence, there exists $j \in \{1, \ldots, m_s\}$ such that

$$\lambda_{j,r}^{(s)} \varphi_s\left(e_j^{(s)}\right)(x) = \sum_{i=1}^{m_s} \lambda_{i,r}^{(s)} \varphi_s\left(e_i^{(s)}\right)(x) \geq \frac{1}{m(n+1)} - \frac{\epsilon}{m(n+1)^2}. \tag{8.22}$$

Since $\lambda_{j,r}^{(s)}$ and $\varphi_s\left(e_j^{(s)}\right)(x)$ are positive numbers less than 1, we obtain

$$\lambda_{j,r}^{(s)} \geq \frac{1}{m(n+1)} - \frac{\epsilon}{m(n+1)^2} \tag{8.23}$$

and

$$\varphi_s\left(e_j^{(s)}\right)(x) \geq \frac{1}{m(n+1)} - \frac{\epsilon}{m(n+1)^2} > \frac{1}{m(n+1)} - \epsilon. \tag{8.24}$$

This immediately shows $x \in W_j^{(s)}$. Moreover, we have

$$\lambda_{j,r}^{(s)} \overset{(8.23)}{>} \frac{1}{m(n+1)} - \frac{\epsilon}{m(n+1)^2} \overset{(8.3)}{>} \frac{5\epsilon}{3}. \tag{8.25}$$

Then, by (8.19), $W_j^{(s)} \cap (X \setminus U_r) = \emptyset$. In other words, $W_j^{(s)} \subseteq U_r$.

This shows that for any $x \in X$, there exists some U_r and $W_j^{(s)}$ such that $x \in W_j^{(s)} \subseteq U_r$. Set

$$\mathcal{W} = \left\{W_i^{(k)} \in \mathcal{W}_0 \;\middle|\; W_i^{(k)} \subseteq U_r \text{ for some } r = 1, \ldots, m\right\}. \tag{8.26}$$

Our previous arguments show that $\mathcal{W}$ is indeed a cover of X that is contained in $\mathcal{W}_0$. Hence, its order is at most n and, by construction, $\mathcal{W}$ refines $\mathcal{U}$. Therefore

$$\dim X \leq \dim_{\mathrm{nuc}} C(X). \tag{8.27}$$

The argument from Section 2.3 actually shows that $\operatorname{dr} C(X) \leq \dim X$. Finally, using that nuclear dimension is always smaller than decomposition rank yields the following,

$$\operatorname{dr} C(X) \leq \dim X \leq \dim_{\mathrm{nuc}} C(X) \leq \operatorname{dr} C(X) \tag{8.28}$$

If we want to include the second countable locally compact case as well, we need to use the following identities regarding a locally compact space X and its one point compactification αX ([**WZ10**, Remark 2.11] [**Pea75**, Corollary 3.5.8]),

$$\dim_{\mathrm{nuc}} C_0(X) = \dim_{\mathrm{nuc}} C(\alpha X), \qquad \text{and} \qquad \dim X = \dim \alpha X. \tag{8.29}$$

We then obtain the following theorem.

THEOREM 8.1. *Let X be a second countably locally compact Hausdorff space. Then*

$$\dim X = \dim_{\mathrm{nuc}} C_0(X) = \operatorname{dr} C_0(X). \tag{8.30}$$

Acknowledgments

The author would like to thank Sam Evington, Jamie Gabe, Adrian Gonzalez-Perez and Stuart White for their comments on an earlier version of this document. The author also thanks the anonymous referee for useful suggestions on an earlier version of this paper.

References

[BBS+19] Joan Bosa, Nathanial P. Brown, Yasuhiko Sato, Aaron Tikuisis, Stuart White, and Wilhelm Winter, *Covering dimension of* C*-*algebras and 2-coloured classification*, Mem. Amer. Math. Soc. **257** (2019), no. 1233, vii+97, DOI 10.1090/memo/1233. MR3908669

[BGSW] Joan Bosa, James Gabe, Aidan Sims, and Stuart White, *The nuclear dimension of* $\mathcal{O}_\infty$*-stable* C*-*algebras*, arXiv:1906.02066, 2019.

[BO08] Nathanial P. Brown and Narutaka Ozawa, *C*-algebras and finite-dimensional approximations*, Graduate Studies in Mathematics, vol. 88, American Mathematical Society, Providence, RI, 2008, DOI 10.1090/gsm/088. MR2391387

[Bra72] Ola Bratteli, *Inductive limits of finite dimensional C*-algebras*, Trans. Amer. Math. Soc. **171** (1972), 195–234, DOI 10.2307/1996380. MR312282

[Cas16] Jorge Castillejos, *Decomposable approximations and coloured isomorphisms for* C*-*algebras*, PhD thesis, University of Glasgow, 2016.

[CE] Jorge Castillejos and Samuel Evington, *Nuclear dimension of simple stably projectionless* C*-*algebras*, Anal. PDE **13** (2020), no. 7, 2205–2240, DOI 10.2140/apde.2020.13.2205. MR4175824

[CET+] Jorge Castillejos, Samuel Evington, Aaron Tikuisis, Stuart White, and Wilhelm Winter, *Nuclear dimension of simple* C*-*algebras*, Invent. Math. **224** (2021), no. 1, 245–290, DOI 10.1007/s00222-020-01013-1. MR4228503

[Čech31] Eduard Čech, *Sur la theorie de la dimension*, *C. R. Acad. Sci., Paris*, 193:976–977, 1931.

[Čech33] Eduard Čech, *Contribution à la théorie de la dimension*, *Čas. Mat. Fys.*, 62:277–291, 1933.

[CE78] Man Duen Choi and Edward G. Effros, *Nuclear C*-algebras and the approximation property*, Amer. J. Math. **100** (1978), no. 1, 61–79, DOI 10.2307/2373876. MR482238

[Cun77] Joachim Cuntz, *Simple C*-algebras generated by isometries*, Comm. Math. Phys. **57** (1977), no. 2, 173–185. MR467330

[DE02] Marius Dadarlat and Søren Eilers, *On the classification of nuclear C*-algebras*, Proc. London Math. Soc. (3) **85** (2002), no. 1, 168–210, DOI 10.1112/S0024611502013679. MR1901373

[Dal00] H. Garth Dales, *Banach algebras and automatic continuity*, London Mathematical Society Monographs. New Series, vol. 24, The Clarendon Press, Oxford University Press, New York, 2000. Oxford Science Publications. MR1816726

[EGM19] Caleb Eckhardt, Elizabeth Gillaspy, and Paul McKenney, *Finite decomposition rank for virtually nilpotent groups*, Trans. Amer. Math. Soc. **371** (2019), no. 6, 3971–3994, DOI 10.1090/tran/7453. MR3917214

[Ell76] George A. Elliott, *On the classification of inductive limits of sequences of semisimple finite-dimensional algebras*, J. Algebra **38** (1976), no. 1, 29–44, DOI 10.1016/0021-8693(76)90242-8. MR397420

[Ell93a] George A. Elliott, *A classification of certain simple C*-algebras*, Operator algebras, mathematical physics, and low-dimensional topology (Istanbul, 1991), Res. Notes Math., vol. 5, A K Peters, Wellesley, MA, 1993, pp. 109–123. MR1259062

[Ell93b] George A. Elliott, *On the classification of C*-algebras of real rank zero*, J. Reine Angew. Math. **443** (1993), 179–219, DOI 10.1515/crll.1993.443.179. MR1241132

[EE93] George A. Elliott and David E. Evans, *The structure of the irrational rotation C*-algebra*, Ann. of Math. (2) **138** (1993), no. 3, 477–501, DOI 10.2307/2946553. MR1247990

[EG96] George A. Elliott and Guihua Gong, *On the classification of C*-algebras of real rank zero. II*, Ann. of Math. (2) **144** (1996), no. 3, 497–610, DOI 10.2307/2118565. MR1426886

[EGL07] George A. Elliott, Guihua Gong, and Liangqing Li, *On the classification of simple inductive limit C*-algebras. II. The isomorphism theorem*, Invent. Math. **168** (2007), no. 2, 249–320, DOI 10.1007/s00222-006-0033-y. MR2289866

[EGLN] George A. Elliott, Guihua Gong, Huaxin Lin, and Zhuang Niu. On the classification of simple amenable C*-algebras with finite decomposition rank, II. arXiv:1507.03437, 2015.

[EGLP96] George A. Elliott, Guihua Gong, Huaxin Lin, and Cornel Pasnicu, *Abelian C^*-subalgebras of C^*-algebras of real rank zero and inductive limit C^*-algebras*, Duke Math. J. **85** (1996), no. 3, 511–554, DOI 10.1215/S0012-7094-96-08520-8. MR1422356

[EN17] George A. Elliott and Zhuang Niu, *The C^*-algebra of a minimal homeomorphism of zero mean dimension*, Duke Math. J. **166** (2017), no. 18, 3569–3594, DOI 10.1215/00127094-2017-0033. MR3732883

[ET08] George A. Elliott and Andrew S. Toms, *Regularity properties in the classification program for separable amenable C^*-algebras*, Bull. Amer. Math. Soc. (N.S.) **45** (2008), no. 2, 229–245, DOI 10.1090/S0273-0979-08-01199-3. MR2383304

[End15] Dominic Enders, *On the nuclear dimension of certain UCT-Kirchberg algebras*, J. Funct. Anal. **268** (2015), no. 9, 2695–2706, DOI 10.1016/j.jfa.2015.01.009. MR3325534

[Eng78] Ryszard Engelking, *Dimension theory*, North-Holland Publishing Co., Amsterdam-Oxford-New York; PWN—Polish Scientific Publishers, Warsaw, 1978. Translated from the Polish and revised by the author; North-Holland Mathematical Library, 19. MR0482697

[GK10] Julien Giol and David Kerr, *Subshifts and perforation*, J. Reine Angew. Math. **639** (2010), 107–119, DOI 10.1515/CRELLE.2010.012. MR2608192

[GLN18] Guihua Gong, Huaxin Lin, and Zhuang Niu, *A classification of finite simple amenable $\mathcal{Z}$-stable C^*-algebras, I: C^*-algebras with generalized tracial rank one* (English, with English and French summaries), C. R. Math. Acad. Sci. Soc. R. Can. **42** (2020), no. 3, 63–450. MR4215379

[GLN19] Guihua Gong, Huaxin Lin, and Zhuang Niu, *A classification of finite simple amenable $\mathcal{Z}$-stable C*-algebras, II: C*-algebras with rational generalized tracial rank one* (English, with English and French summaries), C. R. Math. Acad. Sci. Soc. R. Can. **42** (2020), no. 4, 451–539. MR4215380

[GN43] Israel Gelfand and Mark Neumark, *On the imbedding of normed rings into the ring of operators in Hilbert space* (English, with Russian summary), Rec. Math. [Mat. Sbornik] N.S. **12(54)** (1943), 197–213. MR0009426

[HSWW17] Ilan Hirshberg, Gábor Szabó, Wilhelm Winter, and Jianchao Wu, *Rokhlin dimension for flows*, Comm. Math. Phys. **353** (2017), no. 1, 253–316, DOI 10.1007/s00220-016-2762-0. MR3638315

[HW17] Ilan Hirshberg and Jianchao Wu, *The nuclear dimension of C^*-algebras associated to homeomorphisms*, Adv. Math. **304** (2017), 56–89, DOI 10.1016/j.aim.2016.08.022. MR3558205

[HWZ15] Ilan Hirshberg, Wilhelm Winter, and Joachim Zacharias, *Rokhlin dimension and C^*-dynamics*, Comm. Math. Phys. **335** (2015), no. 2, 637–670, DOI 10.1007/s00220-014-2264-x. MR3316642

[JS99] Xinhui Jiang and Hongbing Su, *On a simple unital projectionless C^*-algebra*, Amer. J. Math. **121** (1999), no. 2, 359–413. MR1680321

[Kir77] Eberhard Kirchberg, *C^*-nuclearity implies CPAP*, Math. Nachr. **76** (1977), 203–212, DOI 10.1002/mana.19770760115. MR512362

[KS18] David Kerr and Gábor Szabó, *Almost finiteness and the small boundary property*, Comm. Math. Phys. **374** (2020), no. 1, 1–31, DOI 10.1007/s00220-019-03519-z. MR4066584

[KW04] Eberhard Kirchberg and Wilhelm Winter, *Covering dimension and quasidiagonality*, Internat. J. Math. **15** (2004), no. 1, 63–85, DOI 10.1142/S0129167X04002119. MR2039212

[Lan73] Christopher Lance, *On nuclear C^*-algebras*, J. Functional Analysis **12** (1973), 157–176, DOI 10.1016/0022-1236(73)90021-9. MR0344901

[Leb11] Henri Lebesgue, *Sur la non-applicabilité de deux domaines appartenant respectivement à des espaces à n et $n+p$ dimensions* (French), Math. Ann. **70** (1911), no. 2, 166–168, DOI 10.1007/BF01461155. MR1511616

[Lin03] Huaxin Lin, *Classification of simple C^*-algebras and higher dimensional noncommutative tori*, Ann. of Math. (2) **157** (2003), no. 2, 521–544, DOI 10.4007/annals.2003.157.521. MR1973053

[Lin04] Huaxin Lin, *Classification of simple C^*-algebras of tracial topological rank zero*, Duke Math. J. **125** (2004), no. 1, 91–119, DOI 10.1215/S0012-7094-04-12514-X. MR2097358

[Lin11] Huaxin Lin, *Asymptotic unitary equivalence and classification of simple amenable C^*-algebras*, Invent. Math. **183** (2011), no. 2, 385–450, DOI 10.1007/s00222-010-0280-9. MR2772085

[Men23] Karl Menger, *Über die Dimensionalität von Punktmengen* (German), Monatsh. Math. Phys. **33** (1923), no. 1, 148–160, DOI 10.1007/BF01705597. Erster Teil. MR1549275

[MS14] Hiroki Matui and Yasuhiko Sato, *Decomposition rank of UHF-absorbing* C*-*algebras*, Duke Math. J. **163** (2014), no. 14, 2687–2708, DOI 10.1215/00127094-2826908. MR3273581

[Mur90] Gerard J. Murphy, *C^*-algebras and operator theory*, Academic Press, Inc., Boston, MA, 1990. MR1074574

[Pat88] Alan L. T. Paterson, *Amenability*, Mathematical Surveys and Monographs, vol. 29, American Mathematical Society, Providence, RI, 1988, DOI 10.1090/surv/029. MR961261

[Pau02] Vern Paulsen, *Completely bounded maps and operator algebras*, Cambridge Studies in Advanced Mathematics, vol. 78, Cambridge University Press, Cambridge, 2002. MR1976867

[Pea75] A. R. Pears, *Dimension theory of general spaces*, Cambridge University Press, Cambridge, England-New York-Melbourne, 1975. MR0394604

[Phi00] N. Christopher Phillips, *A classification theorem for nuclear purely infinite simple C^*-algebras*, Doc. Math. **5** (2000), 49–114. MR1745197

[PV80] M. Pimsner and D. Voiculescu, *Exact sequences for K-groups and Ext-groups of certain cross-product C^*-algebras*, J. Operator Theory **4** (1980), no. 1, 93–118. MR587369

[Rie81] Marc A. Rieffel, *C^*-algebras associated with irrational rotations*, Pacific J. Math. **93** (1981), no. 2, 415–429. MR623572

[Rie83] Marc A. Rieffel, *The cancellation theorem for projective modules over irrational rotation C^*-algebras*, Proc. London Math. Soc. (3) **47** (1983), no. 2, 285–302, DOI 10.1112/plms/s3-47.2.285. MR703981

[Rie90] Marc A. Rieffel, *Noncommutative tori—a case study of noncommutative differentiable manifolds*, Geometric and topological invariants of elliptic operators (Brunswick, ME, 1988), Contemp. Math., vol. 105, Amer. Math. Soc., Providence, RI, 1990, pp. 191–211, DOI 10.1090/conm/105/1047281. MR1047281

[Rør03] Mikael Rørdam, *A simple C^*-algebra with a finite and an infinite projection*, Acta Math. **191** (2003), no. 1, 109–142, DOI 10.1007/BF02392697. MR2020420

[RS02] M. Rørdam, *Classification of nuclear, simple C^*-algebras*, Classification of nuclear C^*-algebras. Entropy in operator algebras, Encyclopaedia Math. Sci., vol. 126, Springer, Berlin, 2002, pp. 1–145, DOI 10.1007/978-3-662-04825-2_1. MR1878882

[RW10] Mikael Rørdam and Wilhelm Winter, *The Jiang-Su algebra revisited*, J. Reine Angew. Math. **642** (2010), 129–155, DOI 10.1515/CRELLE.2010.039. MR2658184

[RSS15] Efren Ruiz, Aidan Sims, and Adam P. W. Sørensen, *UCT-Kirchberg algebras have nuclear dimension one*, Adv. Math. **279** (2015), 1–28, DOI 10.1016/j.aim.2014.12.042. MR3345177

[Sti55] W. Forrest Stinespring, *Positive functions on C^*-algebras*, Proc. Amer. Math. Soc. **6** (1955), 211–216, DOI 10.2307/2032342. MR69403

[SWW15] Yasuhiko Sato, Stuart White, and Wilhelm Winter, *Nuclear dimension and $\mathcal{Z}$-stability*, Invent. Math. **202** (2015), no. 2, 893–921, DOI 10.1007/s00222-015-0580-1. MR3418247

[Sza15] Gábor Szabó, *The Rokhlin dimension of topological $\mathbb{Z}^m$-actions*, Proc. Lond. Math. Soc. (3) **110** (2015), no. 3, 673–694, DOI 10.1112/plms/pdu065. MR3342101

[Sza17] Gábor Szabó, *On the nuclear dimension of strongly purely infinite* C*-*algebras*, Adv. Math. **306** (2017), 1262–1268, DOI 10.1016/j.aim.2016.11.009. MR3581330

[SWZ17] Gábor Szabó, Jianchao Wu, and Joachim Zacharias, *Rokhlin dimension for actions of residually finite groups*, Ergodic Theory Dynam. Systems **39** (2019), no. 8, 2248–2304, DOI 10.1017/etds.2017.113. MR3977342

[Tak64] Masamichi Takesaki, *On the cross-norm of the direct product of C^*-algebras*, Tohoku Math. J. (2) **16** (1964), 111–122, DOI 10.2748/tmj/1178243737. MR165384

[Tik14] Aaron Tikuisis, *Nuclear dimension, $\mathcal{Z}$-stability, and algebraic simplicity for stably projectionless C^*-algebras*, Math. Ann. **358** (2014), no. 3-4, 729–778, DOI 10.1007/s00208-013-0951-0. MR3175139

[TWW17] Aaron Tikuisis, Stuart White, and Wilhelm Winter, *Quasidiagonality of nuclear C^*-algebras*, Ann. of Math. (2) **185** (2017), no. 1, 229–284, DOI 10.4007/annals.2017.185.1.4. MR3583354

[Tom08] Andrew S. Toms, *On the classification problem for nuclear C^*-algebras*, Ann. of Math. (2) **167** (2008), no. 3, 1029–1044, DOI 10.4007/annals.2008.167.1029. MR2415391

[Ury22] Pavel Urysohn, *Les multiplicités cantoriennes*, *C. R. Acad. Sci., Paris*, 175:440–442, 1922.

[Vil98] Jesper Villadsen, *Simple C^*-algebras with perforation*, J. Funct. Anal. **154** (1998), no. 1, 110–116, DOI 10.1006/jfan.1997.3168. MR1616504

[Vil99] Jesper Villadsen, *On the stable rank of simple C^*-algebras*, J. Amer. Math. Soc. **12** (1999), no. 4, 1091–1102, DOI 10.1090/S0894-0347-99-00314-8. MR1691013

[von29] John von Neumann, *Zur allgemeinen Theorie des Maßes*, *Fundam. Math.*, 13:73–116, 1929.

[vN55] John von Neumann, *Mathematical foundations of quantum mechanics*, Princeton University Press, Princeton, 1955. Translated by Robert T. Beyer. MR0066944

[Win03] Wilhelm Winter, *Covering dimension for nuclear C^*-algebras*, J. Funct. Anal. **199** (2003), no. 2, 535–556, DOI 10.1016/S0022-1236(02)00109-X. MR1971906

[Win09] Wilhelm Winter, *Covering dimension for nuclear C^*-algebras. II*, Trans. Amer. Math. Soc. **361** (2009), no. 8, 4143–4167, DOI 10.1090/S0002-9947-09-04602-9. MR2500882

[Win12] Wilhelm Winter, *Nuclear dimension and $\mathcal{Z}$-stability of pure C^*-algebras*, Invent. Math. **187** (2012), no. 2, 259–342, DOI 10.1007/s00222-011-0334-7. MR2885621

[Win14] Wilhelm Winter, *Localizing the Elliott conjecture at strongly self-absorbing C^*-algebras*, J. Reine Angew. Math. **692** (2014), 193–231, DOI 10.1515/crelle-2012-0082. MR3274552

[Win16] Wilhelm Winter, *Classifying crossed product* C*-*algebras*, Amer. J. Math. **138** (2016), no. 3, 793–820, DOI 10.1353/ajm.2016.0029. MR3506386

[Win18] Wilhelm Winter, *Structure of nuclear* C*-*algebras: from quasidiagonality to classification and back again*, Proceedings of the International Congress of Mathematicians—Rio de Janeiro 2018. Vol. III. Invited lectures, World Sci. Publ., Hackensack, NJ, 2018, pp. 1801–1823. MR3966830

[WZ09] Wilhelm Winter and Joachim Zacharias, *Completely positive maps of order zero*, Münster J. Math. **2** (2009), 311–324. MR2545617

[WZ10] Wilhelm Winter and Joachim Zacharias, *The nuclear dimension of C^*-algebras*, Adv. Math. **224** (2010), no. 2, 461–498, DOI 10.1016/j.aim.2009.12.005. MR2609012

[Yin86] Hong Sheng Yin, *A simple proof of the classification of rational rotation C^*-algebras*, Proc. Amer. Math. Soc. **98** (1986), no. 3, 469–470, DOI 10.2307/2046204. MR857943

Department of Mathematics, KU Leuven, 3001 Leuven, Belgium

Current address: Institute of Mathematics, Polish Academy of Sciences, ul. Śniadeckich 8, 00-656 Warszawa, Poland

Email address: jcastillejoslopez@impan.pl

Contemporary Mathematics
Volume **775**, 2021
https://doi.org/10.1090/conm/775/15589

Short survey on the existence of slices for the space of Riemannian metrics

Diego Corro and Jan-Bernhard Kordaß

ABSTRACT. We review the well-known slice theorem of Ebin for the action of the diffeomorphism group on the space of Riemannian metrics of a closed manifold. We present advances in the study of the spaces of Riemannian metrics, and produce a more concise proof for the existence of slices.

1. Introduction

The presence of a group action on a smooth manifold has been a successful tool in the study of the topology of manifolds, as well as their geometry (cf. [**AB15**, **Bre72**, **Kob72**]). In [**Pal61**], Palais showed that given certain topological conditions on the action of the group, one could integrate the normal bundle of an orbit to obtain an equivariant tubular neighborhood called a *slice*. A result guaranteeing its existence is referred to as a *slice theorem*. Such a theorem allows us to study the action of a Lie group G on M locally, as well as the quotient map $\pi\colon M \to M/G$. The required topological condition is automatically satisfied by any compact group, and slices have played an important role in the study of such actions.

Furthermore, the proof of the slice theorem, for the case when the group G and the manifold M are finite-dimensional, is rather geometric. It has been exploited in the so-called *Symmetry Program*, to study and construct, metrics satisfying lower curvature bounds on a given smooth manifold, with a prescribed group contained in the symmetry group (see [**Gro02**, **Gro17**]).

When attempting to reproduce this proof for the case when the dimensions of the acting group $\mathcal{G}$ and the manifold $\mathcal{M}$ are both infinite, a series of technical obstructions arise. Each of these technical points has to be addressed in order to show the existence of a slice. This is the core of Ebin's work in [**Ebi70**] for the action of the group of diffeomorphisms $\mathrm{Diff}(M)$ on the space of Riemannian metrics $\mathcal{R}(M)$ for a closed smooth manifold M. The action is by pulling back Riemannian metrics along diffeomorphisms. To view it as a left action, we set $\phi{\cdot}g := (\phi^{-1})^*g$ for $\phi \in \mathrm{Diff}(M)$ and $g \in \mathcal{R}(M)$. Namely, in [**Ebi70**] the following theorem is proved.

2020 *Mathematics Subject Classification.* Primary 53C, 53C10, 58D17.

Supported by the DFG (281869850, RTG 2229 "Asymptotic Invariants and Limits of Groups and Spaces").

Slice Theorem. *Let M be a closed smooth manifold and consider the Fréchet Lie group of diffeomorphisms* $\operatorname{Diff}(M)$. *Given a Riemannian metric $\gamma \in \mathcal{R}(M)$, there exists a submanifold S_γ containing γ such that the following hold:*

(i) *For any $\phi \in \operatorname{Diff}(M)_\gamma$, we have $\phi \cdot S_\gamma \subset S_\gamma$.*
(ii) *If $\phi \in \operatorname{Diff}(M)$ is a diffeomorphism such that $\phi \cdot S_\gamma \cap S_\gamma \neq \emptyset$, then $\phi \in \operatorname{Diff}(M)_\gamma$.*
(iii) *There exists an open neighborhood U of the identity right-coset in $\operatorname{Diff}(M)/\operatorname{Diff}(M)_\gamma$, and a cross-section $\chi\colon U \to \operatorname{Diff}(M)$, such that the map $F\colon U \times S_\gamma \to \mathcal{R}(M)$ given by*

$$F(u, s) = \chi(u) \cdot s,$$

is a homeomorphism onto an open neighborhood of γ in $\mathcal{R}(M)$.

The slice theorem has acquired recent relevance in Riemannian geometry regarding the study of the *moduli space* $\mathcal{M}(M)$ of Riemannian structures on M. This space is the orbit space $\mathcal{R}(M)/\operatorname{Diff}(M)$, and parametrizes possible Riemannian geometries on M up to isometry. Since the proof of the Slice Theorem is by local means, the slice theorem holds for any open $\operatorname{Diff}(M)$-invariant submanifold of $\mathcal{R}(M)$. Particular examples of such invariant open manifolds are the subsets of Riemannian metrics satisfying some strict lower (resp. upper) curvature bound. For example, we may consider the space of Riemannian metrics with positive sectional curvature $\mathcal{R}^{\sec>0}(M)$. The orbit space $\mathcal{M}^{\sec>0}(M) := \mathcal{R}^{\sec>0}(M)/\operatorname{Diff}(M)$ describes isometry classes of metrics with positive sectional curvature on M. Moreover, we can consider the action of closed subgroups of $\operatorname{Diff}(M)$, such as $\operatorname{Diff}^0(M)$, the subgroup of diffeomorphisms homotopic to the identity. A particular example for a space of metrics with an upper curvature bound is the Teichmüller space (cf. [**FS17**]).

Ebin's approach to prove the Slice Theorem is as follows. He first considers a fixed Riemannian metric and a fixed volume element on M. Then he fixes a degree of differentiability $s \in \mathbb{N} \cup \{\infty\}$, and defines an inner product $(-,-)_s$ on the space of sections $\Gamma^s(S^2T^*M)$. The downside is that the space $\Gamma^s(S^2T^*M)$ is not complete with respect to this inner product. Completing it yields a Sobolev space $H^s(S^2T^*M)$. Furthermore, the topology of this Sobolev space does not depend on the choices of the Riemannian metric, and the volume element on M (see [**Ebi70**, Sec. 3]). Ebin then proceeds to show the existence of a slice for the action of the group of diffeomorphisms on this Sobolev space.

For M a closed smooth manifold, the study of a Riemannian structure for the space $\mathcal{R}(M)$ has greatly advanced in the last 50 years. Namely, Freed and Groisser in [**FG89**], and later Gil-Medrano and Michor in [**GMM91**] have studied the space $\mathcal{R}(M)$ with a canonical non-complete Riemannian structure, given by an L^2-metric (cf. [**KM97**]). They study the Levi-Civita connection of such a metric, and show the existence of an exponential map.

Remark 1.1. The original result in [**Ebi70**] considers only orientable manifolds. In the present proof we believe this hypothesis is not necessary.

The proof we present does not rely on the technical work done in the setting of Sobolev spaces in [**Ebi70**]. Furthermore, the advances in the study of topological group actions gives a clearer picture of the construction used in the proof of the slice. This allows us to describe up to homeomorphism a neighborhood of an orbit of the action of $\mathrm{Diff}(M)$ on $\mathcal{R}(M)$. We observe that the isotropy group of a Riemannian metric $\gamma \in \mathcal{R}(M)$ is the group of isometries $\mathrm{Isom}(\gamma)$ of γ. When M is compact, the group $\mathrm{Isom}(\gamma)$ is a compact Lie group (see [**MS39**]).

THEOREM B. *Let M be a closed smooth manifold, and consider* $\mathrm{Diff}(M)$ *acting on $\mathcal{R}(M)$. For a fixed Riemannian metric γ let S_γ be the slice through γ. Then a closed neighborhood of the orbit* $\mathrm{Diff}(M)(\gamma) \subset \mathcal{R}(M)$ *is homeomorphic to*

$$\mathrm{Diff}(M) \times_{\mathrm{Isom}(\gamma)} S_\gamma.$$

This theorem allows to describe a neighborhood of the orbits, which can be regarded as a topological equivariant tubular neighborhood.

REMARK 1.2. The original approach followed by Ebin relies on the fact that the Fréchet manifold of Riemannian metrics has a graded family of Riemannian metrics generating the topology. This approach has been generalized in [**DR19**], to show the existence of slices for a more general family of Fréchet spaces and Fréchet Lie group actions.

This article is organized as follows. In the first part we present the general theory of smooth actions by finite-dimensional Lie groups on finite-dimensional manifolds. For this setting we give a proof of a slice theorem, to which we will point later when proving a slice theorem for the action of $\mathrm{Diff}(M)$ on $\mathcal{R}(M)$. We recall an infinite-dimensional manifold structure for $\mathcal{R}(M)$, which is modeled on a class of metric spaces called Fréchet spaces. Then we present the L^2-metric on $\mathcal{R}(M)$ studied by Gil-Medrano and Michor in [**GMM91**]. This L^2-metric is $\mathrm{Diff}(M)$-invariant, and has an exponential map. Finally, we present the proof of the Slice Theorem and the Theorem B. These are based on the proofs for the finite-dimensional case. We end this work by stating some consequences of the slice theorem.

2. Preliminaries

CONVENTIONS. We will henceforth use the following conventions:

(1) All manifolds, submanifolds and Lie groups are assumed to be smooth and finite-dimensional, if not stated otherwise.
(2) By M we always denote a closed, finite-dimensional smooth manifold.
(3) We denote by G a Hausdorff topological group; often a finite-dimensional Lie group.
(4) The continuous map describing an action of a group on a space is denoted by μ.
(5) For a smooth vector bundle $E \to M$, we denote by $\Gamma^s(E)$ for $s \in \mathbb{N} \cup \{\infty\}$ the space of sections of differentiability s. Instead of $\Gamma^\infty(E)$, we will write $\Gamma(E)$.

2.1. Group actions. A group G equipped with a topology is called a *topological group*, if the multiplication $m\colon G \times G \to G$, and the inverse $i\colon G \to G$ are continuous maps with respect to this topology. In the present work we will additionally assume that all topological groups are Hausdorff and there is a second countable basis of neighborhoods for the identity $e \in G$. This implies that for the identity element $e \in G$, the set $\{e\}$ is closed. The condition of $\{e\}$ being closed and G having a second countable basis for open sets containing e is equivalent to G being metrizable, and the metric can be chosen left-invariant by [**MZ55**, Sec. 1.22].

Let G be a topological group, and X be a topological space. We say that G *acts on* X from the left, if there exists a continuous map $\mu\colon G \times X \to X$ such that:

(i) $\mu(e, p) = p$ for any $p \in X$;
(ii) $\mu(g, \mu(h, p)) = \mu(gh, p)$ for any $g, h \in G$ and any $p \in X$.

For a fixed $p \in X$ the set $G(p) := \{\mu(g, p) \mid g \in G\} \subset X$ is called the *orbit of* p. The subgroup $G_p := \{g \in G \mid \mu(g, p) = p\}$ is called the *isotropy subgroup* at p, or the *stabilizer* of the action at p. From now on, we will denote $\mu(g, p)$ simply by $g \cdot p$. The subgroup given by intersection of all isotropy groups

$$\bigcap_{p \in X} G_p,$$

is called the *ineffective kernel* of the action. If this subgroup is trivial, the action is called *effective*. Any group action can be turned into an effective action and hence we will assume that all group actions in the present text are effective. Moreover, if at every point $p \in X$ we have $G_p = \{e\}$, we say the action is *free*. An action is *transitive* if the orbit of any point $p \in X$ is the whole space X.

An action of G on X defines an equivalence relation $\sim$ on X. We say $p \sim q$, if and only if, p and q are contained in the same orbit. The quotient space obtained from the action's orbits is called the *orbit space* of the action, and is denoted by X/G. The topology of X induces via the *orbit quotient map* $\pi\colon X \to X/G$ the quotient topology on X/G. For any subset $A \subset X$, we denote its image under π by A^*. For a point $p \in X$ we denote its image as p^*.

An action μ of a topological group G on a topological space X is called *proper*, if the map $A\colon G \times X \to X \times X$ defined as

$$A(g, p) := (g \cdot p, p),$$

is proper, i.e. the preimage of any compact subset of X under the map is a compact subset of $G \times X$. This concept was introduced by Palais in [**Pal61**]. If X is metrizable, then we have the following characterization of proper actions.

PROPOSITION 2.1 (Prop. 3.19 in [**AB15**]). *Let G be a Hausdorff topological group, with second countable basis for open sets containing the identity $e \in G$, and let X be a metrizable topological space. An action $\mu\colon G \times X \to X$ is proper, if and only if, for any sequence $(g_n)_{n\in\mathbb{N}}$ in G and any convergent sequence $(x_n)_{n\in\mathbb{N}}$ in X, such that $(\mu(g_n, x_n))_{n\in\mathbb{N}}$ converges in X, the sequence $(g_n)_{n\in\mathbb{N}}$ admits a convergent subsequence.*

From this characterization of proper actions it easily follows that for a proper action all isotropy groups are closed.

In the particular case when the group G is metrizable via a complete metric, we have the following characterization for a proper action:

PROPOSITION 2.2. *Let G be a topological group with a complete left-invariant metric inducing the topology of G, and X a metric space. An action $\mu\colon G \times X \to X$ is proper, if and only if, for any sequence $(g_n)_{n\in\mathbb{N}}$ in G and any $x_0 \in X$, if the sequence $(\mu(g_n, x_0))_{n\in\mathbb{N}}$ converges to x_0 in X, then $(g_n)_{n\in\mathbb{N}}$ has a convergent subsequence.*

PROOF. It is clear that the conclusion of Proposition 2.1, implies the necessity direction.

Denote by $\rho\colon G \times G \to \mathbb{R}$ the complete left-invariant metric of G. We will show that for any $x \in X$, the isotropy group G_x is compact, i.e. that any sequence $(g_n)_{n\in\mathbb{N}}$ in G_x has a convergent subsequence, with limit in G_x. Since we have ρ an invariant metric, G_x is a metric space. Consider $(g_n)_{n\in\mathbb{N}} \subset G_x \subset G$ an arbitrary sequence. Then for all elements in the sequence, we have that $\mu(g_n, x) = x$. Thus the sequence $(\mu(g_n, x))_{n\in\mathbb{N}}$ is constant and thus it is a convergent sequence. From our hypothesis, we conclude that there exists a convergent subsequence $(g_{n_k})_{k\in\mathbb{N}}$, with limit some $g_0 \in G$. From the continuity of μ, we have that $\mu(g_{n_k}, x) = x$ converges to $\mu(g, x)$. By the uniqueness of limits, we see that $\mu(g, x) = x$. Thus g is an element of G_x, and we conclude that G_x is compact.

Since G_x is compact for any $x \in X$, we can assume, by averaging ρ over G_x, that ρ is G_x-invariant, by right multiplication. This induces a complete metric $\tilde{\rho}$ on the quotient space G/G_x.

Next we show that the action is proper by showing that the characterization given in Proposition 2.1 holds. Consider $((g_n, x_n))_{n\in\mathbb{N}}$ an arbitrary sequence in $G \times X$ such that the sequence $((\mu(g_n, x_n), x_n))_{n\in\mathbb{N}}$ converges in $X \times X$ to (y, x). We will show that there is a convergent subsequence of $(g_n)_{n\in\mathbb{N}}$. Given $\varepsilon > 0$, for sufficiently large $n \in \mathbb{N}$, we have that for the left-invariant metric d on X, the inequality $d(\mu(g_n, x_n), \mu(g_n, x)) = d(x_n, x) < \varepsilon/2$ holds. Since the sequence $(\mu(g_n, x_n))_{n\in\mathbb{N}}$ converges to y, for sufficiently large n we have

$$d(y, \mu(g_n, x)) \leqslant d(y, \mu(g_n, x_n)) + d(\mu(g_n, x_n), \mu(g_n, x)) < \varepsilon.$$

Therefore, for n and m large enough,

$$\varepsilon > d(\mu(g_n, x), \mu(g_m, x)) = d(\mu(g_m^{-1} g_n, x), x).$$

From our hypothesis, there is a convergent subsequence, which we denote again by $g_m^{-1} g_n$ converging to some $g_0 \in G_x$. This implies that for the complete metric $\tilde{\rho}$ on G/G_x, for large indices n and m we have :

$$\tilde{\rho}(g_n G_x, g_m G_x) < \varepsilon.$$

This means, that the sequence $(g_n G_x)_{n\in\mathbb{N}}$ in G/G_x is a Cauchy sequence. Since $\tilde{\rho}$ is complete, there exists a convergent subsequence of $(g_n G_x)_{n\in\mathbb{N}}$ converging to gG_x for some $g \in G$, i.e. for large n we have

$$\begin{aligned} \varepsilon > \tilde{\rho}(g_n G_x, gG_x) &= \inf\{\rho(g_n a_1, g a_2) \mid a_1, a_2 \in G_x\} \\ &= \inf\{\rho(g_n a_1 a_2^{-1}, g) \mid a_1, a_2 \in G_x\} \geqslant 0. \end{aligned}$$

Thus, there is a sequence $(g_n h_n)_{n\in\mathbb{N}}$ in G converging to g. Since G_x is compact, we can find a subsequence $(h_{n_k})_{k\in\mathbb{N}}$ converging to $h \in G_x$. From this, using the triangle inequality, we obtain that the sequence $(g_{n_k})_{k\in\mathbb{N}}$ converges to gh^{-1} in G. □

Furthermore, when G is a smooth manifold, and the group operations $m\colon G \times G \to G$, and $i\colon G \to G$ are smooth we say that G is a *Lie group*. If M is a fixed smooth manifold, and G is a Lie group, then a *smooth action* by G on M is a smooth map $\mu\colon G \times M \to M$ satisfying (i) and (ii) above.

The orbit space of a proper smooth action of a Lie group on a smooth manifold is a second countable Hausdorff space. If the action is not free, the quotient might not be a closed manifold. For example, if we consider the 2-sphere and fix an axis of rotation, we have a smooth action by the circle on the 2-sphere with two fix points. The orbit space is homeomorphic to a closed interval.

THEOREM 2.3. *Let G be a Hausdorff topological group, with a second countable basis for open sets containing the identity element $e \in G$, acting properly on a metric space (X, d), where d is G-invariant.*[1] *Fix $p \in X$, and consider $\mu_p \colon G \to X$ given by $\mu_p(g) = g \cdot p$. Let $\rho \colon G \to G/G_p$ be the quotient map. Then there exists a G-equivariant homeomorphism $\tilde{\mu}_p \colon G/G_p \to G(p)$ onto the orbit through p such that $\tilde{\mu}_p \circ \rho = \mu_p$. Furthermore, the orbit $G(p)$ is a closed subspace of X.*

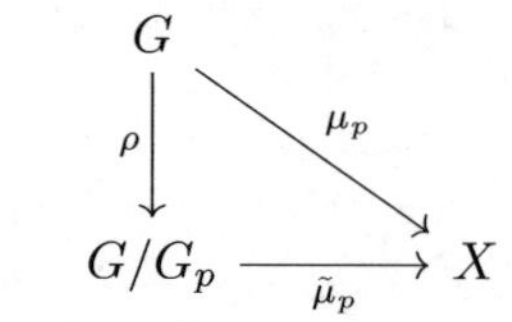

PROOF. Consider $g_1 G_p = g_2 G_p$. Then $g_2^{-1} g_1 \in G_p$, and thus $g_2^{-1} g_1 \cdot p = p$. Therefore, we have $g_1 \cdot p = g_2 \cdot p$. I.e. if $\rho(g_1) = \rho(g_2)$, then $\mu_p(g_1) = \mu_p(g_2)$. Since ρ is a quotient map, there exists a continuous, well-defined map $\tilde{\mu}_p \colon G/G_p \to X$ making the diagram commute. This map is injective, and its image is the orbit $G(p)$. We only have to proof that $\tilde{\mu}_p$ is closed. Observe that since we have an invariant metric, we can talk about convergent sequences. Take $C \subset G/G_p$ closed, and consider $\tilde{\mu}_p(C)$ in X. Consider a sequence $(g_n \cdot p)_{n \in \mathbb{N}}$ in $\tilde{\mu}_p(C)$, with limit $g \cdot p$. From the properness of the G-action we have that the sequence $(g_n)_{n \in \mathbb{N}}$ converges to g in G. Since $\rho^{-1}(C)$ is closed in G, we have that g lies in $\rho^{-1}(C)$, and thus gG_p is an element of C. □

REMARK 2.4.

(1) We used the properness of the action in Theorem 2.3 to show that the inverse of $\tilde{\mu}_p$ is continuous.
(2) In the following section, we will see that for a smooth, proper, effective group action by a Lie group on a smooth manifold M, we can always find an invariant metric on M. In this case, the conclusions of Theorem 2.3 can be strengthened to diffeomorphism for $\tilde{\mu}_p$, and smooth embedded submanifold for $G(p)$ (cf. [**AB15**, Proposition 3.41]).
(3) Moreover, we will see that the existence of a slice yields an improved version of this theorem topologically describing a neighborhood of the orbits.

2.2. The finite-dimensional slice theorem. A group action gives a partition of a smooth manifold M, whose global structure is described by the topological space M/G. Thus, we can attempt to recover information about M from a separate analysis of the orbits, and the orbit space. Proper actions, which where introduced by Palais in [**Pal61**], are a good setting for this decomposition study. This follows from the fact that for a proper smooth action an orbit is an embedded closed submanifold (see [**AB15**, Proposition 3.41]). Actually, the orbit at $p \in M$ is diffeomorphic to G/G_p, and as stated above the orbit space is a second countable Hausdorff topological space.

There are stronger geometric consequences to the properness of a smooth action by a Lie group G on a smooth manifold M. We say that a Riemannian metric γ on M is G-invariant, if for any $g \in G$, the map

$$\mu_g \colon M \to M, \quad p \mapsto g \cdot p,$$

[1] I.e. $d(g \cdot p, g \cdot q) = d(p, q)$ for all $p, q \in X$ and $g \in G$.

is an isometry of γ. The following theorem states that for any proper smooth action by G on M there exists at least one G-invariant metric.

THEOREM 2.5 (Theorem 3.65 in [**AB15**]). *Let G be a Lie group, and let M be a smooth manifold. Further, let $\mu\colon G \times M \to M$ be a proper smooth action. There exists a G-invariant Riemannian metric γ on M such that $\mu_G = \{\mu_g \mid g \in G\}$ is a closed Lie subgroup of* $\mathrm{Isom}(M, \gamma)$.

We note that $\mathrm{Isom}(M, \gamma)$ is a Lie group by the theorem of Myers-Steenrod (see [**MS39**]).

Combining this theorem with the fact that the orbits are closed, the orbit space obtains the structure of a metric space, and the quotient map $\pi\colon M \to M/G$ is a *submetry* with respect to the metrics on M and M/G. This means that the ball of radius δ centered at p in M is mapped onto the ball of radius δ centered at p^* in M/G.

REMARK 2.6. Furthermore by [**Kan05**], the G-invariant metric can be considered complete.

DEFINITION 2.7. Let G be a Lie group acting continuously on a topological manifold X. A *slice* through $p \in X$, is a closed embedded submanifold S_p of X containing p such that:

(i) For any $g \in G_p$, we have $g \cdot S_p \subset S_p$.
(ii) If $g \in G$ is such that $g \cdot S_p \cap S_p \neq \emptyset$, then $g \in G_p$.
(iii) There exists an open neighborhood U of the identity right-coset in G/G_p, and a cross-section $\chi\colon U \to G$, such that the map $F\colon U \times S_p \to X$ given by
$$F(u, s) := \mu(\chi(u), s),$$
is a homeomorphism onto an open neighborhood of p in X.

REMARK 2.8. This term can be naively generalized in several directions. Let $\mathcal{C}$ be a subcategory of manifold objects in a category of topological spaces with finite products and group objects. Let X be an object and G be a group object in $\mathcal{C}$ acting on X internally, i.e. the action $G \times X \to X$ is a morphism in $\mathcal{C}$. For $p \in X$, a *slice* is a closed embedding $i\colon S_p \hookrightarrow X$ with $p \in \mathrm{Im}(i)$ of an object S_p in $\mathcal{C}$ satisfying (i) – (iii) above. If F in (iii) can be strengthened to be a morphism in $\mathcal{C}$, we call it an *internal slice.*

The main theorem states the existence of a slice for the action of $\mathrm{Diff}(M)$ on $\mathcal{R}(M)$ considered in $\mathcal{C}$, the category of Fréchet manifolds.

We will give a proof of the existence of a slice for the case of finite-dimensional Lie groups and smooth manifolds, which we will later retrace in the infinite-dimensional setting.

THEOREM 2.9 (Slice Theorem 3.49 in [**AB15**]). *Consider a proper smooth action by a Lie group G on M, a smooth manifold. For arbitrary $p \in M$ there exists a slice S_p through p.*

PROOF. First we consider the G-invariant Riemannian metric γ given by Theorem 2.5. From the invariance of the metric γ, it follows that if $\lambda\colon I \to M$ is a geodesic then for any $g \in G$, the curve $g \cdot \lambda(t)$ is a geodesic. Next we fix $p \in M$, and consider the normal space $\nu_p M$ to $G(p)$ at p. We can find an open set B contained in the tangent bundle TM, over which the exponential map is a diffeomorphism onto its image. Set $B^\perp$ the intersection of B with the normal bundle $\nu G(p) \to G(p)$. Set $\varepsilon > 0$ small enough, and let V consist of all normal vectors $v \in B^\perp$ with norm less than ε. Take $V_p \subset V$ to be an open ball of radius ε around the origin, contained in the intersection of B with the normal space $\nu_p G(p) \subset T_pM$ of $G(p)$ at p. We claim that the image of V_p under $\exp_p$ is the desired slice through p.

From the fact that the action maps geodesics to geodesics, points (i) and (ii) follow: Take $g \in G_p$ and $s = \exp_p(t_0 X) \in S_p$. Then for the curve $\lambda(t) = \exp_p(tX)$ we have that $g \cdot \lambda(t)$ is a geodesic. Since $\lambda(0) = p$, and $g \in G_p$, we have $g \cdot \lambda(0) = p$. Furthermore from the fact that X is normal to the orbit at p, and the G-invariance of γ, it follows that $g \cdot \gamma$ is normal to the orbit at p. Thus $g \cdot s = g \cdot \lambda(t_0)$ lies in S_p and hence $G_p(S_p) \subset S_p$.

Now consider g an arbitrary element in G, such that for some $s \in S_p$, $g \cdot s \in S_p$. This means that $g \cdot s = \tilde{s}$ for some $\tilde{s} \in S_p$. Assume that $g \cdot p \neq p$. Then we have one point in the image of B under the exponential map exp, with two different foot points: $g \cdot s$ has $g \cdot p$ as base point for exp, and $\tilde{s}$ has p as a base point for exp. Since the exponential map is a diffeomorphism on B, this is a contradiction. Thus $g \cdot p = p$.

To prove point (iii) we need to consider the right action of G_p on G given by the group multiplication. Since this action is free, the fibration $\rho\colon G \to G/G_p$ is a (smooth) G_p-principal bundle (cf. [**AB15**, Corollary 3.38]). Thus there exists an open neighborhood U of the coset $eG_p \in G/G_p$, such that the bundle $\rho|_{\rho^{-1}(U)}\colon \rho^{-1}(U) \to U$ is trivial, i.e. there exists a cross-section $\chi\colon U \to G$ of ρ. By Theorem 2.3 there exists a homeomorphism $\tilde{\mu}_p\colon G/G_p \to G(p)$. Thus we may assume that U is such that $\tilde{U} = \tilde{\mu}_p(U) \subset \exp(V)$. Consider $(g_1G_p, s_1), (g_2G_p, s_2) \in G/G_p \times S_p$ and assume that $F(g_1G_p, s_1) = F(g_2G_p, s_2)$. Then we have that

$$\chi(g_2G_p)^{-1}\chi(g_1G_p) \cdot s_1 = s_2 \in S_p.$$

By point (ii) we have that $\chi(g_2G_p)^{-1}\chi(g_1G_p) \in G_p$. Thus $g_1G_p = g_2G_p$, and from this it follows that $s_1 = s_2$. Therefore, F is injective.

Given $q \in \exp(V)$ there exists a unique vector $v \in V$, normal to the orbit such that $q = \exp(v)$. Furthermore we can assume that for the projection map $\pi\colon V \to G(p)$, we have that $x = \pi(v) \in \tilde{U} \subset G(p)$ (i.e we take $v \in \pi^{-1}(\tilde{U})$). Then, there exists $g \in G$ such that $x = g \cdot p$. Thus for the homeomorphism $\tilde{\mu}_p$ given by Theorem 2.3, we have that $\tilde{\mu}_p^{-1}(x) = gG_p \in U$. Observe that

$$\chi(gH) \cdot p = \mu_p(\chi(gH)) = \tilde{\mu}_p \circ \rho(\chi(gH)) = \tilde{\mu}_p(gH) = g \cdot p = x.$$

Therefore, $\chi(gH)^{-1} \cdot x = p$. We remark that, since q lies in a geodesic normal to $G(p)$, and the Riemannian metric γ on M is G-invariant, then $\chi(gH)^{-1} \cdot q$ lies in a geodesic normal to $G(p)$, which goes through p. Thus $\chi(gH)^{-1} \cdot q$ lies on the slice S_p. From this discussion it follows that $(gG_p, \chi(gH)^{-1} \cdot q) \in U \times S_p$ and

$$F(gG_p, \chi(gH)^{-1} \cdot q) = q.$$

In other words for $q \in \exp(\pi^{-1}(\tilde{U}))$, the inverse of F is given by

$$F^{-1}(q) = \Big(\tilde{\mu}_p^{-1}(\pi(\exp^{-1}(q))), \chi(\tilde{\mu}_p^{-1}(\pi(\exp^{-1}(q)))) \cdot q\Big)$$

Therefore F^{-1} is a continuous function since it is given by continuous functions, and thus F is a homeomorphism. □

The following result depends only on the existence of slices for every point in the space acted on.

THEOREM 2.10. *Let G be a Lie group, X a topological manifold. Assume $\mu\colon G \times X \to X$ is a proper action, such that for any point $p \in X$, a closed slice S_p exists, satisfying* (i), (ii) *and* (iii), *and the isotropy group G_p is compact. Then a closed neighborhood of the orbit $G(p)$ is homeomorphic to*

$$G \times_{G_p} S_p.$$

PROOF. We set $\mathrm{Tub}(G(p)) = \mu(G \times S_p)$. Since the action of G_p leaves S_p invariant, we consider the following action of G_p on $G \times S_\gamma$: For $h \in G_p$ and $(g,s) \in S_p$, we set

$$h \star (g,s) = (gh, \mu(h^{-1}, s)).$$

We denote the orbit space of this action by $G \times_{G_p} S_p$, and we consider $\pi\colon G \times S_p \to G \times_{G_p} S_p$ the orbit projection map. We define $\phi\colon G \times S_p \to \mathrm{Tub}(G(p))$ by $\phi(g,s) = \mu(g,s)$. Thus by definition the map ϕ is continuous and surjective. For (g,s) and $(gh, \mu(h^{-1}, s))$, we have that $\phi(g,s) = \phi(gh, \mu(h^{-1}, s))$. From the fact that π is a quotient map, it follows that there exists a continuous map $\psi\colon G \times_{G_p} S_p \to \mathrm{Tub}(G(p))$ making the following diagram commute

$$\begin{array}{ccc} G \times S_p & & \\ {\scriptstyle\pi}\downarrow & \searrow^{\phi} & \\ G \times_{G_p} S_p & \xrightarrow[\psi]{} & \mathrm{Tub}(G(p)) \end{array}$$

We will show that the map ϕ is an open map. Let $U \subset G/G_p$ be an open neighborhood of the identity coset given by point (iii) of Definition 2.7. For the quotient map $p\colon G \to G/G_p$ the subset $p^{-1}(U) \times S_p \subset G \times S_p$ is open. Consider the map $F\colon U \times S_p \to X$ given by point (iii). Set $N = F(U \times S_p) \subset X$, which is open. Since we have $\chi(U) \subset p^{-1}(U)$ for the cross-section $\chi\colon U \to G$, we have $N \subset \phi(p^{-1}(U) \times S_p)$. Let $(g,s) \in p^{-1}(U) \times S_p$. Since the action of G_p on the fibers of $p\colon G \to G/G_p$ is transitive, there exists $h \in G_p$ such that

$$g = \chi\big(gG_p\big)h.$$

Thus $\phi(g,s) = \phi(\chi\big(gH\big)h, s) = \mu(\chi\big(gH\big)h, s) = \mu(\chi\big(gH\big), \mu(h,s))$. Since $\mu(h,s)$ is an element of S_p by property (i) of the Slice, then $\phi(g,s) \in N$. Thus the image under ϕ of an open set in $G \times S_p$ is open in $\mathrm{Tub}(G(p))$. Therefore ϕ is a quotient map, and ψ is a homeomorphism. □

2.3. Fréchet spaces. Let F be a real vector space. A *seminorm* on F is a real valued function $\|\cdot\|\colon F \to \mathbb{R}$ such that:

(i) $\|f\| \geqslant 0$ for any vector $f \in F$;
(ii) $\|f + g\| \leqslant \|f\| + \|g\|$ for all vectors $f, g \in F$;
(iii) $\|\lambda f\| = |\lambda| \|f\|$ for any vector $f \in F$, and any $\lambda \in \mathbb{R}$.

A family of seminorms $\{\|\cdot\|_\alpha \mid \alpha \in \Lambda\}$ defines a unique topology on F such that a sequence $(f_k)_{k\in\mathbb{N}}$ converges to f if and only if for all α in Λ, the limit of $(\|f_k - f\|_\alpha)_{k\in\mathbb{N}}$ is 0. We call $(F, \{\|-\|_\alpha\})$ with this induced topology a *locally convex topological vector space*. This topology is metrizable if and only if Λ is countable as a set. In this case we say that a sequence $(f_k)_{k\in\mathbb{N}}$ is Cauchy if for any $\varepsilon > 0$ there exists an $N \in \mathbb{N}$, such that for $n, m \geqslant N$ and for all $\alpha \in \Lambda$ we have $\|f_n - f_m\|_\alpha < \varepsilon$. We say that a metrizable locally convex topological vector space is *complete* if for any Cauchy sequence $(f_k)_{k\in\mathbb{N}}$ there exists $f \in F$, such that $(f_k)_{k\in\mathbb{N}}$ converges to f in $(F, \{\|-\|_\alpha\})$. A *Fréchet space* is a complete, metrizable, locally convex topological vector space.

We can define derivatives in Fréchet spaces as follows. Let E and F be Fréchet spaces, and let $U \subset E$ be an open subset. Consider a continuous map $f\colon U \to F$. We say that f is *differentiable at* $x \in U$ in the direction of $v \in E$, if the *differential*

$$Df(x)v := \lim_{t\to 0} \frac{f(x+tv) - f(x)}{t}$$

exists. We say that f is of class $C^1(U)$ if the limit exists for arbitrary pairs (x, v) in $U \times E$, and the map $Df\colon U \times E \to F$ is continuous in both arguments. By considering

$$\lim_{s\to 0} \frac{Df(x+sh)v - Df(x)v}{s},$$

we can define the second derivative $D^2 f\colon U \times E \to F$, of f.

REMARK 2.11. It does not make sense to consider the partial derivative of Df with respect to v, since $Df(x)v$ is linear in v.

By iterating this definition, we can define for $k \in \mathbb{N}$ maps of class C^k, and maps of class C^∞ between Fréchet spaces.

DEFINITION 2.12. Given a Hausdorff topological space $\mathcal{M}$, we say that it is *modeled on a Fréchet space* F in an analogous way to smooth finite-dimensional manifolds. Namely, $\mathcal{M}$ is modeled on F if there exists an atlas $\{(U_i, \phi_i) \mid i \in \mathcal{I}\}$, where each $U_i \subset \mathcal{M}$ is open, and each $\phi_i\colon U_i \to F$ is a homeomorphism onto its image. Furthermore, if $U_{ij} = U_i \cap U_j \neq \emptyset$, the transition map $\phi_i \circ (\phi_j)^{-1}\colon \phi_j(U_{ij}) \to \phi_i(U_{ij})$ is a map of class C^∞ between Fréchet spaces.

REMARK 2.13. Since Fréchet spaces are not necessarily finite-dimensional, a Fréchet manifold is not necessarily finite-dimensional. Furthermore, they differ from Banach manifolds in several aspects. Among the most crucial differences is the absence of an inverse function theorem for arbitrary Fréchet manifolds (cf. [**Ham82**]).

2.4. Compact-open topology. We fix two smooth manifolds M, and N. We denote the set of all smooth maps between M and N by $C^\infty(M, N)$. Let $k \in \mathbb{N} \cup \{\infty\}$. There exists a topology on $C^\infty(M, N)$ with the following property: a sequence of maps in $C^\infty(M, N)$ converges if and only if the first k derivatives converge uniformly on compact subsets of M. This topology is called the *compact-open C^k-topology*[2]. Whereas, if M is a compact manifold, we will also call it the *C^k-topology*. In general, the compact-open C^∞-topology is metrizable (cf. [**KM97**, Corollary 41.12]). This topology is relevant to us by the following result:

THEOREM 2.14. *If M is compact, then for any smooth vector bundle $\pi\colon E \to M$ over M, the space of smooth cross-sections $\Gamma(E)$ with the C^∞-topology induced from $C^\infty(M, E)$ is a Fréchet manifold.*

PROOF. See [**KM97**, Sec. 42]. □

[2]Other common names in the case $k = \infty$ are *weak topology*, e.g. [**Hir94**], or *topology of uniform convergence over compact subsets*, e.g. [**TW15**].

In particular, the *space of Riemannian metrics* $\mathcal{R}(M)$, which is an open subset in $\Gamma(S^2T^*M)$, is a Fréchet manifold.

A topological group G, which is also a Fréchet manifold, is called a *Fréchet Lie group* if the operations of multiplication and taking inverses are smooth in the Fréchet sense.

THEOREM 2.15 (Theorem 43.1 in [**KM97**]). *For a compact smooth manifold M, the group* Diff(M) *of all smooth diffeomorphisms of M is an open Fréchet submanifold of $C^\infty(M,M)$, and composition and inversion are smooth.*

COROLLARY 2.16. *For M a closed smooth manifold, the group of smooth diffeomorphisms* Diff(M) *is a Fréchet Lie group.*

We can define the Lie Algebra of a Fréchet Lie group as the tangent space to the identity element. The Lie bracket is defined via the adjoint representation. The Lie algebra can be identified with the set of left-invariant vector fields on the group (see [**KW09**]). The Lie algebra of Diff(M) is identified in the following theorem.

THEOREM 2.17 (Theorem 43.1 in [**KM97**]). *Let M be a closed smooth manifold. The Lie algebra of the Fréchet Lie group* Diff(M) *is the Lie algebra $\Gamma(TM)$, of all smooth vector fields on M, equipped with the negative of the usual Lie bracket. Furthermore, the exponential map* $\exp\colon \Gamma(TM) \to \mathrm{Diff}(M)$, $V \mapsto \phi^V(1)$ *is given by evaluating thc flow ϕ^V of a vcctor ficld V at timc* 1.

REMARK 2.18. Moreover, for M compact, the Fréchet structure at the tangent space of $\mathrm{Id}_M \in \mathrm{Diff}(M)$ induces a complete left-invariant metric on G, inducing the topology (see Theorem at the end of page 53 in [**Sub85**]).

2.5. Riemannian structure on the space of Riemannian metrics. We can define a Riemannian metric σ on $\mathcal{R}(M)$ as follows. Given a Riemannian metric γ on M, we identify the tangent space of $\mathcal{R}(M)$ at γ with the sections $\Gamma(S^2T^*M)$ (see for example [**KM97**],[**Cla09**]). Consider S and T, two symmetric $(0,2)$-tensors in $\Gamma(S^2T^*M)$. We set

$$\sigma_\gamma(S,T) := \int_M \mathrm{tr}_\gamma(S,T)\,\mathrm{d}\mathrm{vol}(\gamma) = \int_M \sum \gamma^{ij} S_{il}\gamma^{lm}T_{jm}\,\mathrm{d}\mathrm{vol}(\gamma),$$

where $\mathrm{d}\mathrm{vol}(\gamma)$ denotes the Riemannian volume density with respect to γ.

Surprisingly, the topology induced by the Riemannian metric σ on $\mathcal{R}(M)$ is weaker than the topology induced by the Fréchet atlas. For this reason, we say that σ is a *weak* Riemannian structure. One reason why we are not recovering the original topology on $\mathcal{R}(M)$ via σ is that $\mathcal{R}(M)$ equipped with σ is not complete in the sense that each tangent space (which can be identified with $\Gamma(S^2T^*M)$) is not complete with respect to the interior product σ. Its completion is a Sobolev space (cf. [**Ebi70**],[**GMM91**]).

We note that it is still possible to take directional derivatives of σ, and thus we can attempt to define the Levi-Civita connection of σ via the Koszul formula:

$$\begin{aligned}\sigma(\nabla_X Y, Z) =& X\sigma(Y,Z) + Y\sigma(Z,X) - Z\sigma(X,Y)\\ &- \sigma(X,[Y,Z]) - \sigma(Y,[X,Z]) + \sigma(Z,[X,Y]).\end{aligned}$$

The fact that the tangent spaces of $(\mathcal{R}(M),\sigma)$ fail to be complete with respect to σ, implies that $\nabla_X Y$ may, a priori, noi exist. We point out that in case it does exists, the Koszul formula implies it is unique. The possible non-existence of $\nabla_X Y$ has as a consequence the fact that there may be directions in $\mathcal{R}(M)$ where we cannot define the exponential map of σ.

A possible work-around for this problem, is to proceed as in [**Ebi70**], and complete the tangent spaces of $(\mathcal{R}(M),\sigma)$ to obtain Sobolev spaces.

In [**GMM91**], they show that the exponential map, $\exp\colon T\mathcal{R}(M)\to\mathcal{R}(M)$, of the metric σ is actually defined everywhere in $\mathcal{R}(M)$, and for any tangent direction. Furthermore, we have the following theorem.

THEOREM 2.19 (Theorem 45.13 in [**KM97**]). *The mapping* $(\pi,\exp)\colon T\mathcal{R}(M)\to\mathcal{R}(M)\times\mathcal{R}(M)$ *is a diffeomorphism from an open neighborhood of the zero-section in* $T\mathcal{R}(M)$ *onto an open neighborhood of the diagonal in* $\mathcal{R}(M)\times\mathcal{R}(M)$.

This theorem has the following consequence.

COROLLARY 2.20. *Given* $\gamma\in\mathcal{R}(M)$*, there exists an open neighborhood of* γ *in* $\mathcal{R}(M)$ *over which the exponential map is a diffeomorphism.*

2.6. Action of the diffeomorphism group. There is a smooth right action of $\mathrm{Diff}(M)$ on the space of Riemannian metrics $\mathcal{R}(M)$ via pullbacks. Since we are interested in a left action, and the inverse map is smooth in $\mathrm{Diff}(M)$, we consider the left action defined as

$$\mu\colon \mathrm{Diff}(M)\times\mathcal{R}(M)\to\mathcal{R}(M),\quad (\phi,\gamma)\mapsto(\phi^{-1})^*\gamma.$$

Note that this action is linear with respect to the Fréchet structure, i.e. for a fixed $\phi\in\mathrm{Diff}(M)$, the map

$$\mu_\phi\colon\Gamma(S^2T^*M)\supset\mathcal{R}(M)\to\Gamma(S^2T^*M),\quad S\mapsto\mu(\phi,S)$$

is a linear map. We also point out that for a fixed element $\gamma\in\mathcal{R}(M)$, the isotropy group $\mathrm{Diff}(M)_\gamma$ consists of all isometries of (M,γ).

As stated before, for M compact, the group of smooth diffeomorphisms $\mathrm{Diff}(M)$ is a Fréchet Lie group. We recall that there is an identification of the Lie algebra of $\mathrm{Diff}(M)$ with the algebra of smooth vector fields $\Gamma(TM)$. Fix $\gamma\in\mathcal{R}(M)$, and set

$$\mu^\gamma\colon\mathrm{Diff}(M)\to\mathcal{R}(M),\quad \phi\mapsto\mu(\phi,\gamma)$$

with derivative $(\mu^\gamma)_*\colon\Gamma(TM)\to T_\gamma\mathcal{R}(M)$. Furthermore, for a vector field $X\in\Gamma(TM)$ denote by $\phi_t^X\colon M\to M$ the flow of X at time t. In these terms, the derivative is given by

$$(\mu^\gamma)_*(X)=\frac{d}{dt}(\phi_{-t}^X)^*\gamma=\mathcal{L}_{-X}(\gamma).$$

REMARK 2.21. Recall that the Riemannian metric γ induces a Riemannian metric on the bundle of 1-forms (see [**AH11**]), which for $\omega_1,\omega_2\in\Omega^1(M)$ we denote by $\gamma(\omega_1,\omega_2)$, as well.

If X_1 and X_2 are the dual vector fields of the 1-forms ω_1 and ω_2, respectively, we have

$$\gamma(\omega_1,\omega_2)=\gamma(X_1,X_2).$$

With this, for the Riemannian structure σ introduced in Section 2.5 we have the following (cf. [**Bla00**])

$$\sigma(\mathcal{L}_X(\gamma),S)_\gamma=-2\int_M(\nabla_iS^{ij})X_j\,d\mathrm{vol}(\gamma)=2\gamma(\omega,\mathrm{div}\,S).$$

Here $\operatorname{div} S$ is the divergence of S with respect to γ. Thus a symmetric 2-tensor is perpendicular to the orbit of $\operatorname{Diff}(M)$ at γ, if and only if, the divergence of S vanishes. By the work of Berger and Ebin in [**BE69**], the tangent space $\Gamma(S^2T^*M)$ of $\mathcal{R}(M)$ at γ splits as

$$\Gamma(S^2T^*M) \cong T_\gamma \operatorname{Diff}(M)(\gamma) \oplus \operatorname{Ker}(\operatorname{div}). \tag{2.1}$$

From the discussion above, this decomposition is compatible with the inner product σ_γ and we denote $\nu_\gamma \operatorname{Diff}(M)(\gamma) := \operatorname{Ker}(\operatorname{div})$.

REMARK 2.22. The metric σ defined in section 2.5 is invariant under the action of $\operatorname{Diff}(M)$ on $\Gamma(S^2T^*M)$ via pullback. This follows from the fact that the action of $\operatorname{Diff}(M)$ is linear (see [**Ebi70**, p. 20]).

Also, it was proven by Subramaniam in his Ph.D. thesis ([**Sub85**]), that the action of $\operatorname{Diff}(M)$ on $\mathcal{R}(M)$ is proper for a compact action. For the sake of completeness we provide a sketch of the proof here.

THEOREM 2.23. *Let M be a closed smooth manifold. Then the action of* $\operatorname{Diff}(M)$ *on $\mathcal{R}(M)$ is proper.*

PROOF (SEE [**Sub85**, p. 67]). We recall from Remark 2.18 that $\operatorname{Diff}(M)$ has a complete left-invariant metric. Therefore, by Proposition 2.2, the condition of being proper is equivalent to the following statement: If we consider a sequence $(f_n)_{n\in\mathbb{N}}$ of diffeomorphisms of M such that for some fixed Riemannian metric γ, the sequence of Riemannian metrics $(f_n^*\gamma)_{n\in\mathbb{N}}$ converges to γ with respect to the C^∞-topology in $\mathcal{R}(M)$, then there exists a subsequence $(f_{n_k})_{k\in\mathbb{N}}$ converging with respect to the C^∞-topology in $\operatorname{Diff}(M)$.

We will prove that this statement holds in two steps. First we show that with respect to a cover there exists a subsequence converging on each neighborhood with respect to the C^1-topology. Second, we will show that the limit functions agree on the overlaps of the neighborhoods, and that the subsequence converges with respect to the C^∞-topology.

LEMMA 2.24. *Let $\{U_\lambda\}_{1\leq\lambda\leq N}$ be an open cover of (M,γ) by normal coordinates. Then there exists a subsequence $(f_{n_k})_{k\in\mathbb{N}}$ of $(f_n)_{n\in\mathbb{N}}$ converging over each U_λ with respect to the C^1-topology.*

Let m denote the dimension of M. Consider an open cover of (M,γ) by normal coordinates U_λ with $1\leq\lambda\leq N$, each one centered at some point $p_\lambda\in M$. Choose an orthonormal basis $\{V_\lambda^1,\dots,V_\lambda^m\}$ of $T_{p_\lambda}M$ with respect to the metric γ. Since $f_n^*\gamma$ converges uniformly to γ, given $K>1$ we have for all λ, j, and for sufficiently large n

$$\left\|V_\lambda^j\right\|_{f_n^*\gamma} = \left\|(f_n)_*(V_\lambda^j)\right\|_\gamma < K$$

This implies that the vectors V_λ^j are contained in the disk bundle DM_K, which has as fiber the disk of radius K with respect to each metric $f_n^*\gamma$. We note that the total space DM_K is compact since M is compact. We now proceed to construct the subsequence $(f_{n_k})_{k\in\mathbb{N}}$. First we consider the sequences of points $(f_n(p_\lambda))_{n\in\mathbb{N}}$. We begin by considering the point p_1. From the compactness of M, we can find a convergent subsequence $(f_{n_{k_1}})_{k_1\in\mathbb{N}}$ of functions of $(f_n)_{n\in\mathbb{N}}$, such that when we evaluate them at p_1 we have a convergent sequence $(f_{n_{k_1}}(p_1))_{k_1\in\mathbb{N}}$. Denote the limit point q_1. We then evaluate the functions $f_{n_{k_1}}$ at p_2, and use the compactness of M to obtain a new subsequence $(f_{n_{k_2}}(p_2))_{k_2\in\mathbb{N}}$ of $(f_{n_{k_1}}(p_2))_{k_1\in\mathbb{N}}$, with limit q_2. We do this finitely many times to obtain a subsequence $(f_{n_{k_N}})_{k_N\in\mathbb{N}}$ of $(f_n)_{n\in\mathbb{N}}$ such that for all λ the sequence $(f_{n_{k_N}}(p_\lambda))_{k_N\in\mathbb{N}}$ converges to q_λ.

We apply the same process to the subsequence $(f_{n_{k_N}})_{k_N\in\mathbb{N}}$, using the compactness of DM_k, but now with all the bases $\{V_\lambda^1,\dots,V_\lambda^m\}$ and indices λ, to obtain a subsequence $(f_{n_k})_{k\in\mathbb{N}}$ such that for each index λ, the sequence $(f_{n_k}(p_\lambda))_{k\in\mathbb{N}}$ converges to q_λ in M, and for each index $1\leqslant j\leqslant m$, the sequence $((f_{n_k})_*(V_\lambda^j))_{k\in\mathbb{N}}$ converges to $W_\lambda^j\in T_{q_\lambda}M$ in TM.

From now on we consider only the subsequence $(f_{n_k})_{k\in\mathbb{N}}$, and write it as $(f_n)_{n\in\mathbb{N}}$.

So far we have considered arbitrary normal neighborhoods. We now discuss which should be the neighborhoods we should consider. Since we have that the γ-norm of $(f_n^*)_*(V_\lambda^j)$ is less than K, the limit vectors W_λ^j have γ-norm less than $2K$. We will assume from now on that the normal neighborhoods U_λ are normal balls of radius $\delta<2K$ centered at the points p_λ.

Denote by d_n the metric induced on M by the Riemannian metric $f_n^*\gamma$, and by d the metric induced by γ. By definition, since $f_n^*\gamma$ converges to γ with the respect to the C^∞-topology, we have that the metric functions d_n converge pointwise to d. This implies that if q is contained in the ball of radius δ centered at p with respect to d, then for a given ε, and n large enough, q is contained in the $(\delta+\varepsilon)$-ball centered at p with respect to d_n. Furthermore from the energy functional, the minimizing geodesic between p and q with respect to $f_n^*\gamma$ converges to the minimizing geodesic with respect to γ. This also implies that the injectivity radius of $f_n^*\gamma$ converges to the injectivity radius of γ. These observations imply that for a fixed p and ε, if q is contained in the normal ball of radius δ centered at p, then for large n we have that q is contained in the normal ball of radius $(\delta+\varepsilon)$ centered at p.

Take $q\in U_\lambda$. In particular we can write

$$q=\exp_{p_\lambda}^\gamma\left(\sum_j a_{\lambda j}V_\lambda^j\right).$$

Then for large k we have that q is in the normal ball of radius 2δ with respect to $f_n^*\gamma$. Thus we can write

$$q=\exp_{p_\lambda}^{f_n^*\gamma}\left(\sum_j a_{n\lambda j}V_\lambda^j\right).$$

Recall that the minimizing geodesics α_n joining p_λ to q with respect to $f_n^*\gamma$ converge to the minimizing geodesic α joining q to p_λ with respect to γ. Since the vectors $\sum_j a_{n\lambda j}V_\lambda^j$ are the tangent vectors to the geodesics α_n at p_λ, and $\sum_j a_{\lambda j}V_\lambda^j$ is the tangent vector to α at p_λ, then for each λ and j we have that the coefficient $a_{n\lambda j}$ converges to $a_{\lambda j}$.

We now define the limit function $f_\lambda\colon U_\lambda\to M$. For $q\in U_\lambda$ with $q=\exp_{p_\lambda}^\gamma\left(\sum_j a_{\lambda j}V_\lambda^j\right)$ we set

$$f_\lambda(q)=\exp_{q_\lambda}^\gamma\left(\sum_j a_{\lambda j}W_\lambda^j\right).$$

Observe that this function is smooth over U_λ. Using the fact that f_n is an isometry from $(M, f_n^*\gamma)$ to (M,γ), and that $q = \exp_{p_\lambda}^{f_n^*\gamma}\left(\sum_j a_{n\lambda\, j} V_\lambda^j\right)$, we have:

$$f_n(q) = (f_n \circ \exp_{p_\lambda}^{f_n^*\gamma})\left(\sum_j a_{n\lambda\, j} V_\lambda^j\right) = \exp^\gamma_{f_n(p_\lambda)}\left(\sum_j a_{n\lambda\, j} (f_n)_*\left(V_\lambda^j\right)\right).$$

Thus the sequence $(f_n(q))_{n\in\mathbb{N}}$ converges to $\exp^\gamma_{q_\lambda}\left(\sum_j a_{\lambda\, j} W_\lambda^j\right) = f_\lambda(q)$ over U_λ with respect to the C^1-topology.

LEMMA 2.25. *The limit functions f_λ and f_μ agree on the overlaps $U_\lambda \cap U_\mu$ for all λ, μ and thus define a diffeomorphism f on M. Moreover, there exists a subsequence of $(f_n)_{n\in\mathbb{N}}$ converging to f with respect to the C^∞-topology.*

Consider $q \in U_\lambda \cap U_j$. On one side we have that the sequence $(f_n(q))_{n\in\mathbb{N}}$ converges to $f_\lambda(q)$ in M. On the other side we have that the sequence $(f_n(q))_{n\in\mathbb{N}}$ converges to $f_j(q)$ in M. Thus $f_\lambda(q) = f_j(q)$. Since the limits f_λ agree on the overlaps of the open cover $\{U_\lambda\}$ of M, there is a global well-defined function $f\colon M \to M$. This function f is onto, since it is an open map, of a compact space to a connected Hausdorff space. It is also injective: This is true by construction for the restriction of f to each open set U_λ. Consider d_γ the metric on M induced by γ. From the compactness of M, it follows that there exists $L > 0$ such that if $d_\gamma(p,q) < L$, then p and q lie in some U_λ. Take $p \neq q$ in M and assume that $d_\gamma(p,q) > L$. We now consider $d_{f_n^*\gamma}$. Since f_n is an isometry between $(M, d_{f_n^*\gamma})$ and (M, d_γ), we have that $d_{f_n^*\gamma}(p,q) > L$. Since we have that $(f_n^*\gamma)_{n\in\mathbb{N}}$ converges to γ uniformly over M, $(f_n(p))_{n\in\mathbb{N}}$ and $(f_n(q))_{n\in\mathbb{N}}$ converge to $f(p)$ and $f(q)$, respectively. Therefore, $d_\gamma(f(p), f(q)) \geqslant L$ and in particular $f(p) \neq f(q)$. Thus f is injective.

Since by construction f restricted to each U_λ is smooth, f is a diffeomorphism of M. We will show that the sequence $(f_n)_{n\in\mathbb{N}}$ converges to f with respect to the C^∞-topology. To do this, we prove by induction that the sequence $(f_n)_{n\in\mathbb{N}}$ converges to f with respect to the C^k-topology, for all $k \in \mathbb{N}$. The basis of induction is $k = 1$. The statement holds true by construction. Next we assume that the sequence $(f_n)_{n\in\mathbb{N}}$ converges to f with respect to the C^k-topology. This implies that the first partial derivatives of the functions f_n and f_n^{-1} converge to the first derivatives of f and f^{-1} with respect to the C^{k-1}-topology. Let Γ_n and Γ denote the Christoffel symbols of $f_n^*\gamma$ and γ respectively. Since $f_n^*\gamma$ is converging uniformly to γ with respect to the C^∞-topology, for all indices r, s, t we have that ${\Gamma_n}^r_{st}$ converges to Γ^r_{st} with respect to the C^∞-topology. Since each f_n is a diffeomorphism, from the transformation law under change of variable,

$${\Gamma_n}^j_{k\ell} = \sum_{r,s,t} \frac{\partial (f_n^{-1})^j}{\partial x^r} \frac{\partial (f_n)^s}{\partial x^k} \frac{\partial (f_n)^t}{\partial x^\ell} \Gamma^r_{st} + \sum_r \frac{\partial (f_n^{-1})^j}{\partial x^r} \frac{\partial^2 (f_n)^r}{\partial x^k \partial x^\ell},$$

we see that the second partial derivatives of f_n must converges to the second partial derivatives of f with respect to the C^{k-1}-topology. This implies that the sequence $(f_n)_{n\in\mathbb{N}}$ converges to f with respect to the C^{k+1}-topology. □

3. Proofs of the Slice Theorem and Theorem B

We begin by noting that given a slice by the Slice Theorem and properness of the Diff(M) action by Theorem 2.23, the proof of Theorem B is entirely analogous to the proof of Theorem 2.10 in the Fréchet setting.

Our starting point for the proof of the Slice Theorem is the Riemannian metric σ on $\mathcal{R}(M)$ presented in section 2.5, which is defined in local coordinates as

$$\sigma(\alpha, \beta)_\gamma := \int_M \operatorname{tr}(\gamma^{-1}\alpha\gamma^{-1}\beta)\,\mathrm{dvol}(\gamma).$$

We fix a Riemannian metric $\gamma \in \mathcal{R}(M)$. Using the metric σ, and the fact that the orbit Diff$(M)(\gamma)$ is closed, we split the tangent space of $\mathcal{R}(M)$ at γ into two (infinite-dimensional) tangent subspaces: One tangent to the orbit Diff$(M)(\gamma)$, and the normal complement ν_γ Diff$(M)(\gamma)$ with respect to σ_γ (see (2.1)).

REMARK 3.1. Recall that the space normal to the orbit at γ is given by the symmetric 2-forms with 0 divergence with respect to γ (see [**Bla00**, **BE69**]).

We will now follow the proof of Theorem 2.9 (the classical Slice theorem) to show the existence of slices for the action of Diff(M) on $\mathcal{R}(M)$.

Thus we need to show that the exponential map of σ is invariant under the action of Diff(M). Recall by Theorem 2.19 that this exponential map is a local diffeomorphism onto some open subset $U \subset \mathcal{R}(M)$ around γ.

LEMMA 3.2. *The exponential map of σ is invariant under the action of* Diff(M).

PROOF. From the Koszul formula we have uniqueness for the Levi-Civita connection of σ. Since σ is a Diff(M)-invariant Riemannian metric, by the uniqueness, the Levi-Civita connection is Diff(M)-invariant. This implies that the action of Diff(M) respects geodesics. Hence, the exponential map is Diff(M) invariant. See [**FG89**, **KM97**, **GMM91**]. □

With this lemma we can proceed to present the proof of the main theorem.

PROOF OF THE SLICE THEOREM. We observe that we may consider the open neighborhood V to be a small ball around the origin in ν_γ Diff$(M)(\gamma)$. We set $S_\gamma = \exp_\gamma(V)$. We claim this is the desired slice.

We prove that for S_γ defined above, point (i) holds. First, since the metric σ is Diff(M)-invariant, and Isom(γ) is a closed Lie subgroup, then σ is also Isom(γ)-invariant. Recall that $\eta \in S_\gamma$ is, by definition, contained in a geodesic λ, containing γ, and which is normal to the orbit Diff$(M)(\gamma)$. For $f \in$ Isom(γ), from Lemma 3.2, we have that $f \cdot \lambda$ is a geodesic through $f \cdot \gamma = \gamma$. Furthermore, from the invariance of σ under the Isom(γ)-action, and the fact that the geodesic λ is normal to the orbit, we have that $f \cdot \lambda$ is normal to the orbit, as well. Also from the invariance of σ, it follows that the distance of $f \cdot \eta$ to the orbit Diff$(M)(\gamma)$ is the same as the distance of η to the orbit Diff$(M)(\gamma)$. Thus by the definition of S_γ, we have that $f \cdot \eta$ lies in S_γ, i.e. $f \cdot S_\gamma \subset S_\gamma$. This proves point (i).

Point (ii) follows from the fact, due to Theorem 2.19, that the normal exponential map on the neighborhood $V \subset \nu_\gamma$ Diff$(M)(\gamma)$ of a point in the zero section is injective, when V is small enough. We now state how to show point (ii). Take $f \in$ Diff(M), and $\eta, \psi \in S_\gamma$ such that $f \cdot \eta = \psi$ (i.e. assume $f \cdot S_\gamma \cap S_\gamma \neq \emptyset$). Furthermore assume that ψ and η are at a distance less than $\delta > 0$ from γ. Since the action of $\mathrm{Diff}(M)$ is by isometries with respect to σ, it follows that the distance between $f \cdot \gamma$ and $f \cdot \eta = \psi$ is the same distance as the one between γ and η. Thus by the triangle inequality we have that the distance between γ and $f \cdot \gamma$ is less than 2δ. By taking δ small enough this implies that $f \cdot \gamma$ and γ are contained in V. Since η is in S_γ, by the invariance of the Riemannian metric σ, and the invariance of the exponential map of σ, ψ is contained in $S_{f\cdot\gamma}$. From the fact that ψ is contained in S_γ, it follows that ψ is in the image of $\exp_\gamma$. On the other hand, ψ lies in $S_{f\cdot\gamma}$, thus in the image of $\exp_{f\cdot\gamma}$. Since the exponential map is injective on V, we conclude that $\gamma = f \cdot \gamma$.

The proof of point (iii) follows verbatim as in the proof of Theorem 2.9, using the fact that the map $\rho\colon \mathcal{R}(M) \to \mathcal{R}(M)/\mathrm{Diff}(M)$ is a principal fiber bundle. □

REMARK 3.3. The condition of properness for the action of $\mathrm{Diff}(M)$ on $\mathcal{R}(M)$ is necessary for the proof of point (iii), as we need that the orbit $\mathrm{Diff}(M)(\gamma)$ is homeomorphic to the homogeneous space $\mathrm{Diff}(M)/\mathrm{Isom}(\gamma)$.

4. Consequences of the Slice Theorem

For a given compact manifold M, several interesting consequences regarding the action of the diffeomorphism group on the space of Riemannian metrics follow from the Slice Theorem.

The first one is Theorem B, giving a description of a neighborhood of an orbit of $\mathrm{Diff}(M)$ inside $\mathcal{R}(M)$ up to homeomorphism. This result directly follows together with Theorem 2.10.

Another interesting consequence is a statement on the relations of isometry groups of Riemannian metrics which are close to each other in $\mathcal{R}(M)$.

PROPOSITION 4.1 (Theorem 8.1 in [**Ebi70**]). *Let $\gamma \in \mathcal{R}(M)$ be an arbitrary fixed Riemannian metric on M. Then there exists an open neighborhood V of the identity in* $\mathrm{Diff}(M)$*, and an open neighborhood N of γ in $\mathcal{R}(M)$, such that for any $\sigma \in N$, there exists an f in V such that:*

$$f^{-1}\,\mathrm{Isom}(\sigma) f \subset \mathrm{Isom}(\gamma).$$

PROOF. Take V' any open neighborhood of the identity in $\mathrm{Diff}(M)$. Consider U the open neighborhood of the identity coset in $\mathrm{Diff}(M)/\mathrm{Isom}(\gamma)$ given by the Slice Theorem. We recall that there exists a continuous cross-section $\chi\colon U \to \mathrm{Diff}(M)$ of the projection map $\pi\colon \mathrm{Diff}(M) \to \mathrm{Diff}(M)/\mathrm{Isom}(\gamma)$.

Set $U' = U \cap \pi(V)$, which is open in $\mathrm{Diff}(M)/\mathrm{Isom}(\gamma)$. Setting $V = \pi^{-1}(U')$ we get an open neighborhood of the identity in $\mathrm{Diff}(M)$. We point out, that $\chi(U') \subset V$. We consider the homeomorphism $F\colon U \times S_\gamma \to \mathcal{R}(M)$, and set $N = F(U' \times S_\gamma)$.

For $\sigma \in N$, it follows from the definition of F, that

$$\sigma = \chi(u')s,$$

for some $u' \in U'$, and $s \in S_\gamma$. This implies that $s = \chi(u')^{-1}\sigma$. We set $f := \chi(u')$.

Consider $h \in \mathrm{Isom}(\sigma)$ arbitrary. Then by our choices, $(f^{-1} \circ h \circ f)(s) = s$. From point (ii) of the Slice Theorem it follows that $f^{-1} \circ h \circ f \in \mathrm{Isom}(\gamma)$. Since h is arbitrary, it follows that

$$f^{-1}\,\mathrm{Isom}(\sigma) f \subset \mathrm{Isom}(\gamma).$$

□

Let $\mathcal{R}_{\text{triv}}(M)$ be the collection of all Riemannian metrics on M with trivial isometry group. The following Corollary about $\mathcal{R}_{\text{triv}}(M)$ follows from the previous Proposition.

COROLLARY 4.2 (Corollary 8.2 in [**Ebi70**], Theorem 1 in [**Kim87**]). *$\mathcal{R}_{\text{triv}}(M)$ is open in $\mathcal{R}(M)$.*

PROOF. Take $\gamma \in \mathcal{R}_{\text{triv}}(M)$. Then, by the previous proposition there exists an open neighborhood N of γ in $\mathcal{R}(M)$ and V an open neighborhood of the identity in $\text{Diff}(M)$, such that for any σ in N, we have

$$f^{-1}\,\text{Isom}(\sigma) f \subset \text{Isom}(\gamma) = \{Id_M\},$$

for some $f \in V$. This implies that $\text{Isom}(\sigma)$ is trivial. Since $\sigma \in N$ is arbitrary, we conclude that $N \subset \mathcal{R}_{\text{triv}}(M)$. Thus $\mathcal{R}_{\text{triv}}(M)$ is open in $\mathcal{R}(M)$. □

Furthermore, given γ an arbitrary Riemannian metric on a compact n-manifold, we can deform it in such a way that the absolute value of the Ricci curvature increases only on a small neighborhood U of M. Then by increasing the curvature at n points inside U, we get a deformation of γ which is at one point very asymmetrical. I.e. we are able to deform γ into a Riemannian metric in $\mathcal{R}_{\text{triv}}(M)$.

THEOREM 4.3 (Proposition 8.3 in [**Ebi70**]). *The open set $\mathcal{R}_{\text{triv}}(M)$ is dense in $\mathcal{R}(M)$.*

REMARK 4.4. The proof of the previous Theorem does not use the Slice Theorem. It only depends on the topology we consider in $\mathcal{R}(M)$.

This means that a generic metric has trivial isometry group. This agrees with the notion of the principal stratum in the setting of smooth proper Lie group actions on smooth finite-dimensional manifolds. We consider the projection map $p\colon \mathcal{R}(M) \to \mathcal{M}(M)$ from the space of Riemannian metrics to the moduli space. Then by continuity, the set $p(\mathcal{R}_{\text{triv}}(M))$ is open dense in $\mathcal{M}(M)$. Furthermore, from the invariance of the metric σ with respect to $\text{Diff}(M)$, we can define a Riemannian metric on $p(\mathcal{R}_{\text{triv}}(M))$. Thus we obtain the following Corollary:

COROLLARY 4.5. *For a compact manifold M, an open dense set of $\mathcal{M}(M)$ admits a Fréchet, and a Riemannian structure.*

Furthermore, the slice theorem allows us to lift paths from the moduli space, to $\mathcal{R}(M)$.

PROPOSITION 4.6. *Let $\lambda\colon [a,b] \to \mathcal{M}(M)$ be a path. For $\gamma \in \mathcal{R}(M)$ there exists a path $\tilde{\lambda}\colon [a,b] \to \mathcal{R}(M)$ with $\tilde{\lambda}(a) = \gamma$ and $\pi \circ \tilde{\lambda} = \lambda$.*

PROOF. By Theorem B, for any $t \in [a,b]$ there exists an open neighborhood of $\lambda(t)$ homeomorphic to $S_{\gamma_t}/\,\text{Isom}(\gamma_t)$, where γ_t is any metric such that $\pi(\gamma_t) = \lambda(t)$. Since $[a,b]$ is compact, there exist finitely many open neighborhoods $S_{\gamma_{t_i}}/\,\text{Isom}(\gamma_{t_i})$ covering $\lambda[a,b]$. We recall that $\text{Isom}(\gamma_t)$ is a finite-dimensional Lie group. Then from [**Bre72**, Chapter II, Theorem 6.2], there exists such a lift over each open cover. □

Acknowledgment

The authors would like to thank the referee for their helpful comments.

References

[AB15] Marcos M. Alexandrino and Renato G. Bettiol, *Lie groups and geometric aspects of isometric actions*, Springer, Cham, 2015, DOI 10.1007/978-3-319-16613-1. MR3362465

[AH11] Ben Andrews and Christopher Hopper, *The Ricci flow in Riemannian geometry*, Lecture Notes in Mathematics, vol. 2011, Springer, Heidelberg, 2011. A complete proof of the differentiable 1/4-pinching sphere theorem, DOI 10.1007/978-3-642-16286-2. MR2760593

[BE69] M. Berger and D. Ebin, *Some decompositions of the space of symmetric tensors on a Riemannian manifold*, J. Differential Geometry **3** (1969), 379–392. MR266084

[Bla00] David E. Blair, *Spaces of metrics and curvature functionals*, Handbook of differential geometry, Vol. I, North-Holland, Amsterdam, 2000, pp. 153–185, DOI 10.1016/S1874-5741(00)80005-9. MR1736853

[Bre72] Glen E. Bredon, *Introduction to compact transformation groups*, Academic Press, New York-London, 1972. Pure and Applied Mathematics, Vol. 46. MR0413144

[Cla09] Brian Clarke, *The completion of the Manifold of Riemannian Metrics with Respect to its L^2 Metric*, Ph.D. thesis, University of Leipzig, 2009.

[DR19] Tobias Diez and Gerd Rudolph, *Slice theorem and orbit type stratification in infinite dimensions*, Differential Geom. Appl. **65** (2019), 176–211, DOI 10.1016/j.difgeo.2019.03.005. MR3943433

[Ebi70] David G. Ebin, *The manifold of Riemannian metrics*, Global Analysis (Proc. Sympos. Pure Math., Vol. XV, Berkeley, Calif., 1968), Amer. Math. Soc., Providence, R.I., 1970, pp. 11–40. MR0267604

[FG89] Daniel S. Freed and David Groisser, *The basic geometry of the manifold of Riemannian metrics and of its quotient by the diffeomorphism group*, Michigan Math. J. **36** (1989), no. 3, 323–344, DOI 10.1307/mmj/1029004004. MR1027070

[FS17] F. Thomas Farrell and Gangotryi Sorcar, *Teichmüller space of negatively curved metrics on complex hyperbolic manifolds is not contractible*, Sci. China Math. **60** (2017), no. 4, 569–580, DOI 10.1007/s11425-016-0351-8. MR3629483

[GMM91] Olga Gil-Medrano and Peter W. Michor, *The Riemannian manifold of all Riemannian metrics*, Quart. J. Math. Oxford Ser. (2) **42** (1991), no. 166, 183–202, DOI 10.1093/qmath/42.1.183. MR1107281

[Gro02] Karsten Grove, *Geometry of, and via, symmetries*, Conformal, Riemannian and Lagrangian geometry (Knoxville, TN, 2000), Univ. Lecture Ser., vol. 27, Amer. Math. Soc., Providence, RI, 2002, pp. 31–53, DOI 10.1090/ulect/027/02. MR1922721

[Gro17] K. Grove, *A panoramic glimpse of manifolds with sectional curvature bounded from below*, Algebra i Analiz **29** (2017), no. 1, 7–48, DOI 10.1090/spmj/1479; English transl., St. Petersburg Math. J. **29** (2018), no. 1, 3–31. MR3660682

[Ham82] Richard S. Hamilton, *The inverse function theorem of Nash and Moser*, Bull. Amer. Math. Soc. (N.S.) **7** (1982), no. 1, 65–222, DOI 10.1090/S0273-0979-1982-15004-2. MR656198

[Hir94] Morris W. Hirsch, *Differential topology*, Graduate Texts in Mathematics, vol. 33, Springer-Verlag, New York, 1994. Corrected reprint of the 1976 original. MR1336822

[Kan05] Marja Kankaanrinta, *Proper smooth G-manifolds have complete G-invariant Riemannian metrics*, Topology Appl. **153** (2005), no. 4, 610–619, DOI 10.1016/j.topol.2005.01.034. MR2193330

[Kim87] Young Wook Kim, *Semicontinuity of compact group actions on compact differentiable manifolds*, Arch. Math. (Basel) **49** (1987), no. 5, 450–455, DOI 10.1007/BF01194103. MR915919

[KM97] Andreas Kriegl and Peter W. Michor, *The convenient setting of global analysis*, Mathematical Surveys and Monographs, vol. 53, American Mathematical Society, Providence, RI, 1997, DOI 10.1090/surv/053. MR1471480

[Kob72] Shoshichi Kobayashi, *Transformation groups in differential geometry*, Springer-Verlag, New York-Heidelberg, 1972. Ergebnisse der Mathematik und ihrer Grenzgebiete, Band 70. MR0355886

[KW09] Boris Khesin and Robert Wendt, *The geometry of infinite-dimensional groups*, Ergebnisse der Mathematik und ihrer Grenzgebiete. 3. Folge. A Series of Modern Surveys in Mathematics [Results in Mathematics and Related Areas. 3rd Series. A Series of Modern Surveys in Mathematics], vol. 51, Springer-Verlag, Berlin, 2009. MR2456522

[MS39] S. B. Myers and N. E. Steenrod, *The group of isometries of a Riemannian manifold*, Ann. of Math. (2) **40** (1939), no. 2, 400–416, DOI 10.2307/1968928. MR1503467

[MZ55] Deane Montgomery and Leo Zippin, *Topological transformation groups*, Interscience Publishers, New York-London, 1955. MR0073104

[Pal61] Richard S. Palais, *On the existence of slices for actions of non-compact Lie groups*, Ann. of Math. (2) **73** (1961), 295–323, DOI 10.2307/1970335. MR126506

[Sub85] T. N. Subramaniam, *SLICES FOR THE ACTIONS OF SMOOTH TAME LIE GROUPS (FRECHET, GAUGE, DIFFEOMORPHISM, NASH-MOSER INVERSE FUNCTION THEOREM)*, ProQuest LLC, Ann Arbor, MI, 1985. Thesis (Ph.D.)–Brandeis University. MR2634197

[TW15] Wilderich Tuschmann and David J. Wraith, *Moduli spaces of Riemannian metrics*, Oberwolfach Seminars, vol. 46, Birkhäuser Verlag, Basel, 2015. Second corrected printing, DOI 10.1007/978-3-0348-0948-1. MR3445334

Institut für Algebra und Geometrie, Karlsruher Institut für Technologie (KIT), Karlsruhe, Germany

Current address: Instituto de Matemáticas, Oaxaca de Juárez, Universidad Nacional Autónoma de México (UNAM), Mexico

Email address: diego.corro.math@gmail.com

Institut für Algebra und Geometrie, Karlsruher Institut für Technologie (KIT), Karlsruhe, Germany

Current address: Département de mathématiques, Université de Fribourg, Switzerland

Email address: jb@kordass.eu

Contemporary Mathematics
Volume **775**, 2021
https://doi.org/10.1090/conm/775/15590

Model theory and metric convergence II: Averages of unitary polynomial actions

Eduardo Dueñez and José N. Iovino

ABSTRACT. We use model theory of metric structures to prove the pointwise convergence, with a uniform metastability rate, of averages of a polynomial sequence $\{T_n\}$ (in Leibman's sense) of unitary transformations of a Hilbert space. As a special case, this applies to unitary sequences $\{U^{p(n)}\}$ where p is a polynomial $\mathbb{Z} \to \mathbb{Z}$ and U a fixed unitary operator; however, our convergence results hold for arbitrary Leibman sequences. As a case study, we show that the non-nilpotent "lamplighter group" $\mathbb{Z}\wr\mathbb{Z}$ is realized as the range of a suitable quadratic Leibman sequence. We also indicate how these convergence results generalize to arbitrary Følner averages of unitary polynomial actions of any abelian group $\mathbb{G}$ in place of $\mathbb{Z}$.

Introduction

The first result on "mean" convergence of averages was von Neumann's 1932 Mean Ergodic Theorem [**vN32**]:

MEAN ERGODIC THEOREM (MET). *For any unitary operator U on a Hilbert space $\mathcal{H}$ and any $x \in \mathcal{H}$, the sequence $\mathrm{AV}_\bullet(x) = (\mathrm{AV}_n(x) : n \in \mathbb{N})$ of pointwise averages*

$$\mathrm{AV}_n(x) = \frac{1}{n}\sum_{i=1}^{n} U^i(x)$$

converges as $n \to \infty$. The limit is equal to the orthogonal projection of x on the space of vectors fixed by U.

Historically, generalizations of von Neumann's theorem have largely followed a path influenced by a measure-theoretic viewpoint that is completely absent from the formulation above as a statement about convergence in Hilbert spaces. We provide further historical background below. Leaving history and measure theory aside for the moment, one may suggest the following different possible directions of generalization for MET:

(1) Replace the sequence $(U^i : i \in \mathbb{N})$ with a "higher-degree" sequence $(U^{p(i)} : i \in \mathbb{N})$ where p is a fixed polynomial.

2020 *Mathematics Subject Classification.* Primary 37A30; Secondary 03C98, 46Bxx, 28-xx.

Key words and phrases. Mean Ergodic Theorem, PET induction, Leibman sequences, Henson structures.

This research was funded by NSF grant DMS-1500615.

(2) The sequence $(T_i) = (U^{p(i)})$ above necessarily satisfies the commutativity condition $T_i \circ T_j = T_j \circ T_i$ for all i, j. To what extent can such commutativity requirement be removed?
(3) What conditions on a family (T_i) of unitary operators indexed by a semigroup other than $\mathbb{N}$ ensure the pointwise convergence of suitable averages?

Theorem 4 in this manuscript is arguably the most natural generalization of von Neumann's result simultaneously in all three directions above. (For technical reasons, Theorem 4 is proved in the context of (polynomial) actions of *groups* rather than semigroups.) Theorem 1, stated below, is a very particular case of more general results (Theorems 2, 3 and 4). However, it is easiest to formulate and already generalizes MET all the way in direction (1) and beyond.

THEOREM 1 (MET for abelian unitary polynomial actions of $\mathbb{Z}$). *Fix $d \in \mathbb{N}$. Let $\mathcal{H}$ be a Hilbert space, and let $U_0, U_1, \ldots, U_d$ be pairwise-commuting unitary operators on $\mathcal{H}$. For every $x \in \mathcal{H}$, the sequence $\mathrm{AV}_\bullet(x) = (\mathrm{AV}_n(x) : n \in \mathbb{N})$ of averages* [1]

$$\mathrm{AV}_n(x) = \frac{1}{n+1} \sum_{0 \le k \le n} U_0 \circ U_1^k \circ U_2^{\binom{k}{2}} \circ \cdots \circ U_d^{\binom{k}{d}}(x)$$

converges as $n \to \infty$.

In particular, if $p : \mathbb{Z} \to \mathbb{Z}$ is a polynomial of degree at most d and U is a unitary operator on $\mathcal{H}$, then $\left(\sum_{0 \le k \le n} U^{p(n)}(x)/(n+1) : n \in \mathbb{N}\right)$ converges.

Furthermore, there exists a universal metastability rate (depending only on d) that applies uniformly to all sequences of averages of arbitrary x in the unit ball of an arbitrary Hilbert space $\mathcal{H}$ under arbitrary unitary operators $U_0, U_1, \ldots, U_d$ on $\mathcal{H}$.

The notion of *uniformly metastable convergence* above was first introduced in ergodic theory by Tao. It is a main theme of our prior manuscript, but shall presently play a minor role [**DnI17, Tao08, Tao12**].

Taking a step in direction (2), pairwise commutativity is not a necessary assumption; the sequence of averages under a family (T_i) converges provided $i \mapsto T_i$ is a *Leibman polynomial sequence* in the group $\mathrm{U}_{\mathcal{H}}$ of unitary operators on $\mathcal{H}$ (Theorems 2 and 3), but the range of this sequence need not generate an abelian group. The definition of Leibman polynomial sequence (Definition 2.1) is motivated by the familiar fact that degree-d polynomials $\mathbb{R} \to \mathbb{R}$ are characterized as those functions having $(d+1)$-iterated finite differences equal to zero. The same essential definition gives the notion of *Leibman polynomial mapping* from an arbitrary group $\mathbb{G}$ into $\mathrm{U}_{\mathcal{H}}$ [**Lei02**]. Theorem 4 generalizes von Neumann's result in direction (3) for Leibman polynomials $(T_i : i \in \mathbb{G}) \subset \mathrm{U}_{\mathcal{H}}$ on abelian groups $\mathbb{G}$ endowed with a notion of averaging provided by a countable Følner net.

Continuing our historical remarks, the formulation of von Neumann's result above hides its conceptual genesis via the study of convergence of averages of square-integrable functions $f \in \mathscr{L}^2(\Omega)$ on a probability space (Ω, μ) under the action of a measure-preserving transformation T of Ω. In this setting, MET asserts that the sequence $\mathrm{AV}_\bullet(f)$ of averages

$$\mathrm{AV}_n(f) = \frac{1}{n} \sum_{i=1}^{n} f \circ T^i$$

[1] Here, $\binom{k}{j} = k(k-1)\cdots(k-j+1)/j!$ is the j-th binomial coefficient.

converges in $\mathscr{L}^2(\Omega)$ (after all, $f \mapsto f \circ T$ is a unitary transformation of $\mathscr{L}^2(\Omega)$). This particular case of von Neumann's result explains why it is called a convergence result "in mean", i.e., in the mean-square ("$\mathscr{L}^2$") sense. (By contrast, Birkhoff's Ergodic Theorem asserts the almost-everywhere pointwise convergence of the averages $\mathrm{AV}_n(f)$ for any $f \in \mathscr{L}^1(\Omega)$ [**Bir31**].) The $\mathscr{L}^2$ setting entails no loss of generality since every Hilbert space $\mathcal{H}$ is realized as a space of square-integrable functions. However, this viewpoint is artificial for purposes of studying convergence under unitary actions (at least insofar as *simple* actions are concerned, in contrast to multiple actions mentioned below).

Although generalizations of MET in direction (1) seem very natural, we are not aware of direct proofs of Theorem 1, but only of indirect proofs as byproduct of results on mean convergence of "multiple" ergodic averages. Starting in the 1970's, Furstenberg pioneered the ergodic study of actions of *multiple* simultaneous transformations; equivalently, the study of convergence of "multiple averages" of the *product* of two or more measurable bounded functions on a probability space Ω as acted upon by powers of measure-preserving transformations. As an application of multiple averages, Furstenberg obtained a purely ergodic proof of Szemerédi's Theorem on the existence of arbitrary long arithmetic progressions in positive-density subsets of the integers [**Fur77, Sze75**]. However, Furstenberg's seminal results from the seventies did not extend von Neumann's theorem in either of the directions (1)–(3). It was Bergelson who, in 1987, first extended some of Furstenberg's results to multiple ergodic averages of (*plus quam* linear) polynomial powers of a fixed measure-preserving transformation acting on products of functions [**Ber87**]. When specialized to simple measure-preserving actions, Bergelson's results are a step toward generalizing von Neumann's MET in direction (1). However, there is no purely Hilbert-theoretical formulation of Bergelson's weak mixing hypothesis: Even the convergence of pointwise averages of $(U^{p(n)})$ stated in Theorem 1 only follows unconditionally from 2005 results for multiple ergodic averages of Host and Kra, and of Leibman (which depend on no mixing assumptions) [**HK05, Lei05**].

To our knowledge, Walsh's theorem [**Wal12**] on mean convergence of nilpotent ergodic averages is the first result in the literature from which Theorem 1 follows as a corollary. (Pointwise convergence of averages of $(U^n \circ V^{n^2})$ under the assumption $U \circ V = V \circ U$ is a special case of 2009 results of Austin [**Aus15a, Aus15b**].) Thus, Walsh's theorem actually implies the convergence of averages asserted in the more general Theorem 2, but only under the additional explicit hypothesis that (T_i) generates a nilpotent subgroup of $\mathrm{U}_{\mathcal{H}}$. However, our methods do not require a nilpotence hypothesis, but only the more intrinsic property that (T_i) be a Leibman sequence in the sense of Definition 2.11 (or in Leibman's more general sense of *polynomial mapping* used in Theorem 4). In Section 2.2, we construct a quadratic Leibman sequence whose range generates the non-nilpotent "lamplighter group" $\mathbb{Z} \wr \mathbb{Z}$.

Generalizations of Walsh's theorem by Austin and Zorin-Kranich imply steps in direction (3) [**Aus16, ZK16**]. However, Theorems 2, 3 and 4 appear to be new in the general form stated. Nevertheless, given the close relation of our results to others in the existing literature, the main novelty is our "soft" direct approach to proving pointwise convergence of polynomial averages in Hilbert spaces using the framework of Henson metric structures. Our viewpoint is heavily influenced by Tao's outline [**Tao12**] of a nonstandard proof *à la Robinson* of Walsh's theorem (although we use only standard real numbers, and none of Robinson's apparatus as such). A significant part of the manuscript consists of natural definitions and basic results on model-theoretic notions of integration and convergence that parallel classical ones; nevertheless, we capture, refine, and in some cases extend such results in Henson's framework. Section 1 contains the rather long definition of the Henson class of *PET structures over* $\mathbb{Z}$. Section 2 introduces the notion of *Leibman polynomial sequence;* it also exhibits a quadratic Leibman sequence whose range generates the non-nilpotent group $\mathbb{Z} \wr \mathbb{Z}$. In Section 3, we state and prove Theorems 2 and 3 on metastable convergence of polynomial unitary averages for Leibman sequences (over $\mathbb{Z}$), and also explain how Theorem 1 follows as an immediate corollary. In Section 4, we state and prove the most general of our ergodic convergence results in the form of Theorem 4, which generalizes MET in all three directions (1)–(3). A number of foundational results are contained in the Appendix, which bears a close relation to our prior manuscript [**DnI17**]. These results pertain to measure theory and integration of real functions, as well as abstract notions of integration of functions taking values in Banach spaces. In this way we obtain a Dominated Convergence Theorem for notions of integration in an *ad hoc* Henson class of Banach integration frameworks (Theorem 5). We also show that the compactness of Henson's logic implies a Uniform Metastability Principle for convergence in models of any Henson theory (Proposition A.10). Via this principle, all our results on convergence of averages admit refinements to convergence with metastability rates that are universal. These are *gratis* refinements thanks to the model-theoretic approach.

1. PET structures

1.1. Classical PET structures.

NOTATION 1.1. Below we list a number of *formal symbols* $\mathbb{R}, \mathbb{Z}, \mathbb{N}, \mathcal{H}, \dots$ that will eventually become *sort descriptors* for a Henson language of metric structures. However, throughout this subsection, these symbols have the following classical interpretations:

- $\mathbb{R}, \mathbb{Z}, \mathbb{N}$ shall denote the sets of real numbers, integers and naturals.
- $\mathcal{H}$ shall denote a real Hilbert space.
- $\mathfrak{B}$ shall denote the real Banach algebra $\mathfrak{B}(\mathcal{H}, \mathcal{H})$ of bounded operators on $\mathcal{H}$.
- $\mathcal{A}_{\mathbb{Z}}$ shall denote the Boolean algebra of all subsets of $\mathbb{Z}$.
- $\mathfrak{M}$ shall denote the real Banach space of signed finite measures on $\mathbb{Z}$ (i.e., on the measure space $(\mathbb{Z}, \mathcal{A}_{\mathbb{Z}})$).
- $\mathscr{L}^\infty_{\mathbb{Z},\mathbb{R}}$ shall denote the Banach space $\mathscr{L}^\infty(\mathbb{Z}, \mathbb{R})$ of bounded real functions on $\mathbb{Z}$.
- $\mathscr{L}^\infty_{\mathbb{Z},\mathcal{H}}$ shall denote the Banach space $\mathscr{L}^\infty(\mathbb{Z}, \mathcal{H})$ of bounded functions $\mathbb{Z} \to \mathcal{H}$.
- $\mathscr{L}^\infty_{\mathbb{Z},\mathfrak{B}}$ shall denote the Banach space $\mathscr{L}^\infty(\mathbb{Z}, \mathfrak{B})$ of bounded functions $\mathbb{Z} \to \mathfrak{B}$.
- $\mathscr{L}^\infty_{\mathbb{Z},\mathfrak{M}}$ shall denote the Banach space $\mathscr{L}^\infty(\mathbb{Z}, \mathfrak{M})$ of bounded functions $\mathbb{Z} \to \mathfrak{M}$.
- $\mathscr{L}^\infty_{\mathbb{Z}^2,\mathbb{R}}$ shall denote the Banach space $\mathscr{L}^\infty(\mathbb{Z} \times \mathbb{Z}, \mathbb{R})$ of bounded real functions on $\mathbb{Z} \times \mathbb{Z}$.
- $\mathscr{L}^\infty_{\mathbb{Z}^2,\mathcal{H}}$ shall denote the Banach space $\mathscr{L}^\infty(\mathbb{Z} \times \mathbb{Z}, \mathcal{H})$ of bounded functions $\mathbb{Z} \times \mathbb{Z} \to \mathcal{H}$.

- $\mathscr{L}^\infty_{\mathbb{Z}^2,\mathfrak{B}}$ shall denote the Banach space $\mathscr{L}^\infty(\mathbb{Z}\times\mathbb{Z},\mathfrak{B})$ of bounded functions $\mathbb{Z}\times\mathbb{Z}\to\mathfrak{B}$.

From a model-theoretic viewpoint, the sets $\mathbb{R},\mathbb{N},\mathbb{Z},\ldots$ denoted by the formal symbols above are the *sorts* of a metric Henson structure $\mathscr{M}$. (Discrete sorts $\mathbb{N}$, $\mathbb{Z}$, $\mathcal{A}_\mathbb{Z}$ are still viewed as metric spaces endowed with the discrete metric.) In addition, $\mathscr{M}$ is endowed with a number of distinguished elements ("constants") and continuous functions between sorts. The distinguished elements include:

- All elements of $\mathbb{N}$ and $\mathbb{Z}$.
- All rational numbers in $\mathbb{R}$.
- The zero element of each real Banach space above $(\mathcal{H},\mathfrak{B},\mathfrak{M},\mathscr{L}^\infty_{\mathbb{Z},\mathbb{R}},\ldots)$.
- The identity operator $I\in\mathfrak{B}$.
- The zero (empty set $\emptyset$) and unity (improper subset $\mathbb{Z}\subseteq\mathbb{Z}$) of the Boolean algebra $\mathcal{A}_\mathbb{Z}$.

The distinguished functions between sorts include:

- The discrete metric in each the discrete sorts $\mathbb{Z}$, $\mathbb{N}$, $\mathcal{A}_\mathbb{Z}$.
- The operations of addition, subtraction, multiplication, absolute value, and lattice operations (binary minimum and maximum) on $\mathbb{R}$.
- The order $\le$ of $\mathbb{N}$, identified with its characteristic function $[\![\cdot\le\cdot]\!]:\mathbb{N}\times\mathbb{N}\to\{0,1\}$.
- The membership relation from $\mathbb{Z}$ to $\mathcal{A}_\mathbb{Z}$, identified with its characteristic function $[\![\cdot\in\cdot]\!]:\mathbb{Z}\times\mathcal{A}_\mathbb{Z}\to\{0,1\}$.
- The group operations (unary negation, binary addition and subtraction) of $\mathbb{Z}$.
- The operations of union, intersection and complementation on $\mathcal{A}_\mathbb{Z}$.
- The Hilbert space operations (addition, scalar multiplication, and inner product $(x,y)\mapsto x\cdot y$) on $\mathcal{H}$. For convenience, also the norm $\|x\|=\sqrt{x\cdot x}$.
- The operations of addition and scalar product, and the Banach norm $\|\cdot\|$ on each Banach sort $\mathfrak{B},\mathfrak{M},\mathscr{L}^\infty_{X,Y}$.

For $f\in\mathscr{L}^\infty_{X,Y}$, the Banach norm is $\|f\|=\sup_{x\in X}\|f(x)\|$, where $\|f(x)\|$ is the norm of $f(x)$ as an element of Banach sort Y. The Banach norm on $\mathfrak{B}$ is $\|T\|=\sup\{\|T(x)\|:x\in\mathcal{H},\|x\|\le1\}$. The Banach norm on $\mu\in\mathfrak{M}$ is "total variation": Recall that μ has an atomic decomposition $\mu=\sum_{i\in\mathbb{Z}}c_i\delta_i$ where δ_i is the unit mass at i and $c_i=\mu(\{i\})$. With this notation, $\|\mu\|=\sum_i|c_i|$.

(To abbreviate the long list of distinguished functions, above and in what follows we use X to denote either of the "domain" discrete sets $\mathbb{Z}$, $\mathbb{Z}^2$ of the various sorts $\mathscr{L}^\infty$, and Y to denote the "codomain" Banach sorts $\mathbb{R}$, $\mathcal{H}$, $\mathfrak{B}$, $\mathfrak{M}$.)

The list of distinguished functions continues as follows:

- The operations $\mathscr{L}^\infty_{X,\mathcal{H}}\times\mathscr{L}^\infty_{X,\mathcal{H}}\to\mathscr{L}^\infty_{X,\mathbb{R}}$ induced by (pointwise) application of the inner product of $\mathcal{H}$.
- The unary operation of pointwise absolute value $|\cdot|$ and the binary lattice operations (pointwise max and min) on sorts $\mathscr{L}^\infty_{X,\mathbb{R}}$.
- The unary operation $|\cdot|$ of *measure of total variation* and the binary lattice operations ("pointwise" max and min) on $\mathfrak{M}$ (i.e., $|\mu|=\sum_i|a_i|\,\delta_i$, $\max(\mu,\nu)=\sum_i\max\{a_i,b_i\}\delta_i$, and $\min(\mu,\nu)=\sum_i\min\{a_i,b_i\}\delta_i$ if $\mu=\sum_i a_i\delta_i$ and $\nu=\sum_i b_i\delta_i$).
- The operation of pointwise magnitude $|\cdot|:\mathscr{L}^\infty_{X,Y}\to\mathscr{L}^\infty_{X,\mathbb{R}}$, namely $|f|:x\mapsto\|f(x)\|$ for any $f\in\mathscr{L}^\infty_{X,Y}$.

- The unary adjoint operation $T \mapsto T^*$ on $\mathfrak{B}$, and the corresponding induced operations (pointwise adjoint) on sorts $\mathscr{L}^\infty_{X,\mathfrak{B}}$.
- The binary operation $(S,T) \mapsto S \circ T$ of composition on $\mathfrak{B}$, and the corresponding induced operations of pointwise composition on sorts $\mathscr{L}^\infty_{X,\mathfrak{B}}$.
- The inclusions:
  - $\mathbb{Z} \hookrightarrow \mathcal{A}_\mathbb{Z} : i \mapsto \{i\}$.
  - $\mathcal{A}_\mathbb{Z} \hookrightarrow \mathscr{L}^\infty_{\mathbb{Z},\mathbb{R}} : A \mapsto \chi_A$ where χ_A is the characteristic function of the subset $A \subseteq \mathbb{Z}$.
  - $\mathbb{Z} \hookrightarrow \mathfrak{M}$ given by $i \mapsto \delta_i$ (the unit point mass at i).
  - $Y \hookrightarrow \mathscr{L}^\infty_{X,Y}$, with $y \in Y$ identified with the constant function $y(\blacksquare) : x \mapsto y$ in $\mathscr{L}^\infty_{X,Y}$;
  - The right inclusion map $\mathscr{L}^\infty_{\mathbb{Z},Y} \hookrightarrow \mathscr{L}^\infty_{\mathbb{Z}^2,Y}$ whereby $f \in \mathscr{L}^\infty_{\mathbb{Z},Y}$ is identified with $f(\blacksquare,\cdot) : (w,x) \mapsto f(x)$; also, the analogous left inclusion map identifying f with $f(\cdot,\blacksquare) : (w,x) \mapsto f(w)$.
- The function-evaluation maps
  - $(T,x) \mapsto T(x)$ from $\mathfrak{B} \times \mathcal{H}$ to $\mathcal{H}$.
  - $(f,x) \mapsto f(x)$ from $\mathscr{L}^\infty_{X,Y} \times X$ to Y;

  Also, the maps $\mathscr{L}^\infty_{X,\mathfrak{B}} \times \mathscr{L}^\infty_{X,\mathcal{H}} \to \mathscr{L}^\infty_{X,\mathcal{H}}$ induced by pointwise evaluation.
- The partial evaluation maps:
  - Left evaluation $\mathscr{L}^\infty_{\mathbb{Z}^2,Y} \times \mathbb{Z} \to \mathscr{L}^\infty_{\mathbb{Z},Y}$, namely $(F,i) \mapsto F(i,\cdot)$ where $F(i,\cdot) : j \mapsto F(i,j)$.
  - Right evaluation $\mathscr{L}^\infty_{\mathbb{Z}^2,Y} \times \mathbb{Z} \to \mathscr{L}^\infty_{\mathbb{Z},Y}$, namely $(F,j) \mapsto F(\cdot,j)$ where $F(\cdot,j) : i \mapsto F(i,j)$.

(Note that the left evaluation map allows us to identify $\mathscr{L}^\infty_{\mathbb{Z}^2,Y}$ with the space $\mathscr{L}^\infty(\mathbb{Z},\mathscr{L}^\infty_{\mathbb{Z},Y})$ of all bounded functions $\mathbb{Z} \to \mathscr{L}^\infty_{\mathbb{Z},Y}$—thus making a potential sort $\mathscr{L}^\infty(\mathbb{Z},\mathscr{L}^\infty_{\mathbb{Z},Y})$ superfluous. We also have a different identification of $\mathscr{L}^\infty(\mathbb{Z},\mathscr{L}^\infty_{\mathbb{Z},Y})$ with $\mathscr{L}^\infty(\mathbb{Z}^2,Y)$ via right evaluation.)

- The *Følner-measure map* $\sigma : \mathbb{N} \to \mathfrak{M}$, where
$$\sigma_n = \frac{1}{n+1} \sum_{0 \le i \le n} \delta_i \qquad \text{for all } n \in \mathbb{N}.$$
  (σ_n is the average of unit point masses at the points $0,1,2,\dots,n$.)
- The translation action of $\mathbb{Z}$ on $\mathscr{L}^\infty_{\mathbb{Z},Y}$. We regard this action as a function $\mathscr{L}^\infty_{\mathbb{Z},Y} \to \mathscr{L}^\infty_{\mathbb{Z}^2,Y} = \mathscr{L}^\infty(\mathbb{Z},\mathscr{L}^\infty_{\mathbb{Z},Y})$ (with the latter identification by partial evaluation on the left). The action is denoted $f \mapsto {}_\bullet f$ where ${}_\bullet f \in \mathscr{L}^\infty(\mathbb{Z},\mathscr{L}^\infty_{\mathbb{Z},Y})$ is the function $i \mapsto {}_i f$ with ${}_i f \in \mathscr{L}^\infty_{\mathbb{Z},Y}$ the function $j \mapsto f(i+j)$.
- The *shear transformation* $\mathscr{L}^\infty_{\mathbb{Z}^2,Y} \to \mathscr{L}^\infty_{\mathbb{Z}^2,Y}$, namely $F \mapsto \widetilde{F}$ where $\widetilde{F} : (i,j) \mapsto F(i,i+j)$.
- The translation action of $\mathbb{Z}$ on $\mathfrak{M}$, regarded as a mapping $\mathfrak{M} \to \mathscr{L}^\infty_{\mathbb{Z},\mathfrak{M}}$ and denoted $\mu \mapsto {}_\bullet\mu$ where ${}_\bullet\mu \in \mathscr{L}^\infty_{\mathbb{Z},\mathfrak{M}}$ is the mapping $i \mapsto {}_i\mu$, with ${}_i\mu \in \mathfrak{M}$ the measure μ shifted by $-i$, namely
$${}_i\mu = \sum_{j \in \mathbb{Z}} a_j \delta_{j+i} \qquad \text{if } \mu = \sum_{j \in \mathbb{Z}} a_j \delta_j,$$
  which is classically characterized by the property that $\langle f, \mu \rangle = \langle {}_i f, {}_i\mu \rangle$ for all $f \in \mathscr{L}^\infty_{\mathbb{Z},\mathbb{R}}$ and $i \in \mathbb{Z}$.

- The involutions $\mathscr{L}^\infty_{\mathbb{Z}^2,Y} \to \mathscr{L}^\infty_{\mathbb{Z}^2,Y}$ induced by the involution $(i,j) \mapsto (j,i)$ of $\mathbb{Z}^2$.
- The integration operations
  - $\mathscr{L}^\infty_{\mathbb{Z},Y} \times \mathfrak{M} \to Y : (f,\mu) \mapsto \langle f,\mu\rangle = \sum_{i\in\mathbb{Z}} c_i f(i)$ for $\mu = \sum_i c_i\delta_i$.
  - (Left integral) $\mathfrak{M}\times\mathscr{L}^\infty_{\mathbb{Z}^2,Y} \to \mathscr{L}^\infty_{\mathbb{Z},Y} : (\mu, F) \mapsto \langle\!\langle \mu, F\rangle\!\rangle$, where $\langle\!\langle \mu, F\rangle\!\rangle \in \mathscr{L}^\infty_{\mathbb{Z},Y}$ is the function $j \mapsto \langle F(\cdot,j),\mu\rangle = \sum_i c_i F(i,j)$.
  - (Right integral) $\mathscr{L}^\infty_{\mathbb{Z}^2,Y} \times \mathfrak{M} \to \mathscr{L}^\infty_{\mathbb{Z},Y} : (F,\mu) \mapsto \langle\!\langle F,\mu\rangle\!\rangle$, where $\langle\!\langle F,\mu\rangle\!\rangle \in \mathscr{L}^\infty_{\mathbb{Z},Y}$ is the function $i \mapsto \langle F(i,\cdot),\mu\rangle = \sum_j c_j F(i,j)$.
  - $\mathscr{L}^\infty_{\mathbb{Z}^2,Y} \times \mathscr{L}^\infty_{\mathbb{Z},\mathfrak{M}} \to \mathscr{L}^\infty_{\mathbb{Z},Y} : (F,\mu_\bullet) \mapsto \langle\!\langle F,\mu_\bullet\rangle\!\rangle$ where $\langle\!\langle F,\mu_\bullet\rangle\!\rangle \in \mathscr{L}^\infty_{\mathbb{Z},Y}$ is the function $j \mapsto \langle F(\cdot,j),\mu_j\rangle$ (i.e., the operation induced by "pointwise integration" when $\mathscr{L}^\infty_{\mathbb{Z}^2,Y}$ is identified with $\mathscr{L}^\infty(\mathbb{Z},\mathscr{L}^\infty_{\mathbb{Z},Y})$ via left partial evaluation).

For visual convenience, we may use integral notation and write $\int f\,d\mu$ or $\int f(i)\,d\mu(i)$ for $\langle f,\mu\rangle$, and $\int F(i,\cdot)\,d\mu(i)$ for $\langle\!\langle \mu,F\rangle\!\rangle$ (resp., $\int F(\cdot,j)\,d\mu(j)$ for $\langle\!\langle F,\mu\rangle\!\rangle$).

REMARKS 1.2.
- There are redundancies on the list of functions above. For instance, the $\mathscr{L}^2$-norm on $\mathcal{H}$ is implicitly defined by its inner product: $\|x\|^2 = x\cdot x$. As a less trivial example, the action of $\mathbb{Z}$ on $\mathscr{L}^\infty_{\mathbb{Z},Y}$ is obtained from the right inclusion $\mathscr{L}^\infty_{\mathbb{Z},Y} \hookrightarrow \mathscr{L}^\infty_{\mathbb{Z}^2,Y}$ followed by the shear transformation. However, for reasons of exposition we make no effort to present a minimal list of distinguished functions. The model-theoretic approach fundamentally requires that all sorts, functions and constants that are relevant to the problem at hand be part of the structures under study.
- The nonstrict order relations ($\le$ and $\ge$) of $\mathbb{R}$ are the only predicate symbols of a Henson language. However, any discrete predicate P may be identified with a $\{0,1\}$-valued function χ_P (the characteristic function of the truth set of P), so the usual interpretation of $P(x)$ (resp., of $\neg P(x)$) agrees with the interpretation of the Henson formula $\chi_P(x) \ge 1/2$ (resp., of $\chi_P(x) \le 1/2$).

DEFINITION 1.3 (Classical PET structure over $\mathbb{Z}$). A *classical PET structure (over $\mathbb{Z}$)* is a triple $\mathscr{M} = (\mathbf{S},\mathbf{C},\mathbf{F})$ where

$$\mathbf{S} = (\mathbb{R},\mathbb{N},\mathbb{Z},\mathcal{A}_\mathbb{Z},\mathcal{H},\mathfrak{B},\mathfrak{M},\mathscr{L}^\infty_{\mathbb{Z},\mathbb{R}},\mathscr{L}^\infty_{\mathbb{Z},\mathcal{H}},\mathscr{L}^\infty_{\mathbb{Z},\mathfrak{B}},\mathscr{L}^\infty_{\mathbb{Z},\mathfrak{M}},\mathscr{L}^\infty_{\mathbb{Z}^2,\mathbb{R}},\mathscr{L}^\infty_{\mathbb{Z}^2,\mathcal{H}},\mathscr{L}^\infty_{\mathbb{Z}^2,\mathfrak{B}})$$

is a collection of *sorts*, $\mathbf{C}$ is a collection of distinguished elements (*constants*), and $\mathbf{F}$ is a collection of distinguished functions between sorts, provided these sorts, constants and functions are obtained in the manner prescribed by Notation 1.1.

1.2. Abstract PET structures.

DEFINITION 1.4 (Henson signature and language for PET structures over $\mathbb{Z}$). The *Henson signature for PET structures over* $\mathbb{Z}$ consists of three ingredients:

- A collection of formal symbols, called *sort descriptors* (or *sort names*) in one-to-one correspondence with the collection $\mathbf{S}$ of sorts of a classical PET structure. For definiteness, the collection of descriptors is taken to be

$$(\mathbb{R},\mathbb{N},\mathbb{Z},\mathcal{A}_\mathbb{Z},\mathcal{H},\mathfrak{B},\mathfrak{M},\mathscr{L}^\infty_{\mathbb{Z},\mathbb{R}},\mathscr{L}^\infty_{\mathbb{Z},\mathcal{H}},\mathscr{L}^\infty_{\mathbb{Z},\mathfrak{B}},\mathscr{L}^\infty_{\mathbb{Z},\mathfrak{M}},\mathscr{L}^\infty_{\mathbb{Z}^2,\mathbb{R}},\mathscr{L}^\infty_{\mathbb{Z}^2,\mathcal{H}},\mathscr{L}^\infty_{\mathbb{Z}^2,\mathfrak{B}})$$

its members regarded as purely formal symbols.

- A collection of lexical *constant symbols* containing a unique symbol $\mathtt{c}$ for each of the distinguished elements in Definition 1.3, with each such symbol endowed with a sort descriptor s naming that sort to which the element c named by $\mathtt{c}$ belongs per Definition 1.3.
- A collection of lexical *function symbols* containing a unique symbol $\mathtt{f}$ for each of the functions named in Definition 1.3, with each such symbol endowed with a *sort-specification* of the form $s_1 \times \cdots \times s_n \to s_0$ where $s_0, s_1, \dots, s_n$ are sort descriptors chosen in accordance with the specification of the domain (Cartesian product of sorts named by $s_1, \dots, s_n$) and codomain (sort named by s_0) of the function f named by the symbol $\mathtt{f}$.

The *Henson language $\mathcal{L}$ for PET structures over* $\mathbb{Z}$ is the Henson language (of positive bounded formulas) whose signature is the one just described [**HI02**, **Iov14**, **DnI17**].

DEFINITION 1.5 (PET structure over $\mathbb{Z}$). Let $\mathcal{L}$ be the Henson language for PET structures. Let **PET** be the class of all classical PET structures over $\mathbb{Z}$ per Definition 1.3, and let $\mathrm{Th}_{\mathbf{PET}}$ be the $\mathcal{L}$-theory of **PET** in Henson's logic of approximate satisfaction of positive bounded formulas. An *(abstract) PET structure over* $\mathbb{Z}$ is a model of $\mathrm{Th}_{\mathbf{PET}}$.

The class $\overline{\mathbf{PET}}$ of abstract PET structures obviously extends **PET**.

REMARKS 1.6.
- In principle, one may provide an explicit axiomatization in positive bounded Henson formulas of the class **PET**. However, given the large number of sorts and functions in a PET structure this task is impractical. We refer the reader to our prior manuscript in which we provide explicit Henson axiomatizations of certain classes of structures somewhat more general than **PET** [**DnI17**]. Nevertheless, it should be clear that the Henson theory $\mathrm{Th}_{\mathbf{PET}}$ is *uniform* in the sense that it imposes bounds on constants as well as local bounds and local moduli of uniform continuity on distinguished functions. Moreover, $\mathrm{Th}_{\mathbf{PET}}$ obviously is identical to the theory $\mathrm{Th}_{\overline{\mathbf{PET}}}$ of all abstract PET structures.
- The Følner map $\sigma : \mathbb{N} \to \mathfrak{M}$ per Notation 1.1 implies a particular choice of a "notion of averaging" over $\mathbb{Z}$ that is built into $\mathrm{Th}_{\mathbf{PET}}$. Nonequivalent definitions of the PET class over $\mathbb{Z}$ and of $\overline{\mathbf{PET}}$ are obtained by changing this choice (e.g., letting $\sigma_n = 1/(2n+1)\sum_{-n\le i\le n}\delta_i$ in classical structures), but Theorems 2 and 3 on PET structures over $\mathbb{Z}$ remain true under such alternate choice (in fact, they are special cases of the more general Theorem 4).
- If $\mathcal{M}$ is a PET structure, then the $\mathbb{R}$-named sort $\mathbb{R}^{\mathcal{M}}$ of $\mathcal{M}$, under the corresponding operations $+_{\mathbb{R}}^{\mathcal{M}}, -_{\mathbb{R}}^{\mathcal{M}}, \dots,$ is (isomorphic to) the standard real numbers; we shall identify $\mathbb{R}^{\mathcal{M}}$ with $\mathbb{R}$. Correspondingly, the "Hilbert sort" $\mathcal{H}^{\mathcal{M}}$ of $\mathcal{M}$ is a classical real Hilbert space. Typically, the $\mathbb{N}$-named sort $\mathcal{N} = \mathbb{N}^{\mathcal{M}}$ of $\mathcal{M}$ is a proper extension of the set $\mathbb{N}$ of standard natural numbers (when the latter is identified with the set of interpretations $\mathtt{m}^{\mathcal{M}}$ of the constant symbols $\mathtt{m}$ of $\mathcal{L}$, one for each standard natural m) [2], and similarly $\mathcal{Z} = \mathbb{Z}^{\mathcal{M}}$ extends $\mathbb{Z}$ in general. While $\mathfrak{B}^{\mathcal{M}}$ may be identified (via the evaluation map $\mathfrak{B} \times \mathcal{H} \to \mathcal{H}$) with an algebra of bounded operators

[2]The language $\mathcal{L}$ has constants naming only the standard integers and natural numbers, but no nonstandard elements of the sorts $\mathbb{N}^{\mathcal{M}}$, $\mathbb{Z}^{\mathcal{M}}$.

on $\mathcal{H}^{\mathscr{M}}$, it need not contain all bounded operators. The sort $\mathcal{A}_{\mathbb{Z}}^{\mathscr{M}}$ may be identified (via $[\![\cdot \in \cdot]\!]$) with a Boolean algebra of some, but not necessarily all subsets of $\mathbb{Z}^{\mathscr{M}}$, while $(\mathscr{L}_{\mathbb{Z},\mathbb{R}}^{\infty})^{\mathscr{M}}$ may be identified (via the evaluation map) with a space of (not necessarily all) bounded functions $\mathcal{Z} \to \mathbb{R}$.

One of the subtlest differences between classical and abstract PET structures is the fact that $\mathfrak{M}^{\mathscr{M}}$ typically consists of measures that are finitely but not countably additive on $\mathbb{Z}^{\mathscr{M}}$ (in particular, such measures need not have atomic decompositions as in the classical case). Fortunately, this difference turns out not to be critical, at least if one works in saturated PET structures: In this setting, the interplay between sorts $\mathbb{Z}^{\mathscr{M}}$, $\mathcal{A}_{\mathbb{Z}}^{\mathscr{M}}$ and $(\mathscr{L}_{\mathbb{Z},\mathbb{R}}^{\infty})^{\mathscr{M}}$ comes to the rescue via analogues of Loeb measure and Loeb integration [**DnI17**]. Appendix A.2 contains a basic discussion of Loeb structures.

2. Leibman sequences

2.1. Classical Leibman sequences. Leibman introduced the notion of *polynomial sequences* in a group G [**Lei98**]. Leibman's polynomial sequences in G generalize sequences (indexed by $\mathbb{Z}$) of the form $(g^{p(j)} : j \in \mathbb{Z})$ where $p : \mathbb{Z} \to \mathbb{Z}$ is a polynomial and $g \in G$ is fixed. Fix such a sequence $T_\bullet = (T_j : j \in \mathbb{Z})$ where $T_j = g^{p(j)}$. For fixed $i \in \mathbb{Z}$, the sequence $\Delta^i T_\bullet = (T_{i+j} \circ T_j^* : j \in \mathbb{Z})$ of "step-i discrete differences" of $T_\bullet$ is of the form $(g^{q(j)})$ where $q = \Delta^i p : j \mapsto p(i+j) - p(j)$ is a polynomial of degree less than p (or possibly the zero polynomial). This motivates Leibman's recursive definition of polynomial sequence as follows.

DEFINITION 2.1 (Discrete difference and Leibman sequence). Let G be a multiplicative group with identity I. When convenient, the inverse g^{-1} of an element g of G will be denoted g^*. Let $G^{\mathbb{Z}}$ be the group of all $\mathbb{Z}$-sequences $T : j \mapsto T_j$ from $\mathbb{Z}$ into G under the operation of pointwise multiplication induced from G, and endow $G^{\mathbb{Z}}$ with the translation action ${}_iT$ of $\mathbb{Z}$, namely $({}_iT)_j = T_{i+j}$. For $i \in \mathbb{Z}$, the *discrete-difference operator* is the function $T \mapsto \Delta^i T := {}_iT \cdot T^*$ from $G^{\mathbb{Z}}$ to $G^{\mathbb{Z}}$; it is uniquely characterized by the identity

$$(\Delta^i T)_j = T_{i+j} \cdot T_j^* \qquad \text{for all } j \in \mathbb{Z}.$$

(Since $(T^*)_j = (T_j)^*$, parentheses may be omitted without ambiguity.) We will also omit parentheses when writing iterated discrete differences; thus, $\Delta^i \Delta^j T$ means $\Delta^i(\Delta^j T)$.

Let $\mathbb{I}$ denote the constant sequence $j \mapsto I$. Given $d \in \mathbb{N}$, a *Leibman sequence in G of degree at most d* is any $T \in G^{\mathbb{Z}}$ all of whose $(d+1)$-fold iterated discrete differences are trivial, i.e.,

$$\Delta^{i_d} \dots \Delta^{i_1} \Delta^{i_0} T = \mathbb{I} \qquad \text{for all } i_0, i_1, \dots, i_d \in \mathbb{Z}.$$

A Leibman sequence is a Leibman sequence of any degree d; its degree $\deg_{\mathrm{L}} T$ is the least such d. (We define formally $\deg_{\mathrm{L}} \mathbb{I} = -\infty$.) A Leibman sequence T of degree at most 0 is called *translation-invariant* or *constant*; it is of the form $T = g\mathbb{I}$ for some $g \in G$ (i.e., $T_k = g$ for all $k \in \mathbb{Z}$).

REMARKS 2.2.

- The definition of Leibman sequence above is indirect and recursive; it involves only the group structures of $(\mathbb{Z}, +)$ and $(G, \cdot)$, but not the product of $\mathbb{Z}$ as one might otherwise expect from the usual construction of polynomials starting with monomials built from multiplication.

- It can be shown (by an application of the usual method of finite differences) that if T is a Leibman sequence of degree at most d in an *abelian* group G, then there exist $g_0, g_1, \dots, g_d \in G$ such that

$$T_k = g_0 \cdot g_1^k \cdot g_2^{\binom{k}{2}} \cdot \dots \cdot g_d^{\binom{k}{d}} \qquad \text{for all } k \in \mathbb{Z},$$

where $\binom{k}{j} = k(k-1)\cdots(k-j+1)/j!$ is the j-th binomial coefficient. [3] One may regard $g_0, g_1, \dots, g_d$ as the "coefficients" of the Leibman polynomial T. In particular, this abelian setting comprises all families $(g^{p(k)})$ where p is a polynomial $\mathbb{Z} \to \mathbb{Z}$ and $g \in G$ is fixed. Theorem 1 states the convergence of ergodic averages in the abelian case; nevertheless, Theorems 2, 3 and 4 only assume that T is a unitary Leibman sequence per Definition 2.11, but no additional explicit commutativity hypotheses.

- Translations commute with inversion and with discrete differences, i.e., $({}_jT)^* = {}_j(T^*)$ and ${}_j(\Delta^i T) = \Delta^i({}_jT)$. (The latter equality depends on the commutativity of addition on $\mathbb{Z}$.) In particular, Leibman degree is invariant under translation. However, the discrete difference operators do *not* commute with adjoints, so Leibman degree is not invariant under taking adjoints. Correspondingly, T^* need not be a Leibman polynomial if T is.

Leibman sequences of degree at most 1 are easily characterized:

PROPOSITION 2.3. *Given any fixed choice of $a, b \in G$, there exists a unique Leibman sequence T of degree at most 1 satisfying $b = T_0$ and $a = \Delta^1 T_0$, namely $T : k \mapsto a^k b$.*

The straightforward proof of Proposition 2.3 is left to the reader.

2.2. Quadratic Leibman sequences. In this section we characterize classical Leibman sequences that are quadratic, i.e., of degree at most 2. In particular, we construct a quadratic Leibman sequence whose range generates the non-nilpotent "lamplighter" group $\mathbb{Z} \wr \mathbb{Z}$ (Corollary 2.9). Throughout this section, G will denote a multiplicative group with identity I; the inverse g^{-1} of $g \in G$ is denoted g^* when convenient.

In what follows, we fix a quadratic Leibman sequence T. One may suspect that T is uniquely characterized by three constants, say $a = \Delta^1\Delta^1 T_0$, $b = \Delta^1 T_0$ and $c = T_0$; this is easily shown to be true (See Proposition 2.4 below). However, in contrast to Proposition 2.3, the constants a, b, c are not arbitrary: The requirement that they correspond to a *bona fide* Leibman sequence T imposes nontrivial relations among a and b: They must generate a factor of $\mathbb{Z} \wr \mathbb{Z}$ by Propositions 2.4 and 2.8.

PROPOSITION 2.4. *Given a quadratic Leibman sequence T in a group G, the elements $a = \Delta^1\Delta^1 T_0 = T_2 T_1^* T_0 T_1^*$ and $b = \Delta^1 T_0 = T_1 T_0^*$ satisfy the commutation relations*

$$a \cdot {}^{[b^k]}a = {}^{[b^k]}a \cdot a \qquad \textit{for all } k \in \mathbb{Z}, \tag{2.1}$$

[3]See Proposition 2.6 below for the case of sequences in G abelian that are at most quadratic.

where ${}^{[h]}g := hgh^*$ *is the conjugate of* g *by* h. *Conversely, given* $a, b, c \in G$ *such that the above relations hold for* a *and* b, *there exists a unique quadratic Leibman sequence* T *satisfying* $\Delta^1\Delta^1 T_0 = a$, $\Delta^1 T_0 = b$ *and* $T_0 = c$.

Note that the commutation relations (2.1) do not involve c at all.

2.2.1. *Proof of Proposition* 2.4. Free variables $i, j, k, l, m, n, p, x, y, z$ will denote elements of $\mathbb{Z}$ throughout. The iterated discrete differences $\Delta^i T, \Delta^i\Delta^j T, \Delta^i\Delta^j\Delta^k T$ of T will be denoted $\delta(i), \delta(i,j), \delta(i,j,k)$, respectively.

First, let T be a quadratic Leibman sequence; thus, $\delta(i,j,k) = \mathbb{I}$ for all i, j, k, by definition of Leibman degree. It follows that ${}_i\delta(j,k) = \delta(j,k)$, i.e., $\delta(j,k)$ is constant. Denote by ${}^{[S]}R$ the conjugate SRS^* of R by S $(R, S \in G^{\mathbb{Z}})$. Straightforward algebra shows that the "cocycle identity"

$$\delta(j, k+l) = {}_l\delta(j,k) \cdot {}^{[{}_l\delta(k)]}\delta(j,l) \tag{2.2}$$

holds for arbitrary $T \in G^{\mathbb{Z}}$. Under the assumption that T is quadratic, all terms $\delta(\cdot,\cdot)$ in the identity above are constant, so the cocycle identity improves itself to one with an extra free parameter m:

$$\delta(j, k+l) = \delta(j,k) \cdot {}^{[{}_m\delta(k)]}\delta(j,l). \tag{2.3}$$

Let $[\![R, S]\!] = R^*S^*RS$ be the commutator of R, S. Using the cocycle identity (2.3) to expand $\delta(i, j+k+l)$ in two different ways, we find $[\![\delta(i,j), {}_m\delta(k+l)^* \cdot {}_n\delta(k) \cdot {}_p\delta(l)]\!] = \mathbb{I}$. Using the relation $\delta(x)^* = {}_x\delta(-x)$, this identity may be rewritten

$$[\![\delta(i,j), {}_k\delta(x) \cdot {}_l\delta(y) \cdot {}_m\delta(z)]\!] = \mathbb{I} \qquad \text{whenever } x+y+z=0. \tag{2.4}$$

Let H be the subgroup of $G^{\mathbb{Z}}$ generated by all elements ${}_k\delta(x)$, and K the subgroup of H generated by all elements ${}_k\delta(x)\cdot{}_l\delta(y)\cdot{}_m\delta(z)$ with $x+y+z=0$. It follows from the commutation relations (2.4) that K is a subgroup of the centralizer of $\delta(i,j)$ in $G^{\mathbb{Z}}$. It is easy to see that K is a normal subgroup of H, and $x \mapsto \delta(x) \pmod K$ is a homomorphism $\mathbb{Z} \to H/K$. Note that ${}^{[\delta(l)^k]}\delta(m,n) = \delta(l)^k \cdot {}_m\delta(n) \cdot \delta(n)^{-1} \cdot \delta(l)^{-k} \equiv \delta(kl) \cdot \delta(n) \cdot \delta(-n) \cdot \delta(-kl) \equiv \mathbb{I} \pmod K$, so the following commutation identity follows:

$$\delta(i,j) \cdot {}^{[\delta(l)^k]}\delta(m,n) = {}^{[\delta(l)^k]}\delta(m,n) \cdot \delta(i,j). \tag{2.5}$$

Putting $i = j = l = m = n = 1$ (with k arbitrary) and evaluating at 0, we obtain the commutation relations (2.1).

Conversely, let a, b, c satisfy the commutation relations (2.1). We claim that a unique $T \in G^{\mathbb{Z}}$ exists satisfying

$$T_0 = c, \quad \delta(1)_0 = b, \quad \delta(1,1) = a\mathbb{I} \qquad (\text{i.e., } \delta(1,1)_k = a \text{ for all } k \in \mathbb{Z}). \tag{2.6}$$

Let $T_0 = c$ and $T_1 = bc$, so the first two conditions above hold. For $k \geq 0$, the condition $T_{k+2}T^*_{k+1}T_kT^*_{k+1} = \delta(1,1)_k = a$ is equivalent to the forward recurrence $T_{k+2} = aT_{k+1}T^*_kT_{k+1}$, while for $k \leq 0$, it is equivalent to the backward recurrence $T_k = T_{k+1}T^*_{k+2}aT_{k+1}$. Using both recurrences with the initial values $T_0 = c$, $T_1 = bc$, we obtain a unique $T \in G^{\mathbb{Z}}$ satisfying the conditions (2.6).

To prove that T is indeed a quadratic Leibman sequence, it remains to show that $\delta(i,j)$ is constant for all i, j. This is done inductively, starting from (2.6), which implies that $\delta(1,1)$ is constant. The details follow.

LEMMA 2.5. *Let* $A = \delta(1,1) = a\mathbb{I}$ *and* $B = \delta(1)$. *For* $R \in G^{\mathbb{Z}}$, *let the* naive degree $\deg R$ *of* R *be the sum of the exponents of all occurrences of* B *when* R *is* *written as a word in the alphabet A^1, A^{-1}, B^1, B^{-1}. Then naive degree is invariant under translations:* $\deg R = \deg({}_jR)$ *for all* $j \in \mathbb{Z}$. *Define*

$$\{\!\{R\}\!\} := {}^{[R]}A.$$

Let $\langle A, B\rangle$ *be the subgroup of* $G^{\mathbb{Z}}$ *generated by* A *and* B. *The elements* $\{\!\{R\}\!\}$ *for* $R \in \langle A, B\rangle$ *are constant and commute with each other pairwise. Moreover,* $\{\!\{R\}\!\}$ *depends only on* $\deg R$*; in fact,* $\{\!\{R\}\!\} = \{\!\{B^{\deg R}\}\!\}$. [4]

PROOF. Let $\mathcal{W} = \mathcal{W}(X, Y)$ denote a word in four formal symbols X, X^*, Y, Y^*. If $P, Q, R \in G^{\mathbb{Z}}$ are such that the word $\mathcal{W}$ evaluates to R (using the group operation of $G^{\mathbb{Z}}$) under the substitutions $X = P$, $X^* = P^{-1}$, $Y = Q$, $Y^* = Q^{-1}$, we write $R = \mathcal{W}(P, Q)$. For an arbitrary such word $\mathcal{W}$, let $R = \mathcal{W}(A, B)$. The equalities $A = {}_jA$ and ${}_jB = A^jB$ imply

$$ {}_jR = \mathcal{W}({}_jA, {}_jB) = \mathcal{W}(A, A^jB) = \mathcal{W}'(A, B) \tag{2.7}$$

for a new word [5] $\mathcal{W}' = \mathcal{W}'(X, Y)$ using exactly as many of each of the symbols Y, Y^* as $\mathcal{W}$ (but possibly more of X, X^*). It follows that naive degree is invariant under translations. We have ${}_j\{\!\{R\}\!\} = {}_j(RAR^*) = {}_jR \cdot {}_jA \cdot {}_jR^* = \{\!\{{}_jR\}\!\}$ since ${}_jA = A$. Because of the translation invariance of naive degree, once the identity $\{\!\{R\}\!\} = \{\!\{B^{\deg R}\}\!\}$ is proved, it shall follow that $\{\!\{R\}\!\}$ is constant, since ${}_j\{\!\{R\}\!\} = \{\!\{{}_jR\}\!\} = \{\!\{B^{\deg({}_jR)}\}\!\} = \{\!\{B^{\deg R}\}\!\} = \{\!\{R\}\!\}$. By identity (2.7), proving $\{\!\{R\}\!\}\{\!\{S\}\!\} = \{\!\{S\}\!\}\{\!\{R\}\!\}$ for all $R, S \in G^{\mathbb{Z}}$ reduces to showing $\{\!\{R\}\!\}_0 \cdot \{\!\{S\}\!\}_0 = \{\!\{S\}\!\}_0 \cdot \{\!\{R\}\!\}_0$. Abusing notation, define $\{\!\{g\}\!\} = {}^{[g]}a = gag^*$ for $g \in G$. The remainder of the proof thus reduces to proving

(1) $\{\!\{g\}\!\}$ depends only on the naive degree $\deg g$ of $g \in G$—defined as the sum of the exponents of b in an expression $g = \mathcal{W}(a, b)$ of g as a word $\mathcal{W}$ on a^1, a^{-1}, b^1, and b^{-1}—in fact, $\{\!\{g\}\!\} = \{\!\{b^n\}\!\}$ where $n = \deg g$, and

(2) the elements $\{\!\{g\}\!\} = {}^{[g]}a$ for g in the subgroup $\langle a, b\rangle$ generated by a and b commute pairwise.

Since $\{\!\{g\}\!\}$ commutes with $\{\!\{h\}\!\}$ iff $\{\!\{g^*h\}\!\}$ commutes with $\{\!\{I\}\!\} = a$, property (2) follows from

(2') $\{\!\{g\}\!\}$ commutes with a if $g \in \langle a, b\rangle$.

Note that $g \in \langle a, b\rangle$ satisfies properties (1) and (2') iff either one of g^*, ag, a^*g does.

For $m \geq 0$, let $\langle a, b\rangle_m$ be the set of elements of $\langle a, b\rangle$ that are words $\mathcal{W}(a, b)$ using no more than m symbols b, b^*. By induction on m, we prove assertions (1) and (2') for $g \in \langle a, b\rangle_m$. (This will prove the assertions for all $g \in \langle a, b\rangle = \bigcup_m \langle a, b\rangle_m$.) First, $\langle a, b\rangle_0$ consists of powers a^k having naive degree zero. Since a commutes with a^k, it follows that $\{\!\{a^k\}\!\} = {}^{[a^k]}a = a$; thus, assertions (1) and (2') hold for $m = 0$. Next, assume both assertions hold for some fixed $m \geq 0$. Let $g \in \langle a, b\rangle_{m+1}$ be arbitrary. Without loss of generality (possibly multiplying g by

[4] The naive degree need not be well defined as an integer, but it is well defined as an integer modulo N if N is the least positive integer (if any) such that I has an expression as a word of naive degree N (in addition to its expression as the empty word of naive degree 0). Thus, naive degree induces a well-defined notion of degree (modulo N) with respect to which the identity $\{\!\{R\}\!\} = \{\!\{B^{\deg R}\}\!\}$ holds.

[5] To be precise, $\mathcal{W}(A, A^jB)$ is a word $\mathcal{W}'(A, B)$ where $\mathcal{W}'(X, Y)$ is obtained from $\mathcal{W}(X, Y)$ performing the substitutions $Y = X^jY$, $Y^* = Y^*X^{-j}$, where powers X^k, X^{-k} (for $k \geq 0$) are interpreted as the k-words $X \ldots X$ and $X^* \ldots X^*$, respectively.

powers of a or a^* on the left) we may assume that $g = b^{\pm 1}h$ with $h \in \langle a, b\rangle_m$. If $g = b^{\pm 1}h$, then $\deg g = \deg h \pm 1$, and it follows from the inductive hypothesis that $\{\!\{g\}\!\} = \{\!\{b^{\pm 1}h\}\!\} = {}^{[b^{\pm 1}]}\{\!\{h\}\!\} = {}^{[b^{\pm 1}]}\{\!\{b^{\deg h}\}\!\} = \{\!\{b^{\pm 1}b^{\deg h}\}\!\} = \{\!\{b^{\deg g}\}\!\}$. Note that $\{\!\{b^k\}\!\} = {}^{[b^k]}a$ commutes with a by the hypothesis of Proposition 2.4; hence, so does $\{\!\{g\}\!\}$. This shows that assertions (1) and (2') hold for $m+1$, completing the proof of Lemma 2.5. □

Continuing the proof of Proposition 2.4, let $A = \delta(1,1)$ and $B = \delta(1)$ as above, and let $\{\!\{\langle A, B\rangle\}\!\}$ be the subgroup of $\langle A, B\rangle$ generated by elements $\{\!\{R\}\!\}$ with $R \in \langle A, B\rangle$. By Lemma 2.5, $\{\!\{\langle A, B\rangle\}\!\}$ is an abelian group of constants. By induction, one shows first that for all k, l we have ${}_k\delta(l) \in \langle A, B\rangle$, and subsequently that $\delta(k,l) \in \{\!\{\langle A, B\rangle\}\!\}$ (using Lemma 2.5 and the cocycle identity (2.2) to induct on l, then the identity $\delta(i+j,l) = {}_j\delta(i,l)\delta(j,l)$ to induct on k, plus simple manipulations to extend to negative k, l). Thus, $\delta(k,l)$ is constant for all k, l. This implies that $\mathbb{I} = \delta(j,k,l) = \Delta^j\Delta^k\Delta^l T$ for all j, k, l, showing that T is a quadratic Leibman sequence and concluding the proof of Proposition 2.4.

2.2.2. *Some consequences of Proposition* 2.4. First, we give some definitions. Given a sequence $(x_i : i \in \mathbb{Z})$ in any multiplicative group $\mathbb{G}$, there is a natural notion of product $\prod_{i=k}^{l} x_i$ of the terms x_i "as i ranges from k to l"; it is characterized by the properties

(1) $\prod_{i=k}^{k-1} x_i = 1_{\mathbb{G}}$ (the identity of $\mathbb{G}$), and
(2) $\prod_{i=k}^{l+1} x_i = \left(\prod_{i=k}^{l} x_i\right) \cdot x_{l+1}$

for all $k, l \in \mathbb{Z}$. Informally, terms x_i are multiplied left-to-right in succession. For fixed k, one obtains the familiar definitions

$$\prod_{i=k}^{k} x_i = x_k, \quad \prod_{i=k}^{k+1} x_i = x_k x_{k+1}, \quad \ldots, \quad \prod_{i=k}^{k+n} x_i = x_k x_{k+1} \ldots x_{k+n}, \quad \ldots,$$

but also the less familiar

$$\prod_{i=k}^{k-2} x_i = x_{k-1}^{-1}, \quad \prod_{i=k}^{k-3} x_i = x_{k-1}^{-1}x_{k-2}^{-1}, \quad \ldots, \quad \prod_{i=k}^{k-n-1} x_i = x_{k-1}^{-1}x_{k-2}^{-1}\ldots x_{k-n}^{-1}, \quad \ldots.$$

There is a corresponding notion of product ${}^{\mathrm{op}}\prod$ evaluated in the opposite group $\mathbb{G}^{\mathrm{op}}$ of $\mathbb{G}$: iterated products are computed right-to-left instead, namely

(1) ${}^{\mathrm{op}}\prod_{i=k}^{k-1} x_i = 1_{\mathbb{G}}$, and
(2) ${}^{\mathrm{op}}\prod_{i=k}^{l+1} x_i = x_{l+1} \cdot \left({}^{\mathrm{op}}\prod_{i=k}^{l} x_i\right)$

for all $k, l \in \mathbb{Z}$. [6]

PROPOSITION 2.6. *If a, b, c are elements of a group G satisfying the commutation relations* (2.1), *the unique quadratic Leibman sequence T satisfying $\Delta^1\Delta^1 T_0 = a$, $\Delta^1 T_0 = b$ and $T_0 = c$ is given by the expression*

$$T_j = \left[{}^{\mathrm{op}}\prod_{i=1}^{j}(a^{i-1}b)\right] \cdot c \qquad \textit{for all } j \in \mathbb{Z}. \tag{2.8}$$

In particular, if a and b commute, then $T_j = a^{\binom{j}{2}}b^j c$ where $\binom{j}{2} = j(j-1)/2$ for all $j \in \mathbb{Z}$.

[6] One may alternatively define ${}^{\mathrm{op}}\prod_{i=k}^{l} x_i := \left(\prod_{i=k}^{l} x_i^{-1}\right)^{-1}$.

Proof. From the identities $\delta(j+1) = {}_j\delta(1) \cdot \delta(j)$ and ${}_j\delta(1) = \delta(j,1)\delta(1) = \delta(1,1)^j \cdot \delta(1) = A^jB$, we obtain ${}_jT \cdot T^* = \delta(j) = {}^{\mathrm{op}}\prod_{i=1}^j (A^{i-1}B)$. Thus, ${}_jT = {}^{\mathrm{op}}\prod_{i=1}^j (A^{i-1}B) \cdot T$. Equation (2.8) follows evaluating the latter identity at 0.

If a and b commute, then $T_j = \prod_{i=1}^j a^{i-1} \cdot \prod_{i=1}^j b \cdot c = a^{\sum_{i=1}^j (i-1)} \cdot b^j \cdot c$, where $\sum_{i=1}^j (i-1) = j(j-1)/2 = \binom{j}{2}$. (For $j \le 0$, the sum $\sum_{i=1}^j$ is understood in the obvious sense analogous to the definition of $\prod_{i=1}^j$ above.) □

Definition 2.7. The restricted wreath product $\mathbb{Z} \wr \mathbb{Z}$ of $\mathbb{Z}$ with itself is called the *lamplighter group*. It is realized as a group on generators $(\alpha_k : k \in \mathbb{Z})$ and β subject to the relations

$$\alpha_k \alpha_l = \alpha_l \alpha_k \tag{2.9}$$

$$\beta^k \alpha_l = \alpha_{k+l} \beta^k \qquad \text{for all } k, l \in \mathbb{Z}. \tag{2.10}$$

The subgroups $H = \langle \beta \rangle$ and $K = \langle \alpha_k : k \in \mathbb{Z} \rangle$ of $\mathbb{Z} \wr \mathbb{Z}$ are abelian. Together, they generate $\mathbb{Z} \wr \mathbb{Z}$, and it is easy to show that each is its own centralizer in $\mathbb{Z} \wr \mathbb{Z}$; therefore, $\mathbb{Z} \wr \mathbb{Z}$ has trivial center. In particular, $\mathbb{Z} \wr \mathbb{Z}$ is not nilpotent. It is, however, solvable, being a semidirect product of the two abelian groups H, K.

Proposition 2.8. *Given a quadratic Leibman sequence T in a group G, its discrete differences $\Delta^i T, \Delta^i \Delta^j T$ $(i, j \in \mathbb{Z})$ generate a subgroup $\langle \Delta^\circ T \rangle$ of $G^{\mathbb{Z}}$ isomorphic to a factor of $\mathbb{Z} \wr \mathbb{Z}$. Actually, the group $\langle \Delta^\circ T \rangle$ is already generated by $A = \Delta^1 \Delta^1 T$ and $B = \Delta^1 T$. The values $\Delta^i T_m, \Delta^i \Delta^j T_m$ $(i, j, m \in \mathbb{Z})$ of these discrete differences generate a subgroup $\langle \Delta^\circ T \rangle_\bullet$ of G also isomorphic to a factor of $\mathbb{Z} \wr \mathbb{Z}$. In fact, $a = A_0$ and $b = B_0$ already generate $\langle \Delta^\circ T \rangle_\bullet$. Furthermore, there exists a quadratic Leibman sequence such that both $\langle \Delta^\circ T \rangle$ and $\langle \Delta^\circ T \rangle_\bullet$ are isomorphic to $\mathbb{Z} \wr \mathbb{Z}$ itself.*

Proof. The proof of Proposition 2.4 shows that $\langle \Delta^\circ T \rangle$ is generated by A and B. *A fortiori*, $\langle \Delta^\circ T \rangle$ is generated by B and $A_{[k]} := {}^{[B^k]}A$ for $k \in \mathbb{Z}$ (since $A_{[0]} = A$). The special case $i = j = l = m = n = 1$ (with k arbitrary) of equation (2.5) gives the commutation relations

$$A \cdot A_{[k]} = A_{[k]} \cdot A,$$

whence the following relations are easily proved using induction and the definition of $A_{[k]}$:

$$A_{[k]} A_{[l]} = A_{[l]} A_{[k]}$$
$$B^k A_{[l]} = A_{[k+l]} B^k \qquad \text{for all } k, l \in \mathbb{Z}.$$

In general, further relations between A and B may hold; nevertheless, we see that the generators $A_{[k]}$ $(k \in \mathbb{Z})$ and B of $\langle \Delta^\circ T \rangle$ satisfy the defining relations (2.9), (2.10) of $\mathbb{Z} \wr \mathbb{Z}$, so $\langle \Delta^\circ T \rangle$ is isomorphic to a factor of $\mathbb{Z} \wr \mathbb{Z}$. Evaluation at zero is a homomorphism $G^{\mathbb{Z}} \to G$ that restricts to a surjection $\langle \Delta^\circ T \rangle \to \langle \Delta^\circ T \rangle_\bullet$, so $\langle \Delta^\circ T \rangle_\bullet$ is isomorphic to a factor of $\langle \Delta^\circ T \rangle$, and thus of $\mathbb{Z} \wr \mathbb{Z}$, generated by $a = A_0$ and $b = B_0$.

Reciprocally, let $(\alpha_k : k \in \mathbb{Z})$ and β be the canonical generators of $\mathbb{Z} \wr \mathbb{Z}$, i.e., these elements obey only relations implied by (2.9) and (2.10). It follows from Proposition 2.4 that there is a unique quadratic Leibman sequence T in $\mathbb{Z} \wr \mathbb{Z}$ satisfying $\Delta^1 \Delta^1 T_0 = \alpha_0$, $\Delta^1 T_0 = \beta$ and $T_0 = I$. For this sequence T we have $\langle \Delta^\circ T \rangle_\bullet \simeq \mathbb{Z} \wr \mathbb{Z}$, and hence $\langle \Delta^\circ T \rangle \simeq \mathbb{Z} \wr \mathbb{Z}$ also. □

COROLLARY 2.9. *There exists a quadratic Leibman sequence T with $T_0 = I$ whose range generates a non-nilpotent group.*

PROOF. The quadratic Leibman sequence T constructed in the proof of Proposition 2.8 has $T_0 = I$, and its range generates $\mathbb{Z} \wr \mathbb{Z}$, which is not nilpotent. □

REMARK 2.10. Bergelson and Leibman used the lamplighter group to construct counterexamples showing that multiple recurrence and multiple convergence results that hold for ergodic actions generating nilpotent groups do fail for non-nilpotent groups [**BL04**]. In contrast to the case of multiple ergodic averages, all (simple) ergodic convergence results in the present manuscript—including Theorems 2 and 4—hold under the sole hypothesis that the family $T_\bullet$ is a Leibman sequence, which already in the quadratic setting includes cases in which the range of $T_\bullet$ is non-nilpotent, per Corollary 2.9 above.

2.3. Leibman sequences in PET structures.

DEFINITION 2.11 (Discrete difference and abstract Leibman sequence). Let $\mathscr{M}$ be a PET structure over $\mathbb{Z}$. The *discrete-difference operator* is the function $\Delta^\bullet : \mathscr{L}^\infty_{\mathcal{Z},\mathfrak{B}} \to \mathscr{L}^\infty_{\mathcal{Z}^2,\mathfrak{B}}$ uniquely characterized by the identity

$$(\Delta^\bullet T)_{i,j} = T_{i+j} \circ T_j^* \qquad \text{for all } i, j \in \mathcal{Z}.$$

Alternatively, for $i \in \mathcal{Z}$, the left evaluation at i of $\Delta^\bullet T$ is $\Delta^i T = {}_iT \circ T^*$.

Let $\mathbb{I} = I(\blacksquare)$ denote the constant family $i \mapsto I$ in $\mathscr{L}^\infty_{\mathcal{Z},\mathfrak{B}}$. Given $d \in \mathbb{N}$, a *unitary Leibman sequence of degree at most* d is any $T \in \mathscr{L}^\infty_{\mathcal{Z},\mathfrak{B}}$ satisfying

$$\Delta^{i_d} \ldots \Delta^{i_1} \Delta^{i_0} T = \mathbb{I} \qquad \text{for all } i_0, i_1, \ldots, i_d \in \mathcal{Z},$$

that takes values in $\mathrm{U}_{\mathcal{H}}$, i.e., also satisfying $T^* \circ T = \mathbb{I} = T \circ T^*$. A Leibman sequence is a Leibman sequence of any degree d; its degree $\deg_{\mathrm{L}} T$ is the least such d. (We define formally $\deg_{\mathrm{L}} \mathbb{I} = -\infty$.)

REMARKS 2.12.

- Note that abstract Leibman sequences per Definition 2.11 are "internal", i.e., obtained *from elements of* $\mathscr{L}^\infty_{\mathcal{Z},\mathfrak{B}}$—that are otherwise only incidentally regarded as functions $\mathcal{Z} \to \mathfrak{B}$ via evaluation; they may be regarded as taking values in the group

$$G = \mathrm{U}_{\mathcal{H}} = \{U \in \mathfrak{B}_{\mathcal{H}} : U \circ U^* = I = U^* \circ U\}$$

  of (internal) unitary transformations of the Hilbert space $\mathcal{H}$. (For this reason, we sometimes refer to these Leibman sequences as *unitary*.) Accordingly, the defining property of a Leibman sequence amounts to the requirement that $d+1$ discrete-difference operations, *possibly involving nonstandard elements* $j \in \mathcal{Z}$, always transform T into $\mathbb{I}$, i.e., the constant function $i \mapsto I$ *for all* $i \in \mathcal{Z}$, not merely for all $i \in \mathbb{Z}$. The proofs of Theorems 2 and 3 below crucially depend on the richer structure of $\mathcal{Z}$ in saturated PET structures—even if ultimately the results are valid in *all* PET structures, including classical ones whose Leibman sequences are *bona fide* functions $\mathbb{Z} \to \mathrm{U}_{\mathcal{H}} \subset \mathfrak{B}$.
- We have $\Delta^\bullet T = {}_\bullet T \circ (T^*_{\blacksquare,\cdot})$; hence, discrete differentiation is obtained from the $\mathbb{Z}$-action on $\mathscr{L}^\infty_{\mathcal{Z},\mathfrak{B}}$, the right inclusion $\mathscr{L}^\infty_{\mathcal{Z},\mathfrak{B}} \hookrightarrow \mathscr{L}^\infty_{\mathcal{Z}^2,\mathfrak{B}}$, plus the pointwise operations of composition and taking adjoint; thus, $\Delta^\bullet$ may as well be regarded as a distinguished function of any PET structure $\mathscr{M}$.

- The predicate *"T is a unitary Leibman sequence of degree at most d"* is captured by a single Henson formula $\lambda_d(T)$, namely [7]

$$(T \circ T^* = \mathbb{I} = T^* \circ T) \wedge \forall i_d \ldots \forall i_1 \forall i_0 \big[\Delta^{i_d}(\ldots(\Delta^{i_1}(\Delta^{i_0}T))\ldots) = \mathbb{I}\big]. \tag{2.11}$$

- Since any group G is realized as a subgroup of a suitable unitary group $\mathrm{U}_{\mathcal{H}}$, [8] it follows that any classical Leibman sequence is realized as a Leibman sequence in a classical PET structure. In view of Proposition 2.8 and Corollary 2.9, we have instances of pointwise ergodic convergence per Theorem 2 in the setting of quadratic Leibman sequences (T_n) of unitary operators generating a non-nilpotent group of unitary transformations. To our knowledge, this is the first explicit example of pointwise convergence of averages of a non-nilpotent (in fact, not even virtually nilpotent [9]) family of unitary transformations.

3. An ergodic theorem for unitary polynomial actions of $\mathbb{Z}$

Throughout the end of this section, $\mathcal{L}$ will be the Henson language for PET structures over $\mathbb{Z}$. All structures will be in the class $\overline{\textbf{PET}}$ of abstract PET structures over $\mathbb{Z}$.

3.1. The sequence of ergodic averages.

CONVENTION 3.1. Henceforth, the standalone symbols $\mathbb{R}, \mathbb{Z}, \mathbb{N}$ shall denote the usual sets of real, integer and natural numbers. If $\mathcal{M}$ is a PET structure over $\mathbb{Z}$, we shall use interpretation of constants (and the density of $\mathbb{Q}$ in $\mathbb{R}$) to identify $\mathbb{R}$ with the sort $\mathbb{R}^{\mathcal{M}}$, and also $\mathbb{Z}$ and $\mathbb{N}$ with subsets of $\mathcal{Z} = \mathbb{Z}^{\mathcal{M}}$ and $\mathcal{N} = \mathbb{N}^{\mathcal{M}}$, respectively. [10] By an abuse of notation, when the structure $\mathcal{M}$ is clear from context, we may omit the superscript and write $\mathcal{H}$, $\mathfrak{B}$, $\mathcal{A}_{\mathbb{Z}}$ $\mathscr{L}^{\infty}_{\mathbb{Z},\mathbb{R}}$, $\ldots$ to denote the sorts $\mathcal{H}^{\mathcal{M}}$, $\mathfrak{B}^{\mathcal{M}}$, $\mathcal{A}_{\mathbb{Z}}^{\mathcal{M}}$, $(\mathscr{L}^{\infty}_{\mathbb{Z},\mathbb{R}})^{\mathcal{M}}$, $\ldots$ of $\mathcal{M}$.

DEFINITION 3.2 (Ergodic averages). Let $\mathcal{M}$ be a PET structure over $\mathbb{Z}$ and let $T \in \mathscr{L}^{\infty}_{\mathbb{Z},\mathfrak{B}}$. Via the evaluation $\mathscr{L}^{\infty}_{\mathbb{Z},\mathfrak{B}} \times \mathcal{Z} \to \mathfrak{B}$, one may regard T as a function $i \mapsto T_i$. For $n \in \mathcal{N}$, the *n-th average of T* is

$$\mathrm{AV}_n\, T = \langle T, \sigma_n \rangle.$$

The *sequence of averages of T* is $\mathrm{AV}_{\bullet} T = (\mathrm{AV}_n\, T : n \in \mathbb{N})$.

Similarly, for $x \in \mathcal{H}$, the *n-th average of x under T* is $\mathrm{AV}_n T(x)$. The *sequence of averages of x under T* is $\mathrm{AV}_{\bullet} T(x) = (\mathrm{AV}_n T(x) : n \in \mathbb{N})$.

We remark that $\mathrm{Th}_{\mathbf{PET}}$ ensures the validity of the identities

$$\mathrm{AV}_n\, T = \frac{1}{n+1} \sum_{0 \le i \le n} T_i \quad \text{and} \quad \mathrm{AV}_n\, T(x) = \frac{1}{n+1} \sum_{0 \le i \le n} T_i(x)$$

[7]Expressions of the type $x = y$ such as those in (2.11) are not Henson formulas *sensu stricti*, but may be regarded as abbreviations of formulas $\mathrm{d}(x, y) \le 0$ (or $\|y - x\| \le 0$ in Banach sorts).

[8]One may identify G with its faithful homomorphic image under the translation action $G \curvearrowright \mathscr{L}^2(G)$, which realizes G as a group of unitary transformations of $\mathcal{H} = \mathscr{L}^2(G)$.

[9]A group is virtually nilpotent if it has a finite-index nilpotent subgroup.

[10]The identification of $\mathbb{N}$ with a subset of $\mathbb{Z}$ is neither necessary nor beneficial. Theorem 4 below considers ergodic averages relative to Følner nets indexed by any countable directed set $\mathbb{D}$ (in place of $\mathbb{N}$) over an arbitrary abelian group $\mathbb{G}$ (in place of $\mathbb{Z}$).

for (standard) $n \in \mathbb{N}$; however, averages as defined above are non-classical if $n \in \mathcal{N} \setminus \mathbb{N}$. On the other hand, the sequence $(\mathrm{AV}_n T : n \in \mathbb{N})$ has classical terms, and the study of its convergence is purely classical *a priori*.

THEOREM 2 (Poly-MET/$\mathbb{Z}$: Mean Ergodic Theorem for unitary polynomial actions of $\mathbb{Z}$). *Let $\mathcal{M}$ be a PET structure over $\mathbb{Z}$, and let $T \in (\mathscr{L}^\infty_{\mathbb{Z},\mathbb{R}})^{\mathcal{M}}$ be a Leibman sequence of unitary operators on the Hilbert space $\mathcal{H} = \mathcal{H}^{\mathcal{M}}$. For every $x \in \mathcal{H}$, the sequence $\mathrm{AV}_\bullet T(x) = (\mathrm{AV}_n T(x) : n \in \mathbb{N})$ of averages of x under T converges in the norm topology of $\mathcal{H}$.*

Theorem 2 admits the following uniformly metastable strengthening.

THEOREM 3 (Metastable Poly-MET/$\mathbb{Z}$). *Fix $d \in \mathbb{N}$. There exists a universal metastability rate $E^d_\bullet$, depending only on d, that applies uniformly to all sequences $\mathrm{AV}_\bullet T(x)$ of averages of arbitrary x in the unit ball of the Hilbert-space sort $\mathcal{H}$ under any Leibman sequence T in $\mathrm{U}_{\mathcal{H}}$ of degree at most d in any PET structure $\mathcal{M}$ over $\mathbb{Z}$.*

The rest of this section is devoted to proving Theorems 2 and 3.

3.2. Proof preliminaries.

LEMMA 3.3 (Dominated Convergence Theorem in PET structures). *Let $\mathcal{L}$ be the language of PET structures. Let $\mathcal{M}$ be any* saturated *PET structure. Let $\varphi_\bullet = (\varphi_n : n \in \mathbb{N})$ be a bounded sequence in $\mathscr{L}^\infty_{\mathbb{Z},\mathcal{H}}$. For all $x \in \mathbb{Z}^{\mathcal{M}}$ assume that the sequence $\varphi_\bullet(x) = (\varphi_n(x) : n \in \mathbb{N})$ in $\mathcal{H}^{\mathcal{M}}$ is convergent. Then, for arbitrary $\mu \in \mathfrak{M}^{\mathcal{M}}$, the sequence $\langle \varphi_\bullet, \mu\rangle = (\langle \varphi_n, \mu\rangle : n \in \mathbb{N})$ in $\mathcal{H}^{\mathcal{M}}$ is convergent.*

(We will only require the special case of Lemma 3.3 in which φ is of the form $n \mapsto \mathrm{AV}_n T$ with $T \in \mathscr{L}^\infty_{\mathbb{Z},\mathfrak{B}}$.)

PROOF. A saturated PET structure $\mathcal{M}$ is a *Banach integration framework* (with Banach sort $\mathcal{H}^{\mathcal{M}}$ and measure-space sort $\mathbb{Z}^{\mathcal{M}}$) as defined in Appendix A.4. Thus, Lemma 3.3 follows from Theorem 5 whose statement and proof are in Appendix A.5. □

To state the next lemma we need a definition. Let the *reverse difference* operator $\nabla^\bullet : \mathscr{L}^\infty_{\mathbb{Z},\mathfrak{B}} \to \mathscr{L}^\infty_{\mathbb{Z}^2,\mathfrak{B}}$ be the mapping $T \mapsto \nabla^\bullet T$ characterized by the property that $\nabla^\bullet T$ evaluates to the function $(i,j) \mapsto T_j \circ T^*_{i+j}$. We write $\nabla^i T$ to denote the left evaluation of $\nabla^\bullet T$, i.e., $\nabla^i T$ evaluates to the mapping $j \mapsto T_j \circ T^*_{i+j}$. Note that $\nabla^i T$ is the translate by i of the (forward) difference of T with step $-i$, i.e., $\nabla^i T = {}_i\Delta^{[-i]} T$ holds for all $i \in \mathcal{Z}$. Just like the forward difference operator $\Delta^\bullet$, the reverse difference operator $\nabla^\bullet$ is explicitly definable in any PET structure since it is obtained by composing functions of the structure (the $\mathbb{Z}$-action $\mathscr{L}^\infty_{\mathbb{Z},\mathfrak{B}} \to \mathscr{L}^\infty_{\mathbb{Z}^2,\mathfrak{B}}$, the shear map on $\mathscr{L}^\infty_{\mathbb{Z}^2,\mathfrak{B}}$, the pointwise adjoint operation $\mathscr{L}^\infty_{\mathbb{Z},\mathfrak{B}} \to \mathscr{L}^\infty_{\mathbb{Z},\mathfrak{B}}$, the left inclusion $\mathscr{L}^\infty_{\mathbb{Z},\mathfrak{B}} \hookrightarrow \mathscr{L}^\infty_{\mathbb{Z}^2,\mathfrak{B}}$, and the pointwise composition $\mathscr{L}^\infty_{\mathbb{Z}^2,\mathfrak{B}} \times \mathscr{L}^\infty_{\mathbb{Z}^2,\mathfrak{B}} \to \mathscr{L}^\infty_{\mathbb{Z}^2,\mathfrak{B}}$); thus, $\nabla^\bullet$ may as well be considered a distinguished function of any PET structure.

LEMMA 3.4. *Let $\mathcal{M}$ be a PET structure such that $\mathcal{N} = \mathbb{N}^{\mathcal{M}}$ contains a nonstandard natural number $M \in \mathcal{N} \setminus \mathbb{N}$. For every standard natural $n \in \mathbb{N}$ and $T \in \mathscr{L}^\infty_{\mathbb{Z},\mathfrak{B}}$:*

$$(\mathrm{AV}_n T) \circ (\mathrm{AV}_M T)^* = \langle \langle\!\langle \nabla^\bullet T, \sigma_n \rangle\!\rangle, \sigma_M \rangle.$$

In less cryptic notation, the equation above reads:

$$(\mathrm{AV}_n T) \circ (\mathrm{AV}_M T)^* = \int \mathrm{AV}_n(\nabla^i T)\, d\sigma_M(i).$$

Every step of the proof below is justified by an axiom of $\mathrm{Th}_{\mathbf{PET}}$. We prefer to use informal integral notation to make the argument transparent.

PROOF. Note that $(\mathrm{AV}_M T)^* = \mathrm{AV}_M(T^*)$ (in fact, $\langle T, \mu\rangle^* = \langle T^*, \mu\rangle$ for all $\mu \in \mathfrak{M}$). For $m, n \in \mathbb{N}$ we have:

$$\begin{aligned}
(\mathrm{AV}_n T) \circ (\mathrm{AV}_m T)^* &= \int T_j\, d\sigma_n(j) \circ \int T_i^*\, d\sigma_m(i) = \iint T_j \circ T_i^*\, d\sigma_m(i)\, d\sigma_n(j)\\
&= \iint T_j \circ T_{i+j}^*\, d\sigma_m(i+j)\, d\sigma_n(j)\\
&\qquad \text{(letting } i = i + j \text{ in the inner integral)}\\
&= \iint \nabla^i T_j\, d\sigma_m(i+j)\, d\sigma_n(j)\\
&= \iint \nabla^i T_j\, d\sigma_n(j)\, d\sigma_m(i)\\
&\qquad - \iint \nabla^i T_j\, d[\sigma_m - {}_j(\sigma_m)](i)\, d\sigma_n(j),
\end{aligned}$$

where ${}_j(\sigma_m)$ is the translation of σ_m by j. Since $\left\|\nabla^i T_j\right\| = \left\|{}_i\Delta^{[-i]} T_j\right\| \le \|T\|^2$ for all i, j:

$$\left\|(\mathrm{AV}_n T) \circ (\mathrm{AV}_m T)^* - \int \mathrm{AV}_n(\nabla^i T)\, d\sigma_m(i)\right\| \le \|T\|^2 \cdot \max_{0\le j\le m} \|\sigma_m - {}_j(\sigma_m)\|.$$

Given fixed $\epsilon > 0$ and $n \in \mathbb{N}$, let $m_{\epsilon,n}$ be the smallest natural number satisfying $m_{\epsilon,n} \ge 2n/\epsilon$. Clearly, $\|\sigma_m - {}_j(\sigma_m)\| \le 2n/(m+1) \le \epsilon$ for $j \le n \le m+1$. (For m large and n small, this inequality captures the "approximate invariance" of the long interval $\{0, 1, \ldots, m\}$ of $\mathbb{Z}$ under small translations, i.e., the Følner property of the collection of such intervals.) Then we have

(3.1)
$$\left\|(\mathrm{AV}_n T) \circ (\mathrm{AV}_m T)^* - \int \mathrm{AV}_n(\nabla^i T)\, d\sigma_m(i)\right\| \le \epsilon \|T\|^2 \qquad \text{whenever } m \ge m_{\epsilon,n}.$$

Since $M \in \mathcal{N} \setminus \mathbb{N}$ satisfies $M \ge m_{\epsilon,n}$ for all $n \in \mathbb{N}$ and $\epsilon > 0$, the assertion in Lemma 3.4 follows. □

We offer some remarks on the proof of Lemma 3.4 above, which is the crux of our approach to proving Theorem 2. Despite its rather short length, it sheds light on the various sorts and distinguished functions in PET structures. There are no double integrals as such but rather iterated integrals $\mathscr{L}^\infty_{\mathbb{Z}^2,\mathfrak{B}} \to \mathscr{L}^\infty_{\mathbb{Z},\mathfrak{B}}$ and $\mathscr{L}^\infty_{\mathbb{Z},\mathfrak{B}} \to \mathfrak{B}$—the order of the integration (first on the left and then on the right variable, or vice versa) is immaterial as ensured by the PET axiom

$$(\forall \mathfrak{T} \in \mathscr{L}^\infty_{\mathbb{Z}^2,\mathfrak{B}})(\forall \mu, \nu \in \mathfrak{M})\big[\langle\langle\!\langle \mu, \mathfrak{T}\rangle\!\rangle, \nu\rangle = \langle\langle\!\langle \mathfrak{T}, \nu\rangle\!\rangle, \mu\rangle\big].$$

The validity of the substitution $i = i + j$ in the inner integral is justified by the compatibility of the shear transformation on $\mathscr{L}^\infty_{\mathbb{Z}^2,\mathfrak{B}}$ and the action of $\mathbb{Z}$ on $\mathfrak{M}$:

$$(\forall \mathfrak{T} \in \mathscr{L}^\infty_{\mathbb{Z}^2,\mathfrak{B}})(\forall \mu \in \mathfrak{M})\big[\langle\!\langle \mathfrak{T}, \mu\rangle\!\rangle = \langle\!\langle \widetilde{\mathfrak{T}}, {}_\bullet\mu\rangle\!\rangle\big].$$

Other steps in the proof admit similar formal justifications by axioms of PET structures.

LEMMA 3.5. *Let $\mathscr{M}$ be a* saturated *Henson structure with an ordered sort* $(\mathcal{N}, \leq)$ *extending* $(\mathbb{N}, \leq)$*, and let* $\varphi_\bullet$ *be a sequence explicitly defined by a* $\mathcal{L}[S]$*-term* $\varphi(\mathtt{n})$ *of sort* s *(i.e.,* $\varphi_\bullet$ *is the sequence* $(\varphi(n) : n \in \mathbb{N})$ *in* $\mathscr{M}$*, where* $\mathtt{n}$ *is a variable of sort* $\mathcal{N}$ *and* S *is some set of parameters of the universe of* $\mathscr{M}$*). Then every subsequential limit of* φ *is of the form* $\varphi(M)$ *for some* $M \in \mathcal{N} \setminus \mathbb{N}$*. If* $\varphi(M) = \varphi(N)$ *for all* $M \in \mathcal{N} \setminus \mathbb{N}$*, then the sequence* $\varphi_\bullet = (\varphi(n) : n \in \mathbb{N})$ *converges. In such case, the common value* $\varphi(M)$ *is the limit* $\lim_{n\to\infty} \varphi(n)$.

The proof of Lemma 3.5 is a routine application of saturation left to the reader.

LEMMA 3.6. *Let* $\mathcal{X}$*,* $\mathcal{Y}$ *be metric spaces, and let* $\varphi : (x, n) \mapsto \varphi_n(x)$ *be a function from* $\mathcal{X} \times \mathbb{N}$ *to* $\mathcal{Y}$ *such that* $\varphi_n(\cdot) : \mathcal{X} \to \mathcal{Y}$ *is* 1*-Lipschitz for each* $n \in \mathbb{N}$ *(i.e.,* $d(\varphi_n(x), \varphi_n(y)) \leq d(x, y)$ *for* $x, y \in \mathcal{X}$*). Let* S *be a dense subset of* $\mathcal{X}$ *such that* $\varphi_\bullet(x)$ *converges for all* $x \in S$*. Then* $\varphi_\bullet(x)$ *converges for all* $x \in \mathcal{X}$.

We omit the straightforward proof of Lemma 3.6.

3.3. Proof of Theorem 2. For each fixed Leibman degree $d \in \mathbb{N}$, we first prove Theorem 2 for unitary Leibman sequences of degree at most d in any *saturated* PET structure $\mathscr{M}$. The descent argument on the degree d is characteristic of Bergelson's PET induction [**Ber87**].

The assertion is trivial for $T = \mathbb{I}$. If T (is pointwise unitary and) has Leibman degree $\deg_{\mathrm{L}} T = 0$, we have $T_i \circ T_0^* = \Delta^i T_0 = I$, hence $T_i = T_0$ for all $i \in \mathcal{Z}$, so T is constant. Thus, the sequences $\mathrm{AV}_\bullet T$ and $\mathrm{AV}_\bullet T(x)$ are also constant (all terms are equal to T_0 and $T_0(x)$, respectively), so Theorem 2 follows for Leibman polynomials of degree 0.

Assume now that the assertion in Theorem 2 is proved for all $T \in \mathscr{L}^\infty_{\mathbb{Z},\mathfrak{B}}$ having Leibman degree less than some positive integer d. Fix $T \in \mathscr{L}^\infty_{\mathbb{Z},\mathfrak{B}}$ with $\deg_{\mathrm{L}}(T) = d$.

LEMMA 3.7. *If* $x = (\mathrm{AV}_M T)^*(y)$ *for some* $M \in \mathcal{N} \setminus \mathbb{N}$ *and* $y \in \mathcal{H}$*, then* $\mathrm{AV}_\bullet T(x)$ *converges.*

PROOF. Note that $\mathrm{AV}_\bullet T(x)$ is bounded by $\|T\|\,\|x\|$. By Lemma 3.4, we have $\mathrm{AV}_n T(x) = \mathrm{AV}_n T \circ \mathrm{AV}_M T^*(y) = \langle\langle\!\langle \nabla^\bullet T(y), \sigma_n \rangle\!\rangle, \sigma_M \rangle$ for $n \in \mathbb{N}$. For $i \in \mathcal{Z}$ we have $\nabla^i T = {}_i\Delta^{[-i]} T$. By the invariance of Leibman degree under translation and the assumption $\deg_{\mathrm{L}}(T) \leq d$, we have $\deg_{\mathrm{L}}(\nabla^i T) = \deg_{\mathrm{L}}({}_i\Delta^{[-i]} T) < d$, and hence $\mathrm{AV}_\bullet(\nabla^i T)(y)$ is convergent for all $i \in \mathcal{Z}$ by the inductive hypothesis. An application of Lemma 3.3 concludes the proof. □

The space Struct of *structured elements* of $\mathcal{H}$ (relative to T) is the closure of the linear span of all elements of the form $(\mathrm{AV}_M T)^*(y)$ for $M \in \mathcal{N} \setminus \mathbb{N}$ and $y \in \mathcal{H}$. By linearity and Lemmas 3.6 & 3.7, $\mathrm{AV}_\bullet T(x)$ converges for all structured elements x. (The 1-Lipschitz condition follows from the inequalities $\|\mathrm{AV}_n T(x) - \mathrm{AV}_n T(y)\| \leq \|\mathrm{AV}_n T\| \|y - x\|$ and $\|\mathrm{AV}_n T\| = \|\langle T, \sigma_n \rangle\| \leq \|T\| \cdot \|\sigma_n\| = 1 \cdot 1 = 1$.)

The space Rand of *pseudorandom elements* of $\mathcal{H}$ (relative to T) is the orthogonal complement of Struct in $\mathcal{H}$. By the fundamental theorem of linear algebra and the definition of structured elements, an element $x \in \mathcal{H}$ is pseudorandom precisely when $\mathrm{AV}_M T(x) = 0$ for all $M \in \mathcal{N} \setminus \mathbb{N}$. By Lemma 3.5, $\mathrm{AV}_\bullet T(x)$ converges to zero in this case.

Combining the pseudorandom and structured cases using linearity, we deduce that all averages $\mathrm{AV}_\bullet T(x)$ converge. This concludes the inductive step of the proof of Theorem 2 in any saturated PET structure $\mathscr{M}$.

To conclude the proof for any PET structure over $\mathbb{Z}$, let $\mathscr{M}$ be any (not necessarily saturated) PET structure, and let $\widetilde{\mathscr{M}}$ be a saturated elementary $\mathcal{L}$-extension of $\mathscr{M}$ in Henson's logic. For fixed degree $d \in \mathbb{N}$ and $T \in (\mathscr{L}^\infty_{\mathbb{Z},\mathfrak{B}})^{\mathscr{M}}$, the property that T is a unitary Leibman sequence of degree at most d is $\mathcal{L}$-axiomatizable, hence it is true in $\widetilde{\mathscr{M}}$ when T is regarded as an element of $(\mathscr{L}^\infty_{\mathbb{Z},\mathfrak{B}})^{\widetilde{\mathscr{M}}}$. For $x \in \mathcal{H}^{\widetilde{\mathscr{M}}}$ we have proved that $\mathrm{AV}_\bullet T(x)$ converges since $\deg_{\mathrm{L}}(T) \le d$. *A fortiori*, $\mathrm{AV}_\bullet T(x)$ converges for $x \in \mathcal{H}^{\mathscr{M}}$. This concludes the proof of Theorem 2 in full generality.

3.4. Proof of Theorem 3. Let $\widetilde{\mathcal{L}}$ expand the language $\mathcal{L}$ of PET structures with new constants $(\mathtt{T}, \mathtt{x}, \mathtt{y}_n : n < \omega)$, with $\mathtt{T}$ of sort $\mathscr{L}^\infty_{\mathbb{Z},\mathfrak{B}}$, and $\mathtt{x}$ and all $\mathtt{y}_n$ of sort $\mathcal{H}$. For fixed $d \in \mathbb{N}$, consider the $\widetilde{\mathcal{L}}$-theory

$$\Lambda_d = \mathrm{Th}_{\mathbf{PET}} \cup \{\lambda_d(\mathtt{T}), \|\mathtt{x}\| \le 1, \mathtt{y}_n = \mathrm{AV}_n \mathtt{T}(\mathtt{x}) : n < \omega\},$$

where $\lambda_d(\mathtt{T})$ is formula (2.11) stating that the interpretation of $\mathtt{T}$ is Leibman of degree at most d. Note that Λ_d is a uniform theory: $\lambda_d(\mathtt{T})$ implies $\|\mathtt{T}\| \le 1$, hence also $\|\mathtt{y}_n\| \le 1$. Every model $\widetilde{\mathscr{M}}$ of Λ_d is an expansion of a PET structure $\mathscr{M}$ having the form $\widetilde{\mathscr{M}} = (\mathscr{M}, T, x, \mathrm{AV}_\bullet T(x))$. By Theorem 2, all sequences $(c_n^{\widetilde{\mathscr{M}}}) = \mathrm{AV}_\bullet T(x)$ are convergent. An application of Proposition A.10 finishes the proof of Theorem 3.

3.5. Proof of Theorem 1. A classical Leibman sequence $T_\bullet = (T_k : k \in \mathbb{Z})$ with $\deg_{\mathrm{L}}(T) \le d$ in an *abelian* subgroup K of the group $\mathrm{U}_{\mathcal{H}}$ of unitary operators on a Hilbert space $\mathcal{H}$ is easily shown to have the form $T_k = U_0 \circ U_1^k \circ U_2^{\binom{k}{2}} \circ \cdots \circ U_d^{\binom{k}{d}}$ where $U_j = \Delta^j T_0$ and $\binom{k}{j} = k(k-1)\dots(k-j+1)/j!$ are binomial coefficients for $j = 0, 1, \dots, d$. Since the functions $k \mapsto \binom{k}{j}$ $(j = 0, 1, \dots, d)$ are a $\mathbb{Z}$-basis for polynomial mappings $p : \mathbb{Z} \to \mathbb{Z}$ of degree at most d, Theorem 1 is an immediate corollary of Theorems 2 and 3.

4. A Mean Ergodic Theorem for unitary polynomial actions of abelian groups

To formulate our most general result on convergence of averages, we replace $\mathbb{N}$ with an arbitrary countable directed set $(\mathbb{D}, \le)$ and $\mathbb{Z}$ with an arbitrary abelian group $(\mathbb{G}, +)$ endowed with a countable Følner $\mathbb{D}$-net $\mathcal{F}_\bullet = (\mathcal{F}_j : j \in \mathbb{D})$ of nonempty finite subsets of $\mathbb{G}$. These assumptions are sufficient to ensure that the proofs of natural generalizations of Theorems 2 and 3 carry through in this more general context, *mutatis mutandis,* from those given in Section 3.

THEOREM 4 (Poly-MET: Mean Ergodic Theorem for unitary polynomial actions of an abelian group). *Fix an abelian group $(\mathbb{G}, +)$ and a Følner net $\mathcal{F}_\bullet = (\mathcal{F}_j : j \in \mathbb{D})$ of subsets of $\mathbb{G}$, indexed by a countable directed set $(\mathbb{D}, \le)$. Let $\mathcal{H}$ be a Hilbert space. Let $T : \mathbb{G} \to \mathrm{U}_{\mathcal{H}}$ be a polynomial mapping, in Leibman's sense, into the group $\mathrm{U}_{\mathcal{H}}$ of unitary transformations of $\mathcal{H}$. For every $x \in \mathcal{H}$, the $\mathbb{D}$-net $\mathrm{AV}_\bullet T(x) = (\mathrm{AV}_i T(x) : i \in \mathbb{D})$ in $\mathcal{H}$, of averages relative to $\mathcal{F}_\bullet$:*

$$\mathrm{AV}_i T(x) = \frac{1}{\#\mathcal{F}_i} \sum_{g \in \mathcal{F}_i} T_g(x)$$

of x under T, converges in the norm topology of $\mathcal{H}$.

In fact, given fixed choices of $\mathbb{G}$, $\mathbb{D}$, $\mathcal{F}_\bullet$ and $d \in \mathbb{N}$, there exists a rate of metastability

$$E_\bullet^{(\mathbb{G},\mathbb{D},\mathcal{F}_\bullet,d)} = \left(E_{\epsilon,\eta} : \epsilon > 0, \eta \in \prod_{i \in \mathbb{D}} \mathcal{P}^*_{\mathrm{fin}}(\mathbb{D}_{\geq i}) \right)$$

(with $E_{\epsilon,\eta} \in \mathcal{P}^*_{\mathrm{fin}}(\mathbb{D})$ for each ϵ, η) that applies universally to all sequences $\mathrm{AV}_\bullet T(x)$ for any element x in the unit ball of any Hilbert space $\mathcal{H}$ and any Leibman polynomial $T : \mathbb{D} \to \mathrm{U}_{\mathcal{H}}$ of degree at most d.

REMARK 4.1. The definition of Leibman polynomial mapping $T : \mathbb{G} \to \mathrm{U}_{\mathcal{H}}$ is a straightforward generalization of that of Leibman sequence $\mathbb{Z} \to \mathrm{U}_{\mathcal{H}}$ [**Lei02**]. The discrete difference $\Delta^g T$ of T with step $g \in \mathbb{G}$ is the mapping $h \mapsto T_{g+h} \circ T_h^*$. Define $\deg_{\mathrm{L}}(T) \leq 0$ if $\Delta^g T = \mathbb{I}$ (where $\mathbb{I}$ is the constant mapping $g \mapsto I$). Recursively, let $\deg_{\mathrm{L}}(T) \leq d+1$ mean that $\deg_{\mathrm{L}}(\Delta^g T) \leq d$ for all $g \in \mathbb{G}$. Then T is a Leibman mapping if $\deg_{\mathrm{L}}(T) \leq d$ for some d; the least such d is $\deg_{\mathrm{L}}(T)$, although we adopt the convention $\deg_{\mathrm{L}}(\mathbb{I}) = -\infty$.

We only provide an outline of the proof of Theorem 4 since it is formally identical to the arguments in Sections 3.2–3.4. The definition of classical PET structure over $\mathbb{G}$ is completely analogous to that of PET structure over $\mathbb{Z}$ in section 1.1—simply replace all instances of $\mathbb{Z}$ by $\mathbb{G}$ and those of $\mathbb{N}$ by $\mathbb{D}$. The Følner sequence $\mathcal{F}_\bullet$ is captured indirectly via the Følner measure map $\sigma : \mathbb{D} \to \mathfrak{M}$, where

$$\sigma_i = \frac{1}{\#\mathcal{F}_i} \sum_{g \in \mathcal{F}_i} \delta_g \qquad \text{for all } i \in \mathbb{D}.$$

The Henson language $\mathcal{L}$ for the class of classical PET structures over $\mathbb{G}$ is clear. Any model of the theory $\mathrm{Th}^{\mathbb{D},\mathbb{G}}_{\mathbf{PET}}$ of such classical structures is an (abstract) PET structure over $\mathbb{G}$. (Note that the language $\mathcal{L}$ depends on both $\mathbb{G}$ and $\mathbb{D}$; the theory $\mathrm{Th}^{\mathbb{D},\mathbb{G}}_{\mathbf{PET}}$ further depends on the choice of the Følner net $\mathcal{F}_\bullet$.)

Analogues of Theorems 2 and 3 hold in the class of all PET structures over $\mathbb{G}$. The scheme of proof is exactly the same. The countability hypothesis on $\mathbb{D}$ is an essential hypothesis in Theorem 5, which enters the proof via an analogue of Lemma 3.3.

Lemma 3.4 uses the exact same definition of reverse difference $(\nabla^g T)_h = T_h \circ T^*_{g+h}$. Its proof is adapted using the definition of Følner net as we now indicate: By definition, given $\epsilon > 0$ and $g \in \mathbb{G}$ there is $k = k_{g,\epsilon} \in \mathbb{D}$ such that the symmetric difference $\mathcal{F}_l \triangle (g + \mathcal{F}_l)$ has cardinality at most $\epsilon \cdot \#\mathcal{F}_l$ for all $l \geq k$. Thus, letting $K_{i,\epsilon} = \max\{k_{g,\epsilon} : g \in \mathcal{F}_i\}$, we have $\|\sigma_j - {}_g(\sigma_j)\| \leq \epsilon$ for all $g \in \mathcal{F}_i$ provided $j \geq K_{i,\epsilon}$. Whence follows the proof of an analog of Lemma 3.4 stating that $(\mathrm{AV}_j\, T) \circ (AV_K T)^* = \int \mathrm{AV}_j(\nabla^g T)\, d\sigma_K(g)$ holds whenever $j \in \mathbb{D}$ and $K \in \mathbb{D}^{\mathcal{M}}$ satisfies $K \geq i$ for all $i \in \mathbb{D}$.

Lemma 3.5 continues to hold provided one replaces the nonstandard natural numbers M, N with nonstandard elements J, K of $\mathbb{D}^{\mathcal{M}}$ that satisfy $J, K \geq i$ for all standard elements $i \in \mathbb{D}$.

The arguments in Sections 3.3 and 3.4 apply verbatim once the lemmas in Section 3.2 have been adapted, completing the proof of Theorem 4.

Appendix A. A Dominated Convergence Theorem for notions of integration in Banach spaces

This appendix bears a close connection to our prior manuscript on measure, integration and metastable convergence in Henson structures [**DnI17**]. Our main goal is proving Lemma 3.3. Rather than doing so in the specific context of PET structures, we prove a more general result (Theorem 5) about sequences of integrals of functions on a finite measure space taking values in a Banach space. This requires a number of preliminary steps.

A.1. Integration structures. We recall the class of *integration structures* (with underlying finite positive measure) introduced in our earlier manuscript, to which we refer the reader for details [**DnI17**]. These are saturated models of the Henson theory $\mathrm{Th}_\int$ of integration with respect to a positive finite measure on structures with (classical) sorts $\mathbb{R}$, Ω, $\mathcal{A}_\Omega$, $\mathscr{L}^\infty_{\Omega,\mathbb{R}}$ where $\mathbb{R}$ is the set of real numbers, $\mathcal{A}_\Omega$ is a σ-algebra of subsets of Ω, and $\mathscr{L}^\infty_{\Omega,\mathbb{R}}$ is the set of bounded $\mathcal{A}_\Omega$-measurable (everywhere-defined) real functions on Ω. (Here Ω, $\mathcal{A}_\Omega$ are discrete while $\mathbb{R}$, $\mathscr{L}^\infty_{\Omega,\mathbb{R}}$ are real Banach spaces.) This theory contains all Henson formulas involving the functions and distinguished constants below that are valid in such structures:

- *Constants:* Rational numbers $r \in \mathbb{R}$, zero vector in the Banach sort $\mathscr{L}^\infty_{\Omega,\mathbb{R}}$, an arbitrary point ("anchor") $\omega_0 \in \Omega$, the empty set $\emptyset \in \mathcal{A}_\Omega$, and the improper subset $\Omega \in \mathcal{A}_\Omega$.
- *Functions:*
  - Arithmetic operations (addition and multiplication), absolute value and lattice operations (binary min and max) on $\mathbb{R}$;
  - The characteristic function $[\![\cdot \in \cdot]\!] : \Omega \times \mathcal{A}_\Omega \to \{0,1\} \subseteq \mathbb{R}$ of the membership relation $\in$ on $\Omega \times \mathcal{A}_\Omega$;
  - Banach operations (addition, scalar multiplication) and norm on $\mathscr{L}^\infty_{\Omega,\mathbb{R}}$ (namely, $\|f\| = \sup_{x\in\Omega} |f(x)|$ for $f \in \mathscr{L}^\infty_{\Omega,\mathbb{R}}$—note that an almost-everywhere null function f has positive norm per this definition unless $f = 0$ everywhere);
  - The evaluation map $\mathscr{L}^\infty_{\Omega,\mathbb{R}} \times \Omega \to \mathbb{R} : (f,x) \mapsto f(x)$;
  - The Banach lattice operations (binary min and max) on $\mathscr{L}^\infty_{\Omega,\mathbb{R}}$;
  - The unary operation of pointwise absolute value $f \mapsto |f|$ on $\mathscr{L}^\infty_{\Omega,\mathbb{R}}$ where $|f| \in \mathscr{L}^\infty_{\Omega,\mathbb{R}}$ is the function $x \mapsto |f(x)|$;
  - The Boolean algebra operations of union, intersection and relative complement $S \mapsto S^{\mathsf{C}} = \Omega \setminus S$ on $\mathcal{A}_\Omega$;
  - The characteristic-function map $\chi : \mathcal{A}_\Omega \to \mathscr{L}^\infty_{\Omega,\mathbb{R}} : S \mapsto \chi_S$;
  - A positive finite measure μ on Ω;
  - The integration operator $I : \mathscr{L}^\infty_{\Omega,\mathbb{R}} \to \mathbb{R} : f \mapsto \int_\Omega f\,d\mu$.

Let $\mathcal{L}$ be any Henson language including sort symbols $\mathbb{R}, \Omega, \mathcal{A}_\Omega, \mathscr{L}^\infty_{\Omega,\mathbb{R}}$ as well as constant and function symbols matching the lists above, and let $\mathrm{Th}_\int$ be the $\mathcal{L}$-theory of such structures $\mathscr{M} = (\mathbf{S}, \mathbf{F}, \mathbf{C})$, where $\mathbf{S}$ is the list of sorts, $\mathbf{F}$ the collection of distinguished functions, and $\mathbf{C}$ the set of distinguished elements of $\mathscr{M}$. An *(abstract) pre-integration structure* is a model of $\mathrm{Th}_\int$. An *integration structure* is a saturated model of $\mathrm{Th}_\int$. If $\mathscr{M}$ is any pre-integration structure (whether saturated or not), then via interpretation of constants, the membership relation $[\![\cdot \in \cdot]\!]$, and the evaluation map, we may identify $\mathbb{R}^{\mathscr{M}}$ with $\mathbb{R}$, $\mathcal{A}^{\mathscr{M}}_\Omega$ with a Boolean algebra of subsets of $\Omega^{\mathcal{M}}$, and $\mathscr{L}^\infty_{\Omega,\mathbb{R}}$ with a set of functions $\Omega \to \mathbb{R}$. However, $\mathcal{A}_\Omega$ need *not* be a σ-algebra. Accordingly, $\mu^{\mathcal{M}}$ is typically just a *finitely* (not countably) additive measure on $(\Omega^{\mathcal{M}}, \mathcal{A}_\Omega^{\mathcal{M}})$, while elements $f \in \mathscr{L}^\infty_{\Omega,\mathbb{R}}$ are identified with uniformly bounded functions on Ω that may only be *approximately* $\mathcal{A}_\Omega$-measurable. [11] Nevertheless, in earlier work we have shown how the classical (i.e., σ-additive) theory of integration of bounded measurable functions over a finite measure space and the corresponding version of the Dominated Convergence Theorem are recovered essentially verbatim in *saturated* Henson integration structures via an analogous construction to that of Loeb measure in nonstandard analysis [**DnI17**].

A.2. Loeb structures.

DEFINITION A.1 (Loeb structure). Let $\mathrm{Th}_{\mathrm{Loeb}}$ be the reduct of the $\mathcal{L}$-theory $\mathrm{Th}_\int$ of integration structures with a positive measure to the language $\mathcal{L}'$ obtained by removing from $\mathcal{L}$ the symbol for sort $\mathscr{L}^\infty_{\Omega,\mathbb{R}}$ as well as all functions and constants involving $\mathscr{L}^\infty_{\Omega,\mathbb{R}}$ (such as the symbol I for the integral). A model of $\mathrm{Th}_{\mathrm{Loeb}}$ is a *pre-Loeb structure.* (Note that $\mathcal{M}$ may be an $\widetilde{\mathcal{L}}$-structure for a language $\widetilde{\mathcal{L}}$ properly extending the language $\mathcal{L}'$ of Loeb structures, and thus have other sorts, functions and constants prescribed by $\widetilde{\mathcal{L}}$ but not by $\mathcal{L}'$.)

A *Loeb structure* is a saturated pre-Loeb structure.

Note that, for the present discussion, we are requiring the measure $\mu^{\mathcal{M}}$ in a pre-Loeb structure $\mathcal{M}$ to be *positive.*

If $\mathcal{M}$ is any pre-Loeb structure, the *set underlying* a given $A \in \mathcal{A}_\Omega^{\mathcal{M}}$ is

$$[A] = \{x \in \Omega^{\mathcal{M}} : [\![x \in A]\!] = 1\}.$$

We may (externally) identify A with $[A]$ since

$$(\forall A)(\forall B)(A = B \leftrightarrow (\forall x)([\![x \in A]\!] = [\![x \in B]\!]))$$

is a sentence in $\mathrm{Th}_{\mathrm{Loeb}}$. [12]

DEFINITION A.2 (Loeb measure and Loeb-measurable sets). Let $\mathcal{M}$ be a pre-Loeb structure with positive measure $\mu = \mu^{\mathcal{M}}$.

A set $S \subseteq \Omega^{\mathcal{M}}$ is *$\mathcal{A}_\Omega$-measurable* (or just *measurable*) if $S = [A]$ for some $A \in \mathcal{A}_\Omega^{\mathcal{M}}$ (i.e., if "$S \in \mathcal{A}_\Omega^{\mathcal{M}}$"—modulo the identification of $S = [A]$ with A itself).

A set $S \subseteq \Omega^{\mathcal{M}}$ is *μ-measurable* (or *Loeb-measurable (modulo μ)*) if for every $\epsilon > 0$ there exist measurable $A, B \in \mathcal{A}_\Omega^{\mathcal{M}}$ such that $[A] \subseteq S \subseteq [B]$ and $\mu(B - A) \leq \epsilon$.

The *Loeb measure* of a Loeb-measurable set S is

$$\mu_{\mathrm{L}}(S) = \sup\{\mu(A) : A \in \mathcal{A}_\Omega^{\mathcal{M}}, [A] \subseteq S\} = \inf\{\mu(B) : B \in \mathcal{A}_\Omega^{\mathcal{M}}, [B] \supseteq S\}.$$

The *Loeb algebra* of $\mathcal{M}$ is the collection $[\![\mathcal{A}]\!]_\mu$ of all Loeb-measurable subsets of $\Omega^{\mathcal{M}}$.

---

[11] A function f on Ω is *approximately $\mathcal{A}_\Omega$-measurable* if for all rational $r < s$ there exists $A \in \mathcal{A}_\Omega$ such that $f^{-1}((-\infty, r)) \subseteq A \subseteq f^{-1}((-\infty, s])$—a property axiomatizable by countably many Henson formulas in the logic of approximate satisfaction ([**DnI17**], Proposition 4.4).

[12] Although Henson's languages have no conditional connective "$\to$", when P is a discrete predicate (i.e., a term taking only the values $0, 1$), a non-Henson formula such as $P(x) \to \varphi(x)$ can be semantically identified with the Henson formula $(P(x) \leq 1/2) \vee \varphi(x)$. (By contrast, the converse $\varphi(x) \to P(x)$ is not semantically equivalent to a Henson formula in general.) When both P, Q are discrete, a biconditional $P(x) \leftrightarrow Q(x)$ can similarly be rewritten as a Henson formula. The assertion "*R is discrete*" is captured by the Henson formula $(\forall x)(R(x) = 0 \vee R(x) = 1)$, where "$R(x) = r$" is itself an abbreviation for "$(R(x) \leq r) \wedge (R(x) \geq r)$".

Note that $[\![\mathcal{A}]\!]_\mu$ is an *external* collection of subsets of $\Omega^{\mathscr{M}}$. It depends on $\mathscr{M}$ and has no intrinsic definition otherwise. It is easy to check that $[\![\mathcal{A}]\!]_\mu$ is an algebra of sets (i.e., closed under finite unions and intersections as well as complements). In fact, as soon as $\mathscr{M}$ is at least ω-saturated (i.e., types over a countable set of parameters are realized), $[\![\mathcal{A}]\!]_\mu$ is a σ-algebra that is complete for μ_{L} in the sense that any subset of a μ_{L}-null set is itself μ_{L}-null ([**DnI17**], Proposition 3.4). On the other hand, no degree of saturation ensures that $[\![\mathcal{A}]\!]_\mu$ is closed under unions of subfamilies of size ω_1 or more.

A.3. Integration frameworks. We need to introduce the notion of *(real) integration framework*, which generalizes integration structures as presented in section A.1. Roughly speaking, an integration framework is a saturated model of the theory of the operations of integration with respect to arbitrary finite (positive or signed) measures on a measure space.

Consider the reduct $\widetilde{\mathscr{M}}$ of a *classical* pre-integration structure $\mathscr{M}$, obtained by removing from $\mathscr{M}$ the distinguished measure μ and all the functions involving μ (including the integration operator I). Now expand $\widetilde{\mathscr{M}}$ to a structure $\mathscr{M}'$ with a new Banach sort $\mathfrak{M}_\Omega$ containing all finite (signed, real-valued) measures μ on Ω plus the following distinguished functions and constants:

- *Constants:* The zero measure $0 \in \mathfrak{M}_\Omega$;
- *Functions:*
  - Vector space operations of addition and scalar multiplication on $\mathfrak{M}_\Omega$.
  - Banach norm of total variation on $\mathfrak{M}_\Omega$: $\|\mu\| = \sup\{A \in \mathcal{A}_\Omega : |\mu(A)| + |\mu(A^{\mathsf{C}})|\}$;
  - The inclusion maps:
    - $\Omega \hookrightarrow \mathcal{A}_\Omega : x \mapsto \{x\}$,
    - $\Omega \hookrightarrow \mathfrak{M}_\Omega : x \mapsto \delta_x$ (the unit point mass at x);
  - The evaluation map $\mathfrak{M}_\Omega \times \mathcal{A}_\Omega \to \mathbb{R} : (\mu, A) \mapsto \mu(A)$ (which is 1-Lipschitz by definition of the norm $\|\cdot\|$ on $\mathfrak{M}_\Omega$);
  - The *total variation* map $\mathscr{L}^\infty_{\Omega,\mathbb{R}} \to \mathscr{L}^\infty_{\Omega,\mathbb{R}}$: $\mu \mapsto |\mu|$ where $|\mu|$ is the (positive) measure of total variation of μ: $|\mu|(A) = \sup\{|\mu(A \cap B)| + |\mu(A \cap B^{\mathsf{C}})| : B \in \mathcal{A}_\Omega\}$;
  - The integration operator $\langle\cdot,\cdot\rangle : \mathscr{L}^\infty_{\Omega,\mathbb{R}} \times \mathfrak{M}_\Omega \to \mathbb{R} : f \mapsto \langle f, \mu\rangle = \int_\Omega f\, d\mu$.

Given a language $\mathcal{L}$ for pre-integration structures, let $\mathcal{L}'$ be obtained from $\mathcal{L}$ by removing the symbols $\boldsymbol{\mu}, I$ for the distinguished measure and integral operator, and adding a new sort symbol $\mathfrak{M}_\Omega$ as well as new constant and function symbols per the list above.

Definition A.3 (Real integration framework). A *classical (real) pre-integration framework* is any $\mathcal{L}'$-structure $\mathscr{M}'$ as described above. An *(abstract) real pre-integration framework* is any model of the Henson theory $\mathrm{Th}_{\int:\mathbb{R}}$ of classical real pre-integration frameworks. [13]

More generally, any structure $\mathscr{M}$ in a language expanding $\mathcal{L}'$ such that the $\mathcal{L}'$-reduct of $\mathscr{M}$ is a pre-integration framework in the above sense will be called a pre-integration framework.

A *(real) integration framework* is a saturated real pre-integration framework.

[13] It is straightforward to verify that $\mathrm{Th}_{\int:\mathbb{R}}$ is a uniform theory.

A.4. Banach integration frameworks. There is no completely general notion of integration of functions taking values in an arbitrary Banach space $\mathfrak{B}$—not even for bounded functions $F : \Omega \to \mathfrak{B}$ on a finite measure space $(\Omega, \mathcal{A}_\Omega)$. However, it is very natural to require that any such notion of Banach integration should build upon the classical integral of real-valued functions. Our viewpoint is that any reasonable notion of Banach integration must expand a (pre-)integration framework to a *Banach (pre-)integration framework* per Definition A.4 below.

Consider expansions $\mathscr{M}' = (\mathbf{S}', \mathbf{F}', \mathbf{C}')$ of real integration frameworks $\mathscr{M} = (\mathbf{S}, \mathbf{F}, \mathbf{C})$ where $\mathbf{S}' \supset \mathbf{S}$ contains two new sorts $\mathfrak{B}$ and $\mathscr{L}^\infty_{\Omega,\mathfrak{B}}$, while $\mathbf{F}' \supset \mathbf{F}$ and $\mathbf{C}' \supset \mathbf{C}$ contain new functions and symbols as follows:

- Addition, scalar product and norm on $\mathfrak{B}$ making it a real Banach space.
- Addition, scalar product and norm on $\mathscr{L}^\infty_{\Omega,\mathfrak{B}}$ making it a Banach space.
- The zero elements of $\mathfrak{B}$ and $\mathscr{L}^\infty_{\Omega,\mathfrak{B}}$.
- An evaluation map $\mathscr{L}^\infty_{\Omega,\mathfrak{B}} \times \Omega \to \mathfrak{B} : (F, x) \mapsto F(x)$ such that $\|F\| = \sup\{\|F(x)\| : x \in \Omega\}$. (Thus, elements of $\mathscr{L}^\infty_{\Omega,\mathfrak{B}}$ may be identified with functions $\Omega \to \mathfrak{B}$.)
- The operation of multiplication $\mathscr{L}^\infty_{\Omega,\mathbb{R}} \times \mathscr{L}^\infty_{\Omega,\mathfrak{B}} \to \mathscr{L}^\infty_{\Omega,\mathfrak{B}} : (f, F) \mapsto fF$ (such that $(fF)(x) = f(x)F(x)$ for all $x \in \Omega$) under which $\mathscr{L}^\infty_{\Omega,\mathfrak{B}}$ is an $\mathscr{L}^\infty_{\Omega,\mathbb{R}}$-module.
- The inclusion map $\mathfrak{B} \to \mathscr{L}^\infty_{\Omega,\mathfrak{B}} : T \mapsto T(\blacksquare)$ where $T(\blacksquare) \in \mathscr{L}^\infty_{\Omega,\mathfrak{B}}$ is identified via evaluation with the constant function $\Omega \to \mathfrak{B} : x \mapsto T$.
- The pointwise-norm map $|\cdot| : \mathscr{L}^\infty_{\Omega,\mathfrak{B}} \to \mathscr{L}^\infty_{\Omega,\mathbb{R}}$ satisfying $|F|(x) = \|F(x)\|$ for all $F \in \mathscr{L}^\infty_{\Omega,\mathfrak{B}}$, $x \in \Omega$.
- An operation of Banach integration, namely a pairing $\langle\!\langle \cdot, \cdot \rangle\!\rangle : \mathscr{L}^\infty_{\Omega,\mathfrak{B}} \times \mathfrak{M}_\Omega \to \mathfrak{B}$ satisfying the following properties:
  (1) $\langle\!\langle \cdot, \cdot \rangle\!\rangle$ is bilinear;
  (2) $\langle\!\langle \cdot, \cdot \rangle\!\rangle$ is compatible with the integration $\langle \cdot, \cdot \rangle$ of real functions:
    (a) For all $T \in \mathfrak{B}$ and $f \in \mathscr{L}^\infty_{\Omega,\mathbb{R}}$: $\langle\!\langle fT, \mu \rangle\!\rangle = \langle f, \mu \rangle T$.
    (b) For all $F \in \mathscr{L}^\infty_{\Omega,\mathfrak{B}}$: $\|\langle\!\langle F, \mu \rangle\!\rangle\| \leq |\langle |F|, |\mu| \rangle|$.

DEFINITION A.4 (Banach integration framework). A language $\mathcal{L}'$ expanding the language $\mathcal{L}$ of real pre-integration frameworks with the new sort symbols plus symbols for the functions and constants above is called a *language for Banach integration frameworks.*

Let $\mathrm{Th}_{\int:\mathbb{R}}$ be the Henson $\mathcal{L}$-theory of real pre-integration frameworks, and let $\mathrm{Th}_{\int:\mathfrak{B}}$ extend $\mathrm{Th}_{\int:\mathbb{R}}$ with further Henson $\mathcal{L}'$-axioms capturing the properties of new sorts, functions and constants stated above (in semantically equivalent terms, let $\mathrm{Th}_{\int:\mathfrak{B}}$ be the Henson $\mathcal{L}'$-theory of those expansions $\mathscr{M}'$ of real pre-integration frameworks having the properties above). [14]

A *Banach pre-integration framework* is a model of $\mathrm{Th}_{\int:\mathfrak{B}}$. More generally, if $\widetilde{\mathcal{L}}$ is a language extending $\mathcal{L}'$ and $\mathscr{M}$ is an $\widetilde{\mathcal{L}}$-structure whose reduct $\mathscr{M} \restriction \mathcal{L}'$ is a model of $\mathrm{Th}_{\int:\mathfrak{B}}$, we shall still call $\mathscr{M}$ a Banach pre-integration framework.

A *Banach integration framework* is a saturated model of $\mathrm{Th}_{\int:\mathfrak{B}}$.

REMARK A.5. The question whether a real pre-integration framework $\mathscr{M}$ admits an expansion to a Banach pre-integration framework $\mathscr{M}'$ is very delicate. In general, the answer may be negative. However, when $\Omega^{\mathscr{M}}$ is a finite set the answer

[14]The verification that $\mathrm{Th}_{\int:\mathfrak{B}}$ is a uniform theory is routine.

is affirmative: It suffices to let $(\mathscr{L}^{\infty}_{\Omega,\mathfrak{B}})^{\mathscr{M}'}$ be the set of all functions $F : \Omega^{\mathscr{M}} \to \mathfrak{B}^{\mathscr{M}}$, and also let $\langle\!\langle F, \mu\rangle\!\rangle = \sum_{x\in\Omega^{\mathscr{M}}} F(x)\mu(\{x\})$. The remaining ingredients of the expansion are defined in the obvious manner. Similarly, an expansion $\mathscr{M}'$ also exists if the Banach sort $\mathfrak{B}^{\mathscr{M}}$ has finite dimension (using a basis for $\mathfrak{B}^{\mathscr{M}}$, real-valued integration extends to $\mathfrak{B}^{\mathscr{M}}$-valued integration in the straightforward classical fashion).

A.5. A Dominated Convergence Theorem for nets of functions in Banach integration frameworks. In order to formulate a version of the Dominated Convergence Theorem 5 for integration frameworks below, we fix a directed set $(\mathbb{D}, \preceq)$ so we can eventually discuss convergence of nets on it. [15] Classical sequences indexed by the directed set $(\mathbb{N}, \leq)$ of natural numbers are of particular interest. Only *infinite* directed sets are useful as tools to define and study notions of convergence in analysis and topology; on the other hand, critical results such as Theorem 5 depend on the *countability* of the directed set, so we may as well fix an infinite countable directed set $(\mathbb{D}, \preceq)$ for the remainder of the manuscript (this hypothesis will be made explicit whenever needed).

DEFINITION A.6. Fix a directed set $(\mathbb{D} \preceq)$. For $i \in \mathbb{D}$, the *final segment of* $\mathbb{D}$ *starting at* i is $\mathbb{D}_{\succeq i} = \{j \in \mathbb{D} : j \succeq i\}$ (i.e., the set of elements equal to or greater than i in $\mathbb{D}$). A $\mathbb{D}$-net $a_\bullet$ in a metric space (X, d) is any function $\mathbb{D} \to X : i \mapsto a_i$. The *spread of* $a_\bullet$ *from* i is

$$\mathrm{spr}_{\succeq i}(a_\bullet) = \sup_{j,k\succeq i} \mathrm{d}(a_j, a_k).$$

The *oscillation* of $a_\bullet$ is

$$\mathrm{osc}(a_\bullet) = \inf_{i\in\mathbb{D}} \mathrm{spr}_{\succeq i}(a_\bullet).$$

The net $a_\bullet$ *converges* if $\mathrm{osc}(a_\bullet) = 0$.

THEOREM 5 (Dominated Convergence Theorem in Banach integration frameworks). *Fix a* countable *directed set* $\mathbb{D}$. *Let* $\mathscr{M}$ *be any (saturated) Banach integration framework. Let* $\varphi_\bullet$ *be a bounded* $\mathbb{D}$*-net in* $(\mathscr{L}^{\infty}_{\Omega,\mathfrak{B}})^{\mathscr{M}}$. *For every* $x \in \Omega^{\mathscr{M}}$ *and* $\mu \in \mathfrak{M}^{\mathscr{M}}_{\Omega}$, *let* $\varphi_\bullet(x)$ *denote the net* $(\varphi_j(x) : j \in \mathbb{D})$ *and* $\langle\!\langle\varphi_\bullet, \mu\rangle\!\rangle$ *the net* $(\langle\!\langle\varphi_j, \mu\rangle\!\rangle : j \in \mathbb{D})$ *in* $\mathfrak{B}^{\mathscr{M}}$. *Then we have*

$$\mathrm{osc}(\langle\!\langle\varphi_\bullet, \mu\rangle\!\rangle) \leq \|\mu\| \sup_{x\in\Omega^{\mathscr{M}}} \mathrm{osc}(\varphi_\bullet(x)).$$

In particular, if the net $\varphi_\bullet(x)$ *is convergent for all* $x \in \Omega^{\mathscr{M}}$, *then* $\langle\!\langle\varphi_\bullet, \mu\rangle\!\rangle$ *is convergent.*

The proof of Theorem 5 below is an adaptation of our earlier one for real-valued notions of integration ([**DnI17**], Proposition 5.3).

Recall that a collection $\mathcal{F}$ of subsets of a set U is a *(proper) filter on* S if (*i*) $\emptyset \notin \mathcal{F}$, (*ii*) $\mathcal{F}$ is closed under finite intersections, and (*iii*) $\mathcal{F}$ is upward closed: if $A \in \mathcal{F}$ and $A \subseteq B \subseteq U$, then $B \in \mathcal{F}$. A proper filter $\mathcal{F}$ is an *ultrafilter* if $A \in \mathcal{F}$ or $U \setminus A \in \mathcal{F}$ for all $A \subseteq U$.

For an introduction to ultrafilters, ultralimits and ultraproduct constructions in model theory, the reader is referred to Bell and Slomson's monograph [**BS06**].

DEFINITION A.7. If X is any nonempty set, let $\mathcal{P}^*_{\mathrm{fin}}(X)$ be the family of finite nonempty subsets of X. We call a filter $\mathcal{F}$ on $\mathcal{P}^*_{\mathrm{fin}}(X)$ *greedy* if it contains all the sets $X_{\supseteq S} = \{T \in \mathcal{P}^*_{\mathrm{fin}}(X) : T \supseteq S\}$ for all $S \in \mathcal{P}^*_{\mathrm{fin}}(X)$.

[15] I.e., $\preceq$ is a nonstrict partial order on $\mathbb{D}$ such that any two $i, j \in \mathbb{D}$ have an upper bound k.

Note that the collection $\{X_{\supseteq S} : S \in \mathcal{P}^*_{\text{fin}}(X)\}$ is a filter base on $\mathcal{P}^*_{\text{fin}}(S)$ since $X_{\supseteq S} \cap X_{\supseteq T} = X_{\supseteq S \cup T}$. (This means that the collection of subsets of $\mathcal{P}^*_{\text{fin}}(X)$ that are supersets of $X_{\supseteq S}$ for some $S \in \mathcal{P}^*_{\text{fin}}(X)$ is a filter on $\mathcal{P}^*_{\text{fin}}(X)$.) [16] By a routine application of the axiom of choice, greedy ultrafilters on $\mathcal{P}^*_{\text{fin}}(X)$ exist whenever X is nonempty. Observe that the principal ultrafilter generated by a fixed $S \in \mathcal{P}^*_{\text{fin}}(X)$ is greedy precisely when $S = X$; thus, if X is infinite, greedy ultrafilters on $\mathcal{P}^*_{\text{fin}}(X)$ are nonprincipal.

LEMMA A.8. *Let $f : \Omega^{\mathcal{M}} \to \mathbb{R}$ be bounded and (externally) μ_{L}-measurable. Then there exists $\widetilde{f} \in (\mathscr{L}^\infty_{\Omega,\mathbb{R}})^{\mathcal{M}}$ such that $f(x) = \widetilde{f}(x)$ for μ_{L}-almost all $x \in \Omega^{\mathcal{M}}$ and $\inf_x f(x) \le \widetilde{f} \le \sup_x f(x)$.*

PROOF. Let f be a μ_{L}-measurable bounded external function on $\Omega^{\mathcal{M}}$, and let $a = \inf_x f$, $b = \sup_x f$. The assertion is trivial if $a = b$ or if $\|\mu\| = 0$—just take a constant $\widetilde{f}$ in $[a, b]$. Otherwise, we have $a < b$ and, replacing μ with $|\mu|$, we may assume μ to be a positive measure without loss of generality. By definition of Loeb measurability, for rational $r \in [a, b]$ and integer $n \ge 1$ there exist $A^r_n, B^r_n \in \mathcal{A}^{\mathcal{M}}_\Omega$ such that $[A^r_n] \subseteq \{f \le r\}$, $[B^r_n] \subseteq \{f \ge r\}$, and $\mu_{\mathrm{L}}\{f \le r\} - \mu(A^r_n) \le 1/n$, $\mu_{\mathrm{L}}\{f \ge r\} - \mu(B^r_n) \le 1/n$. For fixed r, the sequences (A^r_n), (B^r_n) may be constructed recursively to ensure $A^r_m \subseteq A^r_n$ and $B^r_m \subseteq B^r_n$ for $m \le n$. We may also assume $A^r_m = 1_{\mathcal{A}_\Omega}$ if $r \ge b$, and $B^s_m = 1_{\mathcal{A}_\Omega}$ if $s \le a$.

Let $f^r_n = a \cdot (1 - \chi_{B^r_n}) + r \cdot \chi_{B^r_n}$ and $g^r_n = b \cdot (1 - \chi_{A^r_n}) + r \cdot \chi_{A^r_n}$. Let Q be the set of rational numbers in $[a, b]$. The construction of (A^r_n) and (B^r_n) implies that $f^r_m \le f^r_n \le f \le g^r_n \le g^r_m$ for all $r \in Q$ and $m \le n$. For $I \in \mathcal{P}^*_{\text{fin}}(Q)$ of cardinality n, let $f^I = \max\{f^r_n : r \in I\}$, $g^I = \min\{g^r_n : r \in I\}$. Observe that $f^r_n \le f^I \le f^J \le f \le g^J \le g^I \le g^r_n$ if $I \subseteq J$, $r \in I$ and $\operatorname{card}(I) \ge n$. Since $a < b$ by assumption, Q is infinite countable. Let $\mathcal{U}$ be a greedy ultrafilter on $\mathcal{P}^*_{\text{fin}}(Q)$. By saturation, there are $\widetilde{f}, \widetilde{g} \in (\mathscr{L}^\infty_{\Omega,\mathbb{R}})^{\mathcal{M}}$ realizing the $\mathcal{U}$-ultralimit of the types $\operatorname{tp}_S(f^I, g^I)$ over the set of parameters $S = \{\mu, a, b\} \cup \{A^r_n, B^r_n\}_{r \in Q, n \in \mathbb{N}^*}$. From the construction of $\mathcal{U}$ as a greedy ultrafilter, the definition of ultralimit, and the meaning of realization of a type, it is easy to verify that

$$a \le f^r_n \le \widetilde{f} \le \widetilde{g} \le g^r_n \le b \quad \text{for all } r \in Q \text{ and } n \ge 1.$$

It follows that for fixed $r \in Q$ we have $S^r := \bigcup_n [A^r_n] = \bigcup_n \{g^r_n \le r\} \subseteq \{\widetilde{g} \le r\}$. On the other hand, by construction of A^r_n we have $S^r \subseteq \{f \le r\}$ and $\mu_{\mathrm{L}}(S^r) = \sup_n \mu(A^r_n) = \mu_{\mathrm{L}}\{f \le r\}$. Thus, $S^r \subseteq \{f \le r\} \cap \{\widetilde{g} \le r\}$ and $\mu_{\mathrm{L}}(S^r) = \mu_{\mathrm{L}}\{f \le r\}$; hence, $\{f \le r\}$ is μ_{L}-almost included in $\{\widetilde{g} \le r\}$. By a completely analogous argument, $\{\widetilde{f} \ge r\}$ μ_{L}-almost includes $\{f \ge r\}$. These almost-inclusions for every (rational) $r \in Q$ are easily shown to imply the μ_{L}-a.e. inequalities $\widetilde{g} \le f \le \widetilde{f}$. However, $\widetilde{f} \le \widetilde{g}$, so in fact $\widetilde{f} = f = \widetilde{g}$ (μ_{L}-a.e.) □

LEMMA A.9. *Fix $\mu \in \mathfrak{M}^{\mathcal{M}}_\Omega$ and let $a_\bullet = (a_i : i < \omega)$ be a sequence of external μ_{L}-measurable functions $\Omega^{\mathcal{M}} \to \mathbb{R}$. Then there exist $\sigma, \iota \in (\mathscr{L}^\infty_{\Omega,\mathbb{R}})^{\mathcal{M}}$ such that $\sigma(x) = \sup_{i<\omega} a_i(x)$ and $\iota(x) = \inf_{i<\omega} a_i(x)$ for μ_{L}-almost all $x \in \Omega^{\mathcal{M}}$.*

If the (external) sequence $a_\bullet$ consists of internal functions, i.e., it is a sequence in $(\mathscr{L}^\infty_{\Omega,\mathbb{R}})^{\mathcal{M}}$, then σ, ι may be chosen so $\sigma \ge \sup_{i<\omega} a_i$ and $\iota \le \inf_{i<\omega} a_i$.

[16] Recall that a *filter base* on a set U is a collection $\mathcal{E}$ of subsets of U such that $\emptyset \notin \mathcal{E}$ and $\mathcal{E}$ is downward directed by inclusion in the sense that if $A, B \in \mathcal{E}$ then $A \cap B \supseteq C$ for some $C \in \mathcal{E}$. The *filter $\mathcal{F}$ with base $\mathcal{E}$* is the collection of all subsets of U that are supersets of some $A \in \mathcal{E}$.

PROOF. It is routine to show that $f = \sup_{i<\omega} a_i$ and $g = \inf_{i<\omega} a_i$ are μ_L-measurable, so the first assertion follows from Lemma A.8.

When $a_\bullet$ is a sequence of internal functions, let $\mathcal{U}$ be any nonprincipal ultrafilter on ω and let σ realize the $\mathcal{U}$-ultralimit of the types $\mathrm{tp}_S(a^k)$ over the set of parameters $S = \{\mu\} \cup \{a_i\}_{i<\omega}$, where $a^k = \max\{a_i : i \leq k\}$. (Recall that $\mathscr{L}^\infty_{\Omega,\mathbb{R}}$ is endowed with the binary lattice operation $\max\{a, b\}$, which trivially defines n-ary maximum operations for all $n \geq 1$.) The verification that σ has the required properties is routine. The construction of ι is identical upon replacing "max" by "min". □

PROOF OF THEOREM 5. The asserted inequality evidently holds if $\mu = 0$. Otherwise, using a Jordan decomposition $\mu = \mu_+ - \mu_-$ where $\mu_+ = (|\mu| + \mu)/2$ and $\mu_- = (|\mu| - \mu)/2$ are positive, the proof is easily reduced to the case in which μ is a probability measure, which we assume henceforth.

Choose C such that $\|\varphi_i\| \leq C$ for all i. For $j, k \in \mathbb{D}$ let $\varphi^{j,k} = |\varphi_k - \varphi_j| \in (\mathscr{L}^\infty_{\Omega,\mathbb{R}})^{\mathcal{M}}$. Since $\mathbb{D}$ is countable, Lemma A.9 implies that for each $i \in \mathbb{D}$ there is $\sigma^i \in (\mathscr{L}^\infty_{\Omega,\mathbb{R}})^{\mathcal{M}}$ with $\|\sigma^i\| \leq 2C$ such that σ^i is μ_L-a.e. equal to $\mathrm{spr}_{\succeq i} \varphi_\bullet = \sup_{j,k\succeq i} \varphi^{j,k}$. Similarly, $\mathrm{osc}\, \varphi_\bullet = \inf_i \mathrm{spr}_{\succeq i} \varphi_\bullet$ is μ_L-a.e. equal to $\inf_i \sigma^i$, hence to some $\omega \in (\mathscr{L}^\infty_{\Omega,\mathbb{R}})^{\mathcal{M}}$ with $\|\omega\| \leq 2C$.

Let $s = \sup_x \mathrm{osc}(\varphi_\bullet(x))$ and fix $t > s$. Since $\{x\colon \mathrm{osc}\, \varphi_\bullet(x) \geq t\}$ is empty (by choice of s and t) and $\omega(x) = \mathrm{osc}(\varphi_\bullet(x))$ for μ_L-a.e. x, the set $\{x\colon \omega(x) \geq t\}$ is μ_L-null. Since $\omega = \inf_i \mathrm{spr}_{\succeq i} \varphi_\bullet$ (μ_L-a.e.), we have $\inf_i \mu_L\{x\colon \mathrm{spr}_{\succeq i} \varphi_\bullet(x) \geq t\} = 0$. Thus, for arbitrary fixed $\epsilon > 0$ we have $\mu_L\{x\colon \mathrm{spr}_{\succeq i} \varphi_\bullet(x) \geq t\} \leq \epsilon$ for some $i = i_\epsilon \in \mathbb{D}$. (This depends crucially on the hypothesis that $\mathbb{D}$ is countable.) It follows that for $j, k \succeq i$:

$$\begin{aligned}
|\langle\!\langle \varphi_k - \varphi_j, \mu \rangle\!\rangle| \leq \langle\!\langle |\varphi_k - \varphi_j|, \mu \rangle\!\rangle &= \int |\varphi_k(x) - \varphi_j(x)|\, d\mu_L(x) \\
&= \left(\int_{\{x\colon \mathrm{spr}_{\succeq i} \varphi_\bullet(x) < t\}} + \int_{\{x\colon \mathrm{spr}_{\succeq i} \varphi_\bullet(x) \geq t\}} \right) |\varphi_k(x) - \varphi_j(x)|\, d\mu_L(x) \\
&\leq t \cdot \mu_L\{x\colon \mathrm{spr}_{\succeq i} \varphi_\bullet(x) < t\} + 2C\epsilon \leq t + 2C\epsilon.
\end{aligned}$$

This proves that $\mathrm{spr}_{\succeq i} \langle\!\langle \varphi_\bullet, \mu \rangle\!\rangle \leq t + 2C\epsilon$. As $t > s$ and $\epsilon > 0$ are arbitrary, $\mathrm{osc} \langle\!\langle \varphi_\bullet, \mu \rangle\!\rangle \leq s$. □

A.6. A Uniform Metastability Principle for nets in Henson structures.

PROPOSITION A.10 (Uniform Metastability Principle (UMP)). *Fix a directed set* $(\mathbb{D}, \preceq)$. *Fix a Henson language* $\mathfrak{L}$ *including constants* $(a_j : j \in \mathbb{D})$ *all of a common sort* $\mathbb{S}$. *Let* $\mathfrak{T}$ *be a uniform* $\mathfrak{L}$*-theory such that for every model* $\mathcal{M}$ *of* $\mathfrak{T}$ *the net* $a^{\mathcal{M}}_\bullet = (a^{\mathcal{M}}_j : j \in \mathbb{D})$ *is convergent. Then there exists a metastability rate* $E_\bullet = E^{\mathfrak{T}}_\bullet$ *depending only on* $\mathfrak{T}$ *that applies uniformly to all sequences* $a^{\mathcal{M}}_\bullet$ *in all models* $\mathcal{M}$ *of* $\mathfrak{T}$.

PROOF. ([**DnI17**], Proposition 2.4.) Assume no such rate of metastability exists. Then there exist $\epsilon > 0$ and a sampling $\eta \in \prod_{i\in\mathbb{D}} \mathcal{P}^*_{\mathrm{fin}}(\mathbb{D}_{\succeq i})$ such that for every $S \in \mathcal{P}^*_{\mathrm{fin}}(\mathbb{D})$ there is a model $\mathcal{M} = \mathcal{M}^S_{\epsilon,\eta}$ of $\mathfrak{T}$ such that $a_\bullet = a^{\mathcal{M}}_\bullet$ satisfies $\epsilon \leq \mathrm{spr}_{\eta_i}(a_\bullet) = \max\{\mathrm{d}(a_j, a_k) : j, k \in \eta_i\}$ for all $i \in S$. By the compactness theorem for Henson logic, there is a model $\mathcal{M}$ of $\mathfrak{T}$ such that $a_\bullet = a^{\mathcal{M}}_\bullet$ satisfies

$\operatorname{spr}_{\eta_i}(a_\bullet) \geq \epsilon$ for all $i \in \mathbb{D}$, and hence $\operatorname{osc}(a_\bullet) \geq \epsilon$, contradicting the hypothesis that $a_\bullet^{\mathscr{M}}$ converges. □

Acknowledgments

We thank Xavier Caicedo, Christopher Eagle and Franklin Tall for their encouragement and feedback, as well as the Banff International Research Station for hosting the June 2016 FRG "Topological Methods in Model Theory" where many ideas in the Appendix to this manuscript were first conceived.

References

[Aus15a] Tim Austin, *Pleasant extensions retaining algebraic structure, I*, J. Anal. Math. **125** (2015), 1–36, DOI 10.1007/s11854-015-0001-9. MR3317896

[Aus15b] Tim Austin, *Pleasant extensions retaining algebraic structure, II*, J. Anal. Math. **126** (2015), 1–111, DOI 10.1007/s11854-015-0013-5. MR3358029

[Aus16] Tim Austin, *Non-conventional ergodic averages for several commuting actions of an amenable group*, J. Anal. Math. **130** (2016), 243–274, DOI 10.1007/s11854-016-0036-6. MR3574655

[Ber87] V. Bergelson, *Weakly mixing PET*, Ergodic Theory Dynam. Systems **7** (1987), no. 3, 337–349, DOI 10.1017/S0143385700004090. MR912373

[Bir31] G. D. Birkhoff, *Proof of the ergodic theorem*, Proc. Nat. Acad. Sci. U. S. A. **17** (1931), 656–660.

[BL04] V. Bergelson and A. Leibman, *Failure of the Roth theorem for solvable groups of exponential growth*, Ergodic Theory Dynam. Systems **24** (2004), no. 1, 45–53, DOI 10.1017/S0143385703000427. MR2041260

[BS06] J. L. Bell and A. B. Slomson, *Models and ultraproducts: An introduction*, North-Holland Publishing Co., Amsterdam-London, 1969. MR0269486

[DnI17] Eduardo Dueñez and José Iovino, *Model theory and metric convergence I: Metastability and dominated convergence*, Beyond first order model theory, CRC Press, Boca Raton, FL, 2017, pp. 131–187. MR3729326

[Fur77] Harry Furstenberg, *Ergodic behavior of diagonal measures and a theorem of Szemerédi on arithmetic progressions*, J. Analyse Math. **31** (1977), 204–256, DOI 10.1007/BF02813304. MR498471

[HI02] C. Ward Henson and José Iovino, *Ultraproducts in analysis*, Analysis and logic (Mons, 1997), London Math. Soc. Lecture Note Ser., vol. 262, Cambridge Univ. Press, Cambridge, 2002, pp. 1–110. MR1967834

[HK05] Bernard Host and Bryna Kra, *Nonconventional ergodic averages and nilmanifolds*, Ann. of Math. (2) **161** (2005), no. 1, 397–488, DOI 10.4007/annals.2005.161.397. MR2150389

[Iov14] José Iovino, *Applications of model theory to functional analysis*, Dover Publications, Inc., Mineola, NY, 2014. Revised reprint of the 2002 original; With a new preface, notes, and an updated bibliography. MR3362124

[Lei98] A. Leibman, *Polynomial sequences in groups*, J. Algebra **201** (1998), no. 1, 189–206, DOI 10.1006/jabr.1997.7269. MR1608723

[Lei02] A. Leibman, *Polynomial mappings of groups*, Israel J. Math. **129** (2002), 29–60, DOI 10.1007/BF02773152. MR1910931

[Lei05] A. Leibman, *Convergence of multiple ergodic averages along polynomials of several variables*, Israel J. Math. **146** (2005), 303–315, DOI 10.1007/BF02773538. MR2151605

[Sze75] {*Collection of articles in memory of Juriĭ Vladimirovič Linnik*}, Państwowe Wydawnictwo Naukowe, Warsaw, 1975. Acta Arith. **27** (1975). MR0357289

[Tao08] Terence Tao, *Norm convergence of multiple ergodic averages for commuting transformations*, Ergodic Theory Dynam. Systems **28** (2008), no. 2, 657–688, DOI 10.1017/S0143385708000011. MR2408398

[Tao12] ———, *Walsh's ergodic theorem, metastability, and external Cauchy convergence*, `http://terrytao.wordpress.com/2012/10/25/`, October 2012.

[vN32] John von Neumann, *Proof of the quasi-ergodic hypothesis*, Proc. Nat. Acad. Sci. U. S. A. **18** (1932), no. 1, 70–82.

[Wal12] Miguel N. Walsh, *Norm convergence of nilpotent ergodic averages*, Ann. of Math. (2) **175** (2012), no. 3, 1667–1688, DOI 10.4007/annals.2012.175.3.15. MR2912715

[ZK16] Pavel Zorin-Kranich, *Norm convergence of multiple ergodic averages on amenable groups*, J. Anal. Math. **130** (2016), 219–241, DOI 10.1007/s11854-016-0035-7. MR3574654

Department of Mathematics, The University of Texas at San Antonio, One UTSA Circle, San Antonio, Texas 78249-0664

Email address: eduardo.duenez@utsa.edu

Department of Mathematics, The University of Texas at San Antonio, One UTSA Circle, San Antonio, Texas 78249-0664

Email address: jose.iovino@utsa.edu

Contemporary Mathematics
Volume **775**, 2021
https://doi.org/10.1090/conm/775/15591

A survey on the blow-up method for fast-slow systems

Hildeberto Jardón-Kojakhmetov and Christian Kuehn

Abstract. In this document we review a geometric technique, called *the blow-up method*, as it has been used to analyze and understand the dynamics of fast-slow systems around non-hyperbolic points. The blow-up method, having its origins in algebraic geometry, was introduced to the study of fast-slow systems in the seminal work by Dumortier and Roussarie in 1996, whose aim was to give a geometric approach and interpretation of canards in the van der Pol oscillator. Following Dumortier and Roussarie, many efforts have been performed to expand the capabilities of the method and to use it in a wide range of scenarios. Our goal is to present in a concise and compact form those results that, based on the blow-up method, are now the foundation of the geometric theory of fast-slow systems with non-hyperbolic singularities. Due to their great importance in the theory of fast-slow systems, we cover fold points as one of the main topics. Furthermore, we also present several other singularities such as Hopf, pitchfork, transcritical, cusp, and Bogdanov-Takens, in which the blow-up method has been proved to be extremely useful. Finally, we survey further directions as well as examples of specific applied models, where the blow-up method has been used successfully.

Contents

1. Introduction

A large number of phenomena in nature can be understood as the interaction of simpler sub-processes. There are cases where it can be observed that such sub-processes evolve at distinct time scales. A classical example is found in enzymatic reactions (chemical reactions that are accelerated or catalysed through the addition

Invited survey article.

The first author was supported by a Technical University Foundation Fellowship and by an Alexander von Humboldt Research Fellowship. The second author would like to thank the VolkswagenStiftung for support via a Lichtenberg Professorship+.

of a particular type of protein called enzyme), where it can be often measured, and thus assumed in the mathematical models, that the enzyme's concentration evolves and reaches its final state much faster than any of the other reactants [**BH25**]. Other examples of phenomena where different time scales evolutions may be observed can be found in neuron dynamics, population dynamics, network dynamics (such as social and power networks), biological processes (such as cell division and regulatory processes) [**BL14, Hek10**], among many others. Mathematical models in the form of Ordinary Differential Equations (ODEs) that reflect the presence of multiple time scales are called *fast-slow systems* or *singularly perturbed ODEs*.

From a mathematical perspective, fast-slow systems have been studied for a long time and several techniques including asymptotic and geometric methods [**Eck11, KC12, Kue15, O'M91, Ver05**] have been developed to understand them.

In qualitative terms, the key idea when studying fast-slow systems is to separate the sub-processes acting at the different time scales, understand them, and then try to describe the full dynamics based on the subsystems. The previous rough idea can be made mathematically rigorous and, in fact, is the basis of the geometric analysis of fast-slow systems as we describe in Section 2. However, as we shall detail in the main part of this paper, there are cases in which the fast and slow dynamics cannot be distinguished, and thus separated, in a clear and straightforward manner. At the level of the mathematical model, this obstacle carries several difficulties, the most common is the presence of non-hyperbolic equilibrium points (or singularities), and thus advanced methods of analysis are required in such a case. The presence or absence of singularities determines what type of mathematical technique is suitable for analysis. This article is dedicated to review a technique, called *the blow-up method*, that allows one to study fast-slow systems near nilpotent singularities.

Briefly speaking, the blow-up method is a change of coordinates, see the details in Sections 2.1 and 2.4, with which one transforms the analysis of fast-slow systems around nilpotent singularities to a series of sub-problems with (semi-)hyperbolic singularities, which can be dealt with known methods of dynamical systems. Afterwards, we gather all the information obtained from the sub-problems and draw conclusions on the dynamics of the full fast-slow system. It is important to remark that the blow-up (or also blowing up) method was initially developed for the resolution of singularities of algebraic curves, and more generally of varieties over fields of characteristic zero [**Hir64a, Hir64b**]. The idea behind resolution of singularities consists on a certain coordinate mapping that provides a regular curve from a singular one. We will see that this idea is similar for ordinary differential equations. In the context of dynamical systems, the blow-up method has been used, for example, to study and classify nilpotent singularities of (single-scale) vector fields [**ÁFJ11, BM90, Dum77, Dum91, Tak74**], see also Section 2.1 below. Here we focus on the use of blow-up for fast-slow multiscale systems.

Our goal with this survey is to gather in a concise and compact form the most relevant results in the theory of fast-slow systems for which the blow-up method has been fundamental. The results that we include in the main part of this survey have extended and complemented the general theory of fast-slow systems and are nowadays the foundation of many other studies. Our approach is structured as follows: in Section 2 we provide some necessary mathematical preliminaries, where we sketch Fenichel's theorem and the blow-up method. Later, in Section 3 we present our survey which includes results on generic folds and canards. Then we recall some fundamental results outside the scope of folded singularities, and present several applications, where the blow-up method has been used. We conclude in Section 4 with a summary and an outlook for the blow-up method within the context of fast-slow systems.

Notation

Naturally, different notations are used across the whole body of literature regarding fast-slow systems. In this article we homogenize the notation used for the most important geometric objects that one encounters while studying fast-slow systems. Although the explicit definition of the geometric objects changes depending on the specific problem, their role in the analysis is often similar. Clarifying the way we decide to denote these objects allows us to recycle the notation, thus facilitating the exposition. The reader may refer back to the following conventions throughout:

- $\mathcal{C}_0$ denotes the critical manifold of a given fast-slow system to be specified when appropriate.
- $\mathcal{S}_0$ is a compact subset of $\mathcal{C}_0$. The attracting and repelling parts of $\mathcal{S}_0$ are denoted by $\mathcal{S}_0^{\mathrm{a}}$ and $\mathcal{S}_0^{\mathrm{r}}$ respectively. When further distinction is needed, we add appropriate subscripts.
- $\mathcal{S}_\varepsilon$ denotes a slow manifold, and $\mathcal{S}_\varepsilon^{\mathrm{a}}$ (resp. $\mathcal{S}_\varepsilon^{\mathrm{r}}$) denotes the attracting (resp. repelling) part of $\mathcal{S}_\varepsilon$.
- Σ^{en} is a codimension 1 (that is, the dimension of Σ^{en} is one less as that of the phase-space) *entry* section of the phase-space transverse to $\mathcal{S}_0^{\mathrm{a}}$. Σ^{ex} is used to denote a codimension 1 *exit* section of the phase-space transverse to $\mathcal{S}_\varepsilon^{\mathrm{a}}$ beyond a non-hyperbolic singularity. When more than one exit section is necessary, we add appropriate subscripts.
- Π denotes the transition map between Σ^{en} and Σ^{ex} induced by the flow of a fast-slow system. When more than one transition map is necessary, we add appropriate subscripts.
- $K_\bullet$ denotes a chart of the blow-up space. We distinguish the charts using a subscript depending on the number of charts required for the analysis of a specific singularity.

2. Preliminaries

In this section we provide the mathematical preliminaries needed for the survey given in Section 3. First we describe the main idea of the blow-up method for a single time scale planar vector field. Next, we state a couple of important definitions regarding fast-slow systems. Afterwards, we recall the fundamental Fenichel's theorem, and finish this section with a description of the blow-up method for fast-slow systems.

2.1. The blow-up method. In this section we introduce the blow-up method in its classical context, that is, to desingularize a nilpotent equilibrium point[1] of a planar vector field. For a detailed exposition the reader is referred to [**ÁFJ11**,

[1]We recall that an equilibrium point of a vector field is called nilpotent if the linearization of the vector field at such a point is given by a matrix with only zero eigenvalues.

Dum91] and [**Kue15**, Chapter 7]. Here we shall only treat an example to highlight the main idea of the method. Later, in Section 2.4 we will see how this transformation also fits into the study of fast-slow systems.

Let us consider the planar ordinary differential equation (ODE)

$$\begin{aligned} \frac{\mathrm{d}x}{\mathrm{d}t} &= y \\ \frac{\mathrm{d}y}{\mathrm{d}t} &= x^3 + xy. \end{aligned} \tag{2.1}$$

We note that the origin $(x, y) = (0, 0)$ is a unique equilibrium point and that the linearization of (2.1) at the origin is given by the matrix

$$\begin{bmatrix} 0 & 1 \\ 0 & 0 \end{bmatrix}. \tag{2.2}$$

Thus, the origin is a non-hyperbolic equilibrium point and, moreover, is nilpotent. Our goal is to qualitatively describe the orbits of (2.1) in a small neighbourhood of the origin. However, not only the linearization offers no useful information, but centre manifold reduction [**Car12**] is not suitable since in this case the centre manifold corresponds to the whole phase-space. So, what we are going to use is a suitable change of coordinates, known as *blow-up*, which will induce a new system with only hyperbolic equilibrium points, and therefore can be analyzed by dynamical systems tools.

Let us consider a *weighted polar change of coordinates*

$$\phi : \mathbb{S}^1 \times I \to \mathbb{R}^2, \qquad \phi(\theta, r) = (r\cos\theta, r^2 \sin\theta), \tag{2.3}$$

where $I \subseteq \mathbb{R}$ is an interval containing the origin and $\theta \in [0, 2\pi]$. At the end of this section we clarify the reason to choose a weighted polar change of coordinates, for now let us proceed with the example.

The change of coordinates defined by $(x, y) = (r\cos\theta, r^2\sin\theta)$ defines a new ODE, namely

$$\begin{aligned} \dot{\theta} &= \frac{r\left(1 + \sin\theta - 4\sin^2\theta - \sin^3\theta + \sin^4\theta\right)}{\sin^2\theta + 1} \\ \dot{r} &= \frac{r^2}{\sin^2\theta + 1}\cos\theta\sin\theta\left(\sin\theta - \sin^2\theta + 2\right). \end{aligned} \tag{2.4}$$

Note that the change of coordinates defined by ϕ maps the circle $\mathbb{S}^1 \times \{0\}$ to the origin in the plane[2]. Moreover, since ϕ is a diffeomorphism for $\{r > 0\}$, orbits of (2.1) in a small neighbourhood of the origin correspond to orbits of (2.4) in a small neighbourhood of $\mathbb{S}^1 \times \{0\}$. Note however that (2.4) vanishes along $\mathbb{S}^1 \times \{0\}$. To overcome this we can divide the right-hand side of (2.4) by r. This operation does not change the qualitative properties of the orbits in the region $\mathbb{S}^1 \times \{r > 0\}$. Thus, it shall suffice to study *the desingularized system*

$$\begin{aligned} \dot{\theta} &= \frac{1}{\sin^2\theta + 1}\left(1 + \sin\theta - 4\sin^2\theta - \sin^3\theta + \sin^4\theta\right) \\ \dot{r} &= \frac{r}{\sin^2\theta + 1}\cos\theta\sin\theta\left(\sin\theta - \sin^2\theta + 2\right), \end{aligned} \tag{2.5}$$

which does not vanish any more along $\mathbb{S}^1 \times \{0\}$. The most important fact is that *orbits of* (2.5) *near* $\mathbb{S}^1 \times \{0\}$ *correspond to orbits of* (2.1) *near the origin.*

[2]Equivalently ϕ^{-1} maps the origin in the plane to the circle $\mathbb{S}^1 \times \{0\}$.

It is now straightforward to show that (2.5) has four *hyperbolic* saddle equilibrium points, namely $p_1 = (-\arcsin(\sqrt{2}-1), 0)$, $p_2 = (\arcsin(\sqrt{5}/2 - 1/2), 0)$, $p_3 = (\pi - \arcsin(\sqrt{5}/2 - 1/2), 0)$ and $p_4 = (\pi + \arcsin(\sqrt{2}-1), 0)$. Since the aforementioned equilibrium points are hyperbolic it follows from linear analysis that the phase portrait of (2.5) in a small neighbourhood of $\mathbb{S}^1 \times \{0\}$ is as show in Figure 1.

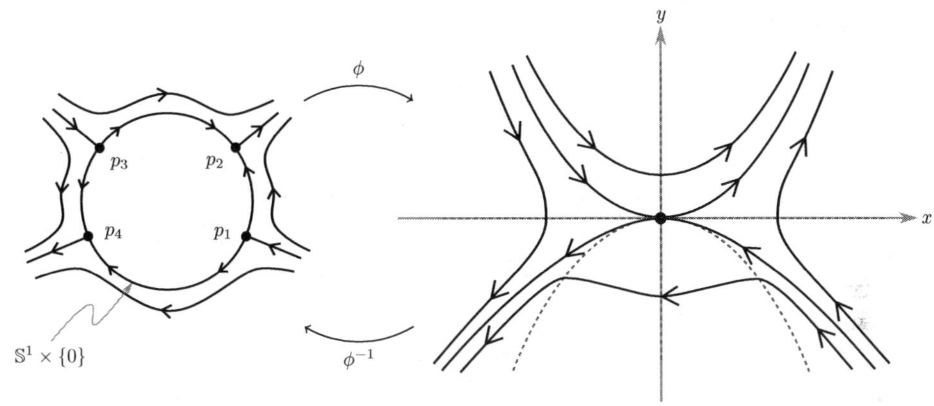

FIGURE 1. Blow-up analysis of (2.1). On the left we show the phase-portrait of (2.5) in a small neighborhood of $\mathbb{S}^1 \times \{0\}$, where four hyperbolic saddle points are found. On the right we show the corresponding orbits of (2.1), where from a qualitative perspective, the circle $\mathbb{S}^1 \times \{0\}$ "blows-down" to the origin and all other orbits of (2.1) are equivalent to orbits of (2.5). To provide more detail on the flow of (2.1) away from the origin we have made use of the corresponding nullclines, shown as dashed-red curves.

We finish this section with some important remarks:

- The procedure we exemplified above is known as *the blow-up method*. In some sense, the transformation ϕ^{-1} "blows the origin up to a circle". The advantage of blowing up is that one obtains a new system which is simpler to analyze. We recall that, in the above example, (2.1) has a nilpotent equilibrium point at the origin while (2.5) has four hyperbolic equilibrium points along $\mathbb{S}^1 \times \{0\}$, which are simpler to study with standard techniques of dynamical systems. Once the blown-up system is understood we then "blow-down" the phase-portrait of (2.5) resulting in a qualitative description of the original system (2.1).
- In the example presented above we have used a weighted version of a polar change of coordinates. Usually one then refers to the transformation as a *quasi-homogeneous* blow-up to emphasize that the weights in the transformation are distinct from 1. The advantage of using a quasi-homogeneous blow-up instead of a homogeneous one is that we can desingularize the origin in just one step. The reader can check that if one uses $(x, y) = (r\cos\theta, r\sin\theta)$ instead of (2.3), the blown-up system then has a pair of nilpotent singularities located at $(\theta, r) = (0, 0)$ and $(\theta, r) = (\pi, 0)$. In turn, the blow-up method can be applied once more to such pair of points. More details are provided in Section 2.4, see also [**Dum91**] and [**Kue15**, Chapter 7].

- It is known, for example, that real-analytic planar and three dimensional vector fields can be desingularized via the blow-up method [**Dum77, Pan02, Pan06**]. In particular, several of the fast-slow systems in this survey fall into this category. The blow-up method has been used in higher dimensional fast-slow systems only in a case-by-case approach.

2.2. Fast-slow systems. A fast-slow system is a singularly perturbed ordinary differential equation (ODE) of the form

$$\begin{aligned} \varepsilon\dot{x} &= f(x,y,\varepsilon), \\ \dot{y} &= g(x,y,\varepsilon), \end{aligned} \tag{2.6}$$

where the over-dot denotes the derivative with respect to *the slow time* τ, $x \in \mathbb{R}^m$ and $y \in \mathbb{R}^n$ denote the fast and the slow variables respectively, and $\varepsilon > 0$ is a small parameter accounting for the time scale separation. System (2.6) is said to be *in standard form*, and we can explicitly see the time scale separation between the slow and fast variables. In some cases, however, fast-slow systems may appear in a more general non-standard form, namely

$$\dot{\zeta} = F(\zeta, \varepsilon), \tag{2.7}$$

where the time scale separation, if any, may not be evident to distinguish, see Section 3.7 for some examples of fast-slow systems which are not in standard form. In this survey we shall adhere to (2.6) but the reader is referred to [**Wec20**] for an exposition of GSPT beyond the standard form.

By defining the fast time $t := \tau/\varepsilon$, we can rewrite (2.6) as an ε-family of ODEs of the form

$$\begin{aligned} x' &= f(x,y,\varepsilon), \\ y' &= \varepsilon g(x,y,\varepsilon), \end{aligned} \tag{2.8}$$

where now the prime $'$ denotes the derivative with respect to *the fast time* t. We assume that $f : \mathbb{R}^m \times \mathbb{R}^n \times \mathbb{R} \to \mathbb{R}^m$ and $g : \mathbb{R}^m \times \mathbb{R}^n \times \mathbb{R} \to \mathbb{R}^n$ are of class C^k, with k sufficiently large. We observe that (2.6) and (2.8) are smoothly equivalent for $\varepsilon > 0$, meaning that their only difference is the time parametrization of the corresponding trajectories.

The term Geometric Singular Perturbation Theory (GSPT) [**Fen79, Jon95**], refers to the collection of geometric techniques with which fast-slow systems can be analyzed. Roughly speaking, the idea of GSPT is to study the limits of (2.6) and (2.8) as $\varepsilon \to 0$, analyze invariant objects in the two limits, and then use perturbation methods to describe the dynamics of (2.6) and (2.8) for $\varepsilon > 0$ sufficiently small. Accordingly, by setting $\varepsilon = 0$ in (2.6) we obtain

$$\begin{aligned} 0 &= f(x,y,0), \\ \dot{y} &= g(x,y,0), \end{aligned} \tag{2.9}$$

which is called the *reduced system* (or *slow subsystem*), and the flow of (2.9) is called *the slow flow*. Structurally, the system (2.9) is a *Constrained Differential Equation* (CDE) [**Tak76**] (or *Differential Algebraic Equation* [**KM06**]). From (2.6) we get for $\varepsilon = 0$ the ODE

$$\begin{aligned} x' &= f(x,y,0), \\ y' &= 0, \end{aligned} \tag{2.10}$$

which is called *the layer equation* (or *fast subsystem*) and where we can view the slow variables y as parameters. The two *singular limit* systems (2.9) and (2.10) are not equivalent any more. However, we observe that the set $\{f(x,y,0)=0\}$ defines the phase-space of solutions of the slow subsystem and the set of equilibrium points of the fast subsystem equation. This leads to the following natural definition:

DEFINITION 2.1 (Critical manifold). The critical manifold is defined as the set

$$\mathcal{C}_0 = \{(x,y) \in \mathbb{R}^m \times \mathbb{R}^n \,|\, f(x,y,0)=0\}\,. \tag{2.11}$$

Although we call $\mathcal{C}_0$ a manifold, strictly speaking, solutions of the equation $f(x,y,0)=0$ do not necessarily define a manifold, that is, they may not be locally diffeomorphic to an Euclidean space. For instance, the solutions of $f(x,y,0)=0$ may self-intersect as in the case of the transcritical and pitchfork singularities in Section 3.5 . However, generically, there are open regions in the phase-space where $\mathcal{C}_0$ is indeed a manifold. So, as a convention in fast-slow systems one keeps referring to $\mathcal{C}_0$ as the critical manifold. An important characteristic that critical manifolds may posses is normal hyperbolicity:

DEFINITION 2.2. A point $p=(x,y)\in\mathcal{C}_0$ is called *hyperbolic* if the matrix $\mathrm{D}_x f(p,0)\in\mathbb{R}^{m\times m}$, where D_x stands for the total derivative with respect to x, has no eigenvalues on the imaginary axis. The critical manifold $\mathcal{C}_0$ is called *normally hyperbolic* if all points $p\in\mathcal{C}_0$ are hyperbolic.

Note that equivalently to Definition 2.2, one can say that $\mathcal{C}_0$ is normally hyperbolic if all its points are hyperbolic equilibrium points of the fast subsystem. We call a hyperbolic point p *attracting* if all eigenvalues have negative real part, *repelling* if all eigenvalues have positive real part, and *saddle-type* if there are negative and positive real part eigenvalues. On the contrary situation, we shall call $p\in\mathcal{C}_0$ *non-hyperbolic* if the matrix $\mathrm{D}_x(p,0)$ has at least one eigenvalue on the imaginary axis. Whether the critical manifold is, or is not, normally hyperbolic will distinguish the tools that are needed for the analysis of the corresponding fast-slow system. In the case of normal hyperbolicity we have Fenichel's theorem (see Section 2.3); in the case of loss of normal hyperbolicity due to the presence of non-hyperbolic points, we may use the blow-up method (see Section 2.4). We emphasize that loss of normal hyperbolicity in fast-slow systems is highly relevant as it is related to dynamic features such as relaxation oscillations, canards, bursting, mixed-mode oscillations, etcétera [**DGK+12, Kue15**].

2.3. Fenichel's theorem. Fenichel's theorem [**Fen79**] (see also [**Jon95, Tik52**]) provides geometric tools and techniques to analyze fast-slow systems with hyperbolic points:

THEOREM 2.3. *Suppose that $\mathcal{S}_0\subseteq\mathcal{C}_0$ is a compact normally hyperbolic submanifold (possibly with boundary) of the critical manifold $\mathcal{C}_0$ of* (2.8). *Then, for $\varepsilon>0$ sufficiently small, the following hold:*

(M1) *There exists a locally invariant manifold $\mathcal{S}_\varepsilon$ diffeomorphic to $\mathcal{S}_0$. Local invariance means that trajectories can enter or leave $\mathcal{S}_\varepsilon$ only through its boundaries.*
(M2) *$\mathcal{S}_\varepsilon$ has Hausdorff distance $\mathcal{O}(\varepsilon)$, as $\varepsilon\to0$, from $\mathcal{S}_0$.*
(M3) *The flow on $\mathcal{S}_\varepsilon$ converges to the slow flow as $\varepsilon\to0$.*
(M4) *$\mathcal{S}_\varepsilon$ is C^k-smooth.*

(M5) $\mathcal{S}_\varepsilon$ *is normally hyperbolic and has the same stability properties, i.e., attracting, repelling or saddle-type, with respect to the fast variables as* $\mathcal{S}_0$.

(M6) $\mathcal{S}_\varepsilon$ *is usually not unique. In regions that remain at a fixed distance from* $\partial\mathcal{S}_\varepsilon$, *all manifolds satisfying items* $(M1)-(M5)$ *above lie at a Hausdorff distance of order* $\mathcal{O}(\exp(-C/\varepsilon))$ *from each other for some* $C>0$, $C\in\mathcal{O}(1)$, *as* $\varepsilon\to 0$.

In the past, one would refer to Fenichel's theorem as *Geometric Singular Perturbation Theory* (GSPT). As a matter of fact, Fenichel's article [**Fen79**] is entitled "Geometric Singular Perturbation Theory for Ordinary Differential Equations". However, nowadays GSPT includes many more geometric techniques such as the blow-up method and fast-slow normal form theory, among others. A few remarks are now in place.

REMARK 2.4.

- The concept of normal hyperbolicity is applicable, under its appropriate modifications, to invariant manifolds in general and not only to sets of equilibria as in Definition 2.2. It turns out, however, that normal hyperbolicity for sets of equilibria is simpler to define. For invariant manifolds in general, normal hyperbolicity is more involved, see [**Fen71, HPS70**] and [**Kue15**, Chapter 2.3].
- Although $\mathcal{S}_\varepsilon$ is usually not unique, any choice of such manifolds is called *"the"* slow manifold. In turn, one should check if the results one obtains are independent or not on the choice of slow manifold. It is often the case that the results do not depend on such a choice. Furthermore, even though the terms "critical manifold" and "slow manifold" are sometimes used interchangeably or even equivalently, it is important to make a distinction between them[3]. One of the reasons for this is that, for example, slow manifolds may be extended beyond a non-hyperbolic point, and such extension may not have a relationship with the critical manifold. Many examples of this situation are shown in Section 4.
- In qualitative terms, Fenichel's theorem shows that, under the assumptions of Theorem 2.3, a fast-slow system without non-hyperbolic points can be regarded as a regular perturbation of its corresponding singular systems near $\mathcal{C}_0$.

The next section describes a geometric technique to locally analyze fast-slow systems around, a particular class of non-hyperbolic points.

2.4. The blow-up method for fast-slow systems. In section 2.1 we sketched the idea of the blow-up method to desingularize nilpotent singularities of planar vector fields. In this section we provide a review on the blow-up method as is nowadays commonly used for the analysis of fast-slow systems with non-hyperbolic singularities. To make the relationship between the blow-up method and fast-slow systems clearer, let us first rewrite the ε-family of vector fields (2.8) on $\mathbb{R}^{m+n}$ as a single vector field on $\mathbb{R}^{m+n+1}$ of the form

$$\begin{aligned} x' &= f(x,y,\varepsilon)\\ y' &= \varepsilon g(x,y,\varepsilon) \\ \varepsilon' &= 0. \end{aligned} \tag{2.12}$$

[3]In this document $\mathcal{C}_0$ is a critical manifold and $\mathcal{S}_0$ is a *compact* subset of $\mathcal{C}_0$. A perturbation of $\mathcal{S}_0$, denoted by $\mathcal{S}_\varepsilon$, is a slow manifold. Note that $\mathcal{C}_\varepsilon$ may not be well-defined since we do not impose compactness to $\mathcal{C}_0$; but see [**Eld13**].

Furthermore, let us assume that the origin $(x, y, \varepsilon) = (0, 0, 0)$ is an equilibrium point and that $\mathrm{D}_x f(0,0,0)$ has all its eigenvalues equal to zero (this is indeed the case in the systems discussed in Section 3). This means that the origin is a nilpotent singularity of (2.12) and, as such, the blow-up method can be adapted to desingularize the origin of (2.12).

REMARK 2.5. It is worth noting that nilpotent singularities are a subset of non-hyperbolic singularities. Thus, not all non-hyperbolic singularities of fast-slow systems may be studied with the blow-up method. However, as we discuss in Section 3, there is a large number of fast-slow systems for which, indeed, the origin is nilpotent. In particular, in all fast-slow systems with one-dimensional fast direction ($x \in \mathbb{R}$), a non-hyperbolic singularity is nilpotent. In other cases where the singularity is non-hyperbolic but not nilpotent, a preliminary transformation may bring a fast-slow system into a suitable form to be analyzed via the blow-up method. For example, a centre manifold reduction may lead to a reduced fast-slow system with a nilpotent singularity [**Dum91**]. In some other cases, as described in Section 3.4, a change of coordinates brings the system into an appropriate form.

Although there are several (equivalent) versions and improvements of the blow-up method, we restrict to the quasihomogeneous case as it is more commonly used nowadays. For further information see [**Kue15**, Chapter 7] and references therein.

Let $X : \mathbb{R}^{m+n+1} \to \mathbb{R}^{m+n+1}$ be the vector field, which in coordinates is defined by (2.12), and let $\mathbb{S}^N$ denote the N-th dimensional sphere[4]. Next, we can formally define the blow-up transformation most commonly used in fast-slow systems:

DEFINITION 2.6 (Quasihomogeneous blow-up). Consider a vector field $X : \mathbb{R}^{m+n+1} \to \mathbb{R}^{m+n+1}$ defined by (2.12) and assume that $X(0) = 0$. Let $\alpha = (\alpha_1, \dots, \alpha_m) \in \mathbb{N}_0^m$, $\beta = (\beta_1, \dots, \beta_m) \in \mathbb{N}_0^n$ and $\gamma \in \mathbb{N}_0$. Let the generalized polar transformation $\phi : \mathbb{S}^{m+n} \times I \to \mathbb{R}^{m+n+1}$ be defined by

$$\phi(\bar{x}, \bar{y}, \bar{\varepsilon}, r) = (r^\alpha \bar{x}, r^\beta \bar{y}, r^\gamma \bar{\varepsilon}) = (x, y, \varepsilon), \tag{2.13}$$

where $(\bar{x}, \bar{y}, \bar{\varepsilon}) = (\bar{x}_1, \dots, \bar{x}_m, \bar{y}_1, \dots, \bar{y}_n, \bar{\varepsilon}) \in \mathbb{S}^{m+n}$, $r \in I$, and $I \subseteq \mathbb{R}$ is an interval containing the origin. Here we use the multi-index notation $r^\alpha \bar{x} = (r^{\alpha_1}\bar{x}_1, \dots, r^{\alpha_m}\bar{x}_m)$, and similarly for $r^\beta \bar{y}$. The *quasihomogeneous* blow-up of the vector field X, denoted as $\bar{X}$, is defined by

$$\bar{X} = \mathrm{D}\phi^{-1}|_{(\bar{x},\bar{y},\bar{\varepsilon},r)} \circ X \circ \phi(\bar{x}, \bar{y}, \bar{\varepsilon}, r). \tag{2.14}$$

We note that ϕ maps the sphere $\mathcal{B}_0 := \mathbb{S}^{m+n} \times \{0\}$ to the origin in $\mathbb{R}^{m+n+1}$, while ϕ^{-1} maps $0 \in \mathbb{R}^{m+n+1}$ to $\mathcal{B}_0$. Hence, the operation ϕ^{-1} is called (quasihomogeneous) blow-up while ϕ is called (quasihomogeneous) blow-down. The word quasihomogeneous reflects the fact that the exponents appearing in (2.13) are not necessarily the same. We omit the term "quasihomogeneous" when all exponents (α, β, γ) are equal to 1.

It follows from (2.14) that $\bar{X}$ and X are conjugate for $r > 0$, meaning that there exists a one-to-one mapping between trajectories of X and trajectories of $\bar{X}$ outside $\mathcal{B}_0$. Moreover, it can be shown that $\bar{X}$ is well defined at $r = 0$ [**Kue15**]. Due to the presence of non-hyperbolic singularities, and depending on the choice of the exponents, it is usually the case that the system denoted by $\bar{X}$ vanishes on $\mathcal{B}_0$.

[4]To use spheres, cylinders, or related spaces as blown-up spaces is often very convenient, yet not necessary [**Kue16**].

In fact, let $j_\ell(X)$ denote the ℓ-jet [**GG12**] of X at the origin. If $j_\ell(X)=0$ for $\ell=0,1,\ldots,k$ and $j_{k+1}(X)\neq 0$, then we define the *desingularized vector field* $\tilde{X}=\frac{1}{r^k}\bar{X}$. Now $\tilde{X}$ does not vanish at $\mathcal{B}_0$. Since $\bar{X}$ and $\tilde{X}$ are smoothly equivalent for $r>0$, all the information obtained from $\tilde{X}$ is equivalent to that of $\bar{X}$ outside $\mathcal{B}_0$. However, since $\tilde{X}$ does not vanish any more along $\{r=0\}$, we may try to infer the dynamics of $\tilde{X}$ for $r>0$ small from the restriction $\tilde{X}|_{\{r=0\}}$. This greatly simplifies the analysis, since usually we find that $\tilde{X}$ has semi-hyperbolic singularities, hyperbolic singularities, or no singularities at all. Finally, due to the equivalences between $X,\bar{X}$, and $\tilde{X}$, we conclude that the flow of $\tilde{X}$ for $r>0$ sufficiently small provides a complete description of the flow of X for $\varepsilon>0$ sufficiently small.

When we study high dimensional problems, say for $m+n>2$, working with polar coordinates can become cumbersome. Then, we rather work in charts that cover the blow-up space. In each of the charts we can define local coordinates and a corresponding *local* vector field. In practice, what we do to define local coordinates in a chart is to fix one of the blow-up coordinates to ±1. This approach is called "directional blow-up". For example, to perform a blow-up in the $\bar{\varepsilon}$-direction we would define new coordinates according to $\phi:\mathbb{R}^{n+m+1}\to\mathbb{R}^{n+m+1}$ given by

$$\phi(\bar{x},\bar{y},\bar{\varepsilon},r)=(r^\alpha\bar{x},r^\beta\bar{y},r^\gamma), \tag{2.15}$$

that is by fixing $\bar{\varepsilon}=1$. Similarly, we can define blow-ups in any of the other directions.

REMARK 2.7. The chart $K:=\{\bar{\varepsilon}=1\}$ is the most important one and it is called *the rescaling chart*, *the family chart* or *the central chart*. The rest of the charts are often referred to as *phase-directional charts*.

Directional blow-ups induce local vector fields on each of the (Euclidean) charts. Once the analysis of the relevant local vector fields is performed, one can overlap suitable regions of the charts and match the flow on such charts via the so-called *matching maps* (or *transition maps*) to describe the dynamics all around $\mathbb{S}^{m+n}\times I$. In particular, this process allows us to track invariant objects, principally centre manifolds [**Car12**], across the blow-up space. A schematic representation of the blow-up map is provided in Figure 2.

In summary, through the blow-up method one attempts to make the analysis of fast-slow systems near non-hyperbolic equilibrium points more accessible. Yet, the analysis of the vector fields obtained after blow-up is still highly non-trivial due to several difficult obstacles such as the nonlinearity of the local vector fields, presence of semi-hyperbolic equilibria, and the appearance of resonances, just to mention a few. It may even happen that, in one of the charts, there is a singularity that is still nilpotent, in which case the blow-up procedure may be applied again. Furthermore, obtaining the right weights (α,β,γ) is not always trivial. One way in which one can obtain such weights is via the Newton Polyhedron [**ÁFJ11, Dum77, Kue15**].

In the next section we are going to review some of the most fundamental and (mathematically) influential works that have established the blow-up method as one preferred technique to study fast-slow systems in the neighborhood of non-hyperbolic points.

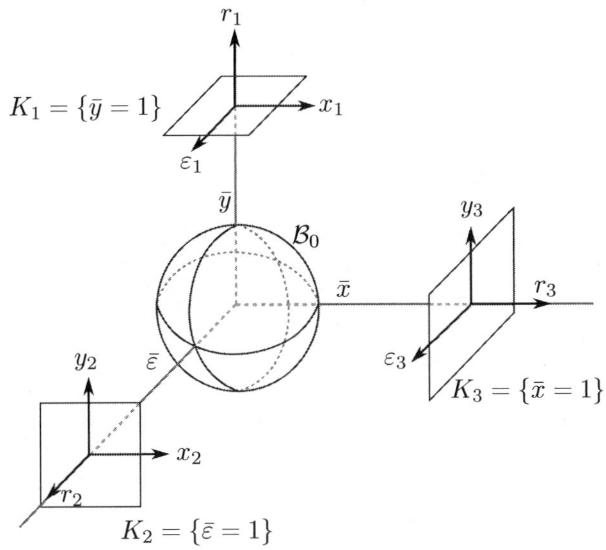

FIGURE 2. Sketch of the blown-up space and of some of the directional charts. In practice, via the blow-up method, we study local vector fields defined in the charts, and then "glue" trajectories and other invariant objects together to describe the dynamics in a small neighborhood of $\mathcal{B}_0$, which in turn provides the dynamics of a fast-slow system around the origin for $\varepsilon > 0$ sufficiently small.

3. Survey

Due to their importance to the theory, we first focus on fold points, folded singularities and canards. Later on, we briefly discuss instances where the blow-up method has been used to study singularities beyond the fold such as Hopf, transcritical, pitchfork, cusp, and Bogdanov-Takens singularities.

3.1. The generic fold.

The generic fold is the first natural situation, where a non-hyperbolic point appears in planar fast-slow systems. The exposition of this section mainly summarizes the results of [**KS01a**].

DEFINITION 3.1 (Generic fold point). Consider (2.6) with $(x, y) \in \mathbb{R}^2$. We say that (x_0, y_0) is a *generic fold point* if

$$f(x_0, y_0, 0) = 0, \qquad \frac{\partial f}{\partial x}(x_0, y_0, 0) = 0, \tag{3.1}$$

hold together with the non-degeneracy conditions

$$\frac{\partial^2 f}{\partial x^2}(x_0, y_0, 0) \neq 0, \qquad \frac{\partial f}{\partial y}(x_0, y_0, 0) \neq 0, \qquad g(x_0, y_0, 0) \neq 0. \tag{3.2}$$

Without loss of generality, we assume that $(x_0, y_0) = (0, 0)$ and we may choose signs of (3.2). Thus a fast-slow system with a generic fold point at the origin can be given as the (local) canonical form [**KS01a**]

$$
\begin{aligned}
x' &= -y + x^2 + \mathcal{O}(\varepsilon, xy, y^2, x^3), \\
y' &= \varepsilon(-1 + \mathcal{O}(x, y, \varepsilon)).
\end{aligned} \tag{3.3}
$$

We observe that the corresponding critical manifold

$$
\mathcal{C}_0 = \left\{(x,y) \in \mathbb{R}^2 \,|\, -y + x^2 + \mathcal{O}(\varepsilon, xy, y^2, x^3) = 0\right\} \tag{3.4}
$$

is locally a parabola as shown in Figure 3.

Let $\mathcal{S}_0^{\mathrm{a}}$ (resp. $\mathcal{S}_0^{\mathrm{r}}$) denote a compact subset of the attracting (resp. repelling) section of the critical manifold $\mathcal{C}_0$. It is known from Fenichel's theorem that a slow manifold $\mathcal{S}_\varepsilon^{\mathrm{a}}$ (resp. $\mathcal{S}_\varepsilon^{\mathrm{r}}$) exists as a smooth perturbation of $\mathcal{S}_0^{\mathrm{a}}$ (resp. $\mathcal{S}_0^{\mathrm{r}}$). However, Fenichel's theorem only guarantees the existence of such manifolds away from the fold point. The main goal of [**KS01a**] is to extend $\mathcal{S}_\varepsilon^{\mathrm{a}}$ beyond the fold point. Let $\rho > 0$ be small and $I \subset \mathbb{R}$ be a suitable interval, and define the sections

$$
\begin{aligned}
\Sigma^{\mathrm{en}} &= \left\{(x,y) \in \mathbb{R}^2 \,|\, x \in I,\, y = \rho^2\right\}, \\
\Sigma^{\mathrm{ex}} &= \left\{(x,y) \in \mathbb{R}^2 \,|\, x = \rho,\, y \in \mathbb{R}\right\}.
\end{aligned} \tag{3.5}
$$

Here I being a suitable interval means that Σ^{en} intersects $\mathcal{S}_0^{\mathrm{a}}$ but not $\mathcal{S}_0^{\mathrm{r}}$, see Figure 3. The following theorem states the properties of the slow manifold $\mathcal{S}_\varepsilon^{\mathrm{a}}$ as it passes through the origin (see Figure 3 for a sketch of the result).

THEOREM 3.2 ([**KS01a**, Theorem 2.1]). *Let $\Pi : \Sigma^{\mathrm{en}} \to \Sigma^{\mathrm{ex}}$ be the transition map for the flow of* (3.3). *Then, there exists $\varepsilon_0 > 0$ such that the following assertions hold for $\varepsilon \in (0, \varepsilon_0]$:*

(F1) *The manifold $\mathcal{S}_\varepsilon^{\mathrm{a}}$ passes through Σ^{ex} at a point $(\rho, h(\varepsilon))$, where $h(\varepsilon) \in \mathcal{O}(\varepsilon^{2/3})$.*
(F2) *The transition Π is a contraction with contraction rate $\mathcal{O}(\exp(-C/\varepsilon))$, where $C > 0$.*

We note the importance of the two statements (F1) and (F2) contained Theorem 3.2 . First of all, the estimate $h(\varepsilon) \in \mathcal{O}(\varepsilon^{2/3})$ indicates that $\mathcal{S}_\varepsilon^{\mathrm{a}}$ is not a regular perturbation of the singular set $\mathcal{S}_0^{\mathrm{a}} \cup \left\{(x,0) \in \mathbb{R}^2 \,|\, x \geq 0\right\}$. In other words, it is not possible to consider an asymptotic series in positive integer weights of ε as an expansion of $\mathcal{S}_\varepsilon^{\mathrm{a}}$ in a small neighborhood of the fold point. Next, the second statement shows that the exponential contraction towards $\mathcal{S}_\varepsilon^{\mathrm{a}}$, which before the fold point is guaranteed by Fenichel's theorem, is not lost after crossing the fold point. We observe that this fact is not a regular perturbation argument derived from the corresponding layer equation, but can be proven via the blow-up analysis.

Let us now point-out the key steps of the proof of Theorem 3.2. The blow-up is defined by

$$
x = \bar{r}\bar{x}, \qquad y = \bar{r}^2\bar{y}, \qquad \varepsilon = \bar{r}^3\bar{\varepsilon}. \tag{3.6}
$$

The weights of the blow-up map follow formally from the quasi-homogeneity [**Dum91**] of the function $-y + x^2$. On a practical level, one can also guess good weights by observing that the scaling $(x, y, t) \mapsto (\varepsilon^{1/3}x, \varepsilon^{2/3}y, \varepsilon^{-1/3}t)$ removes ε to leading-order from (3.3). Using (3.6), the blow-up analysis is carried over three charts, namely

$$
K_1 = \{\bar{y} = 1\}, \quad K_2 = \{\bar{\varepsilon} = 1\}, \quad K_3 = \{\bar{x} = 1\}. \tag{3.7}
$$

The behaviour of the trajectories passing through Σ^{en} as they approach (but do not cross) a small neighbourhood of the origin is studied in chart K_1. The main objects of study in this chart are 2-dimensional centre manifolds associated to semi-hyperbolic equilibrium points. To be more precise, in chart K_1 one can show the existence of a 2-dimensional attracting and a 2-dimensional repelling centre manifolds that are denoted by $\mathcal{M}_1^{\mathrm{a}}$ and $\mathcal{M}_1^{\mathrm{r}}$ respectively. In fact, sections of these centre manifolds correspond to perturbations of $\mathcal{S}_0^{\mathrm{a}}$ and $\mathcal{S}_0^{\mathrm{r}}$, that is to slow manifolds, along normally hyperbolic parts of the critical manifold. The advantage of using the blow-up method is that it allows to track invariant objects (in this case centre manifolds) as they pass through all the charts.

Next, in chart K_2, we study the behaviour of trajectories in a small neighbourhood of the origin. The main object of study in this chart is a Riccati differential equation [**MR80**, 68-72] of the form

$$\begin{aligned} x_2' &= -y_2 + x_2^2, \\ y_2' &= -1, \\ r_2' &= 0 \end{aligned} \tag{3.8}$$

where we use the subscript 2 to indicate local coordinates in the chart K_2. A detailed analysis of (3.8) allows, in particular, to extend $\mathcal{M}_1^{\mathrm{a}}$ found in chart K_1 to an invariant manifold $\mathcal{M}_2^{\mathrm{a}}$ beyond the fold point. The previous is due to the fact that sections of the centre manifold $\mathcal{M}_1^{\mathrm{a}}$ (in chart K_1) can be matched with solutions of (3.8) with r_2 constant (in chart K_2). Afterwards, in chart K_3, we study the dynamics of the trajectories, as they leave a small neighbourhood of the fold point. More specifically, similar to the situation in the previous two charts, via a change of coordinates we can define an invariant object $\mathcal{M}_3^{\mathrm{a}}$ corresponding to $\mathcal{M}_2^{\mathrm{a}}$ but defined in the local coordinates of chart K_3. Then we can pay special attention to trajectories on $\mathcal{M}_3^{\mathrm{a}}$. The main object of study in this chart is a 3-dimensional nonlinear ODE with a resonant hyperbolic saddle at the origin[5]. The fact that the origin is resonant is a major obstacle as it prevents to linearize the nonlinear system in order to have a detailed description of the flow near the origin. In fact, dealing with the aforementioned resonance provides that the next term in the expansion of h is of the order $\mathcal{O}(\varepsilon \ln \varepsilon)$. Once the analysis in these three charts is performed, the transition Π can be described by overlapping the individual transitions in each of the charts via the corresponding matching maps.

We emphasize that the order of the function h was already known in [**MR80**]. In fact, rigorous asymptotic expansions of the function h appear in [**MR80**, equation (16.11) page 106]. It is important to note that, although during the 80's the blow-up method was not known per sé in the context of fast-slow systems, the analysis performed in [**MR80**] starts from a rescaling similar to the blow-up presented above, see (8.1) and (8.2) in [**MR80**]. More recently, in [**VGKS05**, Theorem 2] a rigorous asymptotic expansion of the function h is obtained by combining the blow-up method with matched asymptotic analysis. The main idea of [**VGKS05**] is to compute, in each chart, asymptotic expansions of the centre manifolds $\mathcal{M}_i^{\mathrm{a}}$ mentioned above, and then match such local expansions across the blow-up space.

[5]We recall the definition of resonant equilibrium point: consider an ODE of the form $\dot{x} = Ax$, where $x \in \mathbb{R}^n$, and let A have all its eigenvalues $\lambda_1, \dots, \lambda_n$ with non-zero real part. We say that $x = 0$ is resonant if there is a relation $\lambda_j = \sum_{i=1}^n m_i \lambda_i$, where the m_i's are non-negative integers with $\sum_{i=1}^n m_i \geq 2$.

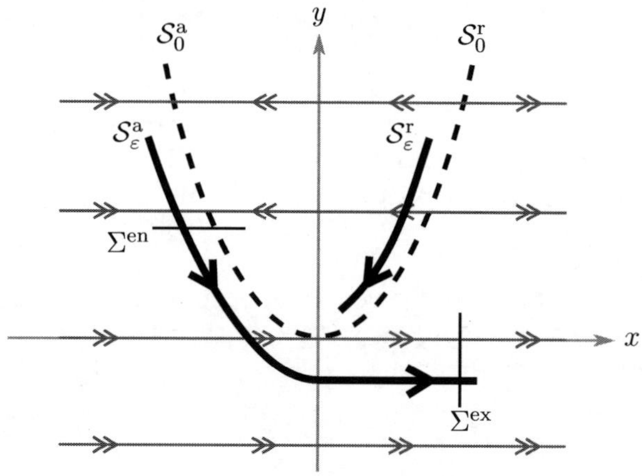

FIGURE 3. Schematic of a fast-slow system near a generic fold point. Up to leading order terms, the critical manifold $\mathcal{C}_0$, shown in dashed, is given by $\mathcal{C}_0 = \left\{(x,y) \in \mathbb{R}^2 \,|\, y = x^2\right\}$. The (blue) lines with double arrows depict the dynamics of the layer equation. Thus, $\mathcal{C}_0$ has an attracting ($\mathcal{S}_0^{\mathrm{a}}$) and a repelling branch ($\mathcal{S}_0^{\mathrm{r}}$). Away from the fold point, Fenichel's theorem shows that, for $\varepsilon > 0$ sufficiently small, $\mathcal{S}_0^{\mathrm{a}}$ and $\mathcal{S}_0^{\mathrm{r}}$ are smoothly perturbed to invariant manifolds (in this case trajectories) $\mathcal{S}_\varepsilon^{\mathrm{a}}$ and $\mathcal{S}_\varepsilon^{\mathrm{r}}$ respectively. The analysis (via the blow-up method) shows that $\mathcal{S}_\varepsilon^{\mathrm{a}}$ can be extended beyond the fold point as depicted in the figure. In particular, one can show that the distance between the x-axis and the intersection $\mathcal{S}_\varepsilon^{\mathrm{a}} \cap \Sigma^{\mathrm{ex}}$ is of order $\mathcal{O}(\varepsilon^{2/3})$.

3.2. Planar canards. In Section 3.1 we have seen that trajectories near a generic fold point first follow the attracting part of the critical manifold, and then follow the fast direction. Perhaps counter-intuitively, for certain one-parameter families of planar fast-slow systems, there are trajectories near a fold singularity that can closely follow the unstable region of the critical manifold for time of order $\mathcal{O}(1)$. Such type of trajectories are called *canards*, see Figure 4 (and the rest of the figures in this Section). In this section we summarize the results of [**DR96**, **KS01c**] and Section 3 of [**KS01a**] dealing with the description of canards in planar fast-slow systems.

REMARK 3.3. The blow-up technique was first introduced for the analysis of fast-slow systems in [**DR96**] to explain, in a geometric way, the canard phenomenon in the van der Pol oscillator. Canards were already explained in [**Ben81**] with the language of non-standard analysis. Although [**DR96**] was the first work describing geometrically the canard phenomenon, we start from [**KS01a**] as it follows several arguments analogous to the generic fold point.

We start by considering a one-parameter family of planar fast-slow systems given by

$$\begin{aligned} x' &= f(x,y,\omega,\varepsilon), \\ y' &= \varepsilon g(x,y,\omega,\varepsilon), \end{aligned} \tag{3.9}$$

where $\omega \in \mathbb{R}$. Just as in the generic fold case of Section 3.1 we may assume that a fold point exists at the origin, say for $\omega = 0$.

DEFINITION 3.4 (Canard point). Consider (3.9). We say that $(x, y) = (0, 0)$ is a *canard point* if

$$f(0,0,0,0) = 0, \quad \frac{\partial f}{\partial x}(0,0,0,0) = 0, \quad g(0,0,0,0) = 0 \tag{3.10}$$

hold together with the non-degeneracy conditions

$$\frac{\partial^2 f}{\partial x^2}(0,0,0,0) \neq 0, \quad \frac{\partial f}{\partial y}(0,0,0,0) \neq 0, \quad \frac{\partial g}{\partial x}(0,0,0,0) \neq 0, \quad \frac{\partial g}{\partial \omega}(0,0,0,0) \neq 0. \tag{3.11}$$

The main difference with the generic fold case of Section 3.1 is that there is an equilibrium point at the origin, which is imposed by $g(0, 0, 0, 0) = 0$. The non-degeneracy conditions imply a transverse intersection at the origin of the nullclines of (3.9). The critical manifold of (3.9) is, up to leading order terms, the same as in Section 3.1. So, we use the same notation and do not repeat its definition here. Similarly to the generic fold case, it can be shown that a planar fast-slow system has a local expression near a canard point given by

$$\begin{aligned} x' &= -yh_1(x, y, \omega, \varepsilon) + x^2 h_2(x, y, \omega, \varepsilon) + \varepsilon h_3(x, y, \omega, \varepsilon), \\ y' &= \varepsilon \left(x h_4(x, y, \omega, \varepsilon) - \omega h_5(x, y, \omega, \varepsilon) + y h_6(x, y, \omega, \varepsilon) \right), \end{aligned} \tag{3.12}$$

where

$$\begin{aligned} h_3 &= \mathcal{O}(x, y, \omega, \varepsilon), \\ h_j &= 1 + \mathcal{O}(x, y, \omega, \varepsilon), \quad j = 1, 2, 4, 5. \end{aligned} \tag{3.13}$$

REMARK 3.5. Up to leading order terms (3.12) reads as

$$\begin{aligned} x' &= -y + x^2, \\ y' &= \varepsilon(x - \omega). \end{aligned} \tag{3.14}$$

Let, for a moment, $\varepsilon > 0$. Then a Hopf bifurcation occurs at $\omega = 0$. Moreover, the norm of the corresponding eigenvalues tends to zero (and to infinity in the slow time scale) as $\varepsilon \to 0$. This situation is known as *singular Hopf bifurcation* [**Bra98, BE86**].

Define the constants

$$\begin{aligned} a_1 &= \frac{\partial h_3}{\partial x}(0,0,0,0), \\ a_5 &= h_6(0,0,0,0), \\ A &= -\frac{\partial h_1}{\partial x}(0,0,0,0) + 3\frac{\partial h_2}{\partial x}(0,0,0,0) - 2\frac{\partial h_4}{\partial x}(0,0,0,0) - 2a_5 \end{aligned} \tag{3.15}$$

and the sections

$$\Sigma^j = \left\{ (x, y) \in \mathbb{R}^2 \,|\, x \in I_j,\, y = \rho^2 \right\}, \qquad j = \text{en}, \text{ex}, \tag{3.16}$$

where $\rho > 0$ and I_j is a suitable small interval, see Figure 4. Let $q_{\text{en},\varepsilon} = \Sigma^{\text{en}} \cap \mathcal{S}^{\text{a}}_{\varepsilon}$, $q_{\text{ex},\varepsilon} = \Sigma^{\text{ex}} \cap \mathcal{S}^{\text{r}}_{\varepsilon}$ and denote by $\Pi : \Sigma^{\text{en}} \to \Sigma^{\text{ex}}$ the transition induced by (3.9). The main result that describes the behavior of the slow manifolds $\mathcal{S}^{\text{a}}_{\varepsilon}$ and $\mathcal{S}^{\text{r}}_{\varepsilon}$ around the canard point is as follows.

THEOREM 3.6 ([**KS01a**, Theorem 3.1]). *Assume that* (3.9) *has a canard point at the origin, i.e. satisfies* (3.10) *and* (3.11). *Assume that the solution* $x_0(t)$ *of the reduced problem connects* $\mathcal{S}_0^{\rm a}$ *to* $\mathcal{S}_0^{\rm r}$. *Then there exists* $\varepsilon_0 > 0$ *and a smooth function* $\omega_c(\sqrt{\varepsilon})$ *defined on* $[0, \varepsilon_0]$ *such that for* $\varepsilon \in (0, \varepsilon_0]$ *the following assertions hold.*

(C1) $\Pi(q_{\text{en},\varepsilon}) = q_{\text{ex},\varepsilon}$ *if and only if* $\omega = \omega_c(\sqrt{\varepsilon})$.

(C2) *The function* ω_c *has the expansion*

$$\omega_c(\sqrt{\varepsilon}) = -\left(\frac{a_1 + a_5}{2} + \frac{1}{8}A\right)\varepsilon + \mathcal{O}(\varepsilon^{3/2}) \tag{3.17}$$

(C3) *The transition map* Π *is defined only for* ω *in an interval around* $\omega_c(\sqrt{\varepsilon})$ *of size* $\mathcal{O}(\exp(-C/\varepsilon))$ *for some* $C > 0$.

(C4) *There is a splitting condition at* $\omega = \omega_c(\sqrt{\varepsilon})$ *given by*

$$\frac{\partial}{\partial \omega}(\Pi(q_{\text{en},\varepsilon}) - q_{\text{ex},\varepsilon})|_{\omega = \omega_c(\sqrt{\varepsilon})} > 0. \tag{3.18}$$

A schematic representation of Theorem 3.6 is given in Figure 4. The most important aspect of Theorem 3.6 is that it tells us that for $\varepsilon > 0$ there is exactly one value of the parameter ω for which $\mathcal{S}_\varepsilon^{\rm a}$ is extended to *exactly* $\mathcal{S}_\varepsilon^{\rm r}$, that is, they coincide as stated in (C1). The previous trajectory is usually called *maximal canard* to distinguish it from others that may follow $\mathcal{S}_\varepsilon^{\rm r}$ for time of order $\mathcal{O}(1)$. Another important point of Theorem 3.6 is its third item (C3), which tells us that canard orbits near planar folds occur only for values of ω within an exponentially small interval. Related to the previous observation we further note that the intersection of $\mathcal{S}_\varepsilon^{\rm a}$ and $\mathcal{S}_\varepsilon^{\rm r}$ is not transversal and thus small C^1-perturbations of the parameter ω destroy the maximal canard; see (C4) and see also Section 3.3.

Let us now point-out the key ingredients of the analysis to prove Theorem 3.6. In many ways, the blow-up analysis of a canard point is similar to that of a generic fold point. One of the differences is that now the parameter ω is included in the blow-up. The appropriate blow-up transformation is defined by

$$x = \bar{r}\bar{x}, \quad y = \bar{r}^2\bar{y}, \quad \varepsilon = \bar{r}^2\bar{\varepsilon}, \quad \omega = \bar{r}\bar{\omega}. \tag{3.19}$$

Due to the definition of the $\Sigma^{\rm en}$ and $\Sigma^{\rm ex}$ sections, the blow-up analysis is performed only in two charts, namely $K_1 = \{\bar{y} = 1\}$ and $K_2 = \{\bar{\varepsilon} = 1\}$. The analysis in chart K_1 for the canard point is analogous to the analysis in chart K_1 for the generic fold point. The only difference is that in the canard case, the consideration of ω in the blow-up map increases by one the dimension of the centre manifolds. Thus, in K_1, the main object of study are 3-dimensional centre manifolds $\mathcal{M}_1^{\rm a}$ and $\mathcal{M}_1^{\rm r}$ that are of centre-stable and centre-unstable type respectively. These centre manifolds correspond to invariant manifolds $\mathcal{M}_2^{\rm a}$ and $\mathcal{M}_2^{\rm r}$ in the chart K_2. In chart K_2 the key object of study is a planar ODE of the form

$$\begin{aligned} x_2' &= -y_2 + x_2^2, \\ y_2' &= x_2, \end{aligned} \tag{3.20}$$

which is obtained after restriction to $\{r_2 = \omega_2 = 0\}$; compare with (3.8). One may check that (3.20) is completely integrable, having a constant of motion denoted by $H(x_2, y_2)$. This means that the trajectories of (3.20) are determined by the level curves of H. Here $(x_2, y_2, r_2, \omega_2)$, as usual, denote local coordinates in K_2.

There is a special curve γ_c given by the solution of $H = 0$. The major importance of γ_c is that it precisely connects $\mathcal{S}_0^{\rm a}$ with $\mathcal{S}_0^{\rm r}$. Naturally in the blow up space $\mathcal{S}_0^{\rm a}$ and $\mathcal{S}_0^{\rm r}$ correspond to $\mathcal{M}_2^{\rm a}|_{r_2=0}$ and $\mathcal{M}_2^{\rm r}|_{r_2=0}$ respectively.

The next step is to investigate how such a connection breaks for $r_2 > 0$, $\omega_2 \neq 0$, which will describe how the manifolds $\mathcal{M}_2^{\rm a}$ and $\mathcal{M}_2^{\rm r}$ are connected. For such a purpose an analysis based on the Melknikov method [**GH13**] is used. In essence, the Melknikov method provides an expression for the distance between $\mathcal{M}_2^{\rm a}$ and $\mathcal{M}_2^{\rm r}$, denoted by $d = d(r_2, \omega_2)$. Therefore, for $\varepsilon > 0$ the manifolds $\mathcal{S}_\varepsilon^{\rm a}$ and $\mathcal{S}_\varepsilon^{\rm r}$ are connected if and only if $d = 0$. Careful estimates of the function d and a blow-down provide the expression for ω_c in Theorem 3.6 and the breaking argument proving (C4). We finalize the summary of the canard point by mentioning that the role of the constant A has to do with the non-degeneracy of the Hopf bifurcation that occurs for $\omega = 0$, see for example [**Kue15**, Section 8.3], and Figure 8.

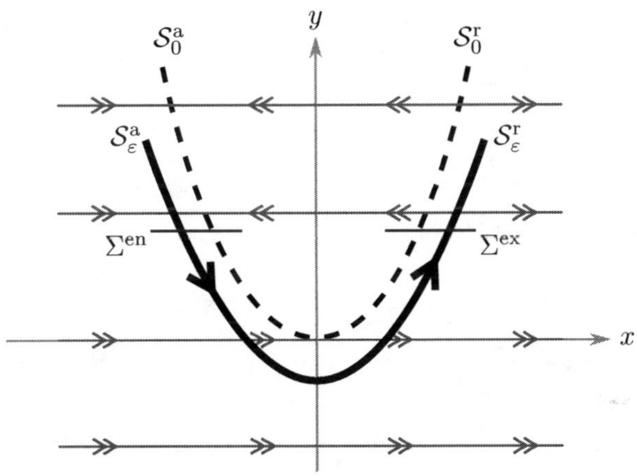

FIGURE 4. Schematic representation of a maximal canard. Note that $\mathcal{S}_\varepsilon^{\rm a}$ extends precisely to $\mathcal{S}_\varepsilon^{\rm r}$ beyond the origin. Trajectories passing through $\Sigma^{\rm en}$ exponentially close to $\mathcal{S}_\varepsilon^{\rm a}$ follow $\mathcal{S}_\varepsilon^{\rm r}$ for time of order $\mathcal{O}(1)$ before being repelled from it.

Next, let us discuss canards in the context of the van der Pol oscillator as they have a particular motivating relevance for the entire subject. Here the main reference is [**DR96**]. To be more precise, [**DR96**] is concerned with *canard cycles*. Canard cycles are closed orbits having "a canard piece". The model of the van der Pol oscillator to be considered is

$$\begin{aligned} x' &= y - \frac{x^2}{2} - \frac{x^3}{3}, \\ y' &= \varepsilon(\omega - x). \end{aligned} \tag{3.21}$$

In the singular limit $\varepsilon = 0$, the phase portrait is as shown in Figure 5 and the critical manifold is a cubic curve given by

$$\mathcal{C}_0 = \left\{ (x,y) \in \mathbb{R}^2 \,|\, y = \frac{x^2}{2} + \frac{x^3}{3} \right\}. \tag{3.22}$$

We note that $\mathcal{C}_0$ has two fold points namely $(x,y) = \left\{ \left(-1, \frac{1}{6}\right), (0,0) \right\}$. Furthermore, there is a canard point at the origin for $\omega = 0$. It is well known [**LDMP77**]

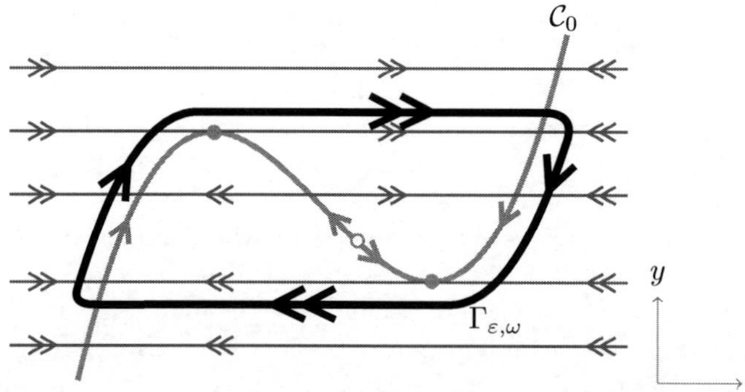

FIGURE 5. The van der Pol oscillator: the critical manifold is the cubic curve $\mathcal{C}_0$; the curve in this case is also called S-shaped curve, which is motivated by viewing it in the (y, x)-plane. The critical manifold has two fold points indicated by filled circles. The singular limit is represented by the blue lines (layer equation) and by the arrows along $\mathcal{C}_0$ (CDE). The CDE has an equilibrium point at $x = \omega$ indicated by an empty circle. As it is described in this Section, the parameter ω is essential to distinguish several types of limit cycles. The curve $\Gamma_{\varepsilon,\omega}$ indicates a typical limit cycle that exists for $\varepsilon > 0$ sufficiently small.

that (3.21) has at most one limit cycle and that when such a limit cycle exists, it is hyperbolic and attracting. Assuming that a limit cycle $\Gamma_{\varepsilon,\omega}$ of (3.21) exists, the question posed by [**DR96**] can be formulated as follows: "what happens to $\Gamma_{\varepsilon,\omega}$ as $\varepsilon \to 0$?" The main contribution of [**DR96**] is the precise asymptotic description of the canard cycles that occur for (3.21). In other words, [**DR96**, Theorem 1] provides asymptotic estimates of the parameter ω for each of the canard cycles shown in Figure 6. All such cycles occur within an exponentially small interval of the parameter ω. This means that the transition between small amplitude canard cycles towards relaxation oscillations occurs for an exponentially small variation of the parameter. This effect is called *canard explosion*. The analysis presented in [**DR96**] is highly non-trivial: besides introducing the blow-up method for fast-slow systems, it uses techniques of nonlinear analysis such as normal form theory, centre manifold theory, asymptotic analysis, Abelian integrals [**Bli33, DR06**], foliations [**CN13**], among others.

Extensions and generalizations of the results of [**DR96**] can be found in [**KS01c**]. The setting of [**KS01c**] is not confined to the van der Pol oscillator, but it applies to general one parameter families of planar fast-slow systems (3.9) having and S-shaped critical manifold (a particular case being the van der Pol oscillator). Similar to [**DR96**] the main result of [**KS01c**] is the proof that the transition between small amplitude periodic orbits to relaxation oscillations occurs within an exponentially small interval of the parameter ω, the precise statement is in [**KS01c**, Theorem 3.3], while results on the stability of the canard cycles are given in [**KS01c**, Theorems 3.4, 3.5 and 3.6]. We briefly summarize the strategy to prove such Theorems as follows. Given an S-shaped critical manifold, one assumes that one of the fold points is generic while the other is a canard point. Then, one

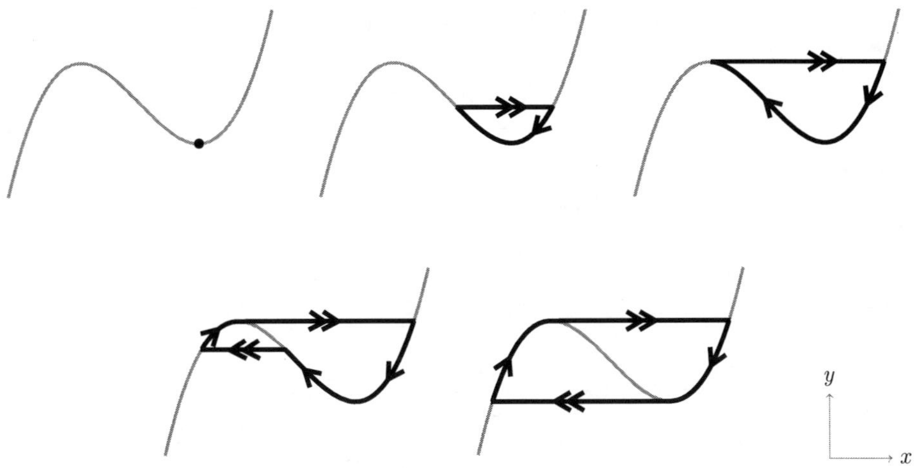

FIGURE 6. Singular canard cycles (or limit periodic sets) giving rise to canard cycles in the van der Pol oscillator, compare with Figure 5. For $\varepsilon > 0$ sufficiently small closed orbits exist as perturbations of these singular cycles.

can locally study the flow near the aforementioned fold points with the techniques reviewed in Section 3.1 and at the beginning of this Section. Afterwards, families of periodic orbits can be constructed by "gluing" together pieces of trajectories in the blow-up space with trajectories of the layer equation and trajectories near normally hyperbolic parts of the critical manifold. A schematic of this construction is shown in Figure 7, while a bifurcation diagram, also representing the canard explosion is shown in Figure 8.

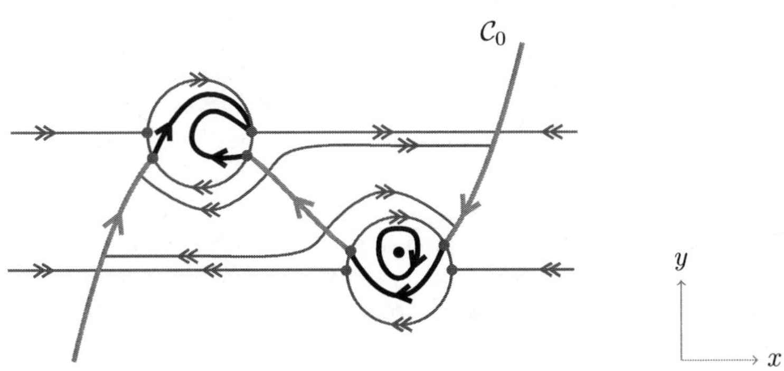

FIGURE 7. Schematic of the families of periodic orbits, parametrized by ω in (3.21), constructed by gluing together trajectories in the blow-up space with trajectories in the singular limit. The flow near the fold and canard points (compare with Figure 5) have been replaced by the flow in the blow-up space. Appropriate combinations of orbits correspond to those depicted in Figure 5. Perturbations of these orbits lead to canard cycles for $\varepsilon > 0$ sufficiently small.

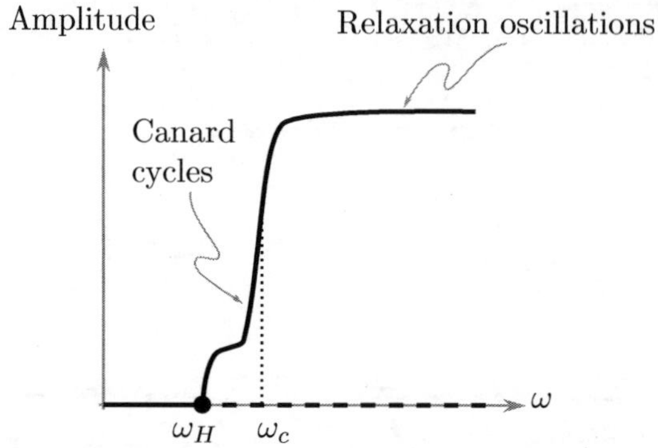

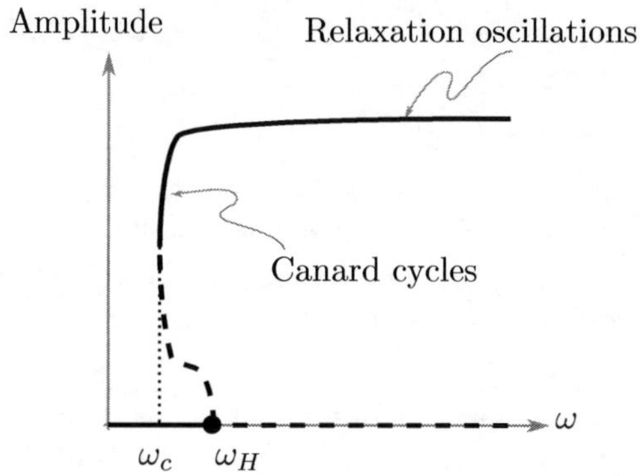

FIGURE 8. Bifurcation diagrams corresponding to a canard explosion for $A < 0$ on the top and $A > 0$ on the bottom (refer to Theorem 3.6). A Hopf bifurcation occurs at $\omega = \omega_H$ and a maximal canard when $\omega = \omega_c$. These diagrams depict the amplitude and stability of the limit cycles $\Gamma_{\varepsilon,\omega}$ for fixed $\varepsilon > 0$ sufficiently small. In particular, we have that for $A < 0$ all canard cycles are stable.

So far, we have seen that a mechanism leading to canards in planar fast-slow systems is a singular Hopf bifurcation. Some higher codimension bifurcations leading to canards have also been studied via the blow-up method. For example, for the Bogdanov-Takens bifurcation, which is of codimension 2, we have e.g. [**DMD11b, Chi11**]. The degenerate case, that is a planar system for which normal hyperbolicity is lost up to arbitrary order, has been dealt with in e.g. [**DMD06**]. Another recent discovery is that canards can also be found in cases where the critical manifold is of codimension zero [**KM19**], that is, in the case when the critical manifold is of the same dimension as the phase-space. Finally, we note that there is strong active research towards proving Hilbert's 16th problem [**Sma98**] based on the tools (and their further generalizations) described in this Section. A non-exhaustive list of references dealing with the cyclicity of canard cycles is: for (fast-slow) Liénard systems [**DMD08, DMD11a, DMDR11, DR07, DR09, HDMD13, HDMD14, Huz16, DPR07**] and non-Liénard systems e.g. [**Huz18**]. A proof of finite cyclicity for fast-slow Darboux systems is available in [**BG16**].

3.3. Folded singularities and canards in higher dimensions. In Section 3.2 we have seen that canards appear in one-parameter families of planar fast-slow systems having a folded critical manifold. However, canards are degenerate in such a case since they exist only for an exponentially small interval of parameter values and disappear after small perturbations of the parameter. This situation is "fixed" in fast slow-systems with at least two slow variables, where canards are generic.

Canards in $\mathbb{R}^3$ were first studied in [**Ben83, Ben90, Ben01**] with techniques from non-standard analysis; see also [**BKR02**] for a different approach in which the critical manifold is assumed to have a generic self-intersection. In this section we review some of the works that, based on the blow-up method, have provided a geometric description of canards in fast-slow systems with two, or more, slow variables.

Let us start with a fast-slow system with two slow and one fast variable given by

$$
\begin{aligned}
x' &= f(x, y_1, y_2, \varepsilon),\\
y_1' &= \varepsilon g_1(x, y_1, y_2, \varepsilon),\\
y_2' &= \varepsilon g_2(x, y_1, y_2, \varepsilon),
\end{aligned}
\tag{3.23}
$$

where $(x, y) = (x, y_1, y_2) \in \mathbb{R}^3$. Analogous to the planar case, the critical manifold

$$
\mathcal{C}_0 = \left\{(x, y_1, y_2) \in \mathbb{R}^3 \,|\, f(x, y_1, y_2, 0) = 0\right\}
\tag{3.24}
$$

is a non-degenerate folded surface if

(3.25)

$$
f(0,0,0,0) = 0,\ \frac{\partial f}{\partial x}(0,0,0,0) = 0,\ \frac{\partial^2 f}{\partial x^2}(0,0,0,0) \neq 0,\ \mathrm{D}_y f(0,0,0,0) \neq 0 \in \mathbb{R}^2.
$$

Without loss of generality we may assume that $\frac{\partial f}{\partial y_1}(0,0,0,0) \neq 0$. Note that there is a fold line along $\ell_f = \left\{(x, y) \in \mathbb{R}^3 \,|\, x = 0,\ y_1 = 0\right\}$ and that the phase-space of the slow subsystem is 2-dimensional. Similar to the planar case, the corresponding critical manifold is a folded surface as shown in Figure 9.

Analogous to the planar case, we say that the origin is a generic fold point if $g_1(0,0,0,0) \neq 0$. In this case (3.23) can be rewritten as

$$
\begin{aligned}
x' &= y_1 + x^2 + \mathcal{O}(\varepsilon x, \varepsilon y_1, \varepsilon y_2, \varepsilon^2, x^3, x y_1 y_2, x^3 y_1),\\
y_1' &= \varepsilon(1 + \mathcal{O}(x, y_1, y_2, \varepsilon)),\\
y_2' &= \varepsilon g_2(x, y_1, y_2, \varepsilon).
\end{aligned}
\tag{3.26}
$$

The analysis of (3.26) as its flow passes near ℓ_f is similar to that of the generic fold in Section 3.1, and the details can be found, for example, in [**SW04**, Theorem 1]. On the other hand, we say that the origin of (3.23) is a *folded singularity* if besides (3.25) we have

$$
\left(\frac{\partial f}{\partial y_1} g_1 + \frac{\partial f}{\partial y_2} g_2\right)\Big|_{(0,0,0,0)} = 0.
\tag{3.27}
$$

If the origin is a folded singularity, then (3.23) can be rewritten as [**SW01**, Proposition 2.1]

$$
\begin{aligned}
x' &= y_1 + x^2 + \mathcal{O}(\varepsilon x, \varepsilon y_1, \varepsilon y_2, \varepsilon^2, x^3, x y_1 y_2, x^3 y_1),\\
y_1' &= \varepsilon(ax + by_2 + \mathcal{O}(y_1, \varepsilon, y_2^2, x y_2, x^2)),\\
y_2' &= \varepsilon(c + \mathcal{O}(x, y_1, y_2, \varepsilon)).
\end{aligned}
\tag{3.28}
$$

Since a folded singularity is necessary for the existence of canards, we now focus on (3.28). Note that due to the leading order terms of (3.28) we have that $\mathcal{C}_0$ is attracting for $x < 0$ and repelling for $x > 0$. The relationship between the flow on the critical manifold and the fold line is of vital importance. A classification of the possible phase-portraits on $\mathcal{C}_0$ near ℓ_f can be found in [**Tak76**, Theorem 5.1] and in [**SW01**, Lemma 2.1]. Such a classification makes use of the so-called *desingularized system* [**Tak76**, Section 4.10] which, up to leading order terms, is given by

$$
\begin{aligned}
\dot{x} &= ax + by_2,\\
\dot{y}_2 &= -2cx.
\end{aligned}
\tag{3.29}
$$

Note that the origin is an equilibrium point of (3.29). Even though the origin is not an equilibrium point of (3.28), it is possible to relate the flow of (3.29) to the flow of (3.28) on $\mathcal{C}_0$. Thus, a classification of the flow of (3.29) is useful and is summarized in Table 1.

Condition	Eigenvalues	Name
$bc < 0$	$\lambda_{1,2} \in \mathbb{R}$, $\lambda_1 < 0 < \lambda_2$	saddle
$0 < 8bc < a^2$	$\lambda_{1,2} \in \mathbb{R}$, $\lambda_1\lambda_2 > 0$	node
$a^2 < 8bc$	$\lambda_{1,2} \in \mathbb{C}$, $\Re(\lambda_{1,2}) \neq 0$	focus
$bc = 0$, $c \neq 0$	$\lambda_1 = 0$, $\lambda_2 = a$	saddle-node type I
$bc = 0$, $b \neq 0$	$\lambda_1 = 0$, $\lambda_2 = a$	saddle-node type II
$8bc = a^2$	$\lambda_{1,2} = \frac{a}{2}$	degenerate node

TABLE 1. Topological Classification of (3.29).

We can see from the above classification that the saddle, node, and focus cases are generic, while the saddle-node and the degenerate node are both degenerate and of codimension 1. From (3.29) one can obtain the flow on $\mathcal{C}_0$ as follows: away from ℓ_f, the trajectories on $\mathcal{C}_0|_{x<0}$ are given by trajectories of (3.29) restricted to $x < 0$, while trajectories on $\mathcal{C}_0|_{x>0}$ are given by reversing the trajectories of (3.29) restricted to $x > 0$, see the details in e.g. [**Tak76**, Section 4.10], and Figure 9 for a couple of examples.

Due to the above classification and relationship between the slow subsystem and the desingularized system, we call the origin of (3.28) a *folded saddle*, *folded node*, *folded focus*, *folded saddle-node*, or *folded degenerate node* according to the coefficients of Table 1. We now have the following definition of canards coming from the flow on $\mathcal{C}_0$.

DEFINITION 3.7. Solutions of the slow subsystem passing through a canard point from an attracting critical manifold to a repelling critical manifold are called *singular canards*. Solutions of the slow subsystem passing through a canard point from a repelling critical manifold to an attracting critical manifold are called *faux canards*.

See, for example Figure 9 (a) where Γ_1 is a singular canard and Γ_2 a faux canard.

It follows from the classification of Table 1 and the resulting reduced flow that there are singular canards for all cases, except for the folded focus case. We can now recall the main result of [**SW01**], stating under which conditions singular canards persist for $\varepsilon > 0$ sufficiently small.

THEOREM 3.8 ([**SW01**, Theorem 4.1]). *Assume* (3.28). *In the folded saddle and in the folded node case singular canards* Γ_1 *perturb to maximal canard solutions for sufficiently small* ε.

For a folded node with $\lambda_1 < \lambda_2 < 0$ *a maximal canard solution corresponding to the weak eigendirection* (Γ_2) *exists for sufficiently small* ε *provided that the ratio* $\mu_2 = \lambda_1/\lambda_2$ *is not a natural number.*

See Figure 9 for a schematic representation of the trajectories Γ_1 and Γ_2 in each case. We shall denote the perturbation of Γ_1 and of Γ_2 by $\Gamma_{1,\varepsilon}$ and $\Gamma_{2,\varepsilon}$, respectively. The idea of the proof is similar to that in Section 3.2. The relevant charts are $K_1 = \{\bar{y}_1 = 1\}$ and $K_2 = \{\bar{\varepsilon} = 1\}$. It is precisely in chart K_2 where transversality of the intersection of the slow manifolds can be shown. We now sketch the main procedure: first of all, the blown up vector field in chart K_2 has (up to appropriate conditions met by the hypothesis of Theorem 3.8) explicit algebraic solutions γ_1, γ_2 [**SW01**, Lemma 4.2]. These are crucial as they connect the repelling and attracting parts of the slow manifolds (equivalently, and analogously to Section 3.2, of centre manifolds in chart K_1). Moreover, using variational arguments, it is further shown by studying a Weber equation (see [**AS65**, Chapter 19] and [**OLBC10**, Chapter 12]) that the invariant manifolds in chart K_2 corresponding to the slow manifolds $\mathcal{S}_\varepsilon^{\mathrm{a}}$ and $\mathcal{S}_\varepsilon^{\mathrm{r}}$ intersect transversaly along the γ_i solutions. Furthermore. the meaning of the ratio μ_2 is quite interesting.

LEMMA 3.9 ([**SW01**, Lemma 4.4]). *Suppose we are in the node scenario and that* $n - 1 < \mu_2 < n$. *Then the slow manifolds* $\mathcal{S}_\varepsilon^{\mathrm{a}}$ *and* $\mathcal{S}_\varepsilon^{\mathrm{r}}$ *twist* $n-1$ *times around the corresponding maximal canard in the neighborhood of the fold curve.*

Regarding the folded node case, it is further shown in [**Wec05**] that one may find more canards bifurcating, upon variation of μ_2, from the weak maximal canard. In terms of the slow manifolds, this means that $\mathcal{S}_\varepsilon^{\mathrm{a}}$ and $\mathcal{S}_\varepsilon^{\mathrm{r}}$ have further transversal intersections, called *secondary maximal canards.* To better grasp what happens in this situation, let us first note that in the folded node case we can differentiate two types of singular canards: a) singular canards Γ_1 and Γ_2 corresponding to the (strong and weak) eigendirections of (3.29) and b) all other singular canards within the shaded region of Figure 9. The latter type of singular canards may also correspond to maximal canards for $\varepsilon > 0$ sufficiently small, this motivates the next definition.

DEFINITION 3.10. All maximal canards not obtained as perturbations of the singular canards Γ_1, Γ_2 using Theorem 3.8 are called *secondary maximal canards.*

Skipping the technicalities, the main result of [**Wec05**] is as follows: "If $\mu_2 \in \mathbb{N}$, then for $\varepsilon > 0$ sufficiently small there are additional secondary maximal canards bifurcating from the weak maximal canard upon variation of the ratio μ_2". The proof of the aforementioned bifurcation result requires a highly non-trivial analysis partially contained in [**Wec05**] and further completed in [**MW17**].

More recently, the folded saddle case has been studied in more detail in [**MW17**]. Interestingly, it is proven that for certain values of the ratio μ_2 there is rotational behaviour of trajectories around the faux canard Γ_2 [**MW17**, Lemma 1]. This rotational behaviour has important consequences, in particular, it is shown that trajectories with distinct rotational behaviour can be identified. The boundaries of sectors of trajectories with the same rotational behaviour are given by the so-called *fast manifolds* (the fast manifolds are defined as the nonlinear stable and unstable fibre bundles of the primary faux canard Γ_2). Furthermore, the transversal intersection of these fast manifolds is responsible for the creation of secondary faux canards. This contrasts with the folded node case where, as described above, the intersection of the slow manifolds of the weak maximal canard is the responsible for the creation of secondary canards. The main analysis of the secondary canards and their bifurcations relies on a variational method and an extension of Melknikov's method.

Visualizing the slow manifolds, especially as they twist in the folded node case, is difficult and the reader is referred to e.g.[**DKO08, GH05**] for further details. We finalize this section by briefly sketching some generalizations and referring to works regarding the degenerate saddle-node cases.

- The canard theory sketched above can be extended to fast-slow systems with $m \geq 2$ slow and $n \geq 1$ fast variables [**Wec12**]. The arguments use the assumption of $n-1$ stable fast directions and centre manifold reduction.
- The codimension 1 folded saddle node cases are studied with the help of the blow-up method in [**KW10**] (FSN-II) and in [**VW15**] (FSN-I). One primary interest in studying such folded singularity is their relation to complex oscillatory motion such as Mixed Mode Oscillations (MMOs) [**DGK$^+$12**] and chaos. Furthermore, folded saddle-nodes appear in many models of applied interest such as the forced van der Pol oscillator [**BEG$^+$03, GWY06, BDG$^+$16**], biochemical reactions [**MSLG98, MS01, Moe02**], and neuron dynamics [**EW09, KPK08, DMW15**], to mention a few. Unfolding such degenerate cases, that is considering one-parameter families of fast-slow systems with folded singularities, provides a much richer geometry. One difference to keep in mind between the two cases is that the FSN-II has an associated singular Hopf bifurcation with two slow variables [**Guc08**], while the FSN-I does not present a Hopf bifurcation but rather a true saddle-node.

3.4. Delayed Hopf bifurcation. In dynamic bifurcation problems, that is where the bifurcation parameter is slowly changed, it can be observed that the transition to instability occurs for parameter values way beyond the prediction of the static bifurcation diagram [**Nei09**]. This effect is known as *delayed loss of stability*. One landmark example of such a phenomenon occurs in the so-called *slow passage through a Hopf bifurcation* [**BER89**]. The aforementioned dynamical behavior is also highly relevant in many applications, e.g. [**SM96, BC11**]; see also [**KV18**] for a recent example of slow passage through a Hopf bifurcation in reaction-diffusion equations.

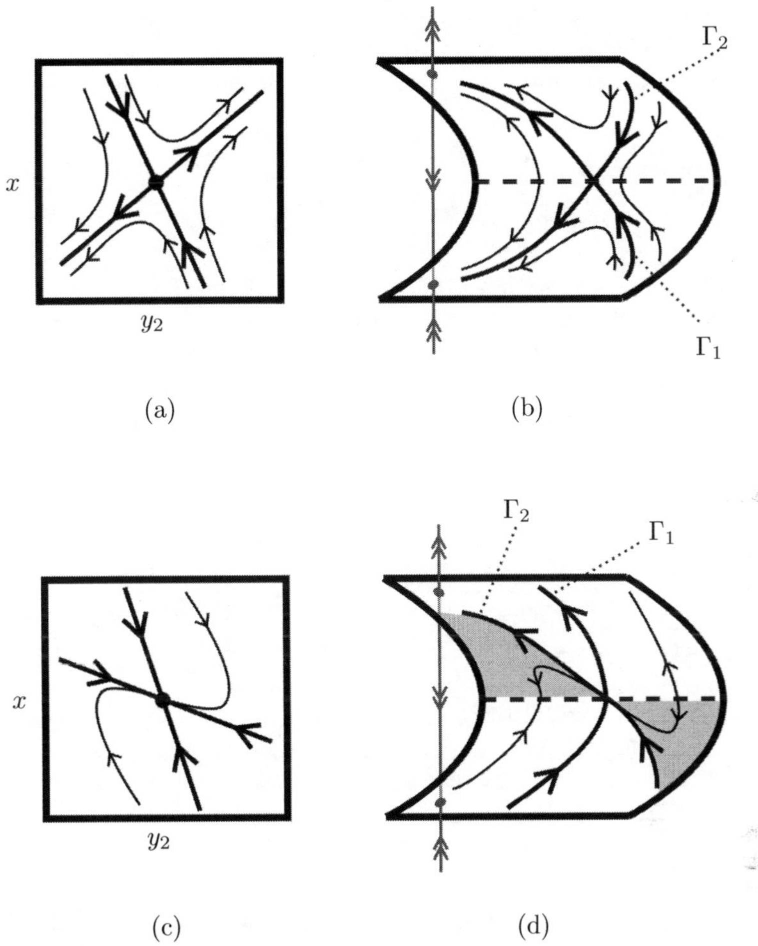

FIGURE 9. Phase-portraits for the folded saddle (first row) and folded node (second row) cases of Table 1. The planar phase-portraits (a),(c) correspond to the flow of the desingularized vector field (3.29), while the phase-portraits (b),(d) are the corresponding reduced flows on $\mathcal{C}_0$. Note that trajectories on the stable (i.e. lower) part of $\mathcal{C}_0$ are given by trajectories of the desingularized system for $x < 0$ while the trajectories on the unstable (i.e. upper) part of $\mathcal{C}_0$ are given by reversing the direction of the trajectories of the desingularized system for $x > 0$. In the saddle case (a),(b) the trajectory Γ_1 is a true canard while Γ_2 is a faux canard. Note that these are the only singular canard trajectories. In the node case (c),(d) Γ_1 is a singular canard tangent to the strong eigendirection while Γ_2 is a singular canard tangent to the weak eigendirection. Observe that, in contrast to the saddle case, in the node case we have many other singular canard trajectories indicated by the shaded region in $\mathcal{C}_0$.

Although the problem of slow passage through a Hopf bifurcation has been long studied, [**HKSW16**] presents a new geometric proof based on the blow-up method, which we now sketch. Consider a fast-slow system (2.6)-(2.8) with $(m, n) = (2, 1)$ given by

$$\begin{aligned} x_1' &= yx_1 - x_2 - x_1(x_1^2 + x_2^2), \\ x_2' &= x_1 + yx_2 - x_2(x_1^2 + x_2^2), \\ y' &= \varepsilon. \end{aligned} \tag{3.30}$$

Note that for $\varepsilon = 0$ we have that (3.30) corresponds to the normal form of a (supercritical) Hopf bifurcation with y the bifurcation parameter and $y = 0$ the bifurcation point. This motivates the following definition.

DEFINITION 3.11. A Hopf bifurcation that occurs in the layer equation of a fast-slow system in which the slow variable acts as the bifurcation parameter is called *Delayed Hopf Bifurcation.*

The idea now is to study the dynamics of (3.30) as the "dynamic parameter" y slowly passes through the Hopf bifurcation point $y = 0$. It is straightforward to see that the critical manifold is

$$\mathcal{C}_0 = \left\{ (x_1, x_2, y) \in \mathbb{R}^3 \,|\, x_1 = 0, x_2 = 0 \right\}, \tag{3.31}$$

i.e., it is just the y-axis. One can further show, for example by changing to polar coordinates, that $\mathcal{C}_0|_{y<0}$ is attracting while $\mathcal{C}_0|_{y>0}$ is repelling, and that the origin is a non-hyperbolic point of the layer equation. The next observation is of crucial importance: the linearization of the layer equation (3.30) at the non-hyperbolic origin is not nilpotent. In fact it is given by $\dot{x} = Jx$, where

$$J = \begin{bmatrix} 0 & -1 \\ 1 & 0 \end{bmatrix}. \tag{3.32}$$

This is a first obstacle to use the blow-up method. Furthermore, note that such linearization corresponds to a fast rotation of the fast variables. We will see below that the main idea of [**HKSW16**] is to "eliminate" such a fast rotation allowing the use of the blow-up method. It is convenient to consider a modified complex version

$$\begin{aligned} z' &= (y + \imath)z - |z|^2 z + \varepsilon k_0, \\ y' &= \varepsilon. \end{aligned} \tag{3.33}$$

System (3.33), with $k_0 = 0$ is obtained from (3.30) via the change of coordinates $z = x_1 + \imath x_2$ with $\imath = \sqrt{-1}$. The term εk_0, $k_0 \neq 0$, is added to break the invariance[6] of $\mathcal{C}_0 = \{z = 0\}$ for $\varepsilon > 0$. It is also clear from (3.33) that $z = 0$ is a stable (resp. unstable) focus for $y < 0$ (resp. $y > 0$) and that at $z = 0$ an exchange of stability occurs. Let $\delta > 0$ be small and, as usual, define

$$\begin{aligned} \mathcal{S}_0^{\mathrm{a}} &= \{(z, y) \in \mathcal{C}_0 \,|\, y < -\delta\}, \\ \mathcal{S}_0^{\mathrm{r}} &= \{(z, y) \in \mathcal{C}_0 \,|\, y > \delta\}, \end{aligned} \tag{3.34}$$

and we denote by $\mathcal{S}_\varepsilon^{\mathrm{a}}$ and $\mathcal{S}_\varepsilon^{\mathrm{r}}$ the corresponding slow manifolds. The manifolds $\mathcal{S}_0^{\mathrm{a}}$ and $\mathcal{S}_0^{\mathrm{r}}$ have analytic extensions to $y = 0$, and the goal is to compute the distance between $\mathcal{S}_\varepsilon^{\mathrm{a}}$ and $\mathcal{S}_\varepsilon^{\mathrm{r}}$ at $y = 0$ for $\varepsilon > 0$ sufficiently small. Let such a distance be

[6] In principle a generic term $\mathcal{O}(\varepsilon)$ can be added instead of εk_0, but the results are similar in such a case.

denoted by $d(\varepsilon)$. The main result of [**HKSW16**] is a new geometric proof of the following well-known theorem.

THEOREM 3.12 ([**HKSW16**, Theorem 3.1] see also [**Nei87, Shi73**]). *Let $k_0 \neq 0$. For sufficiently small values of $\varepsilon > 0$,*

$$d(\varepsilon) \leq \left(\sqrt{\varepsilon}\sqrt{2\pi}\kappa + \mathcal{O}(\varepsilon)\right) \exp\left(-\frac{1}{2\varepsilon}\right), \tag{3.35}$$

where $\kappa > 0$ is bounded away from 0, and is determined in part by $|k_0|$.

The implication of Theorem 3.12 is that, due to the fact that the separation between the slow manifolds $\mathcal{S}_\varepsilon^{\mathrm{a}}$ and $\mathcal{S}_\varepsilon^{\mathrm{r}}$ is exponentially small, solutions starting near $\mathcal{S}_\varepsilon^{\mathrm{a}}$ stay close to $\mathcal{S}_\varepsilon^{\mathrm{r}}$ for time $\mathcal{O}(1/\varepsilon)$ after crossing the bifurcation point. This is observed as a delay in the onset of (large amplitude) oscillations. To prove Theorem 3.12 one first considers y in the complex plane, that is $y = y_R + \imath y_I \in \mathbb{C}$. Next, one chooses a particular path, namely

$$\Gamma_{-1} : y = s - \imath, \qquad s \in (-\infty, \infty). \tag{3.36}$$

The main observation is that, since (3.33) is analytic, by Cauchy's integral theorem it is the equivalent to integrate along Γ_{-1} than to integrate along the real y-axis. The advantage, however, is that along Γ_{-1}, the fast rotation is eliminated, and the resulting non-hyperbolic equilibrium point is nilpotent. So, the fast-slow system (3.33) along Γ_{-1} reads as

$$\begin{aligned} z' &= sz - |z|^2 z + \varepsilon k_0, \\ s' &= \varepsilon, \end{aligned} \tag{3.37}$$

where we can see that $(z, s, \varepsilon) = (0, 0, 0)$ is indeed a nilpotent point. We now proceed to sketch the blow-up analysis. The blow-up map is given by

$$z = r\bar{z}, \quad s = r\bar{s}, \quad \varepsilon = r^2\bar{\varepsilon}. \tag{3.38}$$

The relevant charts are $K_1 = \{\bar{s} = -1\}$, $K_2 = \{\bar{\varepsilon} = 1\}$, and $K_3 = \{\bar{s} = 1\}$. Next, the strategy is similar to the planar fold case: in chart K_1 (resp. K_3) one finds a 2-dimensional centre-stable (resp. centre-unstable) manifold corresponding to $\mathcal{S}_\varepsilon^{\mathrm{a}}$ (resp. $\mathcal{S}_\varepsilon^{\mathrm{r}}$). Then, in chart K_2 one tracks such invariant manifolds. To compute the distance between the manifolds we take advantage of the fact that the blown-up vector field in K_2 can be integrated along appropriate paths. We remark that due to the cubic terms in (3.37) a secondary blow-up is performed. The delayed onset of oscillations due to a slow passage through a Hopf bifurcation can be observed in Figure 10.

3.5. Beyond fold and Hopf singularities. As we have reviewed so far, the understanding of codimension-one fold and Hopf points is of extreme importance due to their genericity in nonlinear fast-slow systems. However, there are many other singularities that have also been studied via the blow-up method, which we summarize here.

Transcritical and pitchfork singularities: Let us consider (2.8) with $(x, y) \in \mathbb{R}^2$ and assume that the layer equation undergoes either a transcritical or a pitchfork bifurcation at the origin where $y \in \mathbb{R}$ is the bifurcation parameter [**KS01b**]. The goal is to understand the dynamics for $\varepsilon > 0$ small in the neighbourhood of the aforementioned singularities.

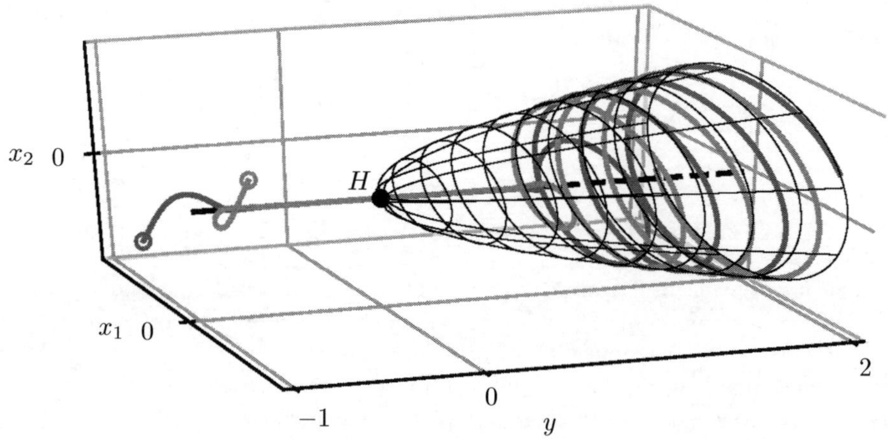

FIGURE 10. Simulation of a slow passage through a Hopf bifurcation of (3.30) with $\varepsilon = 0.05$. The critical manifold is given by $\mathcal{C}_0 = \{(x_1, x_2) = (0,0)\}$, it is attracting (solid black line) for $y < 0$ and repelling (dashed black line) for $y > 0$. The point $H = (0,0,0)$ denotes the bifurcation point and the wire-frame paraboloid represents the amplitude of oscillations according to the y-parameter value. We show two trajectories and observe that the onset of oscillations occurs way beyond the Hopf point.

In the case of the transcritical singularity, it can be shown (Lemma 2.1 in [**KS01b**]) that a canonical form is given as

$$
\begin{aligned}
x' &= x^2 - y^2 + \omega\varepsilon + \mathcal{O}(x^3, x^2y, xy^2, y^3, \varepsilon x, \varepsilon y, \varepsilon^2), \\
y' &= \varepsilon(1 + \mathcal{O}(x, y, \varepsilon)),
\end{aligned} \tag{3.39}
$$

where ω is considered as a continuous parameter. The role of the parameter ω is to distinguish two types of dynamics, as is stated below in Theorem 3.13. The critical manifold $\mathcal{C}_0$ is given by $\mathcal{C}_0 = \{(x,y) \in \mathbb{R}^2 \,|\, x = \pm y + \mathcal{O}(y^2)\}$, which clearly has a self-intersection at the origin. From another perspective, the critical manifold can be regarded as the union of one dimensional manifolds $\{x = y + \mathcal{O}(y^2)\}$ and $\{x = -y + \mathcal{O}(y^2)\}$, which intersect transversally at the origin. Such intersection implies that the origin is a non-hyperbolic equilibrium point of the layer equation. In either case, it is convenient to assume that a compact subset of $\mathcal{C}_0$ around the origin is given by the union of four branches, namely $\mathcal{S}^{\mathrm{a}}_{0,+}$, $\mathcal{S}^{\mathrm{a}}_{0,-}$, $\mathcal{S}^{\mathrm{r}}_{0,+}$, $\mathcal{S}^{\mathrm{r}}_{0,-}$, as shown in Figure 11. Thus, away from the origin, Fenichel's theorem guarantees that the previous critical manifolds persist as slow manifolds $\mathcal{S}^{\mathrm{a}}_{\varepsilon,+}$, $\mathcal{S}^{\mathrm{a}}_{\varepsilon,-}$, $\mathcal{S}^{\mathrm{r}}_{\varepsilon,+}$, $\mathcal{S}^{\mathrm{r}}_{\varepsilon,-}$ for $\varepsilon > 0$ sufficiently small. The idea of [**KS01b**] is to describe the fate of $\mathcal{S}^{\mathrm{a}}_{\varepsilon,-}$ as it crosses a small neighborhood of the origin. In order to state the main result regarding the transcritical singularity, let $\rho > 0$ be small and $I \in \mathbb{R}$ a small interval containing $0 \in \mathbb{R}$. Define the sections

$$
\begin{aligned}
\Sigma^{\mathrm{en}} &= \{(x,y) \in \mathbb{R}^2 \,|\, x = -\rho,\, y + \rho \in I\}, \\
\Sigma^{\mathrm{ex}}_e &= \{(x,y) \in \mathbb{R}^2 \,|\, x = \rho,\, y \in I\}, \\
\Sigma^{\mathrm{ex}}_a &= \{(x,y) \in \mathbb{R}^2 \,|\, x = -\rho,\, y - \rho \in I\}.
\end{aligned} \tag{3.40}
$$

Let $\Pi_e : \Sigma^{\text{ex}} \to \Sigma_e^{\text{ex}}$ and $\Pi_a : \Sigma^{\text{en}} \to \Sigma_a^{\text{ex}}$ denote the corresponding transitions.

THEOREM 3.13 ([**KS01b**]). *Consider* (3.39) *and fix $\omega \neq 1$. There exists $\varepsilon_0 > 0$ such that the following assertions hold for $\varepsilon \in (0, \varepsilon_0]$.*

(T1) *If $\omega > 1$ then the manifold $\mathcal{S}^{\text{a}}_{\varepsilon,-}$ passes through Σ_e^{ex} at a point $(\rho, h(\varepsilon))$ where $h(\varepsilon) \in \mathcal{O}(\sqrt{\varepsilon})$. The section Σ^{en} is mapped by Π_a to an interval containing $\mathcal{S}^{\text{a}}_{\varepsilon,-} \cap \Sigma_a^{\text{ex}}$ of size $\mathcal{O}(\exp(-C/\varepsilon))$, where C is a positive constant.*

(T2) *If $\omega < 1$ then Σ^{en} (including the point $\mathcal{S}^{\text{a}}_{\varepsilon,-} \cap \Sigma^{\text{en}}$) is mapped by Π_e to an interval about $\mathcal{S}^{\text{a}}_{\varepsilon,+}$ of size $\mathcal{O}(\exp(-C/\varepsilon))$, where C is a positive constant.*

See Figure 11 for a schematic of the results of Theorem 3.13. It is important to note that the constant ε_0 tends to 0 as $\omega \to 1$. In fact, the case $\omega = 1$ is degenerate and is not detailed in [**KS01b**]. However, it is argued in Remark 2.2 of [**KS01b**] that canard solutions occur in this case, the treatment being similar to that of [**KS01a**], see also the Example in Section 3 of [**KS01b**]. From another perspective, i.e. not using blow-up, a particular case of this degenerate scenario is investigated in [**Sch85**].

For the proof of Theorem 3.13, one employs a blow-up map defined by $\phi(r, \bar{x}, \bar{y}, \bar{\varepsilon}) = (r\bar{x}, r\bar{y}, r^2\bar{\varepsilon})$. The blow-up analysis is carried out in three charts, namely $K_1 = \{\bar{x} = -1\}$, $K_2 = \{\bar{\varepsilon} = 1\}$ and $K_3 = \{\bar{x} = 1\}$. The analysis in charts K_1 and K_3 is similar and amounts to describing the organization of the dynamics according to two 2-dimensional centre manifolds, and which are tracked along the blow-up space. Another important remark is that in the charts K_1 and K_3 the parameter ω does not play any essential role. In chart K_2, the main object of study is the ODE

$$
\begin{aligned}
x_2' &= x^2 - y^2 + \omega, \\
y_2' &= 1,
\end{aligned}
\tag{3.41}
$$

where as usual we are using (x_2, y_2) to denote local coordinates. The main goal in this chart is to connect the appropriate centre manifolds of charts K_1 and K_3.

A more general scenario compared to the one presented above is treated in [**DM15**], where in brief terms, unfoldings of (3.39) are considered. In fact, since [**DM15**] considers fast-slow systems with a transcritical singularity in its most generic context, more passages as compared to those shown in Figure 11 can occur, [**DM15**, Theorems 5 and 6]. The most important of such passages is, of course, the canard. A consequence of this analysis is the extension of known results on the stability of canard cycles [**DMDR11**] to the case including finite passages through a transcritical singularity [**DM15**, Theorem 4].

Regarding the *pitchfork singularity*, it is shown (Lemma 4.1 of [**KS01b**]) that a canonical form is given by

$$
\begin{aligned}
x' &= x(y - x^2) + \omega\varepsilon + ax^2y + \mathcal{O}(xy^2, y^3, \varepsilon x, \varepsilon y, \varepsilon^2), \\
y' &= \varepsilon(\pm 1 + bx + \mathcal{O}(y, \varepsilon)),
\end{aligned}
\tag{3.42}
$$

where ω is assumed non-zero and $(a, b) \in \mathbb{R}^2$. In this case, the analysis and results depend on the sign in (3.42). Similar to the transcritical case, the critical manifold can be given as the union of four branches $\mathcal{S}^{\text{a}}_{0,\text{t}}$, $\mathcal{S}^{\text{r}}_{0,\text{t}}$, $\mathcal{S}^{\text{a}}_{0,+}$, $\mathcal{S}^{\text{a}}_{0,-}$. Away from a small neighbourhood of the origin and for $\varepsilon > 0$ sufficiently small, the previous four branches persist as the slow manifolds $\mathcal{S}^{\text{a}}_{\varepsilon,\text{t}}$, $\mathcal{S}^{\text{r}}_{\varepsilon,\text{t}}$, $\mathcal{S}^{\text{a}}_{\varepsilon,+}$, $\mathcal{S}^{\text{a}}_{\varepsilon,-}$ respectively, see Figure 11. Let $\rho > 0$ be fixed and $I \subset \mathbb{R}$ be a small interval around $0 \in \mathbb{R}$. Define the sections

$$\begin{aligned} \Sigma^{\mathrm{t}} &= \left\{(x,y)\in\mathbb{R}^2 \,|\, x\in I,\, y=-\rho\right\}, \\ \Sigma^{+} &= \left\{(x,y)\in\mathbb{R}^2 \,|\, x=\rho,\, y-\rho^2\in I\right\}, \\ \Sigma^{-} &= \left\{(x,y)\in\mathbb{R}^2 \,|\, x=-\rho,\, y-\rho^2\in I\right\}, \end{aligned} \tag{3.43}$$

and define transitions $\Pi^t : \Sigma^{\mathrm{t}} \to \Sigma^{+}\cup\Sigma^{-}$ and $\Pi^{\pm} : \Sigma^{\pm} \to \Sigma^{\mathrm{t}}$. The former transition is well defined when the sign in (3.42) is positive while the latter is well defined when the sign in (3.42) is negative. Regarding the pitchfork singularity, the main result is:

THEOREM 3.14 ([**KS01b**]). *Consider* (3.42) *and fix* $\omega \neq 0$. *There exists* $\varepsilon_0 > 0$ *such that the following assertions hold for* $\varepsilon \in (0, \varepsilon_0]$.

(P1) *If the sign in* (3.42) *is positive and* $\omega > 0$ *then* Σ^{t} *(including the point* $\Sigma^{\mathrm{t}}\cap\mathcal{S}^{\mathrm{a}}_{\varepsilon,\mathrm{t}}$*) is mapped by* Π^t *to an interval about* $\Sigma^{+}\cap\mathcal{S}^{\mathrm{a}}_{\varepsilon.+}$ *of size* $\mathcal{O}(\exp(-C/\varepsilon))$, *where* C *is a positive constant.*

(P2) *If the sign in* (3.42) *is positive and* $\omega < 0$ *then* Σ^{t} *(including the point* $\Sigma^{\mathrm{t}}\cap\mathcal{S}^{\mathrm{a}}_{\varepsilon,\mathrm{t}}$*) is mapped by* Π^t *to an interval about* $\Sigma^{-}\cap\mathcal{S}^{\mathrm{a}}_{\varepsilon.-}$ *of size* $\mathcal{O}(\exp(-C/\varepsilon))$, *where* C *is a positive constant.*

(P3) *If the sign in* (3.42) *is negative then* Σ^{+} *and* Σ^{-} *are mapped by* Π^{+} *and* Π^{-}, *respectively, to intervals about* $\mathcal{S}^{\mathrm{a}}_{\varepsilon,\mathrm{t}}\cap\Sigma^{\mathrm{t}}$ *of size* $\mathcal{O}(\exp(-C/\varepsilon))$, *where* C *is a positive constant.*

See Figure 12 for a schematic of the statements of Theorem 3.14. In this case, the blow-up map is defined via $\phi(r,\bar{x},\bar{y},\bar{\varepsilon}) = (r\bar{x}, r^2\bar{y}, r^4\bar{\varepsilon})$. The blow-up analysis is carried out in five charts, namely $K_1 = \{\bar{y}=-1\}$, $K_2 = \{\bar{\varepsilon}=1\}$, $K_3 = \{\bar{y}=1\}$, $K_4 = \{\bar{x}=-1\}$ and $K_5 = \{\bar{x}=1\}$. As in the transcritical case, the idea is to track centre manifolds along the blow-up space. The relevant centre manifolds are first found in charts K_1 and K_3 and are related to normally hyperbolic parts of the critical manifold. Accordingly, $\mathcal{M}^{\mathrm{a}}_{1,\mathrm{t}}$, $\mathcal{M}^{\mathrm{a}}_{3,+}$, $\mathcal{M}^{\mathrm{a}}_{3,-}$ and $\mathcal{M}^{\mathrm{r}}_{3,\mathrm{t}}$ denote centre manifolds (in the blow-up space) associated to the branches $\mathcal{S}^{\mathrm{a}}_{0,\mathrm{t}}$, $\mathcal{S}^{\mathrm{a}}_{0.+}$, $\mathcal{S}^{\mathrm{a}}_{0,-}$ and $\mathcal{S}^{\mathrm{r}}_{0,\mathrm{t}}$ respectively. On the other hand, in charts K_4 and K_5 one finds hyperbolic equilibrium points.

The most important analysis occurs in chart K_2. In chart K_2 one must connect the aforementioned centre manifolds. We sketch the key argument: let $\mathcal{M}^{\mathrm{a}}_{2,\mathrm{t}}$ and $\mathcal{M}^{\mathrm{r}}_{2,\mathrm{t}}$ denote the centre manifolds $\mathcal{M}^{\mathrm{a}}_{1,\mathrm{t}}$ and $\mathcal{M}^{\mathrm{r}}_{3,\mathrm{t}}$, respectively, written in the coordinates of chart K_2. The main result of the analysis in chart K_2 is that the invariant manifolds $\mathcal{M}^{\mathrm{a}}_{2,\mathrm{t}}$ and $\mathcal{M}^{\mathrm{r}}_{2,\mathrm{t}}$ intersect transversally along $(x_2,y_2,r_2) = (0,0,t_2)$, where $t_2 \in (-\infty,\infty)$. The proof of this fact is done via the analysis of a Melknikov integral, see Section 3 in[**KS01b**]. In terms of the slow manifolds, the previous result implies that for $\varepsilon > 0$ sufficiently small and $\omega \neq 0$, the distance between the slow manifolds $\mathcal{S}^{\mathrm{a}}_{\varepsilon,\mathrm{t}}$ and $\mathcal{S}^{\mathrm{r}}_{\varepsilon,\mathrm{t}}$, around the origin, is non-zero. On the other hand, when $\omega = 0$ the slow manifolds $\mathcal{S}^{\mathrm{a}}_{\varepsilon,\mathrm{t}}$ and $\mathcal{S}^{\mathrm{r}}_{\varepsilon,\mathrm{t}}$ are connected.

Cusp singularity: The cusp singularity is more degenerate than the fold. In fact, at the level of classification of singularities of generic smooth functions, the cusp singularity is the next in the list of singularities (or catastrophes) [**Arn90**]. This means that the cusp is of codimension 2, that is, it appears generically in 2-parameter families of smooth functions.

In the context of fast-slow systems, the cusp point was first studied in [**BKK13**]. Given a fast-slow system (2.6) with $(m,n) = (1,2)$, the conditions for having a

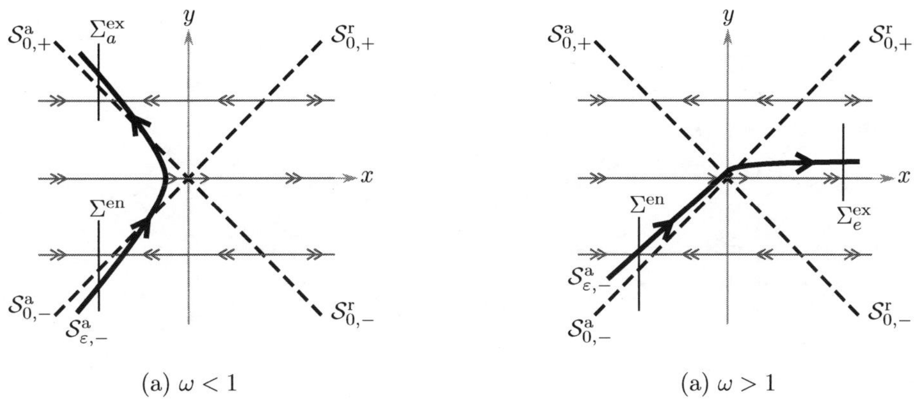

FIGURE 11. Flow of a fast-slow system near a transcritical singularity for $\omega < 1$ on the left and $\omega > 1$ on the right. The case $\omega = 1$ (not shown) corresponds to a maximal canard in which $\mathcal{S}^{a}_{\varepsilon,-}$ and $\mathcal{S}^{r}_{\varepsilon,+}$ coincide.

generic cusp point at the origin are

$$f(0,0,0,0) = 0, \quad \frac{\partial f}{\partial x}(0,0,0,0) = 0, \quad \frac{\partial^2 f}{\partial x^2}(0,0,0,0) = 0 \tag{3.44}$$

and

$$\frac{\partial^3 f}{\partial x^3}(0,0,0,0) \neq 0 \quad \left[\frac{\partial f}{\partial y_1}(0,0,0,0) \quad \frac{\partial f}{\partial y_2}(0,0,0,0)\right] \cdot g(0,0,0,0) \neq 0. \tag{3.45}$$

If a fast-slow system satisfies (3.44)-(3.45), then it can be written as [**BKK13**, Proposition 2.5]

$$\begin{aligned} x' &= -(x^3 + y_2 x + y_1) + \mathcal{O}(\varepsilon, xy_1, x^3 y_2, x^4), \\ y_1' &= \varepsilon(1 + \mathcal{O}(x, y_1, y_2, \varepsilon)), \\ y_2' &= \varepsilon \mathcal{O}(x, y_1, y_2, \varepsilon). \end{aligned} \tag{3.46}$$

We show in Figure 13 the corresponding critical manifold

$$\mathcal{C}_0 = \left\{(x, y_1, y_2) \in \mathbb{R}^3 \,|\, -(x^3 + y_2 x + y_1) + \mathcal{O}(\varepsilon, xy_1, x^3 y_2, x^4) = 0\right\} \tag{3.47}$$

and the reduced flow near the cusp point for a particular choice of signs in (3.45). To state the main theorem of [**BKK13**] let us define the sections

$$\begin{aligned} \Sigma^{en} &= \left\{(x, y_1, y_2) \in \mathbb{R}^3 \,|\, y_1 = -\delta\right\}, \\ \Sigma^{ex} &= \left\{(x, y_1, y_2) \in \mathbb{R}^3 \,|\, y_1 = \delta\right\}, \end{aligned} \tag{3.48}$$

where $\delta > 0$ is small. Furthermore, let $\mathcal{S}_0^-$ and $\mathcal{S}_0^+$ be compact subsets of $\mathcal{C}_0$ satisfying[7]

$$\begin{aligned} \mathcal{S}_{0,-} &= \{(x, y_1, y_2) \in \mathcal{C}_0 \,|\, y_1 < -\delta\}, \\ \mathcal{S}_{0,+} &= \{(x, y_1, y_2) \in \mathcal{C}_0 \,|\, y_1 > \delta\}. \end{aligned} \tag{3.49}$$

[7]Note that both critical manifolds $\mathcal{S}_{0,-}$ and $\mathcal{S}_{0,+}$ are attracting, so we omit the superscript "a".

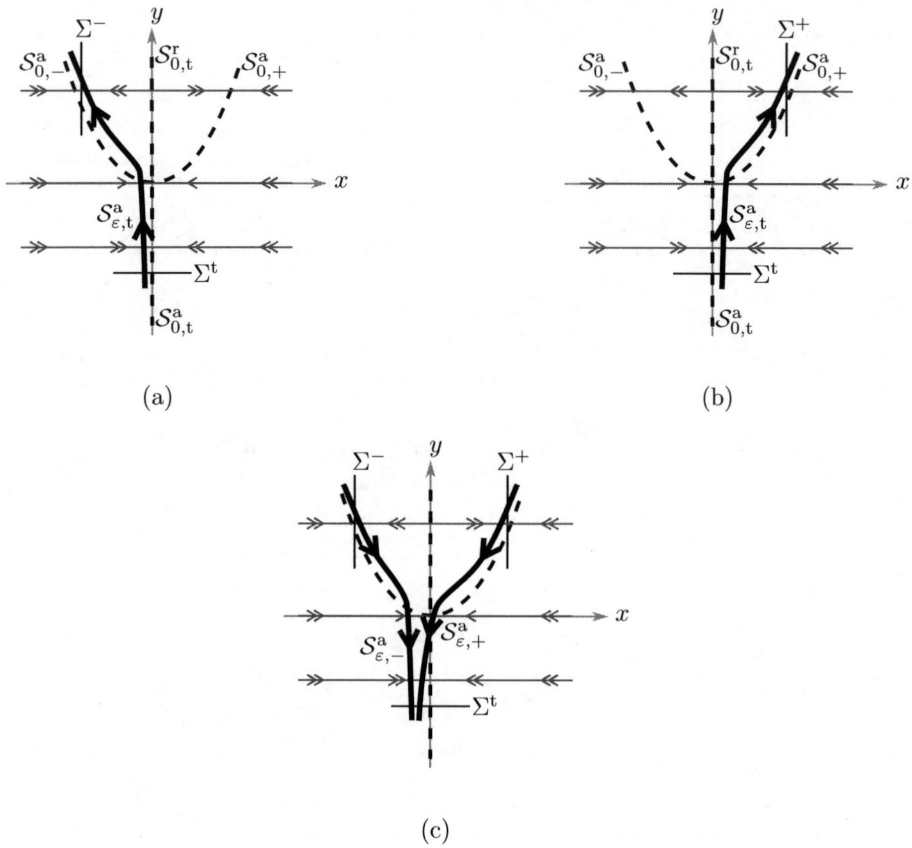

FIGURE 12. Phase-portraits of a fast-slow system near a pitchfork singularity for the cases of Theorem 3.14. Phase-portrait (a) (resp. (b)) corresponds to the case $\omega < 0$ (resp. $\omega > 0$) with the sign of (3.42) positive, while phase-portrait (c) corresponds to the sign of (3.42) negative.

Accordingly we denote by $\mathcal{S}_{\varepsilon,-}$ and $\mathcal{S}_{\varepsilon,+}$ the respective slow manifolds. The main result describes the passage of trajectories starting near the slow manifold $\mathcal{S}_{\varepsilon,-}$ through a neighborhood of the cusp point:

THEOREM 3.15 ([**BKK13**, Theorem 2]). *For system* (3.46) *there exists* $\varepsilon_0 > 0$ *sufficiently small such that for all* $0 < \varepsilon \leq \varepsilon_0$ *the following statements hold:*

(C1) *The transition map* $\Pi : \Sigma^{\text{en}} \to \Sigma^{\text{ex}}$ *induced by the flow of* (3.46) *is a diffeomorphism mapping a rectangular neighborhood of* $\mathcal{S}_{0,-}$ *into* Σ^{out}.

(C2) *The choice of* $\mathcal{S}_{\varepsilon,+}$ *can be made in such a way that* $\Pi(\Sigma^{\text{en}} \cap \mathcal{S}_{\varepsilon,-}) \subset \mathcal{S}_{\varepsilon,+}$.

(C3) *The map* Π *is exponentially contracting in the* x *direction and the derivative in the* y_2 *direction is bounded above and below. More precisely*

$$\left|\frac{\partial \Pi}{\partial x}\right| \in \mathcal{O}(\exp(-C/\varepsilon)) \tag{3.50}$$

for some positive constant C, and there exists a constant $M > 0$ such that

$$\frac{1}{M} \leq \left|\frac{\partial \Pi}{\partial y_2}\right| \leq M. \tag{3.51}$$

The proof of Theorem 3.15 follows the same ideas as for the generic fold case, so we point out some of the differences: first of all, due to the quasihomogeneity of the fast equation, the blow-up map is given by $(x, y_1, y_2, \varepsilon) = (\bar{r}^3\bar{x}, \bar{r}^2\bar{y}_1, \bar{r}\bar{y}_2, \bar{r}^5\bar{\varepsilon})$. These weights can be obtained by checking the quasi-homogeneity of the vector field. Next, due to the geometric properties of the critical manifold, it is now necessary to study the blown-up vector field in five charts, namely: $K_{\pm\bar{y}_1} = \{\bar{y}_1 = \pm 1\}$, $K_{\pm\bar{y}_2} = \{\bar{y}_2 = \pm 1\}$ and $K_{\bar{\varepsilon}} = \{\bar{\varepsilon} = 1\}$. The analysis starts in $K_{-\bar{y}_1}$. Then, depending on the initial conditions one transitions to either $K_{+\bar{y}_2}$, $K_{\bar{\varepsilon}}$, or to $K_{-\bar{y}_2}$. In $K_{+\bar{y}_2}$ we study trajectories passing sufficiently away from the cusp point and along the regular part of $\mathcal{S}_0$. In $K_{\bar{\varepsilon}}$ we study trajectories passing through a small neighborhood of the cusp point. In $K_{-\bar{y}_2}$ we study trajectories passing sufficiently away from the cusp point and across the folded part of $\mathcal{S}_0$. It is important to remark that the flow is regular in the charts $K_{+\bar{y}_2}$, $K_{\bar{\varepsilon}}$ and $K_{-\bar{y}_2}$. Moreover, in $K_{-\bar{y}_2}$ one may invoke the results of the generic fold in Section 3.1. Finally, we further transition from $K_{+\bar{y}_2}$, $K_{+\bar{\varepsilon}}$ and $K_{-\bar{y}_2}$ to $K_{+\bar{y}_1}$ completing the analysis. A schematic of Theorem 3.15 is shown in Figure 13.

Another contribution of [**BKK13**] is the study of trajectories with initial conditions below the cusp point. Interestingly, it can be shown that, under appropriate assumptions, two trajectories with exponentially close initial conditions cross Σ^{ex} algebraically apart, see [**BKK13**, Proposition 3] and Figure 13 for a sketch.

There are some refinements to the results of Theorem 3.15 in the case where the higher order terms of (3.46) are of order $\mathcal{O}(\varepsilon)$. In particular, [**JKBR16**] provides more accurate information on the transition map $\Pi : \Sigma^{\text{en}} \to \Sigma^{\text{ex}}$. The analysis performed in [**JKBR16**] uses a heavy amount of normal form theory, in particular results for quasihomogeneous vector fields [**LS10**], an appropriate normal form for fast-slow systems [**JK15**], as well as for semi-hyperbolic singularities [**Tak71, Bon97, Bon96**], and some other classical normal form techniques [**Che63, Ste58**].

Bogdanov-Takens: Fast-slow systems related to a standard Bogdanov-Takens bifurcation have been studied using the blow-up method in [**DMD11b, Chi11**]. Although the approaches are similar, the concerns and results are different, and here we sketch the most important ones. Let us start from [**DMD11b**] where a planar fast-slow system of the form

$$\begin{aligned} x' &= y, \\ y' &= -xy + \varepsilon(b_0 + b_1 x + x^2 + x^3 G(x, \omega)) + y^2 H(x, y, \omega), \end{aligned} \tag{3.52}$$

where ω denotes parameters in a compact subset of an Euclidean space (any finite dimension is allowed) and $b = (b_0, b_1) \sim 0 \in \mathbb{R}^2$, are considered. It is important to mention that (3.52) is a normal form, implying that the results of [**DMD11b**] apply to a much wider class of planar fast-slow systems which after changes of coordinates can be brought to the form (3.52). Note that for $\varepsilon > 0$, (3.52) indeed corresponds to (a versal unfolding of) the normal form of a Bogdanov-Takens bifurcation [**Bog75**] and [**Tak01**, Theorem 6.2]. The main result of [**DMD11b**] consists on a full description of the bifurcation *independent of* ε, in contrast with [**Bog75, Tak01**] for which "the validity of the results shrinks" as $\varepsilon \to 0$. The blow-up analysis is

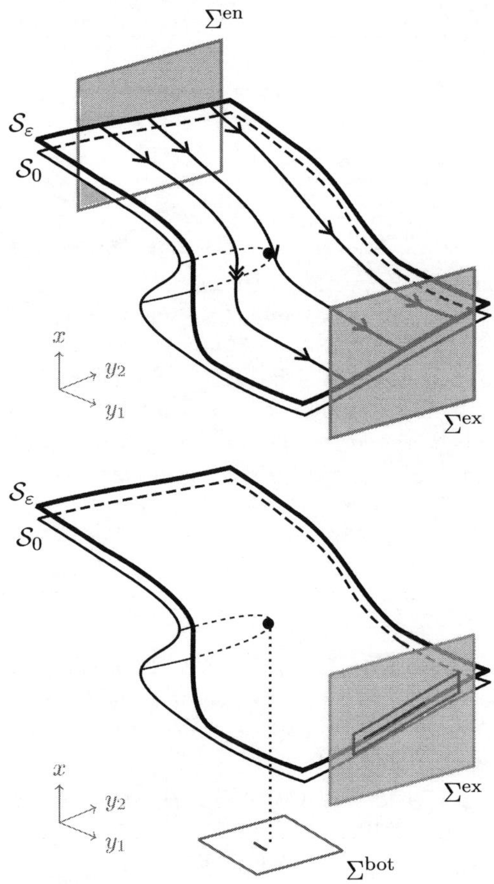

FIGURE 13. Top: schematic representation of the analysis of a fast-slow system near a cusp point (black dot located at the origin). Qualitatively speaking it can be shown that there exists an attracting manifold $\mathcal{S}_\varepsilon$ constructed from continuing $\mathcal{S}_{\varepsilon,-}$ beyond the cusp point (also black dot located at the origin). Bottom: Σ^{bot} is a small rectangle below the cusp point. It can be shown that: a) the image of Σ^{bot} under the flow of (3.46) is an exponentially small strip about $\mathcal{S}_\varepsilon \cap \Sigma^{\text{ex}}$ and b) an exponentially small interval $I_{\text{bot}} \in \Sigma^{\text{bot}}$ (shown in green) is mapped to an algebraically wide interval of order $\mathcal{O}(\varepsilon)$ in Σ^{ex}. The main idea behind this observation is that we can choose trajectories that are exponentially close but that are separated by a repelling slow manifold.

now considerably more involved than in the fold case, and is mostly performed in the family (or central) chart.

In [**Chi11**] a one-parameter family of fast-slow systems presenting a Bogdanov-Takens point is studied, but in this case in a system with two fast and one slow variables of the form

$$
\begin{aligned}
x_1' &= f_1(x_1, x_2, y, \varepsilon, \delta),\\
x_2' &= f_2(x_1, x_2, y, \varepsilon, \delta),\\
y' &= \varepsilon g(x_1, x_2, y, \varepsilon, \delta).
\end{aligned}
\tag{3.53}
$$

Accordingly, the critical manifold is one-dimensional and is supposed to be "S-shaped", similar to the one appearing in the van der Pol oscillator. As such, the critical manifold has two fold points. The main assumption is that such fold points coincide with a Bogdanov-Takens bifurcation point of the layer equation. Furthermore, it is assumed that the stable part of the critical manifold correspond to stable foci of the layer equation, while the unstable part of the critical manifold corresponds to saddle nodes, see Figure 14 for a schematic representation of the critical manifold and the flow of the layer equation.

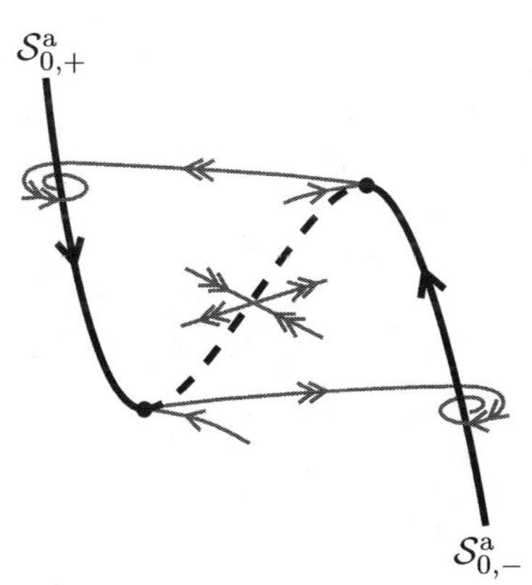

FIGURE 14. Schematic of the critical manifold of (3.53) considered in [**Chi11**], where the critical manifold is a 1-dimensional S-shaped curve. The parameter δ, which controls the strength of attraction of $\mathcal{S}_{0,\pm}^{\mathrm{a}}$, plays a crucial role: for some pairs (ε, δ) (3.53) has a stable periodic orbit, while for other values of (ε, δ) (3.53) exhibits chaotic motion.

The main results of [**Chi11**] prove and describe relaxation oscillations [**Chi11**, Theorem 2] and chaotic motion [**Chi11**, Theorem 3] of (3.53) according to the parameters (δ, ε). The parameter δ is used to control the strength of stability of the stable part of the critical manifold. The core analysis heavily depends on studying a first Painlevé equation (see [**OLBC10**, Chapter 32] and [**Inc56**, Chapter 14.4]) in the blow-up space.

The above examples are particular instances of fast-slow dynamical systems, where singularities beyond the classical fold appear. As it is evident, the analysis of fast-slow systems becomes more complicated, for example, when the number of fast variables increases and/or the degeneracy of the non-hyperbolic points also increases. There are several other situations that fall out of the scope of folded or Hopf-type singularities, see Sections 3.6 and 3.7 for further references.

3.6. Further directions. In this survey we have reviewed research related to the foundations of the blow-up method as an important tool to analyze fast-slow systems. Naturally, there are some particular topics that have not been covered in this survey, and moreover, there are new emerging and exciting research directions, including works that consider problems beyond folded singularities , where the blow-up method is being employed. The following are just some examples of works concerning and/or related to fast-slow dynamics with non-hyperbolic singularities, where the blow-up method has proved useful already:

Unbounded critical manifolds: One usually considers loss of normally due to tangency of the critical manifold and the fast foliation along sets of lower dimension than the critical manifold itself. Clear examples are the fold points and lines reviewed in Sections 3.1, 3.2, 3.3 and cusp points sketched in Section 3.5. However, it may occur that somewhere in the phase-space, the critical manifold aligns with the fast-foliation. These cases have been studied in [**GS09, Kue14, Kri17**].

Codimension 0 critical manifolds: In all examples provided so far, the critical manifold is at least of codimension 1. In other words, the critical manifold is always of lower dimension than the phase-space. For example, for the generic fold in Section 3.1 the critical manifold is of codimension 1 ($\dim(\mathcal{C}_0) = 1$ and phase-space $\mathbb{R}^2$), while in the Bogdanov-Takens case of [**Chi11**], the critical manifold is of codimension 2 ($\dim(\mathcal{C}_0) = 1$ and phase-space $\mathbb{R}^3$). There are cases, however, where the critical manifold may have the same dimension as the phase-space. A trivial but non-generic example is given by the planar fast-slow system

$$\begin{aligned} x' &= 0 \\ y' &= \varepsilon, \end{aligned} \tag{3.54}$$

where it is evident that the critical manifold is the the whole plane. A more elaborate example is given by

$$\begin{aligned} x' &= f(x,y,\varepsilon) \\ y' &= \varepsilon g(x,y,\varepsilon), \end{aligned} \tag{3.55}$$

where

$$f(x,y,\varepsilon) = \begin{cases} h(y)\exp\left(-\frac{1}{x}\right) + \mathcal{O}(\varepsilon), & x > 0 \\ \mathcal{O}(\varepsilon), & x \leq 0, \end{cases} \tag{3.56}$$

where $h(y)$ can be any smooth nonzero function. The function $f(x,y,\varepsilon)$ is smooth and the critical manifold is given by

$$\mathcal{C}_0 = \left\{(x,y) \in \mathbb{R}^2 \,|\, x \leq 0\right\}, \tag{3.57}$$

which is two dimensional. Similarly, many other examples can be created.

In a more relevant context, this situation of codimension 0 critical manifolds has recently been considered in [**KM19**] motivated by hysteresis operators [**KM17**].

Piecewise-linear and non-smooth systems: Canard theory (see Sections 3.2 and 3.3) has been extended to piecewise-linear systems [**DGP$^+$16, DFGK$^+$18**] and in a similar context the blow-up method has been used to analyze non-smooth (fast-slow) systems [**KH15a, KH15b, dMdS18, Jef16, JKSW**], [**JKW20**].

Torus canards: In all the fast-slow systems considered so far, the critical manifold is made of critical points of the fast dynamics. However, in high dimensional problems, the dynamics of the layer equation may converge to, for example, limit cycles instead of equilibrium points. In this context, canard theory has been extended to cases where the fast dynamics has limit cycles as limiting sets, leading to the so-called torus canards [**BDG+16, Vo17**], see also [**JNS18**].

Discrete time systems: The blow-up method has been recently used to study discrete fast-slow systems with a transcritical singularity [**EK18**] directly without using continuous approximations. This represents additional steps to extending the blow-up to the context of maps with singularities; see also [**NS13, NSS09**].

3.7. Applications. The main part of this document has been dedicated to reviewing important theoretical progress on the understanding of fast-slow dynamics using the blow-up method. As such, we have emphasized the usefulness of the method to, for example, explain the canard explosion phenomenon. However, the blow-up method has also been used in interesting particular applications. Here we briefly summarize some examples to illustrate the breadth of different areas, where the blow-up technique can be useful.

In neuroscience: The progress in the theory of fast-slow systems, particularly the understanding of canards, and the related mixed-mode oscillations [**DGK+12**], has had a great impact in neuroscience. As an example, the Fitzugh-Nagumo, the Hodgkin-Huxley, and other related models have been studied from several mathematical perspectives, especially from a multi-time scale approach [**GK09, GK10, DKK13, RW07**]. A recent instance where the blow-up method is instrumental is [**CS18**], where the transition between two distinct oscillatory patters in the Fitzugh-Nagumo system is precisely described.

In systems biology: The analysis of a model for the division cycle of an embryonic cell, known as the *Goldbeter model*, is studied in [**KS16**]. This is the first rigorous analysis of a system based on Michaelis-Menten equations where small Michaelis-Menten constants are considered and play a central role. In its original form, the model under study has no evident time scale separation. After a transformation, an auxiliary and equivalent nonstandard fast-slow model is obtained. By using tools of geometric singular perturbation theory the authors analyze the system using the property that the auxiliary model has three time scales. On a related context, a problem exhibiting relaxation oscillations and depending on two singular parameters is considered in [**KS11**], see also [**MPS20**] and [**TJKSC21**] for further applications in this context.

In chemistry: The *Olsen model* for the peroxidase-oxidase reaction is considered in [**KS15**]. There the model is 4-dimensional. Besides this evident obstacle, the Olsen model is not in standard form (2.6), has several small parameters, and has non-fold singularities. Even more interestingly, the Olsen model has no defined fast and slow variables, but presents three distinct regimes in which there are 1, 2, and 3 fast variables. The blow-up method is essential in this case to understand the dynamics in all the regimes and leads to the description of non-classical relaxation oscillations of the Olsen model.

In engineering: In [**BBK17**], a model describing an earthquake fault is considered. The model being studied is a Hamiltonian system, and possesses an *unbounded critical manifold*, see further information above. The blow-up method is used to show the existence of limit cycles associated to a degenerate Hopf bifurcation. Another recent application is found in [**IPS18**], where a Micro-Electro Mechanical System (MEMS) undergoing a singular effect known as "touchdown" is analyzed. In this case, the blow-up method allows a precise analysis of the steady states of a regularized model, and the description of the dynamics as a couple of small parameters tend to zero. Also, in [**JKK20**] a dynamic consensus network having one slowly varying edge is analyzed.

In control theory: Singular perturbation problems are classical in control theory. The case for which the critical manifold is normally hyperbolic is in fact well understood [**KKO99**]. The case where the critical manifold looses normal hyperbolicity had not received much attention. Recently, the blow-up method has allowed the design of controllers that stabilize non-hyperbolic points (of any degeneracy) of fast-slow systems with one fast variable [**JKS18, JKSdPF19**], the model order reduction of fast-slow systems with hidden/degenerate normally hyperbolic critical manifolds [**JKS17**], and the control of canard cycles [**JKK19**].

4. Summary and outlook

This survey has been devoted to provide a concise recollection of the most influential progress in the blow-up theory of fast-slow systems. As such, we have principally covered folded singularities and in particular how the blow-up method allowed a geometric description of canards, canard cycles, and canard explosion. We have also outlined on the role of the blow-up method when analyzing other singularities beyond the fold, such as in the case of the Hopf, transcritical, pitchfork, cusp, and Bogdanov-Takens points. Moreover, we have briefly reviewed some recent theoretical and applied research in fast-slow systems, where the blow-up method is also one of the main mathematical tools for analysis.

It is evident that the blow-up method enables us to study and understand complicated fast-slow dynamics around non-hyperbolic points. However, it should also be clear that the difficulties do not end once we apply the method. In several cases, the dynamics within the charts are still quite intricate and, for example, one encounters special equations for which their asymptotics are important (e.g. Riccati, Weber, and Panilevé). Furthermore, one must trace invariant manifolds, compute their distance (Melknikov integral), control the obstacle of resonances, or deal with complex and/or complicated path integrals, just to mention a few of the difficulties.

We predict that for many of the theoretical and applied studies of fast-slow systems to come, the blow-up method will likely continue to be one of the techniques of choice when studying such dynamical systems. Furthermore, due to the everyday increase of complexity of systems and their models, the blow-up method will have to be extended and or adapted to such needs. We envision that, for instance, geometric studies of high dimensional systems and/or high codimension singularities are going to increase in relevance. Finally, although the blow-up method has been a great success for deterministic ODE systems, it is at this point a daunting challenge to extend it to a wider range of dynamical systems including partial and stochastic or even stochastic-partial differential equations. Some recent progress in such directions can be found in [**AK20, EK20**].

Acknowledgments

The authors gratefully acknowledge the comments of the anonymous reviewer that helped to improve the content and presentation of this article.

References

[ÁFJ11] María Jesús Álvarez, Antoni Ferragut, and Xavier Jarque, *A survey on the blow up technique*, Internat. J. Bifur. Chaos Appl. Sci. Engrg. **21** (2011), no. 11, 3103–3118, DOI 10.1142/S0218127411030416. MR2871964

[AK20] L. Arcidiacono and C. Kuehn, *Blowing-up Nonautonomous Vector Fields: Infinite Delay Equations and Invariant Manifolds*, arXiv preprint `arXiv:2010.14335` (2020).

[Arn90] V. I. Arnol′d, *The theory of singularities and its applications*, Lezioni Fermiane. [Fermi Lectures], Accademia Nazionale dei Lincei, Rome; Scuola Normale Superiore, Pisa, 1991. MR1122147

[AS65] Milton Abramowitz and Irene A. Stegun (eds.), *Handbook of mathematical functions, with formulas, graphs, and mathematical tables*, Dover Publications, Inc., New York, 1966. MR0208797

[BBK17] Elena Bossolini, Morten Brøns, and Kristian Uldall Kristiansen, *Singular limit analysis of a model for earthquake faulting*, Nonlinearity **30** (2017), no. 7, 2805–2834, DOI 10.1088/1361-6544/aa712e. MR3670008

[BC11] E. Barreto and J. R. Cressman, *Ion concentration dynamics as a mechanism for neuronal bursting*, Journal of Biological Physics **37** (2011), no. 3, 361–373.

[BDG$^+$16] John Burke, Mathieu Desroches, Albert Granados, Tasso J. Kaper, Martin Krupa, and Theodore Vo, *From canards of folded singularities to torus canards in a forced van der Pol equation*, J. Nonlinear Sci. **26** (2016), no. 2, 405–451, DOI 10.1007/s00332-015-9279-0. MR3466229

[BE86] S. M. Baer and T. Erneux, *Singular Hopf bifurcation to relaxation oscillations*, SIAM J. Appl. Math. **46** (1986), no. 5, 721–739, DOI 10.1137/0146047. MR858993

[BEG$^+$03] Katherine Bold, Chantal Edwards, John Guckenheimer, Sabyasachi Guharay, Kathleen Hoffman, Judith Hubbard, Ricardo Oliva, and Warren Weckesser, *The forced van der Pol equation. II. Canards in the reduced system*, SIAM J. Appl. Dyn. Syst. **2** (2003), no. 4, 570–608, DOI 10.1137/S1111111102419130. MR2050243

[Ben81] Éric Benoît, *Chasse au canard. II. Tunnels—entonnoirs—peignes* (French), Collect. Math. **32** (1981), no. 2, 77–97. MR653888

[Ben83] Éric Benoît, *Systèmes lents-rapides dans* $\mathbf{R}^3$ *et leurs canards* (French), Third Schnepfenried geometry conference, Vol. 2 (Schnepfenried, 1982), Astérisque, vol. 109, Soc. Math. France, Paris, 1983, pp. 159–191. MR753147

[Ben90] É. Benoît, *Canards et enlacements* (French), Inst. Hautes Études Sci. Publ. Math. **72** (1990), 63–91 (1991). MR1087393

[Ben01] Éric Benoît, *Perturbation singulière en dimension trois: canards en un point pseudo-singulier nœud* (French, with English and French summaries), Bull. Soc. Math. France **129** (2001), no. 1, 91–113, DOI 10.24033/bsmf.2387. MR1871979

[BER89] S. M. Baer, T. Erneux, and J. Rinzel, *The slow passage through a Hopf bifurcation: delay, memory effects, and resonance*, SIAM J. Appl. Math. **49** (1989), no. 1, 55–71, DOI 10.1137/0149003. MR978825

[BG16] Marcin Bobieński and Lubomir Gavrilov, *Finite cyclicity of slow-fast Darboux systems with a two-saddle loop*, Proc. Amer. Math. Soc. **144** (2016), no. 10, 4205–4219, DOI 10.1090/proc/12678. MR3531173

[BH25] George Edward Briggs and John Burdon Sanderson Haldane, *A note on the kinetics of enzyme action*, Biochemical journal **19** (1925), no. 2, 338.

[BKK13] Henk W. Broer, Tasso J. Kaper, and Martin Krupa, *Geometric desingularization of a cusp singularity in slow-fast systems with applications to Zeeman's examples*, J. Dynam. Differential Equations **25** (2013), no. 4, 925–958, DOI 10.1007/s10884-013-9322-5. MR3138152

[BKR02] A. S. Bobkova, A. Yu. Kolesov, and N. Kh. Rozov, *The "duck survival" problem in three-dimensional singularly perturbed systems with two slow variables* (Russian, with Russian summary), Mat. Zametki **71** (2002), no. 6, 818–831, DOI

10.1023/A:1015812727037; English transl., Math. Notes **71** (2002), no. 5-6, 749–760. MR1933103

[BL14] Jacek Banasiak and Mirosław Lachowicz, *Methods of small parameter in mathematical biology*, Modeling and Simulation in Science, Engineering and Technology, Birkhäuser/Springer, Cham, 2014, DOI 10.1007/978-3-319-05140-6. MR3306891

[Bli33] Gilbert Ames Bliss, *Algebraic functions*, Dover Publications, Inc., New York, 1966. MR0203007

[BM90] Marco Brunella and Massimo Miari, *Topological equivalence of a plane vector field with its principal part defined through Newton polyhedra*, J. Differential Equations **85** (1990), no. 2, 338–366, DOI 10.1016/0022-0396(90)90120-E. MR1054553

[Bog75] R. I. Bogdanov, *Versal deformation of a singular point of a vector field on the plane in the case of zero eigenvalues* (Russian), Funkcional Anal. i Priložen. **9** (1975), no. 2, 63. MR0380878

[Bon96] Patrick Bonckaert, *Partially hyperbolic fixed points with constraints*, Trans. Amer. Math. Soc. **348** (1996), no. 3, 997–1011, DOI 10.1090/S0002-9947-96-01469-9. MR1321568

[Bon97] P. Bonckaert, *Conjugacy of vector fields respecting additional properties*, J. Dynam. Control Systems **3** (1997), no. 3, 419–432, DOI 10.1007/BF02463258. MR1472359

[Bra98] B. Braaksma, *Singular Hopf bifurcation in systems with fast and slow variables*, J. Nonlinear Sci. **8** (1998), no. 5, 457–490, DOI 10.1007/s003329900058. MR1638759

[Car12] Jack Carr, *Applications of centre manifold theory*, Applied Mathematical Sciences, vol. 35, Springer-Verlag, New York-Berlin, 1981. MR635782

[Che63] Kuo-Tsai Chen, *Equivalence and decomposition of vector fields about an elementary critical point*, Amer. J. Math. **85** (1963), 693–722, DOI 10.2307/2373115. MR160010

[Chi11] Hayato Chiba, *Periodic orbits and chaos in fast-slow systems with Bogdanov-Takens type fold points*, J. Differential Equations **250** (2011), no. 1, 112–160, DOI 10.1016/j.jde.2010.09.022. MR2737837

[CN13] César Camacho and Alcides Lins Neto, *Teoria geométrica das folheaç oes* (Portuguese), Projeto Euclides [Euclid Project], vol. 9, Instituto de Matemática Pura e Aplicada, Rio de Janeiro, 1979. MR642831

[CS18] Paul Carter and Björn Sandstede, *Unpeeling a homoclinic banana in the FitzHugh-Nagumo system*, SIAM J. Appl. Dyn. Syst. **17** (2018), no. 1, 236–349, DOI 10.1137/16M1080707. MR3755655

[DFGK+18] Mathieu Desroches, Soledad Fernández-García, Martin Krupa, Rafel Prohens, and Antonio E. Teruel, *Piecewise-linear (PWL) canard dynamics: simplifying singular perturbation theory in the canard regime using piecewise-linear systems*, Nonlinear Systems. Vol. 1, Underst. Complex Syst., Springer, Cham, 2018, pp. 67–86. MR3837319

[DGK+12] Mathieu Desroches, John Guckenheimer, Bernd Krauskopf, Christian Kuehn, Hinke M. Osinga, and Martin Wechselberger, *Mixed-mode oscillations with multiple time scales*, SIAM Rev. **54** (2012), no. 2, 211–288, DOI 10.1137/100791233. MR2916308

[DGP+16] Mathieu Desroches, Antoni Guillamon, Enrique Ponce, Rafael Prohens, Serafim Rodrigues, and Antonio E. Teruel, *Canards, folded nodes, and mixed-mode oscillations in piecewise-linear slow-fast systems*, SIAM Rev. **58** (2016), no. 4, 653–691, DOI 10.1137/15M1014528. MR3567602

[DKK13] Mathieu Desroches, Tasso J. Kaper, and Martin Krupa, *Mixed-mode bursting oscillations: dynamics created by a slow passage through spike-adding canard explosion in a square-wave burster*, Chaos **23** (2013), no. 4, 046106, 13, DOI 10.1063/1.4827026. MR3389775

[DKO08] M. Desroches, B. Krauskopf, and H. M. Osinga, *The geometry of slow manifolds near a folded node*, SIAM J. Appl. Dyn. Syst. **7** (2008), no. 4, 1131–1162, DOI 10.1137/070708810. MR2452674

[DM15] P. De Maesschalck, *Planar canards with transcritical intersections*, Acta Appl. Math. **137** (2015), 159–184, DOI 10.1007/s10440-014-9994-9. MR3343377

[DMD06] Peter De Maesschalck and Freddy Dumortier, *Canard solutions at non-generic turning points*, Trans. Amer. Math. Soc. **358** (2006), no. 5, 2291–2334, DOI 10.1090/S0002-9947-05-03839-0. MR2197445

[DMD08] P. De Maesschalck and F. Dumortier, *Canard cycles in the presence of slow dynamics with singularities*, Proc. Roy. Soc. Edinburgh Sect. A **138** (2008), no. 2, 265–299, DOI 10.1017/S0308210506000199. MR2406691

[DMD11a] P. De Maesschalck and F. Dumortier, *Classical Liénard equations of degree $n \geq 6$ can have $[\frac{n-1}{2}] + 2$ limit cycles*, J. Differential Equations **250** (2011), no. 4, 2162–2176, DOI 10.1016/j.jde.2010.12.003. MR2763568

[DMD11b] P. De Maesschalck and F. Dumortier, *Slow-fast Bogdanov-Takens bifurcations*, J. Differential Equations **250** (2011), no. 2, 1000–1025, DOI 10.1016/j.jde.2010.07.022. MR2737822

[DMDR11] P. De Maesschalck, F. Dumortier, and R. Roussarie, *Cyclicity of common slow-fast cycles*, Indag. Math. (N.S.) **22** (2011), no. 3-4, 165–206, DOI 10.1016/j.indag.2011.09.008. MR2853605

[dMdS18] Paulo R. da Silva and Jaime R. de Moraes, *Piecewise-smooth slow-fast systems*, J. Dyn. Control Syst. **27** (2021), no. 1, 67–85, DOI 10.1007/s10883-020-09480-8. MR4190788

[DMW15] Peter De Maesschalck and Martin Wechselberger, *Neural excitability and singular bifurcations*, J. Math. Neurosci. **5** (2015), Art. 16, 32, DOI 10.1186/s13408-015-0029-2. MR3376904

[DPR07] Freddy Dumortier, Daniel Panazzolo, and Robert Roussarie, *More limit cycles than expected in Liénard equations*, Proc. Amer. Math. Soc. **135** (2007), no. 6, 1895–1904, DOI 10.1090/S0002-9939-07-08688-1. MR2286102

[DR96] Freddy Dumortier and Robert Roussarie, *Canard cycles and center manifolds*, Mem. Amer. Math. Soc. **121** (1996), no. 577, x+100, DOI 10.1090/memo/0577. With an appendix by Cheng Zhi Li. MR1327208

[DR06] Freddy Dumortier and Robert Roussarie, *Abelian integrals and limit cycles*, J. Differential Equations **227** (2006), no. 1, 116–165, DOI 10.1016/j.jde.2005.08.015. MR2233957

[DR07] Freddy Dumortier and Robert Roussarie, *Bifurcation of relaxation oscillations in dimension two*, Discrete Contin. Dyn. Syst. **19** (2007), no. 4, 631–674, DOI 10.3934/dcds.2007.19.631. MR2342266

[DR09] Freddy Dumortier and Robert Roussarie, *Birth of canard cycles*, Discrete Contin. Dyn. Syst. Ser. S **2** (2009), no. 4, 723–781, DOI 10.3934/dcdss.2009.2.723. MR2552119

[Dum77] Freddy Dumortier, *Singularities of vector fields on the plane*, J. Differential Equations **23** (1977), no. 1, 53–106, DOI 10.1016/0022-0396(77)90136-X. MR650816

[Dum91] F. Dumortier, *Local study of planar vector fields: singularities and their unfoldings*, Studies in Mathematical Physics, vol. 2, Elsevier, 1991, 161–241.

[Eck11] Wiktor Eckhaus, *Matched asymptotic expansions and singular perturbations*, North-Holland Publishing Co., Amsterdam-London; American Elsevier Publishing Co., Inc., New York, 1973. North-Holland Mathematics Studies, No. 6. MR0670800

[EK18] Maximilian Engel and Christian Kuehn, *Discretized fast-slow systems near transcritical singularities*, Nonlinearity **32** (2019), no. 7, 2365–2391, DOI 10.1088/1361-6544/ab15c1. MR3957214

[EK20] M. Engel and C. Kuehn, *Blow-up analysis of fast-slow PDEs with loss of hyperbolicity.* arXiv preprint `arXiv:2007.09973` (2020).

[Eld13] Jaap Eldering, *Normally hyperbolic invariant manifolds*, Atlantis Studies in Dynamical Systems, vol. 2, Atlantis Press, Paris, 2013. The noncompact case, DOI 10.2991/978-94-6239-003-4. MR3098498

[EW09] Bard Ermentrout and Martin Wechselberger, *Canards, clusters, and synchronization in a weakly coupled interneuron model*, SIAM J. Appl. Dyn. Syst. **8** (2009), no. 1, 253–278, DOI 10.1137/080724010. MR2481284

[Fen71] Neil Fenichel, *Persistence and smoothness of invariant manifolds for flows*, Indiana Univ. Math. J. **21** (1971/72), 193–226, DOI 10.1512/iumj.1971.21.21017. MR287106

[Fen79] Neil Fenichel, *Geometric singular perturbation theory for ordinary differential equations*, J. Differential Equations **31** (1979), no. 1, 53–98, DOI 10.1016/0022-0396(79)90152-9. MR524817

[GG12] M. Golubitsky and V. Guillemin, *Stable mappings and their singularities*, Springer-Verlag, New York-Heidelberg, 1973. Graduate Texts in Mathematics, Vol. 14. MR0341518

[GH05] John Guckenheimer and Radu Haiduc, *Canards at folded nodes* (English, with English and Russian summaries), Mosc. Math. J. **5** (2005), no. 1, 91–103, DOI 10.17323/1609-4514-2005-5-1-91-103. MR2153468

[GH13] John Guckenheimer and Philip Holmes, *Nonlinear oscillations, dynamical systems, and bifurcations of vector fields*, Applied Mathematical Sciences, vol. 42, Springer-Verlag, New York, 1983, DOI 10.1007/978-1-4612-1140-2. MR709768

[GK09] John Guckenheimer and Christian Kuehn, *Homoclinic orbits of the FitzHugh-Nagumo equation: the singular-limit*, Discrete Contin. Dyn. Syst. Ser. S **2** (2009), no. 4, 851–872, DOI 10.3934/dcdss.2009.2.851. MR2552123

[GK10] John Guckenheimer and Christian Kuehn, *Homoclinic orbits of the FitzHugh-Nagumo equation: bifurcations in the full system*, SIAM J. Appl. Dyn. Syst. **9** (2010), no. 1, 138–153, DOI 10.1137/090758404. MR2595892

[GS09] Ilona Gucwa and Peter Szmolyan, *Geometric singular perturbation analysis of an autocatalator model*, Discrete Contin. Dyn. Syst. Ser. S **2** (2009), no. 4, 783–806, DOI 10.3934/dcdss.2009.2.783. MR2552120

[Guc08] John Guckenheimer, *Singular Hopf bifurcation in systems with two slow variables*, SIAM J. Appl. Dyn. Syst. **7** (2008), no. 4, 1355–1377, DOI 10.1137/080718528. MR2452681

[GWY06] John Guckenheimer, Martin Wechselberger, and Lai-Sang Young, *Chaotic attractors of relaxation oscillators*, Nonlinearity **19** (2006), no. 3, 701–720, DOI 10.1088/0951-7715/19/3/009. MR2209295

[HDMD13] R. Huzak, P. De Maesschalck, and F. Dumortier, *Limit cycles in slow-fast codimension 3 saddle and elliptic bifurcations*, J. Differential Equations **255** (2013), no. 11, 4012–4051, DOI 10.1016/j.jde.2013.07.057. MR3097246

[HDMD14] Renato Huzak, Peter De Maesschalck, and Freddy Dumortier, *Primary birth of canard cycles in slow-fast codimension 3 elliptic bifurcations*, Commun. Pure Appl. Anal. **13** (2014), no. 6, 2641–2673, DOI 10.3934/cpaa.2014.13.2641. MR3248407

[Hek10] Geertje Hek, *Geometric singular perturbation theory in biological practice*, J. Math. Biol. **60** (2010), no. 3, 347–386, DOI 10.1007/s00285-009-0266-7. MR2576546

[Hir64a] Heisuke Hironaka, *Resolution of singularities of an algebraic variety over a field of characteristic zero. I, II*, Ann. of Math. (2) **79** (1964), 109–203; ibid. (2) **79** (1964), 205–326, DOI 10.2307/1970547. MR0199184

[Hir64b] Heisuke Hironaka, *Resolution of singularities of an algebraic variety over a field of characteristic zero. I, II*, Ann. of Math. (2) **79** (1964), 109–203; ibid. (2) **79** (1964), 205–326, DOI 10.2307/1970547. MR0199184

[HKSW16] Michael G. Hayes, Tasso J. Kaper, Peter Szmolyan, and Martin Wechselberger, *Geometric desingularization of degenerate singularities in the presence of fast rotation: a new proof of known results for slow passage through Hopf bifurcations*, Indag. Math. (N.S.) **27** (2016), no. 5, 1184–1203, DOI 10.1016/j.indag.2015.11.005. MR3573755

[HPS70] M. W. Hirsch, C. C. Pugh, and M. Shub, *Invariant manifolds*, Bull. Amer. Math. Soc. **76** (1970), 1015–1019, DOI 10.1090/S0002-9904-1970-12537-X. MR292101

[Huz16] Renato Huzak, *Cyclicity of the origin in slow-fast codimension 3 saddle and elliptic bifurcations*, Discrete Contin. Dyn. Syst. **36** (2016), no. 1, 171–215, DOI 10.3934/dcds.2016.36.171. MR3369218

[Huz18] Renato Huzak, *Canard explosion near non-Liénard type slow-fast Hopf point*, J. Dynam. Differential Equations **31** (2019), no. 2, 683–709, DOI 10.1007/s10884-018-9645-3. MR3951820

[Inc56] E. L. Ince, *Ordinary Differential Equations*, Dover Publications, New York, 1944. MR0010757

[IPS18] Annalisa Iuorio, Nikola Popović, and Peter Szmolyan, *Singular perturbation analysis of a regularized MEMS model*, SIAM J. Appl. Dyn. Syst. **18** (2019), no. 2, 661–708, DOI 10.1137/18M1197552. MR3937919

[Jef16] Mike R. Jeffrey, *Hidden degeneracies in piecewise smooth dynamical systems*, Internat. J. Bifur. Chaos Appl. Sci. Engrg. **26** (2016), no. 5, 1650087, 18, DOI 10.1142/S0218127416500875. MR3506682

[JK15] Hildeberto Jardón-Kojakhmetov, *Formal normal form of A_k slow-fast systems*, C. R. Math. Acad. Sci. Paris **353** (2015), no. 9, 795–800, DOI 10.1016/j.crma.2015.06.009. MR3377675

[JKBR16] H. Jardón-Kojakhmetov, Henk W. Broer, and R. Roussarie, *Analysis of a slow-fast system near a cusp singularity*, J. Differential Equations **260** (2016), no. 4, 3785–3843, DOI 10.1016/j.jde.2015.10.045. MR3434415

[JKK19] H. Jardón-Kojakhmetov and C. Kuehn, *Controlling Canard Cycles*, arXiv preprint `arXiv:1911.11861` (2019), 1–.

[JKK20] Hildeberto Jardón-Kojakhmetov and Christian Kuehn, *On fast-slow consensus networks with a dynamic weight*, J. Nonlinear Sci. **30** (2020), no. 6, 2737–2786, DOI 10.1007/s00332-020-09634-9. MR4170310

[JKS17] Hildeberto Jardón-Kojakhmetov and Jacquelien M. A. Scherpen, *Model order reduction and composite control for a class of slow-fast systems around a non-hyperbolic point*, IEEE Control Syst. Lett. **1** (2017), no. 1, 68–73. MR4208511

[JKS18] H. Jardón-Kojakhmetov and Jacquelien M. A. Scherpen, *Improving the region of attraction of a non-hyperbolic point in slow-fast systems with one fast direction*, IEEE Control Syst. Lett. **2** (2018), no. 2, 296–301. MR4208619

[JKSdPF19] Hildeberto Jardón-Kojakhmetov, Jacquelien M. A. Scherpen, and Dunstano del Puerto-Flores, *Stabilization of a class of slow-fast control systems at non-hyperbolic points*, Automatica J. IFAC **99** (2019), 13–21, DOI 10.1016/j.automatica.2018.10.008. MR3872555

[JKSW] S. Jelbart, K. U. Kristiansen, P. Szmolyan and M. Wechselberger, *Singularly perturbed oscillators with exponential nonlinearities*, arXiv preprint `arXiv:1912.11769`, 2019, 1–.

[JKW20] S. Jelbart, K. U. Kristiansen and M. Wechselberger, *Singularly Perturbed Boundary-Focus Bifurcations*, arXiv preprint `arXiv:2006.06087`, 2020, 1–.

[JNS18] Huiwen Ju, Alexander B. Neiman, and Andrey L. Shilnikov, *Bottom-up approach to torus bifurcation in neuron models*, Chaos **28** (2018), no. 10, 106317, 21, DOI 10.1063/1.5042078. MR3863459

[Jon95] Christopher K. R. T. Jones, *Geometric singular perturbation theory*, Dynamical systems (Montecatini Terme, 1994), Lecture Notes in Math., vol. 1609, Springer, Berlin, 1995, pp. 44–118, DOI 10.1007/BFb0095239. MR1374108

[KC12] J. Kevorkian and J. D. Cole, *Multiple scale and singular perturbation methods*, Applied Mathematical Sciences, vol. 114, Springer-Verlag, New York, 1996, DOI 10.1007/978-1-4612-3968-0. MR1392475

[KH15a] K. Uldall Kristiansen and S. J. Hogan, *On the use of blowup to study regularizations of singularities of piecewise smooth dynamical systems in $\mathbb{R}^3$*, SIAM J. Appl. Dyn. Syst. **14** (2015), no. 1, 382–422, DOI 10.1137/140980995. MR3319843

[KH15b] K. Uldall Kristiansen and S. J. Hogan, *Regularizations of two-fold bifurcations in planar piecewise smooth systems using blowup*, SIAM J. Appl. Dyn. Syst. **14** (2015), no. 4, 1731–1786, DOI 10.1137/15M1009731. MR3404676

[KKO99] Petar Kokotović, Hassan K. Khalil, and John O'Reilly, *Singular perturbation methods in control*, Classics in Applied Mathematics, vol. 25, Society for Industrial and Applied Mathematics (SIAM), Philadelphia, PA, 1999. Analysis and design; Corrected reprint of the 1986 original, DOI 10.1137/1.9781611971118. MR1727138

[KM06] Peter Kunkel and Volker Mehrmann, *Differential-algebraic equations*, EMS Textbooks in Mathematics, European Mathematical Society (EMS), Zürich, 2006. Analysis and numerical solution, DOI 10.4171/017. MR2225970

[KM17] Christian Kuehn and Christian Münch, *Generalized play hysteresis operators in limits of fast-slow systems*, SIAM J. Appl. Dyn. Syst. **16** (2017), no. 3, 1650–1685, DOI 10.1137/17M1110584. MR3690648

[KM19] Christian Kuehn and Christian Münch, *Duck traps: two-dimensional critical manifolds in planar systems*, Dyn. Syst. **34** (2019), no. 4, 584–612, DOI 10.1080/14689367.2019.1575337. MR4020734

[KPK08] Martin Krupa, Nikola Popović, and Nancy Kopell, *Mixed-mode oscillations in three time-scale systems: a prototypical example*, SIAM J. Appl. Dyn. Syst. **7** (2008), no. 2, 361–420, DOI 10.1137/070688912. MR2421110

[Kri17] K. U. Kristiansen, *Blowup for flat slow manifolds*, Nonlinearity **30** (2017), no. 5, 2138–2184, DOI 10.1088/1361-6544/aa6449. MR3639302

[KS01a] M. Krupa and P. Szmolyan, *Extending geometric singular perturbation theory to nonhyperbolic points—fold and canard points in two dimensions*, SIAM J. Math. Anal. **33** (2001), no. 2, 286–314, DOI 10.1137/S0036141099360919. MR1857972

[KS01b] M. Krupa and P. Szmolyan, *Extending slow manifolds near transcritical and pitchfork singularities*, Nonlinearity **14** (2001), no. 6, 1473–1491, DOI 10.1088/0951-7715/14/6/304. MR1867088

[KS01c] M. Krupa and P. Szmolyan, *Relaxation oscillation and canard explosion*, J. Differential Equations **174** (2001), no. 2, 312–368, DOI 10.1006/jdeq.2000.3929. MR1846739

[KS11] Ilona Kosiuk and Peter Szmolyan, *Scaling in singular perturbation problems: blowing up a relaxation oscillator*, SIAM J. Appl. Dyn. Syst. **10** (2011), no. 4, 1307–1343, DOI 10.1137/100814470. MR2854590

[KS15] Christian Kuehn and Peter Szmolyan, *Multiscale geometry of the Olsen model and non-classical relaxation oscillations*, J. Nonlinear Sci. **25** (2015), no. 3, 583–629, DOI 10.1007/s00332-015-9235-z. MR3338451

[KS16] Ilona Kosiuk and Peter Szmolyan, *Geometric analysis of the Goldbeter minimal model for the embryonic cell cycle*, J. Math. Biol. **72** (2016), no. 5, 1337–1368, DOI 10.1007/s00285-015-0905-0. MR3464205

[Kue14] Christian Kuehn, *Normal hyperbolicity and unbounded critical manifolds*, Nonlinearity **27** (2014), no. 6, 1351–1366, DOI 10.1088/0951-7715/27/6/1351. MR3215838

[Kue15] Christian Kuehn, *Multiple time scale dynamics*, Applied Mathematical Sciences, vol. 191, Springer, Cham, 2015, DOI 10.1007/978-3-319-12316-5. MR3309627

[Kue16] Christian Kuehn, *A remark in geometric desingularization of a non-hyperbolic point using hyperbolic space*, J. Phys. Conf. Ser. **727** (2016), 012008, 14, DOI 10.1088/1742-6596/727/1/012008. MR3566837

[KV18] Tasso J. Kaper and Theodore Vo, *Delayed loss of stability due to the slow passage through Hopf bifurcations in reaction-diffusion equations*, Chaos **28** (2018), no. 9, 091103, 9, DOI 10.1063/1.5050508. MR3858264

[KW10] Martin Krupa and Martin Wechselberger, *Local analysis near a folded saddle-node singularity*, J. Differential Equations **248** (2010), no. 12, 2841–2888, DOI 10.1016/j.jde.2010.02.006. MR2644136

[LDMP77] A. Lins, W. de Melo, and C. C. Pugh, *On Liénard's equation*, Geometry and topology (Proc. III Latin Amer. School of Math., Inst. Mat. Pura Aplicada CNPq, Rio de Janeiro, 1976), Springer, Berlin, 1977, pp. 335–357. Lecture Notes in Math., Vol. 597. MR0448423

[LS10] Eric Lombardi and Laurent Stolovitch, *Normal forms of analytic perturbations of quasihomogeneous vector fields: rigidity, invariant analytic sets and exponentially small approximation* (English, with English and French summaries), Ann. Sci. Éc. Norm. Supér. (4) **43** (2010), no. 4, 659–718, DOI 10.24033/asens.2131. MR2722512

[Moe02] J. Moehlis, *Canards in a surface oxidation reaction*, J. Nonlinear Sci. **12** (2002), no. 4, 319–345, DOI 10.1007/s00332-002-0467-3. MR1915940

[MPS20] Zhouqian Miao, Nikola Popović, and Peter Szmolyan, *Oscillations in a cAMP signalling model for cell aggregation—a geometric analysis*, J. Math. Anal. Appl. **483** (2020), no. 1, 123577, 31, DOI 10.1016/j.jmaa.2019.123577. MR4020480

[MR80] E. F. Mishchenko and N. Kh. Rozov, *Differential equations with small parameters and relaxation oscillations*, Mathematical Concepts and Methods in Science and Engineering, vol. 13, Plenum Press, New York, 1980. Translated from the Russian by F. M. C. Goodspeed, DOI 10.1007/978-1-4615-9047-7. MR750298

[MS01] Alexandra Milik and Peter Szmolyan, *Multiple time scales and canards in a chemical oscillator*, Multiple-time-scale dynamical systems (Minneapolis, MN, 1997), IMA Vol. Math. Appl., vol. 122, Springer, New York, 2001, pp. 117–140, DOI 10.1007/978-1-4613-0117-2_5. MR1846575

[MSLG98] A. Milik, P. Szmolyan, H. Löffelmann, and E. Gröller, *Geometry of mixed-mode oscillations in the 3-d autocatalator*, International Journal of Bifurcation and Chaos **8** (1998), no. 03, 505–519.

[MW17] John Mitry and Martin Wechselberger, *Folded saddles and faux canards*, SIAM J. Appl. Dyn. Syst. **16** (2017), no. 1, 546–596, DOI 10.1137/15M1045065. MR3612893

[Nei87] A. I. Neishtadt, *Persistence of stability loss for dynamical bifurcations I*, Differential Equations **23** (1987), 1385–1391.

[Nei09] Anatoly Neishtadt, *On stability loss delay for dynamical bifurcations*, Discrete Contin. Dyn. Syst. Ser. S **2** (2009), no. 4, 897–909, DOI 10.3934/dcdss.2009.2.897. MR2552125

[NS13] Kaspar Nipp and Daniel Stoffer, *Invariant manifolds in discrete and continuous dynamical systems*, EMS Tracts in Mathematics, vol. 21, European Mathematical Society (EMS), Zürich, 2013, DOI 10.4171/124. MR3098463

[NSS09] K. Nipp, D. Stoffer, and P. Szmolyan, *Graph transform and blow-up in singular perturbations*, International Conference on Numerical Analysis and Applied Mathematics, vol. 1, 2009, 861–868.

[OLBC10] F. W. J. Olver, *Airy and related functions*, NIST handbook of mathematical functions, U.S. Dept. Commerce, Washington, DC, 2010, pp. 193–213. MR2655349

[O'M91] Robert E. O'Malley Jr., *Singular perturbation methods for ordinary differential equations*, Applied Mathematical Sciences, vol. 89, Springer-Verlag, New York, 1991, DOI 10.1007/978-1-4612-0977-5. MR1123483

[Pan02] Daniel Panazzolo, *Desingularization of nilpotent singularities in families of planar vector fields*, Mem. Amer. Math. Soc. **158** (2002), no. 753, viii+108, DOI 10.1090/memo/0753. MR1903719

[Pan06] Daniel Panazzolo, *Resolution of singularities of real-analytic vector fields in dimension three*, Acta Math. **197** (2006), no. 2, 167–289, DOI 10.1007/s11511-006-0011-7. MR2296055

[RW07] Jonathan Rubin and Martin Wechselberger, *Giant squid-hidden canard: the 3D geometry of the Hodgkin-Huxley model*, Biol. Cybernet. **97** (2007), no. 1, 5–32, DOI 10.1007/s00422-007-0153-5. MR2326520

[Sch85] Stephen Schecter, *Persistent unstable equilibria and closed orbits of a singularly perturbed equation*, J. Differential Equations **60** (1985), no. 1, 131–141, DOI 10.1016/0022-0396(85)90124-X. MR808261

[Shi73] M. A. Šiškova, *A discussion of a certain system of differential equations with a small parameter multiplying the highest derivatives* (Russian), Dokl. Akad. Nauk SSSR **209** (1973), 576–579. MR0324157

[SM96] P. Strizhak and M. Menzinger, *Slow passage through a supercritical Hopf bifurcation: Time-delayed response in the Belousov–Zhabotinsky reaction in a batch reactor*, The Journal of Chemical Physics **105** (1996), no. 24, 10905–10910.

[Sma98] Steve Smale, *Mathematical problems for the next century*, Math. Intelligencer **20** (1998), no. 2, 7–15, DOI 10.1007/BF03025291. MR1631413

[Ste58] Shlomo Sternberg, *On the structure of local homeomorphisms of euclidean n-space. II*, Amer. J. Math. **80** (1958), 623–631, DOI 10.2307/2372774. MR96854

[SW01] Peter Szmolyan and Martin Wechselberger, *Canards in* $\mathbb{R}^3$, J. Differential Equations **177** (2001), no. 2, 419–453, DOI 10.1006/jdeq.2001.4001. MR1876650

[SW04] P. Szmolyan and M. Wechselberger, *Relaxation oscillations in* $\mathbb{R}^3$, J. Differential Equations **200** (2004), no. 1, 69–104, DOI 10.1016/j.jde.2003.09.010. MR2046318

[TJKSC21] Hadi Taghvafard, Hildeberto Jardón-Kojakhmetov, Peter Szmolyan, and Ming Cao, *Geometric analysis of oscillations in the Frzilator model*, J. Math. Anal. Appl. **495** (2021), no. 1, 124725, 35, DOI 10.1016/j.jmaa.2020.124725. MR4172878

[Tak71] Floris Takens, *Partially hyperbolic fixed points*, Topology **10** (1971), 133–147, DOI 10.1016/0040-9383(71)90035-8. MR307279

[Tak74] Floris Takens, *Singularities of vector fields*, Inst. Hautes Études Sci. Publ. Math. **43** (1974), 47–100. MR339292

[Tak76] Floris Takens, *Constrained equations; a study of implicit differential equations and their discontinuous solutions*, Mathematisch Institut, Rijksuniversiteit, Groningen, 1975. Report ZW-75-03. MR0478236

[Tak01] Floris Takens, *Forced oscillations and bifurcations*, Global analysis of dynamical systems, Inst. Phys., Bristol, 2001, pp. 1–61. MR1858470

[Tik52] A. N. Tihonov, *Systems of differential equations containing small parameters in the derivatives* (Russian), Mat. Sbornik N. S. **31(73)** (1952), 575–586. MR0055515

[Ver05] Ferdinand Verhulst, *Methods and applications of singular perturbations*, Texts in Applied Mathematics, vol. 50, Springer, New York, 2005. Boundary layers and multiple timescale dynamics, DOI 10.1007/0-387-28313-7. MR2148856

[VGKS05] S. van Gils, M. Krupa, and P. Szmolyan, *Asymptotic expansions using blow-up*, Z. Angew. Math. Phys. **56** (2005), no. 3, 369–397, DOI 10.1007/s00033-004-1021-y. MR2142930

[Vo17] Theodore Vo, *Generic torus canards*, Phys. D **356/357** (2017), 37–64, DOI 10.1016/j.physd.2017.06.005. MR3689773

[VW15] Theodore Vo and Martin Wechselberger, *Canards of folded saddle-node type I*, SIAM J. Math. Anal. **47** (2015), no. 4, 3235–3283, DOI 10.1137/140965818. MR3391149

[Wec05] Martin Wechselberger, *Existence and bifurcation of canards in* $\mathbb{R}^3$ *in the case of a folded node*, SIAM J. Appl. Dyn. Syst. **4** (2005), no. 1, 101–139, DOI 10.1137/030601995. MR2136520

[Wec12] Martin Wechselberger, *À propos de canards (Apropos canards)*, Trans. Amer. Math. Soc. **364** (2012), no. 6, 3289–3309, DOI 10.1090/S0002-9947-2012-05575-9. MR2888246

[Wec20] M. Wechselberger, *Geometric singular perturbation theory beyond the standard form*, TSpringer International Publishing, 2020.

Bernoulli Institute, University of Groningen, 9747 AG, Groningen, the Netherlands
Email address: h.jardon.kojakhmetov@rug.nl

Technical University of Munich, Faculty of Mathematics, Research Unit "Multiscale and Stochastic Dynamics", 85748 Garching b. München, Germany
Email address: ckuehn@ma.tum.de

Contemporary Mathematics
Volume **775**, 2021
https://doi.org/10.1090/conm/775/15592

Langlands program and Ramanujan Conjecture: A survey

Luis Alberto Lomelí

ABSTRACT. We present topics in the Langlands Program to graduate students and a wider mathematically mature audience. We study both global and local aspects in characteristic zero as well as characteristic p. We look at modern approaches to the generalized Ramanujan conjecture, which is an open problem over number fields, and present known cases over function fields. In particular, we study automorphic L-functions via the Langlands-Shahidi method. Hopefully, our approach to the Langlands Program can help guide the interested reader into an exciting field of mathematical research.

Introduction

Langlands' functoriality conjectures encompass a vast generalization of classical reciprocity laws that provide crossroads between Number Theory and Representation Theory. We let $\mathbf{G}$ be a connected reductive group and Γ a certain subgroup. With the aid of Einsenstein series, Langlands made the important connection between the theory of Automorphic Forms and Representation Theory. Namely, an automorphic representation occurs as an irreducible constituent of the right regular action on the $\mathscr{L}^2$-space of complex valued functions

$$\mathscr{L}^2(\Gamma\backslash G) = \mathscr{L}^2(\Gamma\backslash G)_{\text{disc}} \oplus \mathscr{L}^2(\Gamma\backslash G)_{\text{cont}},$$

decomposed into discrete and continuous spectra. Initially, the group $\mathbf{G}$ was taken to be defined over the real numbers $\mathbb{R}$, yet we are able to study the spectral decomposition in the general setting of the ring of adèles over a global field k [**68, 82**]. That is, k is either a finite extension of the rational numbers $\mathbb{Q}$ (a number field) or a finite separable extension of the field of functions for the projective line $\mathbb{P}^1$ defined over the finite field $\mathbb{F}_q$ (a function field).

Ubiquitous in modern number theory is the case of $G = \mathrm{SL}_2(\mathbb{R})$, with its principal congruence subgroup $\Gamma = \mathrm{SL}_2(\mathbb{Z})$. Modular forms find themselves inside the discrete spectrum $\mathscr{L}^2(\Gamma\backslash G)_{\text{disc}}$. Of particular importance in the theory is Ramanujan's weight 12 modular form

$$\Delta(z) = q\prod_{n=1}^{\infty}(1-q^n)^{24} = \sum_{n=1}^{\infty}\tau(n)q^n,$$

2020 *Mathematics Subject Classification.* Primary 11F70, 22E50, 22E55.

Key words and phrases. Langlands Program, L-functions, Ramanujan Conjecture.

Partially supported by Project VRIEA-PUCV 039.367 and FONDECYT Grant 1171583.

where $q = \exp(2\pi iz)$ and z is a variable in the hyperbolic upper half plane $\mathfrak{h}$. The coefficients $\tau(n)$ appearing in the Fourier expansion of $\Delta(z)$ give Ramanujan's τ-function. It is a multiplicative function, i.e., $\tau(mn) = \tau(m)\tau(n)$ if $(m,n) = 1$, and it satisfies the recursive relation on primes:

$$\tau(p^{n+1}) = \tau(p)\tau(p^n) - p^{11}\tau(p^{n-1}).$$

The original Ramanujan Conjecture states that

$$|\tau(p)| \leq 2p^{\frac{11}{2}},$$

and was proved by Deligne as a consequence of his work on the Weil conjectures [**28**]. He proved it for modular forms, which is also known as the Ramanujan-Petersson conjecture. The Ramanujan conjecture generalizes to automorphic representations, which we study in this article.

Given a global field k, its ring of adèles $\mathbb{A}_k$ has the correct topology in order for us to do Fourier analysis. Namely, it is a restricted direct product taken over all of the completions over the primes of k, in addition to the arquimedean valuations, giving us that $\mathbb{A}_k$ is a locally compact topological ring. If we let $\mathbf{G} = \mathrm{GL}_1(\mathbb{A}_k)$ and $\Gamma = k^\times$, an automorphic representation in this setting is a Größencharakter $\chi : k^\times \backslash \mathbb{A}_k^\times \to \mathbb{C}^\times$. We then find ourselves in the abelian case of Tate's thesis [**106**]. The abelian case in general provides the base where Langlands functoriality conjectures build upon, namely Class Field Theory [**5**].

We observe that by taking $k = \mathbb{Q}$ and the trivial Größencharakter, we retrieve the Riemann ζ-function from an automorphic L-function for GL_1, i.e.,

$$L^S(s, \mathbb{1}) = \prod_p \left(1 - p^{-s}\right)^{-1} = \sum_{n=1}^{\infty} n^{-s} = \zeta(s),$$

where $S = \{\infty\}$, and we initially have absolute convergence for $\Re(s) > 1$. Tate L-functions were extended to automorphic L-functions for GL_n by Godement and Jacquet [**39**], via an integral representation. Local Rankin-Selberg products for representations of GL_m and GL_n were obtained by Jacquet, Piatetski-Shapiro and Shalika in [**53**]; for remarks and further references on the global theory we refer to [**25**] and [**50**]. After preliminaries in Number Theory and Representation Theory, we present a uniform treatment of these L-functions and products for globally generic representations of the classical groups via the Langlands-Shahidi method.

Now, let $G = \mathrm{GL}_n(\mathbb{A}_k)$ and $\Gamma = \mathbf{Z}(\mathbb{A}_k)\mathrm{GL}_n(k)$, where $\mathbf{Z}$ denotes the center of GL_n. The group GL_n is self dual, i.e., the connected component of its Langlands dual group is ${}^L\mathrm{GL}_n^\circ = \mathrm{GL}_n(\mathbb{C})$. The study of automorphic forms in $\mathscr{L}^2(\Gamma\backslash G)$ leads us to an important non-abelian generalized reciprocity law. Here, a cuspidal automorphic representation π is an irreducible constituent of the right regular action on $\mathscr{L}^2(\Gamma\backslash G)$; furthermore, π has a central character ω, hence we can set $\Gamma_0 = \mathrm{GL}_n(k)$ and study the subspace $\mathscr{L}^2(\Gamma_0\backslash G; \omega)$ of functions transforming via ω under the center, in this case $\mathbf{Z}(\mathbb{A}_k) \cong \mathrm{GL}_1(\mathbb{A}_k)$. The global Langlands conjecture states that there is a correspondence

$$\left\{ \begin{array}{c} \Pi \text{ a cuspidal automorphic} \\ \text{representation of } \mathrm{GL}_n(\mathbb{A}_k) \end{array} \right\} \longleftrightarrow \left\{ \begin{array}{c} \Phi_\Pi : \mathcal{L}_k \to {}^L\mathrm{GL}_n \text{ an} \\ \text{irreducible } L\text{-parameter} \end{array} \right\}$$

agreeing with the abelian GL_1 theory. If k is a global function field, the group $\mathcal{L}_k$ is essentially the Weil group $\mathcal{W}_k$ and we note that global Langlands is a landmark result of V. G. Drinfeld for GL_2 [**30, 31**] and of L. Lafforgue for GL_n [**65**]. The case of a number field k being open, there should be a conjectural locally compact topological group $\mathcal{L}_k$ associated to the absolute Galois group $\mathcal{G}_k$, where $\mathcal{L}_k$ should have the right topology and there should be a morphism $\mathcal{L}_k \to \mathcal{G}_k$ for us to have a one-to-one correspondence between automorphic representations of GL_n and Galois representations, as mentioned by Langlands in [**70**]. The correspondence is in such a way that it respects global L-functions and root numbers, and agrees with the local Langlands correspondence at every place. On the left hand side we have automorphic L-functions and on the right hand side we have Artin L-functions. The correspondence is known to imply the Ramanujan Conjecture for automorphic representations.

When we take the completion of a global field k at one of its absolute values v, we obtain a local field, and all local fields arise in this way. A non-archimedean local field is obtained from a $\mathfrak{p}$-adic valuation, $\mathfrak{p}$ a prime ideal in the ring of integers $\mathcal{O}_k$ of a global field k. We let F denote a non-archimedean local field.

We work with connected quasi-split reductive groups over local and global fields. If the reader is unfamiliar with reductive groups, it is safe to assume on a first reading of this article that $\mathbf{G}$ is a group given by the examples of section § 2.1, which include the classical groups. Given a connected reductive group $\mathbf{G}$ defined over the local field F, we write $G = \mathbf{G}(F)$. We study admissible representations π of G over a complex infinite dimensional vector space V; we refer to the first three sections of [**15**] for basic notions of admissible representations and further references.

The local Langlands correspondence for GL_n provides us with a one to one reciprocity relation

$$\left\{ \begin{array}{c} \pi \text{ an irreducible admissible} \\ \text{representation of } \mathrm{GL}_n(F) \end{array} \right\} \longleftrightarrow \left\{ \begin{array}{c} \phi_\pi : \mathcal{W}'_F \to {}^L\mathrm{GL}_n \\ \text{an } L\text{-parameter} \end{array} \right\},$$

where the Weil group $\mathcal{W}_F$ provides us with the right locally compact group on the Galois side, yet we need L-parameters of the Weil-Deligne group $\mathcal{W}'_F$ or Weil-Deligne representations of $\mathcal{W}_F$, see § 7 of [**15**]. The correspondence is uniquely characterized by the fact that it preserves local L-functions and root numbers for products of representations [**48**].

From the theory of admissible representations of a $\mathfrak{p}$-adic group $G = \mathbf{G}(F)$, an irreducible admissible representation π is supercuspidal if it cannot be obtained as a constituent of a parabolically induced representation

$$\mathrm{Ind}_P^G(\rho).$$

Here, $\mathbf{P}$ is a proper parabolic subgroup of $\mathbf{G}$, with Levi component $\mathbf{M}$ and unipotent radical $\mathbf{N}$; ρ is a representation of M, extended trivially on N; and, we use unitary parabolic induction. For $\mathbf{G} = \mathrm{GL}_n$, the Levi subgroups are all of the form $\prod \mathrm{GL}_{n_i}$, with $n = \sum n_i$, and standard parabolic subgroups are block upper triangular. The Levi subgroups and parabolic subgroups of the classical groups are described in § 2.1. As observed in [**16**], there are two main steps in undesrtanding admissible representations: (i) constructing supercuspidal representations; (ii) studying admissible representations via parabolic induction.

The first examples of supercuspidals were obtained in [**80**], and work using the theory of types of [**16**] gives all supercuspidals for GL_n. Stevens constructs all supercuspidals for the classical groups, assuming $\mathrm{char}(F) \neq 2$ [**103**]. For a general reductive group, the approach of Yu gives tame supercuspidals [**62, 111**]. And we have the supercuspidals of Gross and Reeder, in addition to the work of Kaletha [**56**]. The Kazhdan transfer between close local fields provides an approach to obtain the local Langlands correspondence in characteristic p from known cases in characteristic 0 [**6, 36**].

We observe that studying the space $\mathscr{L}^2(G)$ leads to the Plancherel Theorem much studied by Harish-Chandra [**110**]. For a classical group $\mathbf{G}$, in addition to L-functions and root numbers, we use Harish-Chandra μ-functions in order to provide a characterization of the local Langlands transfer. Tempered representations form an important class of representations of $\mathfrak{p}$-adic reductive groups; we recall that an admissible representation is tempered if its matrix coefficients are in $\mathscr{L}^{2+\varepsilon}(G)$ for every $\varepsilon > 0$.

Ramanujan Conjecture for GL_n. *Let Π be a cuspidal unitary automorphic representation of $\mathrm{GL}_n(\mathbb{A}_k)$, then each local component Π_v is tempered.*

If Π_v is unramified, temperedness implies that its Satake parameters satisfy

$$|\alpha_{j,v}| = 1, \quad 1 \leq j \leq n.$$

A priori, an arbitrary constituent Π_v of a cuspidal automorphic representation is admissible, although unramified for almost every v. Langlands' classification allows us to express general admissible representations in terms of tempered data [**12, 99**]. In order to understand Ramanujan bounds, let us recall the Langlands classification for GL_n. Let ν be the character of a general linear group over a non-archimedean local field F given by $\nu(g) = |\det(g)|_F$. Then, if π is an admissible representation of $\mathrm{GL}_n(F)$, it has a unique Langlands quotient ξ obtained from the induced representation

$$\mathrm{ind}_P^G(\pi_1 \otimes \cdots \otimes \pi_b).$$

Here, each π_i is a representation of $\mathrm{GL}_{n_i}(F)$ of the form

$$\pi_j = \tau_j \cdot \nu^{r_j},$$

with each τ_j tempered, and we can assume that the Langlands parameters are arranged so that $0 \leq r_1 \leq \cdots \leq r_b$.

The Ramanujan conjecture for GL_n states that $r_b = 0$ for every component Π_v of a cuspidal automorphic Π; with the underlying assumption that Π be unitary. If k is a function field, the conjecture was established for GL_2 by Drinfeld [**30**] and GL_n by L. Lafforgue [**65**]. For $\mathbf{G}$ different than GL_n, we have a discussion in § 7.5, since it is no longer true for all cuspidal automorphic representations and it is rephrased for globally generic representations. The Ramanujan conjecture is a result of the author for the classical groups over function fields [**75, 76**].

Let k be a number field. In this open case, we have bounds towards Ramanujan. A first bound, known as the trivial bound is $r_b \leq 1/2$ [**54**]. The best known bounds for arbitrary n, were obtained by Luo, Rudnick and Sarnak in [**78**], expressed by the following formula

$$r_b \leq \frac{1}{2} - \frac{1}{n^2+1}.$$

However, for small n, there are better results. First, for $n = 2$, the best known bound of $r_b \leq 7/64$ was obtained for $k = \mathbb{Q}$ by Kim and Sarnak in [**60**]. For an arbitrary number field k, this Ramanujan bound was proved by Blomer and Brumley in [**8**], where one can also find the best known bounds for n up to 4.

We mention the work by Clozel and Thorne, starting with [**21**], which can lead towards newer Ramanujan bounds for GL_2 over number fields. For GL_n, certain cuspidal automorphic representations are known to satisfy Ramanujan over number fields, where there is the 10 author work on potential automorphy [**1**]. Over function fields, there is the important work by V. Lafforgue, who establishes an automorphic to Galois functorial lift [**66**].

This survey article presents an approach to functoriality in a uniform way for a global field, either characteristic zero or p, via Langlands-Shahidi automorphic L-functions and the Converse Theorem of Cogdell and Piatetski-Shapiro. We do not touch upon the trace formula, where Arthur has established the Endoscopic classification for automorphic representations of the classical groups over number fields, relying on the stabilization of the twisted trace formula completed by Mœglin and Waldspurger in [**83**]. The Fundamental Lemma, proved by Ngo providing a most intricate crucial ingredient.

There are several survey articles available, many of them focusing on number fields. General introductions to the Langlands Program are given by Arthur [**2**], Gelbart [**37**], and a book that may be found useful by graduate students [**7**]. Frenkel touches upon the geometric Langlands Program [**33**]. And, there is much recent progress in the mod-ℓ and the p-adic Langlands Program; a nice read and further references may be found in [**4**]. Furthermore, the reader may find excelent surveys for the Ramanujan Conjecture over number fields by Blomer-Brumley [**9**], Sarnak [**87**], and Shahidi [**95**], for example.

1. Artin L-functions and number theory

Let F be a non-archimedean local field. Thus, F is either an extension of the p-adic numbers $\mathbb{Q}_p$ or it is of characteristic p and of the form $F \simeq \mathbb{F}_q((t))$, the field of Laurent power series with coefficients in the finite field $\mathbb{F}_q$. We let $\mathcal{O}_F$ be the compact open ring of integers of F, it is a discrete valuation ring with a unique maximal compact open prime ideal $\mathfrak{p}_F = (\varpi_F)$, where ϖ_F is a uniformizer. The residue field $\mathcal{O}_F/\mathfrak{p}_F$ is finite with cardinality $q = p^n$. We refer to § 4 of [**86**] and to [**113**] for the classification of locally compact fields and their arithmetic properties.

1.1. The absolute Galois group

$$\mathcal{G}_F = \varprojlim \mathrm{Gal}(E/F),$$

where E/F varies through finite Galois extensions E over F inside a separable algebraic closure $\overline{F}$, is a pro-finite group with the Krull topology, hence compact. We obtain isomorphic Galois groups if we change the separable closure. Initially, we would like to study representations of $\mathcal{G}_F$, however, we study the related locally pro-finite Weil group $\mathcal{W}_F$ in order to obtain the Langlands correspondence. The correspondence will be for certain isomorphism classes of representations, § 4, and we work over a fixed separable algebraic closure $\overline{F}$.

Algebraically, the Weil group $\mathcal{W}_F$ is generated by the Frobenius elements and is normal in $\mathcal{G}_F$. The inertia group $\mathcal{I}_F = \mathrm{Gal}(\overline{F}/F_{\mathrm{nr}})$, where F_{nr} is the maximal unramified extension of F, is embedded inside $\mathcal{W}_F$. We denote the maximal tamely ramified extension of F by E_{t}, it is obtained as the composite of all the cyclic extensions E_n of F_{nr} of degree coprime to p. The wild inertia group $\mathcal{P}_F = \mathrm{Gal}(\overline{F}/E_{\mathrm{t}})$, is the unique pro p-Sylow subgroup of $\mathcal{I}_F$. The topology of $\mathcal{W}_F$ is such that the inertia group $\mathcal{I}_F$ is seen as an open subgroup of $\mathcal{W}_F$ and coincides with the topology of $\mathrm{Gal}(\overline{F}/F_{\mathrm{nr}})$ inside $\mathcal{G}_F$. We express local class field theory in terms of the Artin map in Theorem 1.2, and we now recall the process of "abelianization" and the "Verlagerung" homomorphism.

Let $\mathrm{Der}(\mathcal{W}_F)$ be the derived group and $\overline{\mathrm{Der}(\mathcal{W}_F)}$ its closure, then $\mathcal{W}_F^{\mathrm{ab}} = \mathcal{W}_F/\overline{\mathrm{Der}(\mathcal{W}_F)}$ is the abelianization of $\mathcal{W}_F$. Given a separable extension of non-archimedean local fields E/F, we have the norm map $\mathrm{N}_{E/F} : E^\times \to F^\times$ and

$$\mathrm{Ver}_{E/F} : \mathcal{W}_F^{\mathrm{ab}} \to \mathcal{W}_E^{\mathrm{ab}}$$

is the homomorphism defined by its action on cosets

$$x\,\mathrm{Der}(\mathcal{W}_F) \quad \mapsto \quad \prod_{y\,\in \mathcal{W}_F/\mathcal{W}_E} y_x\,\mathrm{Der}(\mathcal{W}_E),$$

where y_x is defined by taking a section s of the canonical projection $\mathcal{W}_F \to \mathcal{W}_F/\mathcal{W}_E$ and setting $y_x = s(xy)^{-1}xs(y)$; and, by passing to the closure on derived groups $\overline{\mathrm{Der}(\mathcal{W}_F)} \subset \mathcal{W}_F^{\mathrm{ab}}$ and $\overline{\mathrm{Der}(\mathcal{W}_E)} \subset \mathcal{W}_E^{\mathrm{ab}}$.

Weil groups were introduced in [**112**]. An excellent reference for local Weil groups and their representations can be found in Chapter 7 of [**15**]. We will turn towards the global theory, which is existential in nature, in § 1.6.

1.2. Local class field theory. *For every non-archimedean field F there is a reciprocity map*

$$\mathcal{W}_F \to \mathrm{GL}_1(F),$$

sending $\mathcal{I}_F$ *to* $\mathcal{O}_F^\times = \mathrm{GL}_1(\mathcal{O}_F)$, $\mathcal{P}_F$ *to* $1+\mathfrak{p}_F$, *and giving a topological isomorphism*

$$\mathrm{Art} : \mathrm{GL}_1(F) \to \mathcal{W}_F^{\mathrm{ab}},$$

between the abelianization $\mathcal{W}_F^{\mathrm{ab}}$ *of* $\mathcal{W}_F$ *and* $F^\times = \mathrm{GL}_1(F)$. *The Artin map is compatible with finite separable extensions, for which we have commuting diagrams*

$$\begin{array}{ccc} \mathcal{W}_E & \hookrightarrow & \mathcal{W}_F \\ \downarrow & & \downarrow \\ E^\times & \xrightarrow[\mathrm{N}_{E/F}]{} & F^\times \end{array} \qquad \text{and} \qquad \begin{array}{ccc} \mathcal{W}_F^{\mathrm{ab}} & \xrightarrow{\mathrm{Ver}_{E/F}} & \mathcal{W}_E^{\mathrm{ab}} \\ \downarrow & & \downarrow \\ F^\times & \hookrightarrow & E^\times \end{array}$$

1.3. Global fields and valuations. Let k be a global field. Thus, k is either a finite algebraic extension of $\mathbb{Q}$, i.e. a number field, or a finite separable extension of $\mathbb{F}_q(t)$, i.e. a function field. Let $\mathcal{O}_k$ be the ring of integers of k, it is a Dedekind domain.

A valuation v of k can be either non-archimedean or archimedean, the latter possible only in the number field case. We write $\mathrm{Val}(k)$ for the set of valuations of k. After localization, we obtain a local field k_v, where $k_v \simeq \mathbb{R}$ or $\mathbb{C}$ if $v \in \mathrm{Val}(k)$ is archimedean. We write $\mathrm{Val}_{\mathfrak{p}}(k)$ for the set of non-archimedean valuations of k. Every $v \in \mathrm{Val}_{\mathfrak{p}}(k)$ thus leads to a non-archimedean local field k_v, with corresponding ring of integers $\mathcal{O}_v$ and maximal prime ideal $\mathfrak{p}_v$. We write $\mathfrak{p}_v = (\varpi_v)$ for the maximal prime ideal, ϖ_v a uniformizer, and q_v the cardinality of the finite residue field $\mathbb{F}_v = \mathcal{O}_v/\mathfrak{p}_v$.

Let l/k be a finite separable extension, so we have an extension of Dedekind rings $\mathcal{O}_k \subset \mathcal{O}_l$. Given $v \in \mathrm{Val}_{\mathfrak{p}}(k)$, we have that $\mathfrak{p}_v$ factorizes in $\mathcal{O}_l$ in such a way that one obtains all valuations above v

$$\mathfrak{p}_v\mathcal{O}_l = \mathfrak{P}_{w_1}^{e_1}\cdots\mathfrak{P}_{w_g}^{e_g} \qquad \longleftrightarrow$$

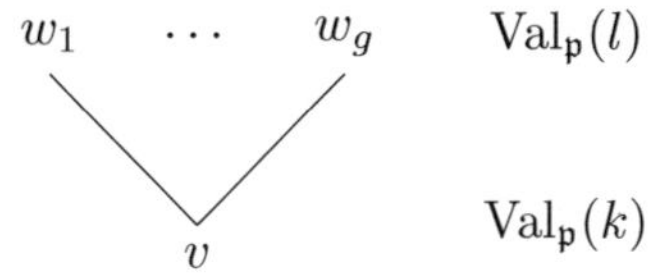

where $w_i \longleftrightarrow \mathfrak{P}_i$, for $1 \leq i \leq g$. We write $w|v$ when $w = w_i$ for some i as above.

In order to help with the transition between global and local notation, we recall that each positive integer e_w is called the ramification index of l_w/k_v, and we also have f_w such that $q_w = q_v^{f_w}$. We say that $v \in \mathrm{Val}_{\mathfrak{p}}(k)$ is unramified if $e_w = 1$ for every $w|v$. It is a known fact that almost every $v \in \mathrm{Val}_{\mathfrak{p}}(k)$ is unramified, i.e., all but finitely many. We further have

$$n = [l:k] = \sum_{w|v}[l_w : k_v] = \sum_{w|v} e_w f_w.$$

If l/k is Galois, then $e_w = e$ and $f_w = f$ for all $w|v$ and we have $n = efg$. For more details see, for example, § 4.4 of [**86**].

1.4. On Artin L-functions. Fix a Galois extension l/k. The Galois group $\mathrm{Gal}(l/k)$ transitively permutes the places $w|v$ and each $\alpha \in \mathrm{Gal}(l/k)$ induces an isomorphism

$$\alpha_w : l_w \simeq l_{\alpha(w)}.$$

The decomposition group is defined by

$$G_w = \{\alpha \in \mathrm{Gal}(l/k) \mid \alpha(w) = w\},$$

for each $w|v$. We have that

$$G_{\beta(w)} = \beta\, G_w \beta^{-1}.$$

If $\alpha \in G_w$, then α induces a k_v-automorphism α_w on l_w, and we have

$$G_w \xrightarrow{\sim} \mathrm{Gal}(l_w/k_v) \twoheadrightarrow \mathrm{Gal}(\mathbb{F}_w/\mathbb{F}_v).$$

The kernel of the last surjective homomorphism is the inertia subgroup $\mathcal{I}_w \subset G_w \subset \mathrm{Gal}(l/k)$. And for almost all $v \in \mathrm{Val}(k)$, we have that $\mathcal{I}_w$ is trivial. Indeed, at unramified places we have

$$\mathrm{Gal}(l_w/k_v) \simeq \mathrm{Gal}(\mathbb{F}_w/\mathbb{F}_v),$$

which is cyclic and generated by

$$\bar{\alpha}_w : x \mapsto x^{q_v}.$$

Now, a Frobenius automorphism for l/k is any element $\alpha_w \in G_w$ with image $\bar{\alpha}_w$ in $\mathrm{Gal}(\mathbb{F}_w/\mathbb{F}_v)$. It is characterized by the property

$$\alpha_w(x) = x^{q_v} \,(\mathrm{mod}\, \mathfrak{P}_w), \; x \in \mathcal{O}_w.$$

We observe that $\alpha_{\beta(w)} = \beta^{-1}\alpha_w\beta$, and hence α_w is determined up to conjugacy. The conjugacy class is by definition the Frobenius element of l/k at the place v $\mathrm{Frob}_{l/k,v}$.

We look at the global Weil group $\mathcal{W}_k$ in § 1.6, which for every Galois extension l/k has an isomorphism

$$\mathcal{W}_k/\mathcal{W}_l \simeq \mathrm{Gal}(l/k).$$

Assume that we have an n-dimensional continuous complex representation (σ, V) of $\mathcal{W}_k$ which factors through $\mathcal{W}_l$, for l/k Galois, and hence has finite image. We thus consider σ as a representation of $\mathrm{Gal}(l/k)$. If we let

$$V_w = V^{\mathcal{I}_w},$$

then $\sigma(\mathrm{Frob}_v)$ is defined on V_w. In this setting, "incomplete" Artin L-functions are given by

$$L(s,\sigma) = \prod_{v \in \mathrm{Val}_{\mathfrak{p}}(k)} \frac{1}{\det(I - \sigma(\mathrm{Frob}_v)|_{V_w} q_v^{-s})},$$

originally absolutely and uniformly convergent when $\Re(s) \geq 1 + \delta$, for some $\delta > 0$.

Artin Conjecture. *Artin L-functions $L(s, \sigma)$ are entire for non-trivial irreducible Galois representations of dimension $n > 1$.*

The conjecture is expected in the setting of § 1.6, where Deligne and Langlands [**27**, **68**] define local root numbers $\varepsilon(s, \sigma_v, \psi_v)$ at every place v of k. Thus, we have a global Artin root number

$$\varepsilon(s,\sigma) = \prod_v \varepsilon(s, \sigma_v, \psi_v),$$

where $\varepsilon(s, \sigma_v, \psi_v) = 1$ at every unramified place. Artin L-functions can be extended to archimedean places v of k, in a way that they are inductive in nature, and the resulting "completed" Artin L-function

$$L(s,\sigma) = \prod_{v \in \mathrm{Val}(k)} L_v(s,\sigma),$$

is in agreement with the previous definition

$$L_v(s,\sigma) = \frac{1}{\det(I - \sigma(\mathrm{Frob}_v)|_{V_w} q_v^{-s})},$$

for $v \in \mathrm{Val}_{\mathfrak{p}}(k)$ and σ factors through $\mathcal{W}_l$ as above. Artin L-functions in general have a meromorphic continuation to the complex plane. With the aid of Brauer theory, they can be proved to satisfy a functional equation

$$L(s,\sigma) = \varepsilon(s,\sigma) L(1-s, \tilde{\sigma}),$$

where $\tilde{\sigma}$ denotes the contragredient representation. This is done by reducing to the case of $n = 1$, studied by Artin and Tate. We discuss later on in § 6 cases where the Artin Conjecture is known.

1.5. Adèles and Idèles. Tate's celebrated thesis [**106**] uses restricted direct products to make a study of Fourier Analysis using valuation vectors, as Chevalley called them, or in modern language the adèles and idèles.

Restricted direct products are defined over a global field k. We assume that we are given a a finite set S_0 consisting of valuations $v \in \mathrm{Val}(k)$, including all archimedean v, and a locally compact topological group G_v for every $v \in \mathrm{Val}(k)$ such that G_v has a compact open subgroup H_v for every $v \notin S_0$. In this situation, we say that a condition is valid for almost all v if there exists a finite set $S \supset S_0$ such that the condition holds for every $v \notin S$. We define the restricted direct product

$$G = \prod{}'(G_v : H_v)$$

as the set of elements of the form (g_v), $g_v \in G_v$, where we have that $g_v \in H_v$ for almost all v.

The open sets of the group G are given by the following basis of neighborhoods of the identity

$$U_S = \prod_{v \in S} U_v \times \prod_{v \notin S} H_v,$$

where $S \supset S_0$ is a finite set of places and, for each $v \in S$ we have an arbitrary open neighborhood of the identity U_v in G_v. We observe that the resulting topology on G makes it a locally compact topological group.

The adèles $\mathbb{A}_k$ of a global field k are defined by taking $S_0 = \mathrm{Val}(k) \setminus \mathrm{Val}_{\mathfrak{p}}(k)$ (the set of archimedean valuations), setting $G_v = k_v$ for every $v \in \mathrm{Val}(k)$ and $H_v = \mathcal{O}_v$ for $v \notin S_0$. The idèles are the group $\mathrm{GL}_1(\mathbb{A}_k)$, and it is not a problem for us define $\mathrm{GL}_n(\mathbb{A}_k)$, for any positive integer n. Indeed, observe that for every $v \in \mathrm{Val}_{\mathfrak{p}}(k)$, the group $\mathrm{GL}_n(\mathcal{O}_v)$ is a maximal compact open subgroup of $\mathrm{GL}_n(k_v)$. Then, $\mathrm{GL}_n(\mathbb{A}_k)$ is given by the restricted direct product

$$\mathrm{GL}_n(\mathbb{A}_k) = \prod{}' \left(\mathrm{GL}_n(k_v) : \mathrm{GL}_n(\mathcal{O}_v)\right).$$

Algebraically $\mathbb{A}_k^\times$ is the multiplicative group of $\mathbb{A}_k$, yet the topology of $\mathrm{GL}_1(\mathbb{A}_k)$ is not the one inherited from $\mathbb{A}_k$ by restriction. Nonetheless, we abuse notation and denote the idèles by $\mathbb{A}_k^\times$, and we make the convention that the topology is the one obtained as a restricted direct product $\mathbb{A}_k^\times = \mathrm{GL}_1(\mathbb{A}_k)$.

Tate's thesis can be seen as an analytic approach to the abelian case studied by Class Field Theory; he reinterprets Hecke's seminal work on L-functions [**45**], now in the adèlic setting. Let $\psi = \otimes \psi_v$ be a non-trivial additive character of $k \backslash \mathbb{A}_k$, allowing us to do Fourier Analysis, and let

$$\chi : k^\times \backslash \mathbb{A}_k^\times \to \mathbb{C}^\times$$

be a Größencharakter of $\mathbb{A}_k^\times = \mathrm{GL}_1(\mathbb{A}_k)$. In [**106**], Tate works over number fields and defines L-functions and root numbers

$$L(s,\chi) = \prod_v L(s,\chi_v) \text{ and } \varepsilon(s,\chi) = \prod_v \varepsilon(s,\chi_v,\psi_v),$$

and establishes their functional equation

$$L(s,\chi) = \varepsilon(s,\chi) L(1-s,\chi^{-1}).$$

Tate's thesis can be studied over function fields as well, and a great reference valid over any global field k can be found in [**86, 113**].

Godement and Jacquet extend Tate's theory of L-functions for Größencharakters χ of $\mathrm{GL}_1(\mathbb{A}_k)$ to automorphic representations Π of $\mathrm{GL}_n(\mathbb{A}_k)$ in [**39**]. We thus have global Godement-Jacquet L-functions and root numbers

$$L(s,\Pi) = \prod_v L(s,\Pi_v) \text{ and } \varepsilon(s,\Pi) = \prod_v \varepsilon(s,\Pi_v,\psi_v),$$

which satisfy the functional equation

$$L(s,\Pi) = \varepsilon(s,\Pi) L(1-s,\tilde{\Pi}).$$

We turn towards automorphic representations in § 2 and we present all the automorphic L-functions that we will need in this article in a uniform way in § 3.

1.6. Weil groups and Artin L-functions. Let k be a global field and let $\mathcal{G}_k$ be its absolute Galois group obtained just as in (1.1) after fixing a separable algebraic closure $\overline{k}$. The global Weil group is a topological group $\mathcal{W}_k$ together with a dense continuous homomorphism $\mathcal{W}_k \to \mathcal{G}_k$, such that for every finite separable extension l/k: $\mathcal{W}_l$ is open in $\mathcal{W}_k$ and maps to $\mathcal{G}_l$ with dense image; there is an isomorphism

$$\mathcal{W}_l^{\mathrm{ab}} \simeq \mathcal{G}_l^{\mathrm{ab}}$$

which is compatible with global abelian class field theory, see § 1.7; and, the projective limit

$$\mathcal{W}_k = \varprojlim (\mathcal{W}_k / \overline{\mathcal{W}_l})$$

taken over all separable l inside $\overline{k}$ is an isomorphism of topological groups. We observe that the constructions of § 1.1, e.g. abelianization and Verlagerung map, are valid in this setting. Furthermore, these constructions also respect isomorphisms of separable field extensions.

In order to take a closer look at Artin L-functions, write $\mathcal{W}_v = \mathcal{W}_{k_v}$, for every $v \in \mathrm{Val}_{\mathfrak{p}}(k)$. Since we mod out by the inertia group $\mathcal{I}_v$, we can define Frob_v on $\mathcal{W}_v$ from the isomorphism

$$\mathrm{Gal}(k_{v,\infty}/k_v) \cong \varprojlim \mathbb{Z}/n\mathbb{Z} = \widehat{\mathbb{Z}}.$$

For each n, there is a unique unramified extension $k_{v,n}$ of k_v with cyclic Galois group $\mathrm{Gal}(k_{v,n}/k_v)$ and $k_{v,\infty}$ is the composite of all $k_{v,n}$.

Artin L-functions at archimedean places v of k can be defined [**7**], and this leads us to "completed" Artin L-functions. In [**27**], Deligne provides a proof that Artin L-functions satisfy the functional equation

$$L(s,\sigma) = \varepsilon(s,\sigma) L(1-s,\tilde{\sigma}),$$

where $\tilde{\sigma}$ denotes the contragredient representation.

The class field theory of Artin-Takagi, gives a very complete description of abelian Galois extensions in connection to the idèles. We next write it in the context of Weil groups, which are constructed cohomologically in Artin-Tate [**5**], and we also note the important references [**20**] and [**107**].

1.7. Global class field theory. *Let k be a global field. For every finite separable extension l/k there is a reciprocity map $\mathcal{W}_l \to \mathrm{GL}_1(\mathbb{A}_k)$ inducing an isomorphism*

$$\mathrm{Art} : \mathcal{C}_l = l^\times \backslash \mathrm{GL}_1(\mathbb{A}_l) \xrightarrow{\sim} \mathcal{W}_l^{\mathrm{ab}}.$$

For abelian Galois extensions l/k it induces an isomorphism

$$\mathrm{Art} : \Gamma \backslash \mathrm{GL}_1(\mathbb{A}_l) \xrightarrow{\sim} \mathrm{Gal}(l/k),$$

for $\Gamma = l^\times \mathrm{N}_{l/k}(\mathrm{GL}_1(\mathbb{A}_l))$, where $\mathrm{N}_{l/k}$ is the global norm. The Artin map is compatible with finite separable extensions, for which we have commuting diagrams

$$\begin{array}{ccc} \mathcal{C}_l & \xrightarrow{\mathrm{N}_{l/k}} & \mathcal{C}_k \\ \downarrow & & \downarrow \\ \mathcal{W}_l^{\mathrm{ab}} & \hookrightarrow & \mathcal{W}_k^{\mathrm{ab}} \end{array} \quad \textit{and} \quad \begin{array}{ccc} \mathcal{C}_k & \hookrightarrow & \mathcal{C}_l \\ \downarrow & & \downarrow \\ \mathcal{W}_k^{\mathrm{ab}} & \xrightarrow[\mathrm{Ver}_{E/F}]{} & \mathcal{W}_l^{\mathrm{ab}} \end{array}$$

.

The reciprocity map preserves one dimensional L-functions. For every $v \in \mathrm{Val}_{\mathfrak{p}}(k)$, *we retrieve the isomorphism*

$$\mathrm{Art}_v : \mathrm{GL}_1(k_v) \xrightarrow{\sim} \mathcal{W}_{k_v}^{\mathrm{ab}},$$

of local class field theory.

2. Reductive groups and representation theory

Reductive groups are at the base of the theory, and were studied in detail by the Chevalley seminar and the Séminaire de Géométrie Algébrique (SGA 3). The scheme theoretic point of view gives a functor $\mathbf{G}$ that allows us to change the field or ring of definition. For practical purposes, as in the theory of Lie groups, becoming familiar with a list of examples works best for familiarizing oneself with the representation theory of reductive groups. For the reader unfamiliar with reductive groups, the examples of § 2.1 suffice on a first reading; for the basic notions of reductive groups we refer to [**102**] and [**81**].

We let $\mathbf{G}$ be a quasi-split connected reductive group defined over a global field k. We take $G = \mathbf{G}(\mathbb{A}_k)$ and $\Gamma = \mathbf{Z}_G(\mathbb{A}_k)\mathbf{G}(k)$, and in general, see Remark 2.1. Langlands made a deep study of Eisenstein series in [**68**], providing an explicit refinement of the decomposition

$$\mathscr{L}^2(\Gamma\backslash G) = \mathscr{L}^2(\Gamma\backslash G)_{\mathrm{disc}} \oplus \mathscr{L}^2(\Gamma\backslash G)_{\mathrm{cont}},$$

where the cuspidal representations live in the discrete spectrum and we have the process of parabolic induction which helps us understand the continuous spectrum. Initially, Langlands works classically for arithmetic quotients of the form $\Gamma\backslash\mathbf{G}(\mathbb{R})$; reformulated adèlicly for number fields in an appendix to the published version of [*loc. cit.*]. Over function fields, Harder made an initial study for split $\mathbf{G}$ in [**41**], which was completed by Morris in [**85**].

Let $\mathbf{P} = \mathbf{MN}$ be a parabolic subgroup of $\mathbf{G}$ with Levi $\mathbf{M}$ and unipotent radical $\mathbf{N}$. We observe that $\mathbf{M}$ is a quasi-split connected reductive group over k and we look at the space of functions

$$\mathscr{L}^2(\mathbf{Z_M}(\mathbb{A}_k)\mathbf{M}(k)\backslash\mathbf{M}(\mathbb{A}_k)).$$

We say that an automorphic representation π of $\mathbf{M}(\mathbb{A}_k)$ is cuspidal if for every f in the space of π and every parabolic subgroup $\mathbf{P}_0 = \mathbf{M}_0\mathbf{N}_0$ of $\mathbf{M}$ we have that

$$\int_{\mathbf{N}_0(k)\backslash\mathbf{N}_0(\mathbb{A}_k)} f(ng)\,dn = 0.$$

Every cuspidal automorphic representation π of $\mathbf{M}(\mathbb{A}_k)$ lives inside the right regular representation of $\mathscr{L}^2(\Gamma\backslash M)$ by definition, as observed in [**69**]. Furthermore, every automorphic representation π of $\mathbf{G}(\mathbb{A}_k)$ arises as a constituent of

$$\mathrm{Ind}_{\mathbf{P}(\mathbb{A}_k)}^{\mathbf{G}(\mathbb{A}_k)}(\rho),$$

for some $\mathbf{P}$, with ρ a cuspidal representation of $\mathbf{M}(\mathbb{A}_k)$, extended to $\mathbf{P}(\mathbb{A}_k)$ by taking it to be trivial on $\mathbf{N}(\mathbb{A}_k)$. Here, Ind denotes normalized parabolic induction, to which we will return locally in § 2.3 and globally in § 2.5. For now, it suffices for us to know that cuspidals are the building blocks of the theory and all automorphic representations π are obtained by this process.

REMARK 2.1. In general, π has a central character ω, hence we can set $\Gamma_0 = \mathbf{G}(k)$ and study the subspace $\mathscr{L}^2(\Gamma_0\backslash G;\omega)$ of functions transforming via ω under the center. In the examples to which we next turn, we have: the general linear group (a), where $\mathbf{Z}(\mathbb{A}_k) \cong \mathrm{GL}_1(\mathbb{A}_k)$; and, the classical groups (b)–(d), which do not have a center.

2.1. Examples. The group GL_1 arises as the maximal torus (and Levi subgroup) $\mathbf{T}$ consisiting of diagonal matrices of the ambient group $\mathbf{G} = \mathrm{SL}_2$. Automorphic represention of $\mathrm{GL}_1(\mathbb{A}_k)$ are Größencharakters

$$\chi : k^\times\backslash\mathbb{A}_k^\times \to \mathbb{C}^\times.$$

As observed by Langlands in [**67**], Eisenstein series in this setting allow us to study L-functions in concordance with the abelian case of Tate's thesis.

Let n be a positive integer. We have in mind general linear groups and split classical groups when working with reductive groups in this article, namely:

$$\left.\begin{array}{ll} \text{(a)} & \mathbf{G}_n = \mathrm{GL}_n \\ \text{(b)} & \mathbf{G}_n = \mathrm{SO}_{2n+1} \\ \text{(c)} & \mathbf{G}_n = \mathrm{Sp}_{2n} \\ \text{(d)} & \mathbf{G}_n = \mathrm{SO}_{2n} \end{array}\right\} \quad \begin{array}{c} \text{with their Levi subgroups} \\ \mathbf{M} \cong \mathrm{GL}_{n_b} \times \cdots \times \mathrm{GL}_{n_1} \times \mathbf{G}_{n_0} \end{array}$$

$n = n_b + \cdots + n_1 + n_0$ and $\mathbf{G_{n_0}}$ a group of the same kind as $\mathbf{G_n}$.

These reductive groups provide us with intuition for the general theory. And, indeed, their Lie algebras are of types A_{n-1}, B_n, C_n and D_n, respectively. For the classical groups, i.e. cases (b)–(d), there are Levi subgroups with $n_0 = 0$ where we interpret $\mathbf{G}_0$ as empty.

In these examples, it is not hard to describe all parabolics $\mathbf{P} = \mathbf{MN}$ explicitly over a field K[1]. For this, let $\mathfrak{M}_{s,t}(K)$ be the set of $s \times t$ matrices with entries in K and let $I_s \in \mathfrak{M}_{s,s}(K)$ denote the identity matrix.

In case (a), we can precisely write the rational points of $\mathbf{M}$ and $\mathbf{N}$ as

$$M = \left\{(X_{ij}) \mid X_{ij} \in \mathfrak{M}_{n_i,n_j}(K), \text{ with } X_{ii} \in \mathrm{GL}_{n_i}(K) \text{ and } X_{ij} = 0 \text{ for } i \neq j\right\},$$

$$N = \left\{(X_{ij}) \mid X_{ij} \in \mathfrak{M}_{n_i,n_j}(K), X_{ii} = I_{n_i} \text{ and } X_{ij} = 0 \text{ for } i > j\right\}.$$

Parabolic subgroups $\mathbf{P}$ of $\mathbf{G}_n$, with $\mathbf{M}$ and $\mathbf{N}$ satisfying the above explicit description for every field K, are the standard parabolics. From the theory of reductive groups, every parabolic subgroup is conjugate to a standard one.

Now assume we are in the case of $\mathbf{G}_n$ a classical group. Let $J_n \in \mathfrak{M}_{n,n}(K)$ be the matrix with i,j entries given by $\delta_{i,n+1-j}$. If $\mathbf{G}_n$ is symplectic, i.e. $\mathbf{G}_n = \mathrm{Sp}_{2n}$, let

$$\Phi_n = \begin{pmatrix} 0 & J_n \\ -J_n & 0 \end{pmatrix}.$$

And we write $\Phi_n = J_{n'}$,where $n' = 2n+1$ or $2n$ for odd or even orthogonal groups, respectively (for symplectic groups, set $n' = 2n$). We can simplify things if we assume $-1 \neq 1$ in K. Let us do so, although the theory is available without this restriction. Then the rational points, $G_n = \mathbf{G}_n(K)$, of a classical group are given by

$$G_n = \left\{g \in \mathrm{SL}_{n'}(K) \mid {}^t g \Phi_n g = \Phi_n\right\}.$$

[1]When K is a global field, there is the well known process of "adelization" giving the rational points over the topological ring $\mathbb{A}_K$. The classical reference is [**114**], and we refer to [**26**] for a discussion à la Grothendieck.

The standard parabolic subgroups $\mathbf{P} = \mathbf{MN}$ of $\mathbf{G}_n$ have rational points given as follows: for $M = \mathbf{M}(K)$ by the following set of block matrices

$$M = \{(X_{ij}) \mid X_{b+1,b+1} \in G_{n_0}, X_{ij} = 0 \text{ for } i \neq j, \text{ and } X_{ii} \in \mathrm{GL}_{n_i}(K),$$
$$X_{b'+1-i,b'+1-i} = J_{n_i}{}^t X_{i,i}^{-1} J_{n_i} \text{ for } 1 \leq i \leq b\},$$

where $b' = 2b+1$ in case (b), and $b' = 2b$ in cases (c) and (d); the group of rational points N of the unipotent radical $\mathbf{N}$ then consists of the corresponding block upper triangular matrices with block identity matrices along the diagonal.

2.2. Orthogonal groups in characteristic 2. While the symplectic group Sp_{2n} can be defined as before regardless of characteristic, the orthogonal groups require quadratic forms; especially SO_{2n}. The previous simplification is possible because quadratic forms are equivalent to bilinear forms if $-1 \neq 1$.

For quadratic forms defined over a ring A, we refer to [**63**]. In particular, let

$$q_{2n+1}(x) = \sum_{i=1}^{n} x_i x_{2n+2-i} + x_{n+1}^2 \text{ and } q_{2n}(x) = \sum_{i=1}^{n} x_i x_{2n+1-i},$$

for $x \in A^{n'}$, where $n' = 2n+1$ or $2n'$, respectively. Then $\mathrm{O}_{n'}$ is the orthogonal group of rank n for the corresponding quadratic form $q_{n'}$.

For the special orthogonal groups, let $\mathbb{Z}_2(A)$ be the additive group of continuous maps $\mathrm{Spec}(A) \to \mathbb{Z}/2\mathbb{Z}$. There is the Dickson invariant, see [*loc. cit.*], for the corresponding quadratic space:

$$\mathrm{O}_{n'}(A) \xrightarrow{D_{q_{n'}}} \mathbb{Z}_2(A) \longrightarrow 1,$$

Then, $\mathrm{SO}_{n'}(A)$ is defined to be the kernel of the epimorphism $D_{q_{n'}}$.

2.3. Admissible representations of $\mathfrak{p}$-adic groups. The first article on $\mathfrak{p}$-adic groups was written by Mautner [**79**], who studied spherical functions of GL_2 and PGL_2. This active area of current research has come a long way since then, and a modern introductory approach in accordance with our exposition can be found in [**15**]. The reader can assume that $\mathbf{G}$ is given by one of the examples of 2.1. However, our framework is in the context of a general connected quasi-split reductive group $\mathbf{G}$, where a reductive group being quasi-split means it has a Borel subgroup over its field of definition.

We fix a Borel subgroup $\mathbf{B} = \mathbf{TU}$ of $\mathbf{G}$. It is a minimal parabolic subgroup, with maximal torus $\mathbf{T}$ and unipotent radical $\mathbf{U}$. In our examples, $\mathbf{B}$ consists of upper triangular matrices and $\mathbf{T}$ of diagonal matrices. We assume parabolic subgroups $\mathbf{P} = \mathbf{MN}$ are standard, by requiring $\mathbf{P} \supset \mathbf{B}$.

Given an algebraic group $\mathbf{H}$ defined over a non-archimedean local field F, we write $H = \mathbf{H}(F)$ for its group of rational points. Then G is a totally disconnected group, or, in other words, it is locally profinite. There exists then a basis for the topology consisting of compact open subgroups (c.o.s.g. for short) K of G.

We study representations (π, V) of G, where V is a complex (often infinite dimensional) vector space. A representation (π, V) is smooth if for every $v \in V$, there is a c.o.s.g. K of G such that $\pi(K)v = v$. A smooth representation (π, V) is admissible if for every c.o.s.g. K the space of fixed vectors V^K has

$$\dim_{\mathbb{C}} V^K < \infty.$$

All representations in the local Langlands conjecture are admissible, and this assumption allows us to define the trace of π. This allows us to define the character of π, sending $g \in G$ to $\operatorname{tr} \pi(g)$.

Given a parabolic subgroup $\mathbf{P} = \mathbf{MN}$ of $\mathbf{G}$, the group $\mathbf{M}$ is itself quasi-split reductive. From the point of view of Haar measures, G and M are unimodular. However, this is no longer the case for P, $\mathbf{P} \neq G$, which has non-trivial modulus character δ_P that allows us to pass between right Haar measures (which we use in all of our integrals) to left Haar measures. Let ρ be an admissible representation of M of finite length, which we extend to one of P by tensoring with the trivial representation of N. Then, parabolic induction defines a representation on G via the right regular action on the space

$$\begin{aligned} \operatorname{Ind}_P^G(\rho) = \{ f : G \to V \,|\, f(mng) = \delta_P^{1/2}(m)\rho(m)f(g), \forall m \in M, n \in N, \\ \exists \text{ c.o.s.g. } K \text{ of } G \text{ such that } f(gk) = f(g), \forall k \in K \} . \end{aligned}$$

When P and G are clear from context, we often simply write $\operatorname{Ind}(\rho)$ and further abuse notation by identifying the induced representation with its space $\operatorname{Ind}(\rho)$. We observe that $\operatorname{Ind}(\rho)$ is also admissible.

Because we normalize by the character $\delta_P^{1/2}$, this parabolic induction is often called unitary parabolic induction. Then $\operatorname{Ind}(\rho)$ is admissible unitary, when the inducing representation ρ is admissible unitary. Without the normalization, parabolic induction does not preserve unitarity. Furthermore, $\operatorname{Ind}(\rho)$ is of finite length, see § 6 of [**19**]. We also have the operation of taking the smooth dual, i.e., the contragredient representation $(\tilde{\rho}, \tilde{V})$, which preserves admissibility and is respected by Ind.

The building blocks of the theory are supercuspidal representations. These are irreducible admissible representations π which cannot be obtained as a subquotient of a parabolically induced representation

$$\operatorname{Ind}_P^G(\rho),$$

with ρ an admissible representation of M and $M \neq G$. In our examples, supercuspidals can be constructed using the theory of types in [**16, 103**].

Let (π, V) be an irreducible admissible representation of G. A matrix coefficient of π is a function defined by

$$c_{v,\tilde{w}}(g) = \langle \pi(g)v, \tilde{w} \rangle \,, \; g \in G,$$

where $v \in V$ and $\tilde{w} \in \tilde{V}$. With matrix coefficients we can define several important classes of representations in the theory.

An irreducible admissible representation π of G is supercuspidal if and only if all of its matrix coefficients have compact support. Discrete series representations are those whose matrix coefficients are in $\mathscr{L}^2(G)$. Tempered representations can be characterized by having their matrix coefficients in $\mathscr{L}^{2+\epsilon}(G)$ for every $\epsilon > 0$.

Langlands classification states that every irreducible admissible representation π of G arises as a Langlands quotient of an induced representation

$$\operatorname{Ind}(\rho \cdot \chi),$$

where ρ is a tempered representation of M and χ is a character in the Langlands situation [**12, 99**]. What we have is the following diagram

$$\left\{\begin{array}{c}\text{supercuspidal}\\ \text{representations}\end{array}\right\} \subset \left\{\begin{array}{c}\text{discrete}\\ \text{series}\end{array}\right\} \subset \left\{\begin{array}{c}\text{tempered}\\ \text{representations}\end{array}\right\} \subset \left\{\begin{array}{c}\text{admissible}\\ \text{representations}\end{array}\right\}$$

Example. It is customary to write ν for the unramified character $|\det(\cdot)|_F$ of a general linear group over F, and we do so. Let δ be a unitary discrete series representation of $\mathrm{GL}_n(F)$. Zelevinsky classification [**115**] tells us that it is a constituent of the induced representation

$$\mathrm{Ind}(\rho\nu^{-\frac{t-1}{2}} \otimes \cdots \otimes \rho\nu^{\frac{t-1}{2}}), \tag{2.1}$$

where ρ is a supercuspidal representation of $\mathrm{GL}_m(F)$, $m|n$, and t is a positive integer. The representation δ is precisely the generic constituent, see § 2.6 and the further examples of § 2.7.

2.4. L-groups and Satake parametrization. Given a quasi-split connected reductive group $\mathbf{G}$, Chevalley's theorem provides the existence of the dual group $\widehat{\mathbf{G}}$, and we take $\widehat{G} = \widehat{\mathbf{G}}(\mathbb{C})$. Given a local or a global field k, we have the Langlands dual group

$$^LG = \widehat{G} \rtimes \mathcal{W}_k,$$

where $\mathcal{W}_k$ is the Weil group. The basic properties of L-groups may be found in [**10**], we only observe that Borel defines them using the Galois group, which can be changed to the Weil group in the discussion.

Let now F be a non-archimedean local field with ring of integers $\mathcal{O}_F$ and residue field $\mathbb{F} = \mathcal{O}_F/\mathfrak{p}_F$, with cardinality q_F. Since we assume $\mathbf{G}$ to be quasi-split over F, there is a Borel subgroup $\mathbf{B}$ of $\mathbf{G}$. If the group is split, then the groups are also defined over $\mathcal{O}_F$. If $\mathbf{G}$ is non-split quasi-split, one can always choose a connected group scheme $\mathbf{G}_0$ over $\mathcal{O}_F$, such that it has the same generic fiber and connected special fiber [**108**]. Then, we can define the Iwahori subgroup as the pre-image in $\mathbf{G}_0(\mathcal{O}_F)$ of the $\mathbb{F}$-points of $\mathbf{B}$. Namely

$$\mathcal{I}_F = \mathrm{Ker}\left\{\mathbf{G}_0(\mathcal{O}_F) \to \mathbf{B}(\mathbb{F})\right\}.$$

Let

$$\mathfrak{Rep}_{\mathrm{nr}}(G) = \left\{\begin{array}{c}\text{isomorphism classes of smooth irreducible}\\ \text{representations } \pi \text{ of } G \text{ such that } \pi^{\mathcal{I}_F} \neq 0\end{array}\right\}.$$

It is well known that this class of representations satisfies the local Langlands correspondence [**88**], a result that is known as the Satake parametrization:

$$\pi \in \mathfrak{Rep}_{\mathrm{nr}}(G) \quad \longleftrightarrow \quad \phi_\pi : \mathcal{W}_F \to {}^LG,$$

where ϕ_π is the local Langlands parameter attached to π. It is represented by

$$\phi_\pi(\mathrm{Frob}_F) = s(\pi),$$

a Satake parameter $s(\pi)$ inside the variety of semi-simple elements $[\widehat{G}^{\mathcal{I}_F} \rtimes \mathrm{Frob}_F]_{\mathrm{ss}}$ modulo conjugation under $\widehat{G}^{\mathcal{I}_F}$. A general construction can be found in [**40**].

DEFINITION 2.2. Let r be an analytic finite dimensional representation of ^LG and let $\pi \in \mathfrak{Rep}_{\mathrm{nr}}(G)$. The corresponding local L-function is defined by

$$L(s, \pi, r) = \frac{1}{\det(I - r(s(\pi))q_F^{-s})}, \quad s \in \mathbb{C}.$$

We also note the fact that unramified representations are obtained via parabolic induction from an unramified character of the maximal torus. This is a basic result of Borel and of Casselman [**18**].

THEOREM 2.3. *Let* $\mathbf{G}$ *be a quasi-split connected reductive group over* F *with Borel subgroup* $\mathbf{B} = \mathbf{TU}$. *If* $\pi \in \mathfrak{Rep}_{\rm nr}(G)$, *then there exists an unramified character* χ *of* T *such that* π *is the unique irreducible generic constituent of*

$$\mathrm{Ind}(\chi).$$

Examples. Our examples of type (a)–(d) are all split, so the Galois action is trivial in the definition of ${}^LG$. Because of this, given our rank n group $\mathbf{G}_n$, the Satake parametrization only depends on the connected component of the Langlands dual group:

$${}^LG_n^\circ = \widehat{G}_n = \begin{cases} \mathrm{GL}_n(\mathbb{C}) & \text{if } \mathbf{G}_n = \mathrm{GL}_n \\ \mathrm{Sp}_{2n}(\mathbb{C}) & \text{if } \mathbf{G}_n = \mathrm{SO}_{2n+1} \\ \mathrm{SO}_{2n+1}(\mathbb{C}) & \text{if } \mathbf{G}_n = \mathrm{Sp}_{2n} \\ \mathrm{SO}_{2n}(\mathbb{C}) & \text{if } \mathbf{G}_n = \mathrm{SO}_{2n} \end{cases}.$$

We observe that the general linear group GL_n and the even special orthogonal group SO_{2n} are self dual.

Fix a classical group $\mathbf{G}_n$ and let $\pi \in \mathfrak{Rep}_{\rm nr}(G_n)$. Then, there exist unramified characters $\chi_1, \ldots, \chi_n$ of $\mathrm{GL}_1(F)$ such that

$$\pi \hookrightarrow \mathrm{Ind}(\chi_1 \otimes \cdots \otimes \chi_n)$$

is the unique irreducible generic constituent. The Satake parameters for $\mathbf{G}_n = \mathrm{GL}_n$ are then given by the diagonal (hence semisimple) matrix

$$s(\pi) = \mathrm{diag}(\chi_1(\varpi_F), \ldots, \chi_n(\varpi_F)) \in \mathrm{GL}_n(\mathbb{C}).$$

If $\mathbf{G}_n = \mathrm{SO}_{2n+1}$ or SO_{2n}, then

$$s(\pi) = \mathrm{diag}(\chi_1(\varpi_F), \ldots, \chi_n(\varpi_F), \chi_n^{-1}(\varpi_F), \ldots, \chi_1^{-1}(\varpi_F)) \in {}^LG_n^\circ.$$

And if $\mathbf{G}_n = \mathrm{Sp}_{2n}$, then

$$s(\pi) = \mathrm{diag}(\chi_1(\varpi_F), \ldots, \chi_n(\varpi_F), 1, \chi_n^{-1}(\varpi_F), \ldots, \chi_1^{-1}(\varpi_F)) \in {}^LG_n^\circ.$$

2.5. Induction and automorphic representations. We use the guidance of Langlands, who observed that the process of parabolic induction allows us to obtain all automorphic representations from cuspidal representations [**69**].

We are in the situation of a quasi-split connected reductive group $\mathbf{G}$, defined over a global field k. And, $\mathbf{P} = \mathbf{MN}$ is a parabolic subgroup. We write G_v for $\mathbf{G}(k_v)$ and similarly for P_v, M_v and N_v. For each $v \in \mathrm{Val}_\mathfrak{p}(k)$, we have the ring of integers $\mathcal{O}_v$ of k_v. The group $K_v = \mathrm{G}(\mathcal{O}_v)$ is defined for almost every $v \in \mathrm{Val}_\mathfrak{p}(k)$, and it is a hyperspecial maximal compact open subgroup of G_v.

Let ρ be a cuspidal automorphic representation of $\mathbf{M}(\mathbb{A}_k)$ extended to $\mathbf{P}(\mathbb{A}_k)$ by taking it to be trivial on $\mathbf{N}(\mathbb{A}_k)$. Every automorphic representation can be written as a restricted tensor product

$$\rho = \otimes' \rho_v,$$

such that $\rho_v \in \mathfrak{Rep}_{\rm nr}(G_v)$ for almost all $v \in \mathrm{Val}(k)$, see [**32**]. The decomposition is made by fixing a family of vectors x_v in the space of ρ_v, fixed by $K_v = \mathbf{M}(\mathcal{O}_v)$, for almost all v. Then we can form the restricted tensor product

$$\mathrm{Ind}_{\mathbf{P}(\mathbb{A}_k)}^{\mathbf{G}(\mathbb{A}_k)} \rho = \otimes' \mathrm{Ind}_{P_v}^{G_v} \rho_v.$$

The restricted tensor product is taken with respect to a family of nonzero vectors f_v^0 in the space of $\mathrm{Ind}_{P_v}^{G_v} \rho_v$. We fix such a family by taking S to be a finite set of places of k, $S \supset \mathrm{Valp}_\infty(k)$, such that ρ_v is unramified for $v \notin S$; we can further assume that $K_v = \mathbf{G}(\mathcal{O}_v)$. We have a unique constituent

$$\pi_v^\circ \text{ of } \mathrm{Ind}_{P_v}^{G_v} \rho_v, \ v \notin S,$$

containing the trivial representation of K_v. Then, for each $v \notin S$, the function f_v^0 is taken in the space of π_v° to be such that $f_v^0(y_v) = x_v$ for all $y_v \in K_v$. Then every function in V_π is of the form

$$f = \otimes_v' f_v, \quad f_v = f_v^0, \ v \notin S',$$

where S' is a set containing S.

The main result of [**68**, **82**], is that a representation π of $\mathbf{G}(\mathbb{A}_k)$ is an automorphic representation if and only if π is a constituent of

$$\mathrm{Ind}_{\mathbf{P}(\mathbb{A}_k)}^{\mathbf{G}(\mathbb{A}_k)} \rho$$

for a parabolic $\mathbf{P} = \mathbf{MN}$ of $\mathbf{G}$ and a cuspidal representation ρ of $\mathbf{M}(\mathbb{A}_k)$. The representation can be expressed as a tensor product

$$\pi = \otimes' \pi_v,$$

where $\pi_v = \pi_v^\circ$ for almost all v.

2.6. Generic representations. Fixing a Borel subgroup $\mathbf{B}$ of $\mathbf{G}$ gives a pinning of the roots, and in particular determines the set of positive roots Φ^+. We use Bourbaki notation, chapter VI of [**13**], when working with root systems. For every $\alpha \in \Phi^+$, there is a quasi-split reductive group $\mathbf{G}_\alpha$ obtained via restriction of scalars from a rank one group, either SL_2 or SU_3. In the former case we have that the unipotent subgroup $N_\alpha = U_\alpha$ is a one parameter additive subgroup and in the latter case we have $N_\alpha = U_\alpha U_{2\alpha}$. Our examples are of split groups, in which the latter possibility does not occur.

Let Δ denote the simple roots. We have the surjective morphism

$$U \twoheadrightarrow U / \prod_{\alpha \in \Phi^+ \setminus \Delta} N_\alpha \simeq \prod_{\alpha \in \Delta} U_\alpha, \tag{2.2}$$

Given a non-trivial additive character $\psi : A \to \mathbb{C}^\times$, on a topological ring A, we extend it to one of $U = \mathbf{U}(A)$ via the morphism (2.2)

$$\psi(u) = \prod_{\alpha \in \Delta} \psi(u_\alpha).$$

In practice, we know how to pass from an arbitrary non-degenerate character χ of U to one of the form just constructed.

Example. For GL_n, we fix the Borel subgroup $\mathbf{B} = \mathbf{TU}$ of upper triangular matrices. The maximal torus $\mathbf{T}$ consists of diagonal matrices and the unipotent radical $\mathbf{U}$ has rational points given by

$$U = \mathbf{U}(A) = \{u = (x_{ij}) \mid x_{ij} \in A, x_{ii} = 1 \text{ and } x_{ij} = 0 \text{ if } i < j\} \subset \mathrm{GL}_n(A).$$

From the non-trivial character $\psi : F \to \mathbb{C}^\times$, we extend ψ to $u \in \mathbf{U}(A)$ by setting

$$\psi(u) = \prod_{i=1}^{n-1} \psi(x_{i,i+1}).$$

It is a good excercise to write the corresponding explicit matrix realization for the classical groups.

Let us now return to the notion of genericity, first, for local representations. Fix a non-archimedean local F and a non-trivial character $\psi : F \to \mathbb{C}^\times$. We say that an irreducible admissible representation (π, V) of $\mathbf{G}(F)$ is generic if there exists a continuous functional

$$\lambda : V \to \mathbb{C},$$

such that

$$\lambda(\pi(u)v) = \psi(u)\lambda(v), \text{ for } u \in U.$$

Such a λ, when it exists, is called a Whittaker functional. It is a result of Shalika [**97**], that generic representations of quasi-split reductive groups satisfy the following multiplicity one property.

THEOREM 2.4. *Let (π, V) be a generic representation of $G = \mathbf{G}(F)$. Let W denote the space of all Whittaker functionals, then*

$$\dim_{\mathbb{C}} W = 1.$$

Now, for a global field k. Let $\pi = \otimes' \pi_v$ be a cuspidal automorphic representation of $\mathbf{G}(\mathbb{A}_k)$ and $\psi : k \backslash \mathbb{A}_k \to \mathbb{C}^\times$ a non-trivial character. We say that π is globally generic if there is a cusp form φ in the space of π such that

$$W_\varphi(g) = \int_{\mathbf{U}(k)\backslash \mathbf{U}(\mathbb{A}_k)} \varphi(ug)\overline{\psi}(u)\, du \neq 0.$$

A globally generic representation $\pi = \otimes' \pi_v$, is locally generic. Indeed, if we write the cusp form as a restricted product $\varphi = \otimes' \varphi_v$, ranging over all places v of k, then we have

$$W_\varphi(g) = \prod_v W_{\varphi_v}(g_v) = \prod_v \lambda_v(\pi_v(g_v)\varphi_v),$$

where each λ_v is a Whittaker functional for π_v.

2.7. Unitary generic representations of the classical groups. Let $\mathbf{G}_n$ be either a general linear group or a classical group defined over a non-archimedean local field F. We have the following result concerning the unitary spectrum.

THEOREM 2.5. *Let π be an irreducible generic unitary admissible representation of $\mathbf{G}_n(F)$, then π arises as the irreducible quotient of a parabolically induced representation*

$$\operatorname{Ind}(\tau_b \nu^{r_b} \otimes \cdots \otimes \tau_1 \nu^{r_1} \otimes \pi_0). \tag{2.3}$$

The induction is with respect to a parabolic subgroup with Levi $\mathbf{M} \cong \mathrm{GL}_{n_b} \times \cdots \times \mathrm{GL}_{n_1} \times \mathbf{G}_{n_0}$. For each $1 \leq i \leq b$, τ_i is a tempered unitary representation of $\mathrm{GL}_{n_i}(F)$, and the Langlands parameters satisfy

$$0 \leq r_1 \leq \cdots \leq r_b < 1.$$

In the case of a general linear group, the representation π_0 of $\mathbf{G}_{n_0}(F)$ is quasi-tempered of the form $\pi_0' \nu^{r_0}$ with π_0' tempered and $r_1 \geq r_0 \geq 0$. In the case of a classical group, π_0 is tempered generic.

The bound $r_b < 1/2$ for GL_n is known as the trivial Ramanujan bound, obtained when viewed from a global perspective [**54**]. For the classical groups, the theorem and a complete description of the generic unitary spectrum in characteristic zero is given in [**73**].

In characteristic p, Theorem 2.5 is an example which can be obtained via the Kazhdan philosophy concerning properties of admissible representations that can be transferred over close local fields [**57**], thanks to the work of Ganapathy [**35**]. Two non-archimedean local fields F and F' are said to be m-close in the sense of Kazhdan, if there is a ring isomorphism

$$\mathcal{O}_F/\mathfrak{p}_F^m \cong \mathcal{O}_{F'}/\mathfrak{p}_{F'}^m.$$

Given F' of characteristic p, it can always be approximated in this way by an F of characteristic zero.

Let $\mathbf{G}$ be a split connected reductive group scheme defined over $\mathbb{Z}$. We write $G = \mathbf{G}(F)$ and $G' = \mathbf{G}'(F)$. Let $\mathcal{I}$ denote the standard Iwahori subgroup scheme of $\mathbf{G}$ and let

$$\mathcal{I}_m = \mathrm{Ker}\left\{\mathcal{I}(\mathcal{O}_F) \to \mathcal{I}(\mathcal{O}_F/\mathfrak{p}_F^m)\right\}.$$

We can rely on the results of Ganapathy, who establishes a refined Kazhdan transfer for Hecke algebras, leading to the following bijection

$$\left\{\begin{array}{c}\text{Isomorphism classes of}\\ \text{irreducible admissible } \pi\\ \text{of } G \text{ with } \pi^{\mathcal{I}_m} \neq 0\end{array}\right\} \xrightarrow{\zeta_m} \left\{\begin{array}{c}\text{Isomorphism classes of}\\ \text{irreducible admissible } \pi'\\ \text{of } G' \text{ with } \pi'^{\mathcal{I}'_m} \neq 0\end{array}\right\}.$$

With this approach, it is proved in § 5 of [**76**] that unitarity is preserved by the Kazhdan transfer and Theorem 2.5 is established in positive characteristic.

2.8. Intertwining operators and Plancherel measures. Harish-Chandra studied the Plancherel theorem over the reals as well as the $\mathfrak{p}$-adics. We look at μ-factors which make an appearence in the local Langlands conjecture (see § 4.2), in addition to their original role in the Plancherel theorem [**110**].

Fix a pair of quasi-split reuductive groups $(\mathbf{G}, \mathbf{M})$ defined over a non-archimedean local field F, with $\mathbf{M}$ a maximal Levi subgroup of $\mathbf{G}$. We let

$$w_0 = w_l w_{l,M}, \tag{2.4}$$

where w_l and $w_{l,M}$ are the longest Weyl group elements of G and M, respectively. We have that conjugation by w_0 defines a Levi subgroup $\mathbf{M}_{w_0}$. Given a representation ξ on M, we let $w_0(\xi)$ denote the representation on M_{w_0} defined by $w_0(\xi)(m') = \xi(w_0^{-1} m' w_0)$.

Let $X_{\mathrm{nr}}(\mathbf{M})$ be the group of unramified characters $\chi : M \to \mathbb{C}$; it is a complex algebraic variety. Given an irreducible admissible representation (π, V) of M, and a $\chi \in X_{\mathrm{nr}}(\mathbf{M})$ we have an induced representation

$$\mathrm{I}(\chi, \pi) = \mathrm{Ind}_P^G(\pi \cdot \chi).$$

Furthermore, we have an intertwining operator

$$\mathrm{A}(\chi, \pi, w_0) : \mathrm{I}(\chi, \pi) \to \mathrm{I}(w_0(\chi), w_0(\pi)),$$

which is defined via the principal value integral

$$\mathrm{A}(\chi, \pi, w_0) f(g) = \int_{N_{w_0}} f(w_0^{-1} n g)\, dn.$$

It extends to a rational operator on $\chi \in X_{\mathrm{nr}}(\mathbf{M})$, see § IV of [**110**].

Harish-Chandra observed the connection between Plancherel measures and intertwining opertors, where the composite of two intertwining operators defines

$$\mu(\chi, \rho, \psi)^{-1} = A(w_0(\chi), w_0(\pi), \psi) \circ A(\chi, \pi, \psi),$$

giving a scalar-valued rational function $\mu(\chi, \pi, \psi)$ on the variable $\chi \in X_{\mathrm{nr}}(\mathbf{M})$.

Examples. We note that every unramified character $\chi \in X_{\mathrm{nr}}(\mathrm{GL}_n)$ is of the form $\chi = \nu^s$, as s ranges through $\mathbb{C}$. In case (a) of examples 2.1, we obtain functions of one complex variable by taking the representation $\rho = \tau \otimes \tilde{\pi}$ of $M = \mathrm{GL}_m(F) \times \mathrm{GL}_n(F)$ and setting

$$\mathrm{I}(s, \rho) = \mathrm{Ind}_P^G(\tau\nu^{s/2} \otimes \tilde{\pi}\nu^{-s/2}).$$

For the classical groups, i.e., cases (b)–(d), we take the representation $\rho = \tau \otimes \tilde{\pi}$ of $M = \mathrm{GL}_m(F) \times \mathrm{G}_n$ and set

$$\mathrm{I}(s, \rho) = \mathrm{Ind}_P^G(\tau\nu^s \otimes \tilde{\pi}).$$

With these normalizations, the intertwining operator is of the form

$$\mathrm{A}(s, \rho, w_0) : \mathrm{I}(s, \rho) \to \mathrm{I}(-s, \tilde{w}_0(\rho)).$$

While $M_{w_0} \neq M$ in general for GL_l, it is the case that $M_{w_0} = M$ for the classical groups. We also observe that we have rational functions

$$\mu(s, \rho, \psi) \in \mathbb{C}(q^{-s}).$$

2.9. Partial L-functions. Let $\mathbf{G}$ be a quasi-split connected reductive group defined over a global field k. It is a basic property that an automorphic representation π of $\mathbf{G}(\mathbb{A}_k)$, has a decomposition

$$\pi = \otimes' \pi_v,$$

so that there is a finite set S of places of k (containing the archimedean places in the case of a number field), for which $\pi_v \in \mathfrak{Rep}_{\mathrm{nr}}(\mathbf{G}(k_v))$, for $v \notin S$ [**32**].

DEFINITION 2.6. Let r be an analytic finite dimensional representation of ${}^LG$. Let $\pi = \otimes'\pi_v$ be an automorphic representation of $\mathbf{G}(\mathbb{A}_k)$ and S a finite set of places of k such that $\pi_v \in \mathfrak{Rep}_{\mathrm{nr}}(\mathbf{G}(k_v))$, for all $v \notin S$. Then, partial L-functions are defined by

$$L^S(s, \pi, r) = \prod_{v \notin S} L(s, \pi_v, r_v).$$

An basic property of partial automorphic L-functions is that $L^S(s, \pi, r)$ converges for $\Re(s) \gg 0$, see [**10**]. However, it is much more difficult to obtain a meromorphic continuation and functional equation. These last two properties, we address in many cases via the Langlands-Shahidi method, to which we turn next.

3. The Langlands-Shahidi method

3.1. Fix a pair of quasi-split connected reductive groups $(\mathbf{G}, \mathbf{M})$, with $\mathbf{M}$ a maximal Levi subgroup of $\mathbf{G}$. We call $(\mathbf{G}, \mathbf{M})$ a Langlands-Shahidi pair of quasi-split groups. Locally, the Langlands-Shahidi method applies to:

(i) a generic representation π of M;
(ii) a constituent r of the adjoint action of ${}^LM$ on ${}^L\mathfrak{n}$, where $\mathfrak{n}$ is the Lie algebra of the unipotent radical of $\mathbf{N}$;

(iii) a non-trivial additive character ψ.

Given a local triple (π, r, ψ), with $M = \mathbf{M}(F)$, the Langlands-Shahidi machinery produces rational functions

$$\gamma(s, \pi, r, \psi) \in \mathbb{C}(q^{-s}),$$

of a complex variable $s \in \mathbb{C}$. Furthermore, if π is tempered, then the γ-factor can be written as a product of a monomial $\varepsilon(s, \pi, r, \psi)$ and a ratio of two monic polynomias on q^{-s}

$$\gamma(s, \pi, r, \psi) = \varepsilon(s, \pi, r, \psi) \frac{L(1 - s, \tilde{\pi}, r)}{L(s, \pi, r)}. \tag{3.1}$$

This definig relationship is such that for $\pi \in \mathfrak{Rep}_{\mathrm{nr}}(M)$ it agrees with the definition given by the Satake correspondence, where we have $\varepsilon(s, \pi, r, \psi) = 1$. It is extended from tempered generic to all generic π via Langlands classification and a multiplicativity property, explained in § 3.3.

Now, a global triple (π, r, ψ), with $M = \mathbf{M}(\mathbb{A}_k)$, would be one with π a globally generic cuspidal automorphic representation of M, r an irreducible constituent of the adjoint action of ${}^LM$ on ${}^L\mathfrak{n}$, and $\psi = \otimes' \psi_v$ an additive character of $k \backslash \mathbb{A}_k$. We let S be a finite set of places of k such that π_v, k_v and ψ_v are unramified for $v \notin S$, were we have the partial L-function $L^S(s, \pi, r)$ defined in the previous section.

The Langlands-Shahidi machinery proves the functional equation for global triples (π, r, ψ)

$$L^S(s, \pi, r) = \prod_{v \in S} \gamma(s, \pi_v, r_v, \psi_v) L^S(1 - s, \tilde{\pi}, r). \tag{3.2}$$

Furthermore, with the local theory in place, we have completed L-functions and global root numbers

$$L(s, \pi, r) = \prod_v L(s, \pi_v, r_v) \text{ and } \varepsilon(s, \pi, r) = \prod_v \varepsilon(s, \pi_v, r_v, \psi_v). \tag{3.3}$$

Examples. Suppose we are in cases (a)–(d) of examples 2.1. The Langlands-Shahidi method leads us to automorphic L-functions for Rankin-Selberg products

$$L(s, \pi \times \tau),$$

where π and τ are globally generic cuspidal automorphic representations of $\mathbf{G}_n(\mathbb{A})$ and $\mathrm{GL}_m(\mathbb{A}_k)$, respectively. The underlying analytic representation of ${}^LM$ is given by $r = \rho_n \otimes \rho_m$, where for every positive integer l, we let ρ_l denote the standard representation of $\mathrm{GL}_l(\mathbb{C})$. This is one of two possible irreducible constituents of the adjoint action that arise in cases (b)–(d), the case (a) giving only Rankin-Selberg products.

We note that for GL_n, all cuspidal automorphic representations are globally generic. This is no longer the case for the classical groups, as we mention in § 7.5, however, locally the tempered L-packet conjecture of Shahidi is known for the classical groups [**3, 36**], which allows us to obtain all local L-functions from the generic case.

In cases (b)–(d) we also have the related L-functions $L(s, \tau, r)$ for a globally generic cuspidal automorphic representation τ of $\mathrm{GL}_m(\mathbb{A}_k)$, and we can have $r = \rho_m$, $r = \mathrm{Sym}^2 \rho_m$ or $\wedge^2 \rho_m$. These arise in the Siegel Levi case of $\mathbf{M} \cong \mathrm{GL}_m$ as a maximal Levi subgroup of a classical group $\mathbf{G}_m$ of rank m.

3.2. Induced representations and genericity. Fix a Langlands-Shahidi pair of quasi-split groups $(\mathbf{G}, \mathbf{M})$. We consider the cases of $A = F$, a non-archimedean local field, or $A = \mathbb{A}_k$, the ring of adèles of a global field k. We write $G = \mathbf{G}(A)$ for the rational points, and similarly for M.

Given a non-trivial characrter ψ of A, we first use equation (2.2) to obtain a character ψ_M of U_M, we then require the following relation on a character ψ of U:

$$\psi(u) = \psi_M(w_0^{-1} u w_0),\ u \in U_M = U \cap M,$$

with w_0 as in (2.4).

Let $\mathbf{P} = \mathbf{MN}$ be a maximal parabolic subgroup of $\mathbf{G}$. The group $\mathbf{M}/\mathbf{Z_G}$ has a one dimensional split center, hence $\mathrm{Hom}_F(\mathbf{M}/\mathbf{Z_G}, \mathrm{GL}_1)$ is a rank one free $\mathbb{Z}$-module. Let

$$\delta = \det(\mathrm{Ad}_{\mathbf{M}}|_{\mathfrak{n}}) \in \mathrm{Hom}_F(\mathbf{M}/\mathbf{Z_G}, \mathrm{GL}_1).$$

Now, in order to have L-functions of a complex variable, we let α be the simple root such that the maximal parabolic $\mathbf{P}$ corresponds to the subset of simple roots $\theta = \Delta - \{\alpha\}$. Let δ_α be the unramified character of M given by

$$\delta_\alpha : M \xrightarrow{\delta^l} \mathrm{GL}_1(A) \xrightarrow{|\cdot|_A} \mathbb{C}^\times,\ l = \langle \delta, \alpha^\vee \rangle^{-1}, \tag{3.4}$$

where $\langle \cdot, \cdot \rangle$ is the perfect paring between roots and co-roots.

Let π be a generic representation of M. We write

$$\mathrm{I}(s, \pi) = \mathrm{I}(\delta_\alpha^s, \pi),\ s \in \mathbb{C},$$

with δ_α as in equation (3.4); its corresponding space is denoted by $\mathrm{V}(s, \pi)$.

Examples. Suppose we are in the case of a maximal Levi subgroup

$$\mathbf{M} \cong \mathrm{GL}_m \times \mathbf{G}_n \subset \mathbf{G}_l,$$

$l = m + n$, and both m, $n > 0$. If we are given ψ-generic representations τ of GL_m and π of G_n, then we form the ψ_M-generic representation

$$\rho = \tau \otimes \tilde{\pi} \text{ of } M.$$

The use of the contragredient representation $\tilde{\pi}$ is made in order to obtain Rankin-Selberg product L-functions from the local Langlands-Shahidi method. In all of these cases we have

$$\mathrm{I}(s, \rho) = \mathrm{I}((\tau \cdot \nu^s) \otimes \tilde{\pi}),\ s \in \mathbb{C}.$$

The following result follows from the theory of Whittaker models for $\mathfrak{p}$-adic reductive groups; a proof and further references can be found in § 2 of [**76**].

THEOREM 3.1. *Let (F, π, ψ) be a local triple in the Langlands-Shahidi setting. Then the induced representation $\mathrm{I}(s, \pi)$ is also generic. It has a Whittaker functional $\lambda_\psi(s, \pi)$ which is a polynomial on $\{q^s, q^{-s}\}$.*

3.3. Local Langlands-Shahidi method. Let $(\mathbf{G}, \mathbf{M})$ be a Langlands-Shahidi pair of quasi-split groups defined over a non-archimedean local field F. Given a local triple (π, r, ψ), the Langlands-Shahidi method produces a local coefficient $C_\psi(s, \pi, w_0)$, which combined with an induction step produces γ-factors, L-functions and ε-factors.

From Theorem 3.1, we have a Whittaker functional $\lambda_\psi(s, \pi)$ on $\mathrm{I}(s, \pi)$, and from § 2.8 we have an intertwining operator

$$\mathrm{A}(s, \pi, w_0) : \mathrm{I}(s, \pi) \to \mathrm{I}(w_0(\delta_\alpha^s), w_0(\pi)),$$

where in our examples, the unramified character $w_0(\delta_\alpha^s)$ corresponds to $-s$. It is known from the general theory that, inducing from an irreducible admissible σ gives $\mathrm{Ind}(\sigma)$ reducible if and only if $w_0(\sigma) \cong \sigma$ [**100, 110**]; in $\mathrm{I}(s,\pi)$ we have $\sigma = \pi \cdot \delta_\alpha^s$. And multiplicity one, Theorem 2.4, tells us that an irreducible admissible representation of G is generic if and only if its space of Whittaker functionals is one dimensional.

We can then define the Langlands-Shahidi local coefficient via the formula

$$\lambda_\psi(s,\pi) = C_\psi(s,\pi,w_0)\,\lambda_\psi(\delta_\alpha^s, w_0(\pi)) \circ \mathrm{A}(s,\pi,w_0).$$

We have seen that the intertwining operator is rational and the Whittaker functional is a Laurent polynomial, we conclude that

$$C_\psi(s,\pi,w_0) \in \mathbb{C}(q^{-s}).$$

Let us now look into the adjoint action R of ${}^LM$ on ${}^L\mathfrak{n}$, which has a finite number $m(R)$ of irreducible components. In our examples, we actually have $m(R) = 1$ or 2. Then, the following result, see for example Lemma 2.4 of [**77**], tells us that we can inductively reduce to the case where the adjoint representation is irreducible.

LEMMA 3.2. *Let* $(\mathbf{G},\mathbf{M})$ *be a Langlands-Shahidi pair of quasi-split groups. Let* $R = \oplus_{i=1}^{m(R)} r_i$ *be the adjoint action of* ${}^LM$ *on* ${}^L\mathfrak{n}$. *For each* $i > 1$, *there exists a pair* $(\mathbf{G}_i,\mathbf{M}_i)$ *such that the corresponding adjoint action of* ${}^LM_i$ *on* ${}^L\mathfrak{n}_i$ *decomposes as*

$$R' = \bigoplus_{j=1}^{m(R')} r_j' \text{ with } m(R') < m(R),$$

and $r_{m(R)} = r_1'$.

The upshot is that we are able to obtain individual γ-factors from the local coefficient and Lemma 3.2, since $\gamma(s,\pi,R,\psi)$ is some $C_\psi(s,\pi,w_0)$ when R is irreducible. This leads towards local L-functions and ε-factors, as observed in § 3.1. They satisfy a number of local properties, and we highlight two of them starting with their compatibility with Artin γ-factors, proved in [**58**] for non-archimedean local fields of characteristic zero and extended to characteristic p in [**77**]. And, it is also valid for real groups [**96**].

THEOREM 3.3. *Let* (F,π,ψ) *be a local triple such that* $\pi \in \mathfrak{Rep}_{\mathrm{nr}}(M)$. *Let*

$$\phi_\pi : \mathcal{W}_F' \to {}^LM$$

be the Langlands parameter corresponding to π *via the Satake parametrization, see* § 2.4. *Then*

$$\gamma(s,\pi,r,\psi) = \gamma(s, r\circ\phi_\pi,\psi).$$

Local γ-factors satisfy a very interesting multiplicativity property, depending on cocycle relations of Weyl group elements. For our examples, we can make this property explicit as follows.

Multiplicativity. Let $(\mathbf{G},\mathbf{M})$ be a pair such that $\mathbf{G} = \mathbf{G}_l$ is a general linear group or a classical group of rank $l = m+n$, with $\mathbf{M} \cong \mathrm{GL}_m \times \mathbf{G}_n$ and both m, $n \geq 1$. Let τ be a generic representation of $\mathrm{GL}_m(F)$, such that it is a constituent of

$$\mathrm{Ind}(\tau_1 \otimes \cdots \otimes \tau_e), \tag{3.5}$$

where the τ_i's are generic representations of $\mathrm{GL}_{m_i}(F)$, $m = m_1 + \cdots m_e$. Let π be a generic representation of $\mathbf{G}_n(F)$, such that it is a constituent of

$$\mathrm{Ind}(\pi_1 \otimes \cdots \otimes \pi_f \otimes \pi_0),$$

where the π_j's are generic representations of $\mathrm{GL}_{n_j}(F)$, $1 \leq j \leq f$, and π_0 is a generic representation of $\mathbf{G}_{n_0}(F)$, $n = n_1 + \cdots + n_f + n_0$, with $\mathbf{G}_{n_0}$ a classical group of the same type as $\mathbf{G}_n$.

The adjoint representation R factors into two irreducible constituents r_1 and r_2. The first constituent is $r_1 = \rho_m \otimes \tilde{\rho}_n$, hence we obtain $\gamma(s, \tau \otimes \tilde{\pi}, r_1, \psi) = \gamma(s, \tau \times \pi, \psi)$, a Rankin-Selberg γ-factor. If $\mathbf{G}_l = \mathrm{GL}_l$, then

$$\gamma(s, \tau \times \pi, \psi) = \prod_{i,j} \gamma(s, \tau_i \times \pi_j, \psi)$$

And, if $\mathbf{G}_l$ is a classical group, then

$$\gamma(s, \tau \times \pi, r, \psi) = \prod_{i=1}^{e} \gamma(s, \tau_i \times \pi_0, \psi) \prod_{i=1}^{e} \prod_{j=1}^{f} \gamma(s, \tau_i \times \pi_j, \psi) \gamma(s, \tau_i \times \tilde{\pi}_j, \psi).$$

Now, the second constituent of R is given by

$$r_2 = \begin{cases} \mathrm{Sym}^2 \rho_m & \text{if } \mathbf{G} = \mathrm{SO}_{2l+1} \\ \wedge^2 \rho_m & \text{if } \mathbf{G} = \mathrm{SO}_{2l} \text{ or } \mathrm{Sp}_{2l}. \end{cases} \tag{3.6}$$

In these cases, multiplicativity involves only τ of $\mathrm{GL}_m(F)$ given as a constituent of the induced representation of (3.5), and we have the following formula

$$\gamma(s, \tau, r_2, \psi) = \prod_{i=1}^{e} \gamma(s, \tau_i, r_{m_i}, \psi) \prod_{i<j} \gamma(s, \tau_i \times \tau_j, \psi),$$

where r_{m_i} is either $\mathrm{Sym}^2 \rho_{m_i}$ or $\wedge^2 \rho_{m_i}$, of the same kind as r_2.

3.4. Global Langlands-Shahidi L-functions. The global triples we consider (π, r, ψ) require that $\pi = \otimes' \pi_v$ be globally generic. In particular, we can assume that π is globally ψ-generic. By definition, there is a cusp form φ in the space of π such that

$$W_{M,\varphi}(m) = \int_{\mathbf{U}_M(k) \backslash \mathbf{U}_M(\mathbb{A}_k)} \varphi(um) \overline{\psi}(u) \, du \neq 0.$$

One extends φ to an automorphic function

$$\Phi : \mathbf{M}(k) \mathbf{U}(\mathbb{A}_k) \backslash \mathbf{G}(\mathbb{A}_k) \to \mathbb{C},$$

as in § I.2.17 of [**82**]. We thus form a function in the space of the globally induced representation $\mathrm{I}(s, \pi)$, by setting

$$\Phi_s = \Phi \cdot \delta_\alpha^s, \; s \in \mathbb{C}.$$

The corresponding Eisenstein series is defined by

$$E(s, \Phi, g, \mathbf{P}) = \sum_{\gamma \in \mathbf{P}(k) \backslash \mathbf{G}(k)} \Phi_s(\gamma g),$$

which initially converges absolutely for $\Re(s) \gg 0$. For the globally generic representation π of $\mathbf{M}(\mathbb{A}_k)$, the Fourier coefficient of the Eisenstein series $E(s, \Phi, g, \mathbf{P})$ is given by

$$E_\psi(s, \Phi, g, \mathbf{P}) = \int_{\mathbf{U}(k)\backslash\mathbf{U}(\mathbb{A}_k)} E(s, \Phi, ug, \mathbf{P})\overline{\psi}(u)\, du.$$

Eisenstein series were studied in great generality by Langlands in [**68**], they have a meromorphic continuation and satisfy a functional equation. An argument of Harder [**41**], shows that in the case of function fields, Eisenstein series and their Fourier coefficients are rational functions in the variable q^{-s}. The following can be found in [**94**] for number fields, and in [**76**] for function fields.

THEOREM 3.4. *Each Langlands-Shahid L-function $L(s, \pi, r_i)$ converges absolutely for $\Re(s) \gg 0$ and has a meromorphic continuation to the complex plane. If k is a function field, then it is a rational function in q^{-s}.*

In the previous theorem, we have completed L-functions $L(s, \pi, r_i)$. Going from partial L-functions $L^S(s, \pi, r_i)$, to completed ones, is a very delicate process. The Langlands-Shahidi method first makes the connection between Eisenstein series and Whittaker functionals

$$E_\psi(s, \Phi, g, \mathbf{P}) = \prod_v \lambda_{\psi_v}(s, \pi_v)(\mathrm{I}(s, \pi_v)(g_v)f_{s,v}).$$

Then a careful computation at unramified places gives the connection to a product of partial L-functions

$$E_\psi(s, \Phi, g, \mathbf{P}) = \prod_{v \in S} \lambda_{\psi_v}(s, \pi_v)(\mathrm{I}(s, \pi_v)(g_v)f_{s,v}) \prod_{i=1}^{m(R)} L^S(1 + is, \pi, r_i)^{-1}.$$

What we can obtain by arguing in this way, is a first form of the functional equation.

THEOREM 3.5 (Crude Functional Equation). *Let π be a globally ψ-generic cuspidal automorphic representation of $\mathbf{M}(\mathbb{A})$, let $R = \oplus_{i=1}^{m(R)} r_i$ be the adjoint representation. Then*

$$\prod_{i=1}^{m(R)} L^S(is, \pi, r_i) = \prod_{v \in S} C_{\psi_v}(s, \pi_v, w_0) \prod_{i=1}^{m(R)} L^S(1 - is, \tilde{\pi}, r_i).$$

The induction result, Lemma 3.2, allows us to refine this into individual functional equations for each i, $1 \leq i \leq m(R)$, where we obtain local γ-factors at places v in S:

$$L^S(s, \pi, r_i) = \prod_{v \in S} \gamma(s, \pi_v, r_{i,v}, \psi_v) L^S(1 - s, \tilde{\pi}, r_i).$$

Where this global process is the motivation behind the local inductive definition of γ-factors from local coefficients.

Since local ε-factos are trivial for $v \notin S$, we can define global ε-factors:

$$\varepsilon(s, \pi, r_i) = \prod_v \varepsilon(s, \pi_v, r_{i,v}, \psi_v).$$

We already know that partial L-functions initially converge on a right half plane, and we now define completed L-functions:

$$L(s, \pi, r_i) = \prod_v L(s, \pi_v, r_{i,v}),$$

where we added finitely many rational or meromorphic $L(s, \pi_v, r_{i,v})$ to the product defining $L^S(s, \pi, r_i)$. The Langlands-Shahidi method thus produces the following form of the functional equation.

THEOREM 3.6 (Functional Equation). *Let (π, r, ψ) be a global Langlands-Shahidi triple, then*

$$L(s, \pi, r) = \varepsilon(s, \pi, r) L(1 - s, \tilde{\pi}, r).$$

4. Local Langlands

4.1. GL_n. Let F be a non-archimedean local field and n a positive integer. Let $\mathscr{A}_F(n)$ be the set of isomorphism classes of supercuspidal representations of $\mathrm{GL}_n(F)$ and let $\mathscr{G}_F(n)$ be the set of isomorphism classes of irreducible Frob-semisimple n-dimensional representations of $\mathcal{W}_F'$.

THEOREM 4.1 (Local Langlands for GL_n). *For every non-archimedean local field F and every integer $n \geq 1$, there is a correspondence*

$$\mathcal{L}_{\mathrm{loc}} : \mathscr{A}_F(n) \to \mathscr{G}_F(n),$$

where we write $\sigma_\pi = \mathcal{L}_{\mathrm{loc}}(\pi)$. The maps are bijective, they satisfy and are uniquely characterized by the following properties:

(i) *$\mathcal{L}_{\mathrm{loc}}$ is the reciprocity map* Art *when $n = 1$ and*

$$\det(\sigma_\pi) = \omega_\pi \circ \mathrm{Art}.$$

(ii) *If $\pi \in \mathscr{A}_F(n)$ and $\chi \in \mathscr{A}_F(1)$, we have that*

$$\mathcal{L}_{\mathrm{loc}}(\pi \cdot \chi) = \sigma_\pi \otimes \chi \text{ and } \tilde{\sigma}_\pi = \mathcal{L}_{\mathrm{loc}}(\tilde{\pi}).$$

(iii) *$\mathcal{L}_{\mathrm{loc}}$ preserves and is uniquely determined by the property that it preserve local factors for every $\pi \in \mathscr{A}_F(m)$ and $\tau \in \mathscr{A}_F(n)$*

$$L(s, \pi \times \tau) = L(s, \sigma_\pi \otimes \sigma_\tau)$$
$$\varepsilon(s, \pi \times \tau, \psi) = \varepsilon(s, \sigma_\pi \otimes \sigma_\tau, \psi),$$

where $\psi : F \to \mathbb{C}^\times$ is a non-trivial character.

The local Langlands conjecture was proved by Kutzko for GL_2 [**64**] followed by Henniart for GL_3 [**47**]. Over local function fields, it is a result of Laumon, Rapoport and Stuhler [**74**] for every n. For finite extensions of $\mathbb{Q}_p$, it was established independently by Harris-Taylor and Henniart [**44,49**], and more recently by Scholze [**90**].

The uniqueness criterion presented in Theorem 4.1 using Rankin-Selberg local factors is due to Henniart [**48**], and originally appeared as an appendix to [**74**]; it is valid in any characteristic. However, we observe that Scholze's proof presents an alternative uniqueness criterion that makes the result of Harris-Taylor independent of Henniart's result. Namely, in [**90**], Scholze constructs functions h and $f_{\tau,h}$ satisfying a trace relationship

$$\mathrm{Tr}(f_{\tau,h}|\pi) = \mathrm{Tr}(\tau|\mathcal{L}_{\mathrm{loc}}(\pi))\mathrm{Tr}(h|\pi)$$

and proves that this condition uniquely characterizes the generalized reciprocity map $\mathcal{L}_{\mathrm{loc}}$. In addition, the map respects local L-functions and root numbers. The method of proof in this approach relies on powerful methods of Harris [**42**] and Harris-Taylor [**44**], based on Shimura varieties.

Local factors satisfy a multiplicativity property, where we observe that tempered representations for $\mathrm{GL}_n(F)$ are generic [**115**]. In particular, supercuspidal representations of general linear groups are generic. Compatibility with the Langlands classification, enables the passage from tempered to admissible representations.

(iv) (Multiplicativity). *Assume π is the unique generic constituent of*

$$\mathrm{Ind}(\pi_1 \otimes \cdots \otimes \pi_b),$$

where each π_i is a supercuspidal representation of $\mathrm{GL}_{n_i}(F)$. Let τ be a supercuspidal representation of $\mathrm{GL}_m(F)$. Then

$$L(s, \pi \times \tau) = \prod_i L(s, \pi_i \times \tau),$$

$$\varepsilon(s, \pi \times \tau, \psi) = \prod_i \varepsilon(s, \pi_i \times \tau, \psi).$$

(v) (Langlands quotient). *If π is a subquotient of*

$$\mathrm{Ind}(\pi_1 \otimes \cdots \otimes \pi_b),$$

with each π_i an irreducible quasi-tempered representation of $\mathrm{GL}_{n_i}(F)$, then

$$\sigma_\pi = \mathcal{L}_{\mathrm{loc}}(\pi) = \sigma_{\pi_1} \oplus \cdots \oplus \sigma_{\pi_b}.$$

And we have compatibility with local L-functions and ε-factors for pairs.

4.2. Classical groups. Given a classical group $\mathbf{G}$, if F is a non-archimedean local field, we write $G = \mathbf{G}(F)$. Let $\mathscr{A}_F(\mathbf{G})$ be the set of irreducible admissible representations of $G = \mathbf{G}(F)$. Let ${}^L\mathscr{G}_F(\mathbf{G})$ be the set of L-parameters

$$\phi : \mathcal{W}'_F \to {}^LG,$$

where the homomorphism ϕ is taken up to $\widehat{G}$ conjugacy for SO_{2n+1} and Sp_{2n}, in our examples, cases (b) and (c), and up to $\mathrm{O}_{2n}(\mathbb{C})$-conjugacy in case (d).

An L-parameter $\phi \in {}^L\mathscr{G}_F(\mathbf{G})$ is discrete if its image does not factor through a parabolic subgroup of ${}^LG$. We say that $\phi \in {}^L\mathscr{G}_F(\mathbf{G})$ is bounded if its image in ${}^LG$ projects onto a bounded subset of $\widehat{G}$.

THEOREM 4.2. *For every split classical group $\mathbf{G}$ and every non-archimedean local field F there is a finite-to-one map*

$$\mathcal{L}_{\mathrm{loc}} : \mathscr{A}_F(\mathbf{G}) \to {}^L\mathscr{G}_F(\mathbf{G}).$$

Writing $\phi_\pi = \mathcal{L}_{\mathrm{loc}}(\pi)$ for the corresponding Langlands parameter of a representation π, we have that

(i) *π is a discrete series representation iff ϕ_π is a discrete L-parameter.*
(ii) *π is a tempered representation iff ϕ_π is a bounded L-parameter.*
(iii) *Assume π is the unique Langlands quotient of*

$$\mathrm{Ind}_P^G(\tau \cdot \eta),$$

where τ is tempered and η is in the Langlands situation. Then ϕ_π factors through ${}^LM$, i.e.,

$$\phi_\pi : \mathcal{W}'_F \xrightarrow{\phi_\tau} {}^LM_F \to {}^LG_F,$$

where $\phi_\tau = \mathcal{L}_{\mathrm{loc}}(\tau)$.

(iv) *Assume* π *is a generic representation of* G *and* τ *is a supercuspidal representation of* $\mathrm{GL}_n(F)$, *then*

$$L(s, \pi \times \tau) = L(s, \phi_\pi \otimes \phi_\tau)$$
$$\varepsilon(s, \pi \times \tau, \psi) = \varepsilon(s, \phi_\pi \otimes \phi_\tau, \psi).$$

(v) *For* π *irreducible admissible non-generic and* τ *supercuspidal of* $\mathrm{GL}_n(F)$, *we have*

$$\mu(s, \pi \times \tau, \psi) = \gamma(s, \phi_\pi^\vee \otimes \phi_\tau, \psi)\, \gamma(-s, \phi_\pi \otimes \phi_\tau^\vee, \overline{\psi})$$
$$\gamma(2s, r \circ \phi_\tau, \psi)\, \gamma(-2s, r^\vee \circ \phi_\tau, \overline{\psi}).$$

Furthermore, $\mathcal{L}_{\mathrm{loc}}$ *is characterized by properties (i)–(v).*

The Langlands correspondence for the classical groups in characteristic zero, both local and global, is due to Arthur [**3**], who provides an endoscopic classification via the trace formula, depending on the stabilization of the twisted trace formula proved by Mœglin and Waldspurger in a series of papers culminating in [**83**].

Under certain restrictions on the residual characteristic, the Langlands correspondence for local function fields was proved by Ganapathy-Varma [**36**]. Their method relies on the Kazhdan philosophy, on transferring information for representations of $\mathfrak{p}$-adic groups from characteristic 0 to characteristic p, which is also established by Aubert, Baum, Plymen and Solleveld in [**6**]. We refer to § 7 of [**34**], for a description of the local Langlands correspondence over function fields for the classical groups and a general proof depending on a working hypothesis.

4.3. Tempered L-packet conjecture. In contrast to the case of general linear groups, tempered representations of classical groups are not necessarily generic. However, we do have the tempered L-packet conjecture, Conjecture 4.3 below. It is known in characteristic zero [**3**], and is part of the work of Ganapathy-Varma [**36**] on the local Langlands correspondence for the split classical groups in characteristic p.

Given an L-parameter ϕ in ${}^L\mathscr{G}_F(\mathbf{G})$, there is an L-packet Π_ϕ which is a finite multi-set of representations π in $\mathscr{A}_F(\mathbf{G})$ sharing in common arithmetic information. We are interested in tempered L-packets, which arise from bounded L-parameters. In this case, the definitions of Langlands and Arthur packets agree.

CONJECTURE 4.3. *If an* L*-packet contains a tempered element, then all of its elements are tempered. Every tempered* L*-packet* Π_ϕ *of* G *contains a representation* π_0 *which is generic.*

The importance of the tempered L-packet conjecture is that it reduces the study of local γ-factors, L-functions and ε-factors to the case of generic representations. Furthermore, we observe that Conjecture 4.3 can be phrased for $\mathbf{G}$ a connected quasi-split reductive group.

4.4. On Local Langlands for reductive groups. The work of Arthur and Mœglin-Waldpurger for the split classical groups was extended to quasi-split unitary groups by Mok [**84**], and recent work of Kaletha [**56**] gives foundations for the general case of inner forms. Ongoing work of Fargues and Scholze aims towards the local Langlands conjecture when $\mathrm{char}(F) = 0$. The work of Genestier-Lafforgue

[**38**], now proves the local correspondence in characteristic p for connected reductive groups. The work of V. Lafforgue in characteristic p, and that of Scholze in characteristic zero, give the semisimple part of the L-parameter.

We only mention here that Base Change has played a crucial role in the local Langlands correspondence, starting with its proof for GL_n. The article by Harris [**43**], provides a Base Change approach towards local Langlands and related open questions.

5. Global Langlands

Let k be a global field with Weil group $\mathcal{W}_k$. Let $\mathscr{A}_k(n)$ be the set of isomorphism classes of cuspidal automorphic representations of $\mathrm{GL}_n(\mathbb{A}_k)$ and let $\mathscr{G}_k(n)$ be the set of isomorphism classes of irreducible continuous degree n representations of $\mathcal{W}_k$ which are unramified almost everywhere.

CONJECTURE 5.1 (Global Langlands for GL_n). *For every $n \geq 2$, there is a correspondence*

$$\mathcal{L}_{\mathrm{glob}} : \mathscr{A}_k(n) \to \mathscr{G}_k(n).$$

Writing $\Sigma = \mathcal{L}_{\mathrm{glob}}(\Pi)$, we have that

$$L(s, \Pi) = L(s, \Sigma)$$

and at every place v of k we have that

$$\Sigma_v = \mathcal{L}_{\mathrm{loc}}(\Pi_v).$$

The maps $\mathcal{L}_{\mathrm{glob}}$ are bijective and uniquely determined.

The conjecture is known over function fields, where Drinfelds' Shtuka technique is the key to functoriality [**30**]. L. Lafforgue makes a tour de force in his very important work for GL_n [**65**]. However, it is an open problem over number fields. Cases where there are global representations $\Pi \in \mathscr{A}_k(n)$ and $\Sigma \in \mathscr{G}_k(n)$ which correspond to each other, are discussed in § 6.

5.1. Converse Theorem.

An important technique in proving that a representation of GL_n is automorphic is produced by the Converse Theorem of Cogdell and Piatetski-Shapiro [**24**]. For this we have a twisted version [**22**], where we fix a Größencharakter character η and a finite set S of places of a global field k, including all archimedean places.

Given an integer $N \geq 2$, consider the set $\mathcal{T}(S; \eta)$ consisting of $\tau = \tau_0 \otimes \eta$ such that: $\tau_0 \in \mathscr{A}_k(m)$, where $\tau_{0,v}$ is unramified for $v \notin S$, and m is an integer ranging from $1 \leq m \leq \max\{1, N-2\}$.

THEOREM 5.2. *Let $\Pi = \otimes' \Pi_v$ be an irreducible admissible representation of $\mathrm{GL}_N(\mathbb{A}_k)$ whose central character ω_Π is invariant under $k^\times$ and whose L-function $L(s, \Pi)$ is absolutely convergent for $\Re(s) \gg 0$. Suppose that for every $\tau \in \mathcal{T}(S; \eta)$ the L-function $L(s, \Pi \times \tau)$ is* nice, *i.e., it satisfies:*

(i) *The functional equation*

$$L(s, \Pi \times \tau) = \varepsilon(s, \Pi \times \tau) L(1-s, \widetilde{\Pi} \times \widetilde{\tau}).$$

(ii) *$L(s, \Pi \times \tau)$ and $L(s, \widetilde{\Pi} \times \widetilde{\tau})$ are entire on $s \in \mathbb{C}$.*

(iii) *$L(s, \Pi \times \tau)$ and $L(s, \widetilde{\Pi} \times \widetilde{\tau})$ are bounded on vertical strips away from poles.*

Then there exists an automorphic representation Π' *of* $\mathrm{GL}_N(\mathbb{A}_k)$ *such that* $\Pi_v \cong \Pi'_v$ *for all* $v \notin S$.

REMARK 5.3. If k is a function field, we have the condition that $L(s, \Pi \times \tau)$ and $L(s, \widetilde{\Pi} \times \widetilde{\tau})$ be polynomials in the variable q^{-s}. This stronger condition, replaces conditions (ii) and (iii) in the above formulation of the Converse Theorem.

5.2. Functoriality for the classical groups. We restrict ourselves to generic representations, since it leads to the Ramanujan Conjecture. In what follows, we assume generic representations are irreducible. We observe that a possible general approach to functoriality for (not necessarily generic) irreducible admissible representations, is by means of the trace formula [**3**]. A treatment of Arthur's trace formula and endoscopic classification would require a separate survey article.

We now proceed to give an overview of the results of [**23, 75, 76**] on the globally generic functorial lift for the classical groups, Theorem 5.4. Let $\mathbf{G}_n$ be a split classical group of rank n defined over a global field k; $n \geq 2$ if $\mathbf{G}_n = \mathrm{SO}_{2n}$. Langlands functoriality is for $\mathbf{G}_n$ to $\mathbf{H}_N$, given by the following table.

$\mathbf{G}_n$	${}^L G_n \hookrightarrow {}^L H_N$	$\mathbf{H}_N$
SO_{2n+1}	$\mathrm{Sp}_{2n}(\mathbb{C}) \times \mathcal{W}_k \hookrightarrow \mathrm{GL}_{2n}(\mathbb{C}) \times \mathcal{W}_k$	GL_{2n}
Sp_{2n}	$\mathrm{SO}_{2n+1}(\mathbb{C}) \times \mathcal{W}_k \hookrightarrow \mathrm{GL}_{2n+1}(\mathbb{C}) \times \mathcal{W}_k$	GL_{2n+1}
SO_{2n}	$\mathrm{SO}_{2n}(\mathbb{C}) \times \mathcal{W}_k \hookrightarrow \mathrm{GL}_{2n}(\mathbb{C}) \times \mathcal{W}_k$	GL_{2n}

Table 1

Locally, we let $\mathscr{A}_F^{\mathrm{gen}}(\mathbf{G}_n)$ be the set of isomorphism classes of generic representations of $\mathbf{G}_n(F)$, when $\mathbf{G}_n$ is a classical group; and, we write $\mathscr{A}_F^{\mathrm{gen}}(\mathrm{GL}_n)$, when we are in the case of a general linear group. Given Theorems 4.1 and 4.2, we have a map

$$\mathcal{L}_{\mathrm{loc}}^{\mathrm{gen}} : \mathscr{A}_F^{\mathrm{gen}}(\mathbf{G}_n) \to \mathscr{A}_F^{\mathrm{gen}}(\mathrm{GL}_N).$$

Let us adapt the notation of § 4 to this setting, now writing $\Pi_0 = \mathcal{L}_{\mathrm{loc}}^{\mathrm{gen}}(\pi_0)$. Then the map is such that for every supercuspidal $\tau_0 \in \mathscr{A}_F^{\mathrm{gen}}(\mathrm{GL}_m)$ and every non-trivial character $\psi_0 : F \to \mathbb{C}^\times$ there is equality of γ-factors

$$\gamma(s, \Pi_0 \times \tau_0, \psi_0) = \gamma(s, \pi_0 \times \tau_0, \psi_0).$$

And the map respects L-functions and ε-factors

$$L(s, \Pi_0 \times \tau_0) = L(s, \pi_0 \times \tau_0)$$
$$\varepsilon(s, \Pi_0 \times \tau_0, \psi_0) = \varepsilon(s, \pi_0 \times \tau_0, \psi_0).$$

Globally, we let $\mathscr{A}_k^{\mathrm{gen}}(\mathbf{G}_n)$ be the set of isomorphism classes of globally generic cuspidal automorphic representations of $\mathbf{G}_n(\mathbb{A}_k)$, when $\mathbf{G}_n$ is a classical group. We let $\mathscr{A}_k^{\mathrm{gen}}(\mathrm{GL}_n)$ be the set of isomorphism classes of globally generic automorphic (not necessarily cuspidal) representations of $\mathrm{GL}_n(\mathbb{A}_k)$, when we are in the case of a general linear group.

THEOREM 5.4 (Classical groups). *There is a map*

$$\mathcal{L}_{\mathrm{gen}} : \mathscr{A}_k^{\mathrm{gen}}(\mathbf{G}_n) \to \mathscr{A}_k^{\mathrm{gen}}(\mathrm{GL}_N).$$

Writing $\Pi = \mathcal{L}_{\mathrm{gen}}(\pi)$, *we have that*

$$L(s, \pi \times \tau) = L(s, \Pi \times \tau)$$

for every $\tau \in \mathscr{A}_k^{\mathrm{gen}}(\mathrm{GL}_m)$. *The representation* Π *has trivial central character and can be expressed as an isobaric sum*

$$\Pi = \Pi_1 \boxplus \cdots \boxplus \Pi_d,$$

where each $\Pi_i \in \mathscr{A}_k^{\mathrm{gen}}(\mathrm{GL}_{N_i})$ *is unitary self-dual cuspidal and* $\Pi_i \not\cong \Pi_j$ *for* $i \neq j$. *Furthermore*

$$\Pi_v = \mathcal{L}_{\mathrm{loc}}^{\mathrm{gen}}(\pi_v)$$

at every place v *of* k.

Let us give a sketch of proof and then make a note on the image of functoriality. Let $\pi = \otimes' \pi_v \in \mathscr{A}_k^{\mathrm{gen}}(\mathbf{G}_n)$. We construct a candidate lift $\Pi' = \otimes' \Pi'_v$, which will be a priori just admissible. We take S to include all places where ramification may occur.

Initially, we understand the local Langlands correspondence for unramified representations π_v of $\mathbf{G}_n(k_v)$ and Π'_v of $\mathrm{GL}_N(k_v)$. Let these representations have L-parameters ϕ_{π_v} and $\phi_{\Pi'_v}$, they are then related by the following commutative diagram

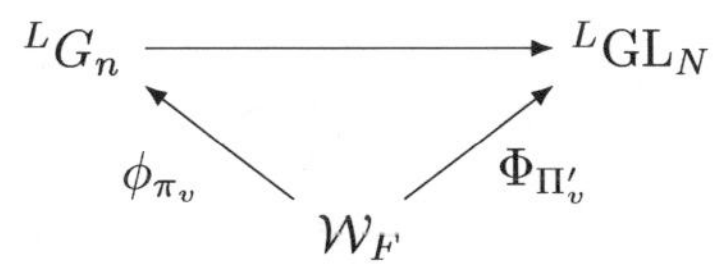

possible via the Satake correspondence. Also, at archimedean places, we have compatibility with Artin factors from the work of Shahidi [**96**].

The construction is such that global L-functions agree

$$L(s, \pi \times \tau) = L(s, \Pi' \times \tau)$$

for $\tau = \tau_0 \otimes \eta \in \mathcal{T}(S;\eta)$. And, at finite places $v \in S$, we have that $\tau = \tau_0 \otimes \eta \in \mathcal{T}(S;\eta)$ has $\tau_{0,v}$ unramified. Hence, $\tau_{0,v}$ for $v \in S$ is an irreducible constituent of an unramified principal series represnetation

$$\mathrm{Ind}(\chi_{1,v} \otimes \cdots \otimes \chi_{m,v}),$$

so that each $\chi_{i,v} : F^\times \to \mathbb{C}^\times$ is an unramified character. Then, we have

$$\gamma(s, \pi_v \otimes \tau_v, \psi_v) = \prod_{i=1}^{m} \gamma(s, \pi_v \otimes (\chi_{i,v} \cdot \eta_v), \psi_v),$$

We can always take a Größencharakter, such that η_v is highly ramified at these places. Hence, the product on the right hand side of this last equation becomes stable.

By the Langlands-Shahidi method the global L-functions $L(s, \pi \times \tau)$ are nice, hence, so are $L(s, \Pi' \times \tau)$. An application of the Converse Theorem, then gives an automorphic representation $\Pi = \otimes \Pi_v$ of $\mathrm{GL}_N(\mathbb{A}_k)$ obtained from π in such a way that $\Pi_v = \mathcal{L}_{\mathrm{loc}}^{\mathrm{gen}}(\pi_v)$ for $v \notin S$. From the classification of automorphic forms for GL_N [**54**], Π is the unique generic subquotient of a globally induced representation of the form $\mathrm{Ind}(\Pi_1 \otimes \cdots \otimes \Pi_d)$ with each Π_i a cuspidal representation of $\mathrm{GL}_{N_i}(\mathbb{A}_k)$. We write this as an isobaric sum

$$\Pi = \Pi_1 \boxplus \cdots \boxplus \Pi_d,$$

The representation Π is unitary, what we need to show is that each Π_i is unitary.

For this, we have the character ν, obtained by composing the determinant with the absolute value. We have used it explicitly over local fields, and we will now also use ν to denote its global version on $\mathbf{Z}(\mathbb{A}_k) = \mathrm{GL}_1(\mathbb{A}_k)$. We write $\Pi_i = \nu^{r_i}\Pi_{i,0}$ with each $\Pi_{i,0}$ unitary and $t_d \geq \cdots \geq t_1$, $t_1 \leq 0$. Because Π is unitary, we cannot have $t_1 > 0$. Then

$$\begin{aligned} L(s, \pi \times \widetilde{\Pi}_1) &= L(s, \Pi \times \widetilde{\Pi}_1) \\ &= \prod_i L(s + t_i, \Pi_i \times \widetilde{\Pi}_1). \end{aligned}$$

An important analytic property, for Langlands-Shahidi L-functions for the classical groups, is that $L(s, \pi \times \widetilde{\Pi}_1)$ is holomorphic for $\mathrm{Re}(s) > 1$. However, $L(s+t_1, \Pi_1 \times \widetilde{\Pi}_1)$ has a simple pole at $s = 1 - t_1 \geq 1$. This pole carries through to a pole of $L(s, \pi \times \widetilde{\Pi}_1)$, which gives a contradiction unless $t_1 = 0$. A recursive argument shows that all the t_i's must be zero.

Also, every time we have $\Pi_i \cong \Pi_j$, we add to the multiplicity of the pole at $s = 1$ of the L-function

$$L(s, \pi \times \tilde{\Pi}_i) = \prod_j L(s, \Pi_j \times \tilde{\Pi}_i). \tag{5.1}$$

However, $L(s, \pi \times \tilde{\Pi}_j)$ has at most a simple pole at $s = 1$. Hence, we must have $\Pi_i \not\cong \Pi_j$, for $i \neq j$.

To show that each Π_i is self-dual, we argue by contradiction and assume $\Pi_i \not\cong \tilde{\Pi}_i$. The representation $\sigma = \tilde{\Pi}_i \otimes \tilde{\tau}$ of $\mathbf{M}_i(\mathbb{A}_k)$ satisfies $w_0(\sigma) \not\cong \sigma$. Then, the L-function $L(s, \pi \times \tilde{\Pi}_i)$ has no poles. Now, we have that a pole appears on the right hand side of equation (5.1) from the L-function $L(s, \Pi_i \times \tilde{\Pi}_i)$. This is a contradiction, unless Π_i is self-dual.

We now go through the classification of generic representations for the classical groups. First for supercuspidal representations, which uses a globalization argument discussed in the following section. Then, there is the Mœglin-Tadić classification for discrete series. The standard module conjecture for tempered representations and the Langlands classification. A crucial step is the stability property of γ-factors under twists by highly ramified characters.

5.3. A note on the image of functoriality. We observe that the argument found in § 7 of [**76**] is valid over any global field. That is, the argument provides a self conatined proof concerning the image of functoriality over number fields of [**23**] and [**101**], with the added note that symmetric and exterior square L-functions for cuspidal unitary automorphic representations are non-vanishing for $\Re(s) > 1$, Proposition 7.4. For function fields, this non-vanishing is proved in § 6 of [**76**].

6. Globalization methods

Throughout this section assume that F is a non-archimedean local field and k is a global field such that $k_{v_0} \simeq F$ at a place v_0. We discuss globalization methods that take a representation π over a non-archimedean local field F, for example, a supercuspidal representation of $\mathrm{GL}_n(F)$, and produce global representations Π and Σ that correspond to each other and $\Pi_{v_0} \simeq \pi$.

DEFINITION 6.1. Let Π be a cuspidal automorphic representation of $\mathrm{GL}_n(\mathbb{A}_k)$ and let Σ be an n-dimensional ℓ-adic representation of $\mathcal{W}_k$ on the Galois side. We say that the representations correspond to each other if

$$\Sigma = \mathcal{L}_{\mathrm{glob}}(\Pi),$$

as in Conjecture 5.1.

Note that if Π and Σ correspond to each other, then by definition

$$\Sigma_v = \mathcal{L}_{\mathrm{loc}}(\Pi_v),$$

at every place v of k.

6.1. GL_1 and monomial representations for GL_n. Already present in Jacquet-Langlands' work [**52**], is the case of monomial representations, which provide a first family of cases where the global Langlands correspondence for GL_2 is known. Locally, this arises from the tame case, and we have this for GL_n, for example, when p does not divide n. We begin with a basic lemma, which is a version of the Grundwald-Wang theorem of class field theory [**5**].

LEMMA 6.2. *Let $\chi_0 : \mathrm{GL}_1(F) \to \mathbb{C}^\times$ be a character. There exists a Größencharakter $\chi : k^\times\backslash\mathrm{GL}_1(\mathbb{A}_k) \to \mathbb{C}^\times$ such that:*

(i) $\chi_{v_0} \simeq \chi_0$;
(ii) χ_v *is unramified if* $v \notin \{v_0, \infty\}$;
(iii) *if* $\mathrm{char}(F) = p$, *then* χ_∞ *is tame.*

Then, by global class field theory, there is a 1-dimensional representation $\sigma = \mathrm{Art}(\chi)$ of $\mathcal{W}_k$ that corresponds to χ.

From the theory of types [**16**], we have level zero supercuspidal representations of $\mathrm{GL}_n(F)$. A level zero supercuspidal π corresponds to a tamely ramified complex representation σ of $\mathcal{W}_F$ via the local Langlands correspondence. More precisely, there exists an unramified extension E/F of degree n and a character $\eta_0 : \mathcal{W}_E \to \mathbb{C}^\times$, such that

$$\sigma = \mathrm{Ind}_{\mathcal{W}_E}^{\mathcal{W}_F}(\eta_0).$$

For these representations we have the following globalization result.

PROPOSITION 6.3. *Let π be a level zero supercuspidal representation of $\mathrm{GL}_n(F)$. Then there exist global representations Π of $\mathrm{GL}_n(\mathbb{A}_k)$ and Σ of $\mathcal{W}_k$ that correspond to each other and $\Pi_{v_0} \simeq \pi$.*

To see this, we need the next basic lemma from Algebraic Number Theory.

LEMMA 6.4. *If E/F is a finite separable extension, then there exists an extension l/k, a place v_0 of k such that $F \cong k_{v_0}$, and an F-isomorphism $E \simeq l \otimes_k k_{v_0}$.*

Thus, given the extension l/k, an application Lemma 6.2 gives a Größencharakter $\eta : l^\times\backslash\mathrm{GL}_1(\mathbb{A}_l) \to \mathbb{C}^\times$ with $\eta_{v_0} \simeq \eta_0$. Then, we can set

$$\Sigma = \mathrm{Ind}_{\mathcal{W}_l}^{\mathcal{W}_k}(\eta).$$

Notice that, using Lemma 6.2, we may choose an η that reduces to cases where the local Langlands correspondence is known. For example, at finite unramified places we have the Satake parametrization § 2.4. Hence

$$\Sigma_v = \mathcal{L}_{\mathrm{loc}}(\Pi_v),$$

at every place v of k.

6.2. On GL_n over number fields. The global Langlands conjecture is an open problem over a number field k, even for GL_2. However, there are important results of Langlands and Tunnell on base change, that allow us to obtain automorphic representations of $\mathrm{GL}_2(\mathbb{A}_k)$ that correspond to Galois representations.

We have the following refinement of globalizing Lemma 6.4 on the Galois side (Lemma 4.13 of [**27**], Lemma 3.6 of [**47**]).

LEMMA 6.5. *Let E/F be a finite Galois extension of non-archimedean local fields. Then there exist a global field k, a finite Galois extension l/k, with a place v_0 of k, and an isomorphism η of k_{v_0} onto of F inducing an isomorphism of $k_{v_0} \otimes_k l$ onto E. In particular, l has only one place w_0 above v_0 and the decomposition subgroup of* $\mathrm{Gal}(l/k)$ *at w_0 is itself, and identifies with* $\mathrm{Gal}(E/F)$ *via η.*

A local Galois representation σ of $\mathrm{Gal}(E/F)$, globalizes to a representation Σ of $\mathrm{Gal}(l/k)$. Local Galois groups being solvable, the image of Σ in the projective linear group is dihedral, A_4 or S_4 so there is by [**52, 72, 109**] a cuspidal automorphic representation Π of $\mathrm{GL}_2(\mathbb{A}_l)$ so that

$$\Sigma \longleftrightarrow \Pi,$$

correspond to each other.

For GL_n, we have a globalization theorem of Harris [**42**] and Harris-Taylor [**44**], more recently refined by Scholze [**90**] and Scholze-Shin [**91**]. These results, lead towards proofs of the local Langlands correspondence, by globalizing supercuspidal representations where there is a corresponding Galois representation. If we begin with a supercuspidal representation π of $\mathrm{GL}_n(F)$, then there exists a number field k, with a place v_0 of k, and an automorphic representation Π of $\mathrm{GL}_n(\mathbb{A}_k)$ for which there is a global n-dimensional ℓ-adic representation Σ on the Galois side, such that

$$\Pi_{v_0} \cong \pi \quad \text{and} \quad \Sigma_v = \mathcal{L}_{\mathrm{loc}}(\Pi_v),$$

at all places v of k, i.e., the representations correspond to each other via local Langlands for GL_n.

The work of many mathematicians aims at establishing a large class of Galois representations which correspond to an automorphic representation under global Langlands. We refer to the work of Clozel and Thorne, starting with [**21**], where several basic definitions can be found. And, the awe-inspiring 10 author effort [**1**] for potential automorphy.

Let l be an imaginary CM field with a totally real subfield k, where $c \in \mathrm{Gal}(l/k)$ denotes conjugation. An automorphic representation Π of $\mathrm{GL}_n(\mathbb{A}_l)$ is regular algebraic, conjugate self-dual, cuspidal (RACSDC), if for every place v of l such that $v \mid \infty$, the archimedean Langlands correspondence leads to a sum of pairwise distinct algebraic characters (regular algebraicity); $\Pi^c \cong \tilde{\Pi}$ (conjugate self-dual); and, Π is cuspidal. Similar notions are RAECSDC, for essentially conjugate self-dual representations, and RAESDC, for essentially self-dual representations.

To summarize, such representations Π of $\mathrm{GL}_n(\mathbb{A}_l)$ will have a corresponding global Galois representation after semisimplification

$$\Pi \longleftrightarrow \Sigma,$$

such that

$$\Sigma_v = \mathcal{L}_{\mathrm{loc}}(\Pi_v),$$

at every place v of k.

6.3. Function fields. The global field k in characteristic p, arises as the field of functions of a smooth connected projective curve X over the finite field $\mathbb{F}_q$. Each closed point of X, gives a valuation v of k. This permits the use of methods from algebraic geometry and étale cohomology, where we obtain the the Galois group of k from the fundamental group of X.

In the article [**57**], Katz shows that the morphism

$$\operatorname{Spec}\left(\mathbb{F}_q((t^{-1}))\right) \to \operatorname{Spec}\left(\mathbb{F}_q\left[t,t^{-1}\right]\right)$$

induces a functor

$$\left\{\begin{array}{c}\text{finite 'special' étale}\\ \text{coverings of } \operatorname{Spec}\left(\mathbb{F}_q\left[t,t^{-1}\right]\right)\end{array}\right\} \longrightarrow \left\{\begin{array}{c}\text{finite étale coverings}\\ \text{of } \operatorname{Spec}\left(\mathbb{F}_q((t^{-1}))\right)\end{array}\right\},$$

which is in fact an equivalence of categories. Here, a finite 'special' étale cover is one of the punctured disk at two places v_0 and v_∞, and is such that it is tame at v_∞ and is unramifed for $v \notin \{v_0, v_\infty\}$. On the right hand side, we have the fundamental group corresponding to the Galois group of F, which connects to the global Galois group of k on the left hand side. If we begin with a local representation σ_0 on the right hand side, then the equivalence of categories proves the existence of a global Galois representation Σ, whose properties we summarize in Lemma 6.6.

We observe that when working with the étale topology, we need to pass between ℓ-adic and complex representations. For this, we fix a prime number ℓ, distinct from p, and an algebraic closure $\overline{\mathbb{Q}}_\ell$ of the field $\mathbb{Q}_\ell$ of ℓ-adic numbers. We do this by fixing an isomorphism $\iota : \overline{\mathbb{Q}}_\ell \simeq \mathbb{C}$.

LEMMA 6.6. *Let σ_0 be an irreducible n-dimensional Weil-Deligne representation of $\mathcal{W}_F$. Then, there is a rational function field k with a set of two places $S = \{v_0, v_\infty\}$, $k_{v_0} \simeq F$, and an irreducible continuous n-dimensional representation Σ of $\mathcal{W}_k$ such that*

(i) $\Sigma_{v_0} \simeq \sigma_0$;
(ii) Σ_v *is unramified for* $v \notin S$;
(iii) Σ_{v_∞} *is tame.*

The result with prescribed ramification is for rational function fields. For a general function field, Katz proves that the representation can be globalized, yet there are finitely many other places where ramification can occur. Remark 6.8 below sketches a solution to the general problem with possibly one other tame place.

Now, working purely on the representation theoretic side, we obtain the corresponding globalization theorem for representations of GL_n in [**50**]. The argument is generalized to a connected reductive group $\mathbf{G}$ in [**34**], but introduces possibly finitely many tamely ramified places; it is refined in [**77**] to the following form.

LEMMA 6.7. *Let π be a supercuspidal representation of $\mathbf{G}(F)$. There exists a set of two places $S = \{v_0, v_\infty\}$, $k_{v_0} \cong F$, and a cuspidal automorphic representation $\Pi = \otimes'_v \Pi_v$ of $\mathbf{G}(\mathbb{A}_k)$ satisfying the following properties:*

(i) $\Pi_{v_0} \cong \pi$;
(ii) Π_v *has an Iwahori fixed vector for* $v \notin S$;
(iii) Π_{v_∞} *is a constituent of a tamely ramified principal series;*
(iv) *if π is generic, then Π is globally generic.*

REMARK 6.8. The global Langlands correspondence for $\mathbf{G} = \mathrm{GL}_n$ over a function field k [**65**], tells us that the cuspidal automorphic representation Π of the previous Theorem corresponds to a Galois representation Σ. Because Π and Σ correspond to each other at every v, this extends Lemma 6.6 on the Galois side to any k.

7. Ramanujan Conjecture

We now turn towards Generalized Ramanujan for the general linear group. Then we note results that are possible for reductive groups, thanks to the principle of Langlands Functoriality. In particular, in the case of function fields, the Ramanujan Conjecture is valid for our examples (a)–(d).

7.1. The general linear group. Let F be a non-archimedean local field. An admissible unitary representation π of $\mathrm{GL}_n(F)$ arises as a subquotient of a parabolically induced representation

$$\pi \hookrightarrow \mathrm{Ind}(\delta_0\nu^{r_0} \otimes \cdots \otimes \delta_b\nu^{r_b}), \tag{7.1}$$

where the δ_i's are discrete series representations of $\mathrm{GL}_{n_i}(F)$, $n = n_0 + \cdots + n_b$, and the Langlands parameters satisfy $0 \leq r_0 \leq \cdots \leq r_b$.

CONJECTURE 7.1 (Ramanujan Conjecture for GL_n). *Let Π be a cuspidal unitary automorphic representation of $\mathrm{GL}_n(\mathbb{A}_k)$, then each local component Π_v is tempered. If Π_v is unramified, temperedness implies that its Satake parameters satisfy*

$$|\alpha_{j,v}| = 1, \quad 1 \leq j \leq n.$$

Over function fields, it is a theorem of Drinfeld for $n = 2$ [**30**] and L. Lafforgue for any n [**65**]. However, it is an open problem over number fields, even in the case of GL_2 and $k = \mathbb{Q}$.

7.2. Ramanujan bounds. Given a cuspidal (unitary) automorphic representation $\Pi = \otimes' \Pi_v$ of $\mathrm{GL}_n(\mathbb{A}_k)$, the Ramanujan Conjecture states that for every place v of k, the representation Π_v is isomorphic to a representation given by the form of equation (7.1) with $r_b = 0$; this exact bound is only known for a function field k.

A first bound that results from the study of L-functions for Rankin-Selberg products is what is known as the "trival" bound of $r_b \leq 1/2$ [**54**]. For arbitrary n and any number field, it is Luo, Rudnick and Sarnak who established in [**78**] the bound

$$r_b \leq \frac{1}{2} - \frac{1}{n^2+1}.$$

It is interesting that this is still the best bound for general n, although breakthrough results have been obtained for $n \leq 4$.

There is the observation, made by Langlands that the automorphy of symmetric powers implies the Ramanujan Conjecture. The following conjecture is a very important open problem over number fields; it follows from the work of L. Lafforgue over function fields.

CONJECTURE 7.2 (Automorphy of symmetric powers). *Given a cuspidal automorphic representation Π of $\mathrm{GL}_n(\mathbb{A}_k)$, then the representation $\mathrm{Sym}^m(\Pi)$ is also automorphic for every m.*

The first breakthrough results in symmetric powers were obtained in the work of Kim and Shahidi, who use the Langlands-Shahidi method and establish the automorphy of the symmetric cube in [**61**], in addition to the Ramanujan bound of $1/9$. Then, Kim carries their work further to establish symmetric fourth [**59**] and with Sarnak [**60**], they have the current best bound towards Ramanujan when $k = \mathbb{Q}$, namely

$$r_b \leq \frac{7}{64}.$$

Due to the presence of an infinite group of roots of unity in an arbitrary number field k, the matching result for $\mathrm{GL}_2(\mathbb{A}_k)$ had to wait until the work of Blomer and Brumley [**8**]. In addition to extending the Kim-Sarnak bound to arbitrary k, they obtain the record bounds of $r_b \leq 5/14$ for $n = 3$ and $r_b \leq 9/22$ for $n = 4$.

Blomer and Brumley, build on the result of Kim-Sarnak, using methods from Analytic Number Theory to study the series

$$L^S(s, \pi, \mathrm{Sym}^2) = \sum_{\mathfrak{a}} \lambda_{\pi, \mathrm{Sym}^2}(\mathfrak{a}) \mathcal{N}(\mathfrak{a})^{-s},$$

where the sum ranges over ideals $\mathfrak{a} \subset \mathcal{O}_k$ with $(\mathfrak{a}, S) = 1$, and the $\lambda_{\pi,\mathrm{Sym}^2}(\mathfrak{a})$'s are Fourier coefficients. They work with Deligne's bounds for hyper-Kloosterman sums to bound test functions and employ a Vonoroi summation formula with S-adic computations.

THEOREM 7.3. *If $L(s, \pi, \mathrm{Sym}^2)$ converges absolutely for $\Re(s) > 1$, then*

$$r_b \leq \frac{1}{2} - \frac{1}{m+1}, \text{ for } m = \frac{n(n+1)}{2}.$$

We make a note, due to their appearence in several important places, concerning $L(s, \pi, \mathrm{Sym}^2)$ and the more general L-functions $L(s, \pi, \mathrm{Sym}^2 \boxtimes \chi)$.

7.3. On twisted symmetric and exterior square L-functions. In addition to their appearance in the bounds towards Ramanujan, twisted symmetric and exterior square L-functions also play an important role in determining the image of functoriality. We make note that taking a trivial character suffices for the classical groups, non-trivial χ make an appearance for general spin groups.

PROPOSITION 7.4. *Let τ be a cuspidal (unitary) automorphic representation of $\mathrm{GL}_n(\mathbb{A}_k)$ and χ a unitary Größencharakter. Let $r = \mathrm{Sym}^2$ or $\wedge^2$. Then $L(s, \tau, r \boxtimes \chi)$ is holomorphic except for possible poles appearing at $s = 0$ or $s = 1$; it is non-vanishing for $\Re(s) > 1$.*

Twisted symmetric and exterior square L-functions are available both via an integral representation and the Langlands-Shahidi method. However, the proof of this proposition was recently completed by Takeda [**104, 105**], who builds upon results of Bump and Ginzburg [**14**], using an integral representation for the partial L-functions $L^S(s, \tau, \mathrm{Sym}^2 \boxtimes \chi)$. It is instructive to recall that non-vanishing is obtained first for the Rankin-Selberg L-function. Indeed, Corollary 5.8 of [**54**] states that $L(s, \tau \times (\tau \cdot \chi))$, which is holomorphic and non-vanishing for $\Re(s) > 1$. We then use the relationship

$$L(s, \tau \times (\tau \cdot \chi)) = L(s, \tau, \mathrm{Sym}^2 \boxtimes \chi) L(s, \tau, \wedge^2 \boxtimes \chi).$$

This well known relationship is proved by a local-to-global argument, where we use [**94**] over number fields and Lemma 6.7 over function fields.

We observe that exterior square L-functions were studied by Jacquet-Shalika, who produce an integral representation in [**55**]. Furthermore, it is possible to show that the integral representation approach and that of the Langlands-Shahidi method are compatible. While Takeda's result is for partial L-functions, at the remaining possibly ramified places, local L-functions are non-zero and holomorphic for $\Re(s) > 1$. To see this, we use the tempered L-function conjecture, proved in [**46**], which gives holomorphy of local L-functions for $\Re(s) > 0$ if π is tempered, then, Langlands classification and the trivial Ramanujan bound $r_b \leq 1/2$ for GL_n.

7.4. Cases towards Ramanujan for GL_n**.** The globalization techinques discussed in § 6.2, give cases where the Ramanujan Conjecture holds. There is work of Shin [**98**] and Caraiani [**17**]; the series of papers by Clozel-Thorne, starting with [**21**], which have applications up to symmetric 8th powers; and, the current 10 author work in progress [**1**].

THEOREM 7.5. *If* Π *is a cuspidal unitary automorphic form of* $\mathrm{GL}_n(\mathbb{A}_k)$ *constructed as in* § 6.2 *and has corresponding Galois representation* Σ *of* $\mathcal{W}_k$. *Then* Π *satisfies the Ramanujan Conjecture.*

7.5. G a reductive group. If $\mathbf{G}$ is a connected reductive group over a global field k, then the Ramanujan conjecture no longer holds for all cuspidal automorphic representations π of $\mathbf{G}(\mathbb{A}_k)$. Counterexamples were produced by Howe and Piatetski-Shapiro for $\mathbf{G} = \mathrm{Sp}_4$ and U_3, via the θ-correspondence [**51**]. However, if one adds the genericity condition, then it is expected to be true.

CONJECTURE 7.6 (Generalized Ramanujan Conjecture). *Let* $\Pi = \otimes' \Pi_v$ *be a globally generic cuspidal (unitary) automorphic representation of* $\mathbf{G}(\mathbb{A}_k)$. *Then each* Π_v *is tempered. If* Π_v *is unramified, then its Satake parameters satisfy*

$$|\alpha_{j,v}| = 1, \quad 1 \leq j \leq n.$$

The globally generic lift from the classical groups to GL_N provides us with a way to establish new cases of the Generalized Ramanujan Conjecture over function fields thanks to the exact Ramanujan bounds for GL_N of L. Lafforgue. Over a number field, we can import the best known bounds for GL_N, discussed in the previous section, to the classical groups [**3, 23**]. Over function fields, we have the Ramanujan conjecture for the classical and unitary groups over funtion fields [**75, 77**].

THEOREM 7.7. *Let* $\mathbf{G}$ *be either* SO_{2n+1}, Sp_{2n} *or* SO_{2n} *defined over a function field* k. *Then, a globally generic cuspidal automorphic representation* Π *of* $\mathbf{G}_n(\mathbb{A}_k)$, *satisfies the Ramanujan Conjecture.*

For a general connected reductive group over a function field k, V. Lafforgue has established a very important automorphic to Galois lift, associating a representation Σ to a cuspidal automorphic representation Π of $\mathbf{G}(\mathbb{A}_k)$ [**66**]. This construction agrees at every unramified place with the Satake parametrization. We also note recent work of Sawin and Templier, who prove under certain conditions a general Ramanujan conjecture over function fields [**89**].

Acknowledgments

The author thanks the organizers of the reunions: Mexican Mathematicians in the World 2016 and 2018; and the institutions CIMAT, Guanajuato, and Banff, Casa México Oaxaca, where the conferences were held; a rich atmosphere for academic exchange was provided and the idea for this article came to be. He is grateful to IISER, Pune, and TIFR, Mumbay, for an invitation to conduct research in January-February 2017, where a first draft of this article was written. He would like to thank A.-M. Aubert, G. Henniart, L. Lafforgue, V. Lafforgue, M. Mishra, D. Prasad, F. Shahidi, S. Varma and A. Zenteno for mathematical communications and helpful discussions. Many thanks are due to the referee, for very uselful observations.

References

[1] P. Allen, F. Calegari, A. Caraiani, T. Gee, D. Helm, B. Le Hung, H. Newton, P. Scholze, R. Taylor, and J. Thorne, *Potential automorphy over CM fields*, preprint. `arXiv:1812.09999`.

[2] James Arthur, *The principle of functoriality*, Bull. Amer. Math. Soc. (N.S.) **40** (2003), no. 1, 39–53, DOI 10.1090/S0273-0979-02-00963-1. Mathematical challenges of the 21st century (Los Angeles, CA, 2000). MR1943132

[3] ———, *The Endoscopic Classification of Representations: Orthogonal and Symplectic Groups*, Colloquium Publications of the AMS, Vol. 61, 2013, Providence, RI.

[4] Anne-Marie Aubert, *Around the Langlands program*, Jahresber. Dtsch. Math.-Ver. **120** (2018), no. 1, 3–40, DOI 10.1365/s13291-017-0171-8. MR3740530

[5] Emil Artin and John Tate, *Class field theory*, AMS Chelsea Publishing, Providence, RI, 2009. Reprinted with corrections from the 1967 original, DOI 10.1090/chel/366. MR2467155

[6] Anne-Marie Aubert, Paul Baum, Roger Plymen, and Maarten Solleveld, *The local Langlands correspondence for inner forms of* SL_n, Res. Math. Sci. **3** (2016), Paper No. 32, 34, DOI 10.1186/s40687-016-0079-4. MR3579297

[7] D. Bump, J. W. Cogdell, E. de Shalit, D. Gaitsgory, E. Kowalski, and S. S. Kudla, *An introduction to the Langlands program*, Birkhäuser Boston, Inc., Boston, MA, 2003. Lectures presented at the Hebrew University of Jerusalem, Jerusalem, March 12–16, 2001; Edited by Joseph Bernstein and Stephen Gelbart. MR1990371

[8] Valentin Blomer and Farrell Brumley, *On the Ramanujan conjecture over number fields*, Ann. of Math. (2) **174** (2011), no. 1, 581–605, DOI 10.4007/annals.2011.174.1.18. MR2811610

[9] Valentin Blomer and Farrell Brumley, *The role of the Ramanujan conjecture in analytic number theory*, Bull. Amer. Math. Soc. (N.S.) **50** (2013), no. 2, 267–320, DOI 10.1090/S0273-0979-2013-01404-6. MR3020828

[10] A. Borel, *Automorphic L-functions*, Automorphic forms, representations and L-functions (Proc. Sympos. Pure Math., Oregon State Univ., Corvallis, Ore., 1977), Proc. Sympos. Pure Math., XXXIII, Amer. Math. Soc., Providence, R.I., 1979, pp. 27–61. MR546608

[11] A. Borel, *Automorphic L-functions*, Automorphic forms, representations and L-functions (Proc. Sympos. Pure Math., Oregon State Univ., Corvallis, Ore., 1977), Proc. Sympos. Pure Math., XXXIII, Amer. Math. Soc., Providence, R.I., 1979, pp. 27–61. MR546608

[12] A. Borel and N. Wallach, *Continuous cohomology, discrete subgroups, and representations of reductive groups*, 2nd ed., Mathematical Surveys and Monographs, vol. 67, American Mathematical Society, Providence, RI, 2000, DOI 10.1090/surv/067. MR1721403

[13] N. Bourbaki, *Éléments de mathématique. I: Les structures fondamentales de l'analyse. Fascicule XI. Livre II: Algèbre. Chapitre 4: Polynomes et fractions rationnelles. Chapitre 5: Corps commutatifs* (French), Actualités Scientifiques et Industrielles [Current Scientific and Industrial Topics], No. 1102, Hermann, Paris, 1959. Deuxième édition. MR0174550

[14] Daniel Bump and David Ginzburg, *Symmetric square L-functions on* $\mathrm{GL}(r)$, Ann. of Math. (2) **136** (1992), no. 1, 137–205, DOI 10.2307/2946548. MR1173928

[15] Colin J. Bushnell and Guy Henniart, *The local Langlands conjecture for* $GL(2)$, Grundlehren der Mathematischen Wissenschaften [Fundamental Principles of Mathematical Sciences], vol. 335, Springer-Verlag, Berlin, 2006, DOI 10.1007/3-540-31511-X. MR2234120

[16] Colin J. Bushnell and Philip C. Kutzko, *The admissible dual of* $\mathrm{GL}(N)$ *via compact open subgroups*, Annals of Mathematics Studies, vol. 129, Princeton University Press, Princeton, NJ, 1993, DOI 10.1515/9781400882496. MR1204652

[17] Ana Caraiani, *Local-global compatibility and the action of monodromy on nearby cycles*, Duke Math. J. **161** (2012), no. 12, 2311–2413, DOI 10.1215/00127094-1723706. MR2972460

[18] P. Cartier, *Representations of p-adic groups: a survey*, Automorphic forms, representations and L-functions (Proc. Sympos. Pure Math., Oregon State Univ., Corvallis, Ore., 1977), Proc. Sympos. Pure Math., XXXIII, Amer. Math. Soc., Providence, R.I., 1979, pp. 111–155. MR546593

[19] W. Casselman, *Introduction to the theory of admissible representations of p-adic reductive groups*, unpublished notes.
[20] J. W. S. Cassels and A. Fröhlich (eds.), *Algebraic number theory*, 2nd ed., London Mathematical Society, London, 2010. Papers from the conference held at the University of Sussex, Brighton, September 1–17, 1965; Including a list of errata. MR3618860
[21] Laurent Clozel and Jack A. Thorne, *Level-raising and symmetric power functoriality, I*, Compos. Math. **150** (2014), no. 5, 729–748, DOI 10.1112/S0010437X13007653. MR3209793
[22] J. W. Cogdell, H. H. Kim, I. I. Piatetski-Shapiro, and F. Shahidi, *On lifting from classical groups to* GL_N, Publ. Math. Inst. Hautes Études Sci. **93** (2001), 5–30, DOI 10.1007/s10240-001-8187-z. MR1863734
[23] J. W. Cogdell, H. H. Kim, I. I. Piatetski-Shapiro, and F. Shahidi, *Functoriality for the classical groups*, Publ. Math. Inst. Hautes Études Sci. **99** (2004), 163–233, DOI 10.1007/s10240-004-0020-z. MR2075885
[24] J. W. Cogdell and I. I. Piatetski-Shapiro, *Converse theorems for* GL_n, Inst. Hautes Études Sci. Publ. Math. **79** (1994), 157–214. MR1307299
[25] ———, *Remarks on Rankin-Selberg convolutions*, (published in [H. Hida, D. Ramakrishnan and F. Shahidi, editors, Contributions to Automorphic Forms, Geometry and Number Theory, John Hopkins University Press, 2004, 255-278]).
[26] B. Conrad, *Weil and Grothendieck approaches to adelic points*, notes.
[27] P. Deligne, *Les constantes des équations fonctionnelles des fonctions L* (French), Modular functions of one variable, II (Proc. Internat. Summer School, Univ. Antwerp, Antwerp, 1972), Springer, Berlin, 1973, pp. 501–597. Lecture Notes in Math., Vol. 349. MR0349635
[28] ———, *La conjecture de Weil. I & II*, Pub. Math. IHÉS **43** (1974), 273-307, & **52** (1980), 137-252.
[29] *Schémas en groupes. III: Structure des schémas en groupes réductifs* (French), Lecture Notes in Mathematics, Vol. 153, Springer-Verlag, Berlin-New York, 1970. Séminaire de Géométrie Algébrique du Bois Marie 1962/64 (SGA 3); Dirigé par M. Demazure et A. Grothendieck. MR0274460
[30] V. G. Drinfel′d, *Langlands' conjecture for* GL(2) *over functional fields*, Proceedings of the International Congress of Mathematicians (Helsinki, 1978), Acad. Sci. Fennica, Helsinki, 1980, pp. 565–574. MR562656
[31] ———, *The proof of Petersson's conjecture for* GL(2) *over a global field of characteristic p*, Functional Analysis and its Applications **22** (1988), 23-43.
[32] D. Flath, *Decomposition of representations into tensor products*, Automorphic forms, representations and L-functions (Proc. Sympos. Pure Math., Oregon State Univ., Corvallis, Ore., 1977), Proc. Sympos. Pure Math., XXXIII, Amer. Math. Soc., Providence, R.I., 1979, pp. 179–183. MR546596
[33] Edward Frenkel, *Langlands program, trace formulas, and their geometrization*, Bull. Amer. Math. Soc. (N.S.) **50** (2013), no. 1, 1–55, DOI 10.1090/S0273-0979-2012-01387-3. MR2994994
[34] Wee Teck Gan and Luis Lomelí, *Globalization of supercuspidal representations over function fields and applications*, J. Eur. Math. Soc. (JEMS) **20** (2018), no. 11, 2813–2858, DOI 10.4171/JEMS/825. MR3861809
[35] Radhika Ganapathy, *The local Langlands correspondence for* GSp_4 *over local function fields*, Amer. J. Math. **137** (2015), no. 6, 1441–1534, DOI 10.1353/ajm.2015.0045. MR3432266
[36] Radhika Ganapathy and Sandeep Varma, *On the local Langlands correspondence for split classical groups over local function fields*, J. Inst. Math. Jussieu **16** (2017), no. 5, 987–1074, DOI 10.1017/S147474801500033X. MR3709003
[37] Stephen Gelbart, *An elementary introduction to the Langlands program*, Bull. Amer. Math. Soc. (N.S.) **10** (1984), no. 2, 177–219, DOI 10.1090/S0273-0979-1984-15237-6. MR733692
[38] Vincent Lafforgue, *Chtoucas pour les groupes réductifs et paramétrisation de Langlands globale* (French), J. Amer. Math. Soc. **31** (2018), no. 3, 719–891, DOI 10.1090/jams/897. MR3787407
[39] Roger Godement and Hervé Jacquet, *Zeta functions of simple algebras*, Lecture Notes in Mathematics, Vol. 260, Springer-Verlag, Berlin-New York, 1972. MR0342495
[40] Thomas J. Haines, *On Satake parameters for representations with parahoric fixed vectors*, Int. Math. Res. Not. IMRN **20** (2015), 10367–10398, DOI 10.1093/imrn/rnu254. MR3455870

[41] G. Harder, *Chevalley groups over function fields and automorphic forms*, Ann. of Math. (2) **100** (1974), 249–306, DOI 10.2307/1971073. MR563090

[42] Michael Harris, *Supercuspidal representations in the cohomology of Drinfel'd upper half spaces; elaboration of Carayol's program*, Invent. Math. **129** (1997), no. 1, 75–119, DOI 10.1007/s002220050159. MR1464867

[43] ———, *Incorrigible representations*, preprint. `arXiv:1811.05050`.

[44] Michael Harris and Richard Taylor, *The geometry and cohomology of some simple Shimura varieties*, Annals of Mathematics Studies, vol. 151, Princeton University Press, Princeton, NJ, 2001. With an appendix by Vladimir G. Berkovich. MR1876802

[45] Erich Hecke, *Mathematische Werke* (German), Vandenhoeck & Ruprecht, Göttingen, 1959. Herausgegeben im Auftrage der Akademie der Wissenschaften zu Göttingen. MR0104550

[46] Volker Heiermann and Eric Opdam, *On the tempered L-functions conjecture*, Amer. J. Math. **135** (2013), no. 3, 777–799, DOI 10.1353/ajm.2013.0026. MR3068402

[47] Guy Henniart, *La conjecture de Langlands locale pour* GL(3) (French, with English summary), Mém. Soc. Math. France (N.S.) **11-12** (1984), 186. MR743063

[48] Guy Henniart, *Caractérisation de la correspondance de Langlands locale par les facteurs ϵ de paires* (French, with English and French summaries), Invent. Math. **113** (1993), no. 2, 339–350, DOI 10.1007/BF01244309. MR1228128

[49] ———, *Une preuve simple des conjectures de Langlands pour* $\mathrm{GL}(n)$ *sur un corps p-adique*, Invent. Math. **139** (2000), no. 2, 439–456.

[50] Guy Henniart and Luis Lomelí, *Uniqueness of Rankin-Selberg products*, J. Number Theory **133** (2013), no. 12, 4024–4035, DOI 10.1016/j.jnt.2013.05.015. MR3165629

[51] R. Howe and I. I. Piatetski-Shapiro, *A counterexample to the "generalized Ramanujan conjecture" for (quasi-) split groups*, Automorphic forms, representations and L-functions (Proc. Sympos. Pure Math., Oregon State Univ., Corvallis, Ore., 1977), Proc. Sympos. Pure Math., XXXIII, Amer. Math. Soc., Providence, R.I., 1979, pp. 315–322. MR546605

[52] H. Jacquet and R. P. Langlands, *Automorphic forms on* GL(2), Lecture Notes in Mathematics, Vol. 114, Springer-Verlag, Berlin-New York, 1970. MR0401654

[53] H. Jacquet, I. I. Piatetskii-Shapiro, and J. A. Shalika, *Rankin-Selberg convolutions*, Amer. J. Math. **105** (1983), no. 2, 367–464, DOI 10.2307/2374264. MR701565

[54] H. Jacquet and J. A. Shalika, *On Euler products and the classification of automorphic representations. I*, Amer. J. Math. **103** (1981), no. 3, 499–558, DOI 10.2307/2374103. MR618323

[55] Hervé Jacquet and Joseph Shalika, *Exterior square L-functions*, Automorphic forms, Shimura varieties, and L-functions, Vol. II (Ann Arbor, MI, 1988), Perspect. Math., vol. 11, Academic Press, Boston, MA, 1990, pp. 143–226. MR1044830

[56] Tasho Kaletha, *Rigid inner forms of real and p-adic groups*, Ann. of Math. (2) **184** (2016), no. 2, 559–632, DOI 10.4007/annals.2016.184.2.6. MR3548533

[57] Nicholas M. Katz, *Local-to-global extensions of representations of fundamental groups* (English, with French summary), Ann. Inst. Fourier (Grenoble) **36** (1986), no. 4, 69–106. MR867916

[58] C. David Keys and Freydoon Shahidi, *Artin L-functions and normalization of intertwining operators*, Ann. Sci. École Norm. Sup. (4) **21** (1988), no. 1, 67–89. MR944102

[59] Henry H. Kim, *Functoriality for the exterior square of* GL_4 *and the symmetric fourth of* GL_2, J. Amer. Math. Soc. **16** (2003), no. 1, 139–183, DOI 10.1090/S0894-0347-02-00410-1. With appendix 1 by Dinakar Ramakrishnan and appendix 2 by Kim and Peter Sarnak. MR1937203

[60] Henry H. Kim, *Functoriality for the exterior square of* GL_4 *and the symmetric fourth of* GL_2, J. Amer. Math. Soc. **16** (2003), no. 1, 139–183, DOI 10.1090/S0894-0347-02-00410-1. With appendix 1 by Dinakar Ramakrishnan and appendix 2 by Kim and Peter Sarnak. MR1937203

[61] Henry H. Kim and Freydoon Shahidi, *Functorial products for* $\mathrm{GL}_2 \times \mathrm{GL}_3$ *and the symmetric cube for* GL_2, Ann. of Math. (2) **155** (2002), no. 3, 837–893, DOI 10.2307/3062134. With an appendix by Colin J. Bushnell and Guy Henniart. MR1923967

[62] Ju-Lee Kim, *Supercuspidal representations: an exhaustion theorem*, J. Amer. Math. Soc. **20** (2007), no. 2, 273–320, DOI 10.1090/S0894-0347-06-00544-3. MR2276772

[63] Max-Albert Knus, *Quadratic and Hermitian forms over rings*, Grundlehren der Mathematischen Wissenschaften [Fundamental Principles of Mathematical Sciences], vol. 294, Springer-Verlag, Berlin, 1991. With a foreword by I. Bertuccioni, DOI 10.1007/978-3-642-75401-2. MR1096299

[64] Philip Kutzko, *The Langlands conjecture for* Gl_2 *of a local field*, Ann. of Math. (2) **112** (1980), no. 2, 381–412, DOI 10.2307/1971151. MR592296

[65] Laurent Lafforgue, *Chtoucas de Drinfeld et correspondance de Langlands* (French, with English and French summaries), Invent. Math. **147** (2002), no. 1, 1–241, DOI 10.1007/s002220100174. MR1875184

[66] Vincent Lafforgue, *Chtoucas pour les groupes réductifs et paramétrisation de Langlands globale* (French), J. Amer. Math. Soc. **31** (2018), no. 3, 719–891, DOI 10.1090/jams/897. MR3787407

[67] Robert P. Langlands, *Euler products*, Yale University Press, New Haven, Conn.-London, 1971. A James K. Whittemore Lecture in Mathematics given at Yale University, 1967; Yale Mathematical Monographs, 1. MR0419366

[68] ———, *On the functional equations satisfied by Eisenstein series*, Lecture Notes Math. **544**, Springer, Berlin, 1976.

[69] ———, *On the notion of an automorphic representation*, Proc. Symp. Pure Math., **33**, part 1 (1979), 203-207.

[70] ———, *Automorphic representations, Shimura varieties, and motives. Ein Märchen*, Proc. Symp. Pure Math., **33**, part 2 (1979), 205-246.

[71] ———, *On the functional equation of Artin's L-functions*, mimeographed notes, Yale University.

[72] ———, *Base Change for* GL(2), Ann. Math. Studies, vol. 96, Princeton Univ. Press, New Jersey, 1980.

[73] Erez Lapid, Goran Muić, and Marko Tadić, *On the generic unitary dual of quasisplit classical groups*, Int. Math. Res. Not. **26** (2004), 1335–1354, DOI 10.1155/S1073792804132832. MR2046512

[74] G. Laumon, M. Rapoport, and U. Stuhler, *$\mathcal{D}$-elliptic sheaves and the Langlands correspondence*, Invent. Math. **113** (1993), no. 2, 217–338, DOI 10.1007/BF01244308. MR1228127

[75] Luis Alberto Lomelí, *Functoriality for the classical groups over function fields*, Int. Math. Res. Not. IMRN **22** (2009), 4271–4335, DOI 10.1093/imrn/rnp089. MR2552304

[76] Luis Alberto Lomelí, *Rationality and holomorphy of Langlands-Shahidi L-functions over function fields*, Math. Z. **291** (2019), no. 1-2, 711–739, DOI 10.1007/s00209-018-2100-7. MR3936087

[77] ———, *The Langlands-Shahidi method over function fields*, `arXiv:1507.03625`.

[78] Wenzhi Luo, Zeév Rudnick, and Peter Sarnak, *On the generalized Ramanujan conjecture for* GL(n), Automorphic forms, automorphic representations, and arithmetic (Fort Worth, TX, 1996), Proc. Sympos. Pure Math., vol. 66, Amer. Math. Soc., Providence, RI, 1999, pp. 301–310. MR1703764

[79] F. I. Mautner, *Spherical functions over $\mathfrak{P}$-adic fields. I*, Amer. J. Math. **80** (1958), 441–457, DOI 10.2307/2372794. MR93558

[80] F. I. Mautner, *Spherical functions over $\mathfrak{b}$-adic fields. II*, Amer. J. Math. **86** (1964), 171–200, DOI 10.2307/2373039. MR166305

[81] J. S. Milne, *Algebraic groups*, Cambridge Studies in Advanced Mathematics, vol. 170, Cambridge University Press, Cambridge, 2017. The theory of group schemes of finite type over a field, DOI 10.1017/9781316711736. MR3729270

[82] Colette Mœglin and Jean-Loup Waldspurger, *Décomposition spectrale et séries d'Eisenstein* (French, with English summary), Progress in Mathematics, vol. 113, Birkhäuser Verlag, Basel, 1994. Une paraphrase de l'Écriture. [A paraphrase of Scripture]. MR1261867

[83] ———, *Stabilisation de la formule des traces tordue X: stabilisation spectrale*, preprint. `arXiv:1412.2981`.

[84] Chung Pang Mok, *Endoscopic classification of representations of quasi-split unitary groups*, Mem. Amer. Math. Soc. **235** (2015), no. 1108, vi+248, DOI 10.1090/memo/1108. MR3338302

[85] L. E. Morris, *Eisenstein series for reductive groups over global function fields. II. The general case*, Canadian J. Math. **34** (1982), no. 5, 1112–1182, DOI 10.4153/CJM-1982-080-6. MR675683

[86] Dinakar Ramakrishnan and Robert J. Valenza, *Fourier analysis on number fields*, Graduate Texts in Mathematics, vol. 186, Springer-Verlag, New York, 1999, DOI 10.1007/978-1-4757-3085-2. MR1680912

[87] Peter Sarnak, *Notes on the generalized Ramanujan conjectures*, Harmonic analysis, the trace formula, and Shimura varieties, Clay Math. Proc., vol. 4, Amer. Math. Soc., Providence, RI, 2005, pp. 659–685. MR2192019

[88] Ichirô Satake, *Theory of spherical functions on reductive algebraic groups over $\mathfrak{p}$-adic fields*, Inst. Hautes Études Sci. Publ. Math. **18** (1963), 5–69. MR195863

[89] W. Sawin and N. Templier, *On the Ramanujan conjecture for automorphic forms over function fields I. Geometry*, preprint. arXiv:1805.12231.

[90] Peter Scholze, *The local Langlands correspondence for* GL_n *over p-adic fields*, Invent. Math. **192** (2013), no. 3, 663–715, DOI 10.1007/s00222-012-0420-5. MR3049932

[91] Peter Scholze and Sug Woo Shin, *On the cohomology of compact unitary group Shimura varieties at ramified split places*, J. Amer. Math. Soc. **26** (2013), no. 1, 261–294, DOI 10.1090/S0894-0347-2012-00752-8. MR2983012

[92] Jean-Pierre Serre, *Cours d'arithmétique* (French), Collection SUP: "Le Mathématicien", vol. 2, Presses Universitaires de France, Paris, 1970. MR0255476

[93] ———, *Cohomologie Galoisienne*, 5ème éd. révisée, Lecture Notes Math. **5**, Springer-Verlag, Berlin, 1997.

[94] Freydoon Shahidi, *Eisenstein series and automorphic L-functions*, American Mathematical Society Colloquium Publications, vol. 58, American Mathematical Society, Providence, RI, 2010, DOI 10.1090/coll/058. MR2683009

[95] ———, *On the Ramanujan Conjecture for quasisplit groups*, Asian J. Math **8** (2004), 813-836.

[96] Freydoon Shahidi, *Local coefficients as Artin factors for real groups*, Duke Math. J. **52** (1985), no. 4, 973–1007, DOI 10.1215/S0012-7094-85-05252-4. MR816396

[97] J. A. Shalika, *The multiplicity one theorem for* GL_n, Ann. of Math. (2) **100** (1974), 171–193, DOI 10.2307/1971071. MR348047

[98] Sug Woo Shin, *Galois representations arising from some compact Shimura varieties*, Ann. of Math. (2) **173** (2011), no. 3, 1645–1741, DOI 10.4007/annals.2011.173.3.9. MR2800722

[99] Allan J. Silberger, *The Langlands quotient theorem for p-adic groups*, Math. Ann. **236** (1978), no. 2, 95–104, DOI 10.1007/BF01351383. MR507262

[100] Allan J. Silberger, *Introduction to harmonic analysis on reductive p-adic groups*, Mathematical Notes, vol. 23, Princeton University Press, Princeton, N.J.; University of Tokyo Press, Tokyo, 1979. Based on lectures by Harish-Chandra at the Institute for Advanced Study, 1971–1973. MR544991

[101] David Soudry, *On Langlands functoriality from classical groups to* GL_n (English, with English and French summaries), Astérisque **298** (2005), 335–390. Automorphic forms. I. MR2141707

[102] T. A. Springer, *Linear algebraic groups*, 2nd ed., Progress in Mathematics, vol. 9, Birkhäuser Boston, Inc., Boston, MA, 1998, DOI 10.1007/978-0-8176-4840-4. MR1642713

[103] Shaun Stevens, *The supercuspidal representations of p-adic classical groups*, Invent. Math. **172** (2008), no. 2, 289–352, DOI 10.1007/s00222-007-0099-1. MR2390287

[104] Shuichiro Takeda, *The twisted symmetric square L-function of* $\mathrm{GL}(r)$, Duke Math. J. **163** (2014), no. 1, 175–266, DOI 10.1215/00127094-2405497. MR3161314

[105] Shuichiro Takeda, *On a certain metaplectic Eisenstein series and the twisted symmetric square L-function*, Math. Z. **281** (2015), no. 1-2, 103–157, DOI 10.1007/s00209-015-1476-x. MR3384864

[106] J. T. Tate, *Fourier analysis in number fields, and Hecke's zeta-functions*, Algebraic Number Theory (Proc. Instructional Conf., Brighton, 1965), Thompson, Washington, D.C., 1967, pp. 305–347. MR0217026

[107] J. Tate, *Number theoretic background*, Automorphic forms, representations and L-functions (Proc. Sympos. Pure Math., Oregon State Univ., Corvallis, Ore., 1977), Proc. Sympos. Pure Math., XXXIII, Amer. Math. Soc., Providence, R.I., 1979, pp. 3–26. MR546607

[108] J. Tits, *Reductive groups over local fields*, Automorphic forms, representations and L-functions (Proc. Sympos. Pure Math., Oregon State Univ., Corvallis, Ore., 1977), Proc. Sympos. Pure Math., XXXIII, Amer. Math. Soc., Providence, R.I., 1979, pp. 29–69. MR546588

[109] Jerrold B. Tunnell, *On the local Langlands conjecture for $GL(2)$*, Invent. Math. **46** (1978), no. 2, 179–200, DOI 10.1007/BF01393255. MR476703

[110] J.-L. Waldspurger, *La formule de Plancherel pour les groupes p-adiques (d'après Harish-Chandra)* (French, with French summary), J. Inst. Math. Jussieu **2** (2003), no. 2, 235–333, DOI 10.1017/S1474748003000082. MR1989693

[111] Jiu-Kang Yu, *Construction of tame supercuspidal representations*, J. Amer. Math. Soc. **14** (2001), no. 3, 579–622, DOI 10.1090/S0894-0347-01-00363-0. MR1824988

[112] André Weil, *Sur la théorie du corps de classes* (French), J. Math. Soc. Japan **3** (1951), 1–35, DOI 10.2969/jmsj/00310001. MR44569

[113] ———, *Basic Number Theory*, 3rd ed., Springer-Verlag, 1974.

[114] ———, *Adeles and Algebraic Groups*, Progress in Math. **23**, Birkhäuser, Boston, 1982.

[115] A. V. Zelevinsky, *Induced representations of reductive $\mathfrak{p}$-adic groups. II. On irreducible representations of* $\mathrm{GL}(n)$, Ann. Sci. École Norm. Sup. (4) **13** (1980), no. 2, 165–210. MR584084

Instituto de Matemáticas, Pontificia Universidad Católica de Valparaíso, Blanco Viel 596, Cerro Barón, Valparaíso, Chile

Email address: luis.lomeli@pucv.cl

Contemporary Mathematics
Volume **775**, 2021
https://doi.org/10.1090/conm/775/15593

The least quadratic non-residue

Kevin McGown and Enrique Treviño

ABSTRACT. For a prime number p, we say a is a quadratic non-residue modulo p if there is no integer x such that $x^2 \equiv a \bmod p$. The problem of bounding the least quadratic non-residue modulo p has a rich mathematical history. Moreover, there have been recent results, especially concerning explicit estimates. In this survey paper we give the history of the problem and explain many of the main achievements, giving explicit versions of these results in most cases. The paper is intended as a self-contained collection of the main ideas that have been used to attack the problem.

1. Introduction

The sequence 2, 5, 8, 11, ... contains no perfect squares. This is for the simple reason that $x^2 \equiv 0, 1 \bmod 3$. We say that 2 is a quadratic non-residue modulo 3. In general we say a is a *quadratic non-residue* modulo n if there is no x such that $\gcd(x,n) = 1$ and $x^2 \equiv a \bmod n$. If such an x exists and $\gcd(x,n) = 1$, then we say a is a *quadratic residue.* We will concentrate on the case where $n = p$ an odd prime number. (Throughout p will always denote an odd prime.) The Legendre symbol is defined as

$$\left(\frac{a}{p}\right) = \begin{cases} 1 & \text{if there exists } x \not\equiv 0 \bmod p \text{ such that } x^2 \equiv a \bmod p \\ -1 & \text{if there is no } x \text{ such that } x^2 \equiv a \bmod p \\ 0 & \text{if } x \equiv 0 \bmod p. \end{cases}$$

Notice that a is a quadratic non-residue modulo p if and only if $\left(\frac{a}{p}\right) = -1$.

Our main object of study in this paper is the following question: How large is the least quadratic non-residue modulo p? We will denote the least quadratic non-residue n_p. Our goal is to provide (in one place), the proofs of the main "classical" results on this problem and to survey the more recent explicit results. In section 2 we will describe the history and some heuristics on the size of n_p. In section 3 we will prove the Pólya–Vinogradov theorem. In section 4 we will give a clear and relatively short proof of an explicit Burgess inequality. In section 5 we will mention recent explicit results and we will include a couple of new theorems (Theorems 5.4 and 5.6) to showcase the techniques in the paper. In section 6, we will discuss a classical result of Linnik on n_p for most primes p. Finally, in section 7 we will prove an explicit version of a theorem of Ankeny under the Generalized Riemann Hypothesis.

2. History and heuristics

Let's first try to get some intuition on how big n_p is. First note that n_p is always prime. This is because if $n_p = ab$ for $1 < a, b < n_p$, then since n_p is the least quadratic non-residue, that means a, b are quadratic residues, but then $m_1^2 \equiv a \bmod p$ and $m_2^2 \equiv b \bmod p$, so $(m_1 m_2)^2 \equiv ab = n_p \bmod p$. But that means n_p is a quadratic residue. Therefore, n_p is prime. Now, let's try to see how often $n_p = 2$ (the smallest prime). A famous result of Gauss is that 2 is a quadratic non-residue whenever $p \equiv 3, 5 \bmod 8$ and 2 is a quadratic residue whenever $p \equiv 1, 7 \bmod 8$. Since primes are equidistributed among the coprime residue classes (this is due to Dirichlet's theorem on primes in arithmetic progressions), then $n_p = 2$ half of the time. In other words if we let x be a real number and $\pi(x)$ is the number of primes less than or equal to x, then

$$\lim_{x\to\infty} \frac{|\{p \le x \,|\, n_p = 2\}|}{\pi(x)} = \frac{2}{4} = \frac{1}{2}.$$

Using quadratic reciprocity, we can see that $n_p = 3$, whenever $p \equiv 7, 17 \bmod 24$. There are 8 residue classes coprime with 24 and each of them has the same number of primes asymptotically (by Dirichlet's Theorem), so

$$\lim_{x\to\infty} \frac{|\{p \le x \,|\, n_p = 3\}|}{\pi(x)} = \frac{2}{8} = \frac{1}{4}.$$

Let $p_1 = 2, p_2 = 3, \ldots, p_k =$ the k-th prime, one can prove (using quadratic reciprocity, the Chinese remainder theorem, and Dirichlet's theorem) that

$$\lim_{x\to\infty} \frac{|\{p \le x \,|\, n_p = p_k\}|}{\pi(x)} = \frac{1}{2^k}.$$

The above suggests the following heuristic: Suppose that $k \approx \log_2 x$, then $\frac{1}{2^k} \approx \frac{1}{x}$. Therefore, the "expected number" of primes with $n_p = p_k$ is 1. If k is much larger, then we wouldn't expect any, and if k is smaller we would expect some. Therefore, it seems that n_p can be as large as $p_{\log_2 x}$ which is (by the Prime Number Theorem) asymptotically $\log_2 x \log\log x$. Table 1 contains the list of smallest primes p satisfying $n_p = p_k$, it also compares to $\log p \log\log p$ to see how good an estimate it is.[1]

The heuristic above suggests that $n_p \le C \log p \log\log p$ for some constant C. It also suggests that $n_p \ge C \log p \log\log p$ for infinitely many p (and some constant C). Evidence in the direction of these conjectures is that Graham and Ringrose [**14**], in 1990, proved that $n_p \ge C \log p \log\log\log p$ for infinitely many p and some constant C.[2] In fact, assuming the Generalized Riemann Hypothesis (GRH), Montgomery [**33**, Theorem 13.5], in 1971, proved precisely the lower bound heuristic, namely that $n_p \ge C \log p \log\log p$ for infinitely many p and some constant C. For the upper bound, assuming GRH, Ankeny [**1**], in 1952, proved $n_p \le C \log^2 p$ for some constant C. Bach [**2**], in 1985, made the constant explicit, showing $n_p \le 2\log^2 p$. Finally in 2015, Lamzouri, Li and Soundararajan [**21**] improved this to $n_p \le \log^2 p$. However, we cannot prove anything remotely close to that upper bound unconditionally. The

[1] Salié in 1965 [**41**], using a table just like the first two columns of Table 1 (but with one extra row), disproved two conjectures: that the first column is increasing and that all of its elements greater than 3 are 7 mod 8.

[2] In 1949, Fridlender [**12**] and Salié [**40**], independently proved that $n_p \ge C \log p$ for some constant C.

p	n_p	$\log p \log\log p$
3	2	0.10
7	3	1.30
23	5	3.58
71	7	6.18
311	11	10.03
479	13	11.23
1559	17	14.67
5711	19	18.66
10559	23	20.63
18191	29	22.40
31391	31	24.20
422231	37	33.18
701399	41	40.00
366791	43	32.68
3818929	47	41.20

TABLE 1. The smallest primes p satisfying that $n_p = p_k$ for $k = 1, 2, \ldots, 15$, and a comparison with $\log p \log\log p$. It is worth noting that not all of the p in the table are 7 mod 8.

first breakthrough in getting a good upper bound for the least quadratic non-residue was done by Vinogradov in 1918 (see pages 53–57 of [**50**]). He came up with the following strategy using Dirichlet characters. For a positive integer q, a function $\chi : \mathbb{Z} \to \mathbb{C}$ is called a Dirichlet character modulo q if

(1) $\chi(nm) = \chi(n)\chi(m)$ for all $n, m \in \mathbb{Z}$, i.e., χ is completely multiplicative;
(2) $\chi(n + kq) = \chi(n)$ for all $n, k \in \mathbb{Z}$, i.e., χ is periodic modulo q;
(3) $\chi(n) = 0$ if and only if $(n, q) > 1$.

χ is called a Dirichlet character modulo q in honor of Dirichlet's introduction of such functions in 1839 for his glorious proof of his namesake theorem regarding primes in arithmetic progressions.

Vinogradov then proved what is now known as the Pólya–Vinogradov inequality (Pólya discovered this independently), which is

THEOREM 2.1 (Pólya–Vinogradov inequality). *Let χ be a non-principal*[3] *Dirichlet chararcter modulo q. Let N be any positive integer, then there exists an absolute constant C such that*

$$\left|\sum_{n=1}^{N} \chi(n)\right| \le C\sqrt{q}\log(q).$$

REMARK 2.2. Vinogradov's proof works with $C = 1$ when q is prime. The best current estimates for C can be found in [**13**] and [**16**].

Now, the Legendre symbol is a Dirichlet character modulo p (for any odd p), and it's not the principal character (since there are $\frac{p-1}{2} > 0$ quadratic non-residues

[3]The principal Dirichlet character modulo q is the one defined as $\chi(n) = 1$ when $\gcd(n, q) = 1$ and $\chi(n) = 0$ otherwise.

modulo p for any odd prime p). Therefore, if $p > N > C\sqrt{p}\log(p)$, then

$$\sum_{n=1}^{N}\left(\frac{n}{p}\right) \leq C\sqrt{p}\log p < N.$$

But if the sum is less than N, that means it cannot consist of only 1's. Therefore, one of the values is -1 (since $p > N$, none of the values is 0). This means that (for large enough p)

$$n_p \leq C\sqrt{p}\log p.$$

REMARK 2.3. There is an elementary proof that $n_p < \sqrt{p}+1$. The proof is as follows. Suppose that $n_p = q \geq \sqrt{p}+1$. Let $k = \lceil \frac{p}{q} \rceil$. Since $q \geq \sqrt{p}+1$, then $\frac{p}{q} < \sqrt{p}$, so $\lceil \frac{p}{q} \rceil < \frac{p}{q}+1 < \sqrt{p}+1$. Therefore $k < q$, which implies k is a quadratic residue. Furthermore, $p < kq < p+q$, therefore kq is also a quadratic residue. But since q is a quadratic non-residue, and k is a quadratic residue, then kq is a quadratic non-residue. Contradiction!

Vinogradov then came up with a very clever idea which is now known as Vinogradov's trick. To understand the following we will need a little bit of notation. We say that the function $f(n)$ is *little oh* of the function $g(n)$, and write $f(n) = o(g(n))$ if $\lim_{n\to\infty} f(n)/g(n) = 0$.

THEOREM 2.4 (Vinogradov's trick). *Let $\chi(n) = \left(\frac{n}{p}\right)$ be the Legendre symbol of n modulo p for p prime. Suppose that x is a large real number such that*

$$\sum_{1\leq n\leq x} \chi(n) = o(x). \tag{1}$$

Let $\varepsilon > 0$ and let $y = x^{\frac{1}{\sqrt{e}}+\varepsilon}$. Then, for large enough x, there exists $n \leq y$ such that $\chi(n) = -1$.

PROOF. We may assume $x < p$ since the sum is 0 for $x = p$ and χ is periodic modulo p. Observe that since χ is totally multiplicative, then $\chi(n) = -1$ implies n has a prime divisor q satisfying $\chi(q) = -1$. Now suppose that for all $q \leq y$, $\chi(q) = 1$. Then

$$\sum_{1\leq n\leq x} \chi(n) = \sum_{1\leq n\leq x} 1 - 2\sum_{\substack{1\leq n\leq x\\ \chi(n)=-1}} 1 \geq \lfloor x \rfloor - 2\sum_{\substack{y<q\leq x\\ \chi(q)=-1}}\sum_{n\leq\frac{x}{q}} 1.$$

Therefore

$$\sum_{n\leq x}\chi(n) \geq \lfloor x\rfloor - 2\sum_{y<q\leq x}\left\lfloor\frac{x}{q}\right\rfloor \geq x - 1 - 2x\sum_{y<q\leq x}\frac{1}{q}.$$

Merten's theorem says that

$$\sum_{y<q\leq x}\frac{1}{q} = \log\log x - \log\log y + O\left(\frac{1}{\log y}\right)$$

and therefore there is a $\delta > 0$ such that for large enough x we have

$$\sum_{y<q\leq x}\frac{1}{q} \leq \log\log x - \log\log y + \delta = -\log\left(\frac{1}{\sqrt{e}}+\varepsilon\right) + \delta < \frac{1}{2}.$$

Therefore, there exists a constant $C > 0$ such that $\sum_{n\leq x}\chi(n) \geq Cx$, which contradicts (1). □

Translating this to our interests, from the Pólya–Vinogradov inequality we have that for $x = \sqrt{p}(\log p)^2$, equation (1) is satisfied. But then that means that for any $\varepsilon > 0$, for a large enough p (depending on ε),

$$n_p \leq p^{\frac{1}{2\sqrt{e}}+\varepsilon}.$$

One interesting thing about the Pólya–Vinogradov inequality is that it does not depend on the number of summands, the upper bound of $\sqrt{p}\log p$ works regardless of how many terms N are summed. One heuristic for how Dirichlet characters behave is to model them as if they were uniformly random on the unit circle. The Central Limit Theorem from probability would then suggest that a constant multiple of $\sqrt{N}$ works as an upper bound. But characters are not random, so Vinogradov conjectures that for any $\varepsilon > 0$,

$$\left|\sum_{n=1}^{N} \chi(n)\right| \leq \sqrt{N} q^{\varepsilon}.$$

From this heuristic, he predicts:

CONJECTURE 2.5 (Vinogradov). *Given $\varepsilon > 0$, we have $n_p \leq p^{\varepsilon}$ for sufficiently large primes p.*

In the 1940s, Linnik introduced the large sieve in order to prove that $n_p \ll p^{\varepsilon}$ holds for most primes ([**25**], [**26**]); this will be discussed in section 6. A big breakthrough in the direction of Vinogradov's conjecture came from Burgess in a series of papers in the early 1960s ([**6**], [**8**], [**7**]), where he proved a slightly weaker version of the following inequality, which can be found in [**18**]:

THEOREM 2.6 (Burgess inequality). *Let χ be a non-principal character* mod q, *where $q > 1$ is prime, and r is a positive integer. Then, there exists a constant C, such that*

$$\left|\sum_{n=1}^{N} \chi(n)\right| \leq C N^{1-\frac{1}{r}} q^{\frac{r+1}{4r^2}} (\log q)^{1/r},$$

where the constant C depends only on ε and r.

REMARK 2.7. Burgess proved a slightly weaker inequality, but it also holds when q is cubefree, as well as for any q when $r = 1, 2, 3$.

REMARK 2.8. When $r = 1$, one recovers the Pólya–Vinogradov inequality. When $r = 2$ one gets the desired $\sqrt{N}$ term, at the cost of a big power of q. As r is larger, the power of q is smaller, so as $r \to \infty$, the bound tends to the trivial inequality.

What does this imply for the least quadratic non-residue modulo p? If $\chi(n) = 1$ for all $n \leq N$, then by the Burgess inequality one would have

$$N = \left|\sum_{n \leq N} \chi(n)\right| \leq C N^{1-\frac{1}{r}} q^{\frac{r+1}{4r^2}} (\log p)^{1/r},$$

and consequently

$$N \leq C^r q^{\frac{1}{4}+\frac{1}{4r}} \log p.$$

For $\delta > 0$ and q large enough , we can see that $N = q^{1/4+\delta}$ satisfies (1). Therefore, given $\varepsilon > 0$, for all large enough p, we have

$$n_p \leq p^{\frac{1}{4\sqrt{e}}+\varepsilon}.$$

No significant unconditional improvements have been done on this result from the 1960s. The modern improvements are in terms of making the constants explicit. These results will be described in section 5.

3. The Pólya–Vinogradov inequality

Let $q \in \mathbb{Z}^+$. Recall that a function $\chi : \mathbb{Z} \to \mathbb{C}$ is called a Dirichlet character modulo q if

(1) $\chi(nm) = \chi(n)\chi(m)$ for all $n, m \in \mathbb{Z}$;
(2) $\chi(n + kq) = \chi(n)$ for all $n, k \in \mathbb{Z}$;
(3) $\chi(n) = 0$ if and only if $(n, q) > 1$.

There is a one-to-one correspondence between Dirichlet characters $\chi : \mathbb{Z} \to \mathbb{C}$ and group homomorphisms $\chi : (\mathbb{Z}/q\mathbb{Z})^\star \to \mathbb{C}^\star$. The principal character mod q is defined by

$$\chi_0(n) = \begin{cases} 1 & \text{if } (n, q) = 1 \\ 0 & \text{if } (n, q) > 1. \end{cases}$$

As mentioned before, the Legendre symbol $\chi(n) = (n/p)$ is a Dirichlet character modulo p. Since this is our main focus, we will simplify our exposition by considering only Dirichlet characters modulo p. Our goal in this section is to prove the Pólya–Vinogradov inequality for prime moduli.

REMARK 3.1. Let p be an odd prime. Choose a primitive root g modulo p; that is, $(\mathbb{Z}/p\mathbb{Z})^\star = \langle g \rangle$. For $a = 0, 1, \dots, p-2$, we can define a Dirichlet character χ_a by the mapping $g^\nu \mapsto \zeta_{p-1}^{a\nu}$, where $\zeta_{p-1} = e^{\frac{2\pi i}{p-1}}$. This gives all Dirichlet characters mod p. In fact, the map $g^\nu \mapsto \chi_\nu$ gives an isomorphism between $(\mathbb{Z}/p\mathbb{Z})^\star$ and the group all Dirichlet characters modulo p.

EXAMPLE. There are 6 Dirichlet characters modulo 7. Choosing $g = 5$, and using that $\zeta_6 = -\zeta_3^2$, we get the following table

n	0	1	2	3	4	5	6
$\chi_0(n)$	0	1	1	1	1	1	1
$\chi_1(n)$	0	1	ζ_3^2	$-\zeta_3$	ζ_3	$-\zeta_3^2$	-1
$\chi_2(n)$	0	1	ζ_3	ζ_3^2	ζ_3^2	ζ_3	1
$\chi_3(n)$	0	1	1	-1	1	-1	-1
$\chi_4(n)$	0	1	ζ_3^2	ζ_3	ζ_3	ζ_3^2	1
$\chi_5(n)$	0	1	ζ_3	$-\zeta_3^2$	ζ_3^2	$-\zeta_3$	-1

To simplify our notation going forward, let $e(x) = e^{2\pi i x}$. For example $\zeta_{p-1} = e^{\frac{2\pi i}{p-1}} = e(\frac{1}{p-1})$.

DEFINITION 3.2. If χ is a Dirichlet character modulo p, then the Gauss sum associated to χ is defined as

$$\tau(\chi) = \sum_{n=1}^{p} \chi(n) e(n/p).$$

For $a = 0, 1, \ldots, p-1$, we also define $\tau_a(\chi) = \sum_{n=1}^{p} \chi(n)e(an/p)$. Notice that $\tau(\chi) = \tau_1(\chi)$.

The following lemma is a consequence of the fact that if $a \not\equiv 0 \bmod p$, then $a, 2a, \ldots, pa \bmod p$ is a permutation of $0, 1, 2, \ldots, p-1$.

LEMMA 3.3.

$$\tau_a(\chi) = \begin{cases} \chi(a^{-1})\tau(\chi) & \textit{if } \chi \neq \chi_0, \textit{ and } a \neq 0 \\ 0 & \textit{if } \chi \neq \chi_0, \textit{ and } a = 0 \\ 0 & \textit{if } \chi = \chi_0, \textit{ and } a \neq 0 \\ p-1 & \textit{if } \chi = \chi_0, \textit{ and } a = 0. \end{cases}$$

We are now ready to prove a classical result due to Gauss.

THEOREM 3.4. *If χ is a non-principal Dirichlet character modulo p, then*

$$|\tau(\chi)| = \sqrt{p}.$$

PROOF. The idea is to evaluate the sum

$$S = \sum_{a=0}^{p-1} \tau_a(\chi)\overline{\tau_a(\chi)}$$

in two different ways. On the one hand, $|\tau_a(\chi)| = |\tau(\chi)|$ when $a \neq 0$ and therefore $S = (p-1)|\tau(\chi)|^2$. On the other hand, we can write

$$S = \sum_a \sum_{n,m} \chi(n)\overline{\chi}(m)e(a(n-m)/p)$$

and push the $\sum_a$ to the inside to find that $S = p(p-1)$. □

First, we will prove a result where the evaluation of a Dirichlet character at n is written in terms of a Gauss sum.

LEMMA 3.5. *If χ is a non-principal Dirichlet character modulo p, then*

$$\chi(n) = \frac{1}{\tau(\overline{\chi})} \sum_{a=1}^{p} \overline{\chi}(a)\, e(an/p)\,. \tag{2}$$

PROOF.

$$\tau(\overline{\chi})\chi(n) = \sum_{m=1}^{p} \overline{\chi}(m)\chi(n)e(m/p).$$

When $n \not\equiv 0 \bmod p$, then $m \equiv an \bmod p$ has a unique solution $a \bmod p$. But then $\overline{\chi}(m)\chi(n) = \overline{\chi}(an)\chi(n) = \overline{\chi}(a)$, and $e(m/p) = e(an/p)$. Therefore

$$\sum_{m=1}^{p} \overline{\chi}(m)\chi(n)e(m/p) = \sum_{a=1}^{p} \overline{\chi}(a)e(an/p).$$

If $n \equiv 0 \bmod p$, then the left hand side of (2) is 0 and the right hand side can be computed (using Remark 3.1)

$$\frac{1}{\tau(\overline{\chi})} \sum_{a=1}^{p} \overline{\chi}(a) = \frac{1}{\tau(\overline{\chi})} \sum_{a=1}^{p} e\left(\frac{-a\nu}{p-1}\right) = 0.$$

□

We now have all the ingredients to prove the Pólya–Vinogradov inequality.

THEOREM 3.6 (Pólya–Vinogradov). *If χ is a non-principal Dirichlet character modulo p, then*

$$\left|\sum_{H\le n<H+N} \chi(n)\right| \le \sqrt{p}\log p\,.$$

PROOF. First, observe that

$$\frac{1}{p}\sum_{a=0}^{p-1} e(ax/p)e(-an/p) = \begin{cases} 1 & \text{if } x\equiv n \bmod p \\ 0 & \text{otherwise.}\end{cases}$$

Let $\mathbb{F}_p$ be the field with p elements $\{0,1,2,\ldots,p-1\}$. Then $\sum_{H\le n<H+N}\chi(n)$ is equal to

$$\begin{aligned}
&\sum_{x\in\mathbb{F}_p}\sum_{H\le n<H+N}\chi(x)\left(\frac{1}{p}\sum_{a=0}^{p-1} e(ax/p)e(-an/p)\right)\\
&=\frac{1}{p}\sum_{a=0}^{p-1}\sum_{x\in\mathbb{F}_p}\chi(x)e(ax/p)\sum_{H\le n<H+N} e(-an/p)\\
&=\frac{1}{p}\sum_{a=0}^{p-1}\tau_a(\chi)\sum_{H\le n<H+N} e(-a/p)^n=\frac{1}{p}\sum_{a=1}^{p-1}\tau_a(\chi)\frac{e(-a(H+N)/p)-e(-aH/p)}{1-e(-a/p)}.
\end{aligned}$$

Thus

$$\left|\sum_{H\le n<H+N}\chi(n)\right|\le\frac{1}{p}\sum_{a=1}^{p-1}|\tau_a(\chi)|\frac{2}{|1-e(-a/p)|}\le\frac{4}{p}\sum_{a=1}^{(p-1)/2}|\tau_a(\chi)|\frac{1}{|1-e(-a/p)|}.$$

We have

$$|1-e(-a/p)|\ge\frac{4a}{p},$$

for $a=1,\ldots(p-1)/2$ and we have $|\tau_a(\chi)|=\sqrt{p}$, so

$$\left|\sum_{H\le n<H+N}\chi(n)\right|\le\sqrt{p}\sum_{a=1}^{(p-1)/2}\frac{1}{a}.$$

To conclude the proof, we need only show that $\sum_{a=1}^{(p-1)/2}\frac{1}{a}\le\log p$. Note that for $p=3,5,7$, one can verify this manually.[4] For $p\ge 11$, we know $(p-1)/2\ge 5$. Since $1/t$ is a decreasing function, for an integer $x\ge 5$, we have

$$\sum_{a\le x}\frac{1}{a}\le 1+\frac{1}{2}+\frac{1}{3}+\frac{1}{4}+\frac{1}{5}+\int_5^x\frac{1}{t}\,\mathrm{d}t=\frac{137}{60}+\log(x)-\log(5).$$

Therefore

$$\sum_{a=1}^{(p-1)/2}\frac{1}{a}\le\frac{137}{60}+\log(p-1)-\log(2)-\log 5<\log p.$$

□

[4]The case $p=2$ is omitted, since the statement of the theorem is vacuously true; there are no non-principal characters modulo 2.

4. The Burgess inequality

There are many explicit estimates for the Burgess inequality in the literature. For example, in [**47**], Treviño proves

THEOREM 4.1. *Suppose χ is a non-principal Dirichlet character modulo a prime $p \geq 10^7$. Let r be a positive integer, and let N and H be integers with $H \geq 1$. Then*

$$\left| \sum_{n \in (N, N+H]} \chi(n) \right| \leq 2.74\, H^{1-\frac{1}{r}} p^{\frac{r+1}{4r^2}} (\log p)^{\frac{1}{r}}.$$

Booker [**5**] has better constants for quadratic characters under certain restrictions on the range of H with respect to p. For characters of any order, there is work of McGown [**28**] and Treviño [**47**] that gets explicit estimates for different ranges of H with respect to p where the estimate gains a power of $\log p$, by getting an exponent of $1/(2r)$ as opposed to the exponent of $1/r$. Recent work of Kerr, Shparlinski, and Hung Yau [**19**] is able to improve the exponent of $\log p$ to $1/(4r)$ but without finding the constants explicitly. More recent work, by de la Bretèche and Munsch [**11**], given $H \leq p^{\frac{1}{2}+\frac{1}{4r}}$, improves the exponent of $\log p$ further to $(\delta_0 + o(1))/(2r)$, where $\delta_0 \approx 0.16656$.

For this paper, we want to show the techniques to prove explicit estimates on the Burgess inequality without aiming at the best possible constants. To simplify the proof, we have an exponent of $3/(2r)$ for $\log p$ instead of an exponent of $1/r$ or $1/(2r)$.

Our main goal this section will be proving the following explicit Burgess inequality

THEOREM 4.2. *Suppose χ is a non-principal Dirichlet character modulo a prime $p \geq 10^{11}$. Let $N, H \in \mathbb{Z}$ with $H \geq 1$. Fix a positive integer $r \geq 2$. Then*

$$\left| \sum_{n \in (N, N+H]} \chi(n) \right| < 5\, H^{1-\frac{1}{r}} p^{\frac{r+1}{4r^2}} (\log p)^{\frac{3}{2r}}.$$

4.1. An important upper bound. To be able to prove Theorem 4.2, we need to get an upper bound on a particular character sum. To do that, we use the following character sum estimate, first given by Weil as a consequence of his deep work on the Riemann hypothesis for function fields (see [**51**]). It is also proved as Theorem 2C' in [**43**] using an elementary method due to Stepanov (see [**44**]), which was later extended by both Bombieri (see [**4**]) and Schmidt (see [**42**]).

LEMMA 4.3. *Let χ be a non-principal Dirichlet character to the prime modulus p, having order n. Let $f(x) \in \mathbb{Z}[x]$ be a polynomial with m distinct roots which is not an n-th power in $\mathbb{F}_p[x]$, where $\mathbb{F}_p$ denotes the finite field with p elements. Then*

$$\left| \sum_{x \in \mathbb{F}_p} \chi(f(x)) \right| \leq (m-1)\, p^{1/2}.$$

Let $S(\chi, h, r)$ be defined as

$$S(\chi, h, r) = \sum_{x \bmod p} \left| \sum_{b=1}^{h} \chi(x+b) \right|^{2r}.$$

We will require bounds which follow from an explicit version of Stirling's formula (for example, see [**38**]):

$$\left(\frac{2r}{e}\right)^r < \frac{(2r)!}{2^r r!} < \sqrt{2}\left(\frac{2r}{e}\right)^r .$$

The following Lemma is due to Treviño, building on work of Burgess, Norton, and Booker [**5, 6, 8, 36, 47, 48**]. Some of the following exposition appears in applications of the lemma in [**27, 32**].

LEMMA 4.4. *Suppose χ is any non-principal Dirichlet character to the prime modulus p. If $r, h \in \mathbb{Z}^+$, then*

$$S(\chi, h, r) < \frac{(2r)!}{2^r r!} p h^r + (2r-1)p^{1/2}h^{2r} .$$

PROOF. First we claim that we may assume, without loss of generality, that $h < p$ and $r < (e/2)h$. We commence by observing that $h = p$ implies $S(\chi, h, r) = 0$, in which case there is nothing to prove. We see that $h > p$ implies $S(\chi, h - p, r) = S(\chi, h, r)$, which allows us to inductively bring h into the range $0 < h < p$. Additionally, we notice that if $r \geq (e/2)h$, then the theorem is trivial since in this case we would have

$$S(\chi, h, r) \leq h^{2r} p \leq \left(\frac{2r}{e}\right)^r h^r p < \frac{(2r)!}{2^r r!} h^r p .$$

This establishes the claim.

Now, to begin the proof proper, we observe that

$$S(\chi, h, r) = \sum_{1 \leq m_1, \ldots, m_{2r} \leq h} \; \sum_{x=0}^{p-1} \chi(x+m_1)\ldots\chi(x+m_r)\overline{\chi}(x+m_{r+1})\ldots\overline{\chi}(x+m_{2r}) .$$

Define

$$\mathcal{M} := \{\mathbf{m} = (m_1, \ldots, m_{2r}) \mid 1 \leq m_1, \ldots, m_{2r} \leq h\} .$$

We can rewrite the above as

$$S(\chi, h, r) = \sum_{\mathbf{m} \in \mathcal{M}} \sum_{x \in \mathbb{F}_p} \chi(f_{\mathbf{m}}(x)) ,$$

where

$$f_{\mathbf{m}}(x) = (x+m_1)\ldots(x+m_r)(x+m_{r+1})^{n-1}(x+m_{2r})^{n-1} ,$$

and n denotes the order of χ. If $f_{\mathbf{m}}(x)$ is not an n-th power mod p, then by Lemma 4.4 we have

$$\left| \sum_{x \in \mathbb{F}_p} \chi(f_{\mathbf{m}}(x)) \right| \leq (2r-1)\sqrt{p} .$$

Otherwise, we employ the trivial bound of p.

It remains to count the number of exceptions – that is, the number of $\mathbf{m} \in \mathcal{M}$ such that $f_{\mathbf{m}}(x)$ is an n-th power mod p. If $n = 2$, it is easy to see that the number of exceptions is bounded above by $(2r-1)(2r-3)\ldots(3)(1) = (2r)!/(2^r r!)$ simply by pairing each m_j with a duplicate. When $n > 2$, the counting problem is much more difficult. To avoid this difficulty, Burgess in [**6**] counts the number of $\mathbf{m} = (m_1, \ldots, m_{2r}) \in \mathcal{M}$ such that each m_j is repeated at least once and arrives at the expression $(4r)^{r+1}h^r$.

Treviño in [**48**] shows that the number of exceptions is bounded above by the quantity

$$c_r(h,n) = \sum_{d=0}^{\lfloor \frac{r}{n} \rfloor} \left(\frac{w!}{d!(k!)^d} \right)^2 \frac{h^{r-(n-2)d}}{(r-nd)!} ;$$

moreover, under the condition $r \leq 9h$, he shows that $c_r(h,n)$ is a decreasing function of n and hence $c_r(h,n) \leq c_w(h,2) = (2r)!/(2^r r!)h^r$. But since we have $r < (e/2)h$ in the context of our proof, this condition is automatic. □

4.2. Proof of Theorem 4.2. Throughout this section, χ will denote a Dirichlet character modulo an odd prime p and N, H will be integers with $0 \leq N < p$ and $1 \leq H < p$. The latter assumption is justified as reducing N and H modulo p leaves the sum in Theorem 4.2 unchanged. Defining the quantities

$$S_\chi(H) := \sum_{n \in (N, N+H]} \chi(n), \quad E(H) := H^{1-\frac{1}{r}} p^{\frac{r+1}{4r^2}} (\log p)^{\frac{3}{2r}},$$

we seek a bound of the form $S_\chi(H) < C\, E(H)$, for some constant C.

Fix $A \in \mathbb{Z}$ with $1 < A < p$. For $x \in \mathbb{F}_p$, we define $\nu_A(x)$ to be the number of ways we can write $x \equiv \overline{a}n \pmod{p}$, where $a \in [1, A]$ is a prime and $n \in (N, N+H]$ is an integer. Here the notation $\overline{a}$ denotes a multiplicative inverse of a modulo p.

LEMMA 4.5. *Suppose $|S_\chi(H_0)| \leq C\, E(H_0)$ for all $H_0 < H$. Fix $H_0 = AB < H$. Then*

$$|S_\chi(H)| \leq \frac{1}{\pi(A)B} \sum_{x \in \mathbb{F}_p} \nu_A(x) \left| \sum_{1 \leq b \leq B} \chi(x+b) \right| + 2C\, E(H_0).$$

PROOF. Applying a shift $n \mapsto n+h$ with $1 \leq h \leq H_0$ gives

$$S_\chi(H) = \sum_{n \in (N, N+H]} \chi(n+h) + 2C\theta E(H_0).$$

(The letter θ will denote a complex number with $|\theta| \leq 1$, possibly different each time it appears.) We set $h = ab$ in the above, and average over all primes $a \in [1, A]$ and all integers $b \in [1, B]$. This gives

$$S_\chi(H) = \frac{1}{\pi(A)B} \sum_{a,b}{}' \sum_{n \in (N, N+H]} \chi(n+ab) + 2C\theta E(H_0),$$

where $\sum'$ here indicates that we are summing over all primes $a \in [1, A]$ and all integers $b \in [1, B]$. Rearranging the sum in the above expression yields

$$\sum_{a,b}{}' \sum_{n \in (N, N+H]} \chi(n+ab) = \sum_{1 \leq a \leq A}{}' \sum_{n \in (N, N+H]} \chi(a) \sum_{1 \leq b \leq B} \chi(\overline{a}n + b),$$

and hence

$$\left| \sum_{a,b}{}' \sum_{n \in (N, N+H]} \chi(n+ab) \right| \leq \sum_{x \in \mathbb{F}_p} \nu_A(x) \left| \sum_{1 \leq b \leq B} \chi(x+b) \right|.$$

The result follows. □

LEMMA 4.6. *Suppose $a_1 \neq a_2$ are prime and $b \in \mathbb{Z}$. Then the number of integer solutions $(x,y) \in \mathbb{Z}^2$ to the equation $a_1 x - a_2 y = b$ with $x, y \in (N, N+H]$ is at most*

$$\frac{H}{\max\{a_1, a_2\}} + 1 .$$

PROOF. Let Q denote the number of solutions to $a_1 x - a_2 y = b$ with $x, y \in (N, N+H]$. We will show $Q \leq H/a_2 + 1$. It will immediately follow from the same argument that $Q \leq H/a_1 + 1$ as well; indeed, just multiply both sides of the equation by -1 and interchange the roles of x and y. Suppose we have two solutions (x,y) and (x',y'). It follows that $a_1(x-x') = a_2(y-y')$, and since $a_1 \neq a_2$ are prime, we see that a_2 divides $x - x'$ which implies $|x-x'| \geq a_2$. Then, there are at most $H/a_2 + 1$ choices for x, but given x, y is fixed. The result follows. □

LEMMA 4.7. *Fix $A \in \mathbb{Z}$ with $1 < A < p$. If $2AH \leq p$, then*

$$\sum_{x \in \mathbb{F}_p} \nu_A(x)^2 < \pi(A) H \left(\frac{5}{3} + \frac{2\pi(A)}{H} \right) .$$

PROOF. Define S to be the set of all quadruples (a_1, a_2, n_1, n_2) with

$$a_1 n_2 \equiv a_2 n_1 \pmod{p},$$

where $a_1, a_2 \in [1, A]$ are prime and $n_1, n_2 \in (N, N+H]$ are integers. We observe that $\#S = \sum_{x \in \mathbb{F}_p} \nu_A(x)^2$. Suppose $(a_1, a_2, n_1, n_2) \in S$ with $a_1 = a_2$. Then we have $n_1 \equiv n_2 \pmod{p}$ and hence $n_1 = n_2$ since $n_1, n_2 \in (N, N+H]$ and $H \leq p$. Thus there are exactly $\pi(A)H$ solutions of this form.

Now we treat the remaining cases. Let $(a_1, a_2, n_1, n_2) \in S$ with $a_1 \neq a_2$. Then $a_1 n_2 - a_2 n_1 = kp$ for some k. It is an exercise to verify that a_1 and a_2 determine k. Now Lemma 4.6 tells us that there are at most $H/\max\{a_1, a_2\} + 1$ choices of (n_1, n_2) for each fixed (a_1, a_2). Thus the number of elements in S with $a_1 \neq a_2$ is bounded above by

$$2 \sum_{\substack{a_2 \leq A \\ a_2 \text{ prime}}} \sum_{\substack{a_1 < a_2 \\ a_1 \text{ prime}}} \left(\frac{H}{a_2} + 1 \right) \;<\; 2H \sum_{\substack{a \leq A \\ a \text{ prime}}} \frac{\pi(a) - 1}{a} + 2 \sum_{\substack{a \leq A \\ a \text{ prime}}} (\pi(a) - 1) .$$

It follows that

$$\sum_{x \in \mathbb{F}_p} \nu_A(x)^2 < \pi(A) H \left(1 + \frac{2}{\pi(A)} \sum_{a \leq A} \frac{\pi(a) - 1}{a} + \frac{2}{\pi(A) H} \sum_{a \leq A} (\pi(a) - 1) \right) .$$

Finally we observe

$$\sum_{\substack{a \leq A \\ a \text{ prime}}} (\pi(a) - 1) \leq \pi(A)^2 , \qquad \sum_{\substack{a \leq A \\ a \text{ prime}}} \frac{\pi(a) - 1}{a} < \frac{\pi(A)}{3}.$$

The second inequality can be confirmed manually for $A \leq 49$ and then one can use $\pi(a) < a/3$ for $a \geq 50$, which follows from (3.6) of [**39**], namely $\pi(a) \leq 1.3a/\log a$ for $a > 1$. The Lemma follows. □

We are now ready to prove the Burgess inequality.

PROOF OF THEOREM 4.2. We may assume

$$C^r p^{\frac{1}{4}+\frac{1}{4r}} \log p \leq H \leq p^{\frac{1}{2}+\frac{1}{4r}} \log p\,; \tag{3}$$

otherwise, the result follows from either $|S_\chi(H)| \leq H$ or $|S_\chi(H)| \leq p^{1/2} \log p$. We will prove the result by induction on H. We assume that $|S_\chi(H_0)| \leq CE(H_0)$ for all $H_0 < H$. We choose an integer H_0 with

$$\frac{H}{d+1} < H_0 \leq \frac{H}{d}\,,$$

for which we can write $H_0 = AB$ with $A, B \in \mathbb{Z}^+$, where

$$B = \left\lceil \frac{2}{d} p^{\frac{1}{2r}} (\log p)^2 \right\rceil.$$

Accomplishing this is possible provided $H/d - H/(d+1) > B$. Assuming $d = 10$ and $p \geq 10^{10}$ we find $B > 100$ and therefore the desired inequality follows from the estimate

$$H \geq (1.01)2(d+1) p^{\frac{1}{2r}} (\log p)^2\,,$$

which is implied by

$$C^r p^{\frac{1}{4}-\frac{1}{4r}} \geq 2.02(d+1) \log p\,;$$

the latter condition holds upon setting $C = 5$ and assuming $p \geq 10^{11}$. Observe that our choice of B leads to $2AH \leq 2H^2/(dB) \leq p$. At this point, we give upper and lower bounds on A. Observe that

$$A \leq \frac{H}{dB} \leq \frac{1}{2} p^{\frac{1}{2}-\frac{1}{4r}} (\log p)^{-1}\,,$$

and hence

$$\log A + 1 \leq \frac{1}{2} \log p. \tag{4}$$

We also have

$$A > \frac{H}{(d+1)B} \geq \frac{C^r d}{(2.02)(d+1)} p^{\frac{1}{4}-\frac{1}{4r}} (\log p)^{-1} > 10\,.$$

Applying Lemma 4.5 and our inductive hypothesis, we have

$$|S_\chi(H)| \leq \frac{1}{\pi(A)B} \sum_{x \in \mathbb{F}_p} \nu_A(x) \left| \sum_{1 \leq b \leq B} \chi(x+b) \right| + \frac{2C}{d^{1-\frac{1}{r}}} E(H)\,. \tag{5}$$

In order to bound the sum above, we apply Hölder's inequality to the functions $\nu_A(x)^{1-\frac{1}{r}}$, $\nu_A(x)^{\frac{1}{r}}$, and $\left|\sum_{1 \leq b \leq B} \chi(x+b)\right|$ using the Hölder exponents $(1-1/r)^{-1}$, $2r$, and $2r$ respectively; this yields:

$$\sum_{x \in \mathbb{F}_p} \nu_A(x) \left| \sum_{1 \leq b \leq B} \chi(x+b) \right|$$
$$\leq \left(\sum_{x \in \mathbb{F}_p} \nu_A(x) \right)^{1-\frac{1}{r}} \left(\sum_{x \in \mathbb{F}_p} \nu_A(x)^2 \right)^{\frac{1}{2r}} \left(\sum_{x \in \mathbb{F}_p} \left| \sum_{1 \leq b \leq B} \chi(x+b) \right|^{2r} \right)^{\frac{1}{2r}}.$$

We bound each of the three sums above in turn. Clearly, one has

$$\sum_{x\in\mathbb{F}_p} \nu_A(x) = \pi(A)H\,.$$

We invoke Lemma 4.7 to conclude that

$$\sum_{x\in\mathbb{F}_p} \nu_A(x)^2 \leq 2\pi(A)H\,.$$

Indeed, using (3.6) of [**39**], we have $\pi(A) \leq 1.3\,A/\log A$ for $A > 1$ and therefore

$$\frac{\pi(A)}{H} \leq \frac{1.3A}{H\log A} \leq \frac{1.3}{dB\log A} \leq \frac{1.3}{2p^{\frac{1}{2r}}(\log p)^2\log A} < 0.1\,.$$

To bound the third sum, we apply Lemma 4.4; this gives

$$\sum_{x\in\mathbb{F}_p}\left|\sum_{1\leq b\leq B}\chi(x+b)\right|^{2r} \leq B^{2r}p^{1/2}\left(\sqrt{2}\left(\frac{2r}{eB}\right)^r p^{1/2}+(2r-1)\right) \leq 2rB^{2r}p^{1/2}\,.$$

The final inequality follows from $\sqrt{2}\left(\frac{2r}{eB}\right)^r p^{1/2} \leq 1$ whenever $r \leq 0.1\log^2 p$. But from (3) we may assume that $r < \frac{1}{4}\log p \leq 0.1\log^2 p$. All together, this gives

$$\sum_{x\in\mathbb{F}_p}\nu_A(x)\left|\sum_{1\leq b\leq B}\chi(x+b)\right| \leq (\pi(A)H)^{1-\frac{1}{r}}(2\pi(A)H)^{\frac{1}{2r}}\left(2rB^{2r}p^{1/2}\right)^{\frac{1}{2r}}\,.$$

Therefore

$$\frac{1}{\pi(A)B}\sum_{x\in\mathbb{F}_p}\nu_A(x)\left|\sum_{1\leq b\leq B}\chi(x+b)\right| \quad\leq\quad (4r)^{\frac{1}{2r}}H^{1-\frac{1}{r}}p^{\frac{1}{4r}}\left(\frac{H}{\pi(A)}\right)^{\frac{1}{2r}}\,.$$

Using (3.5) of [**39**] and some simple computation, provided $A \geq 3$ and $A \in \mathbb{Z}$, we have $\pi(A) \geq A/(1+\log A)$; using this, together with (4), we can estimate

$$\frac{H}{\pi(A)} \leq \frac{H(\log A+1)}{A} \leq (d+1)B(\log A+1) \leq 1.01\frac{d+1}{d}\cdot p^{\frac{1}{2r}}(\log p)^3\,.$$

Therefore

$$\left(\frac{H}{\pi(A)}\right)^{\frac{1}{2r}} \leq \left(1.01\frac{d+1}{d}\right)^{\frac{1}{2r}} p^{\frac{1}{4r^2}}(\log p)^{\frac{3}{2r}},$$

which leads to

$$\frac{1}{\pi(A)B}\sum_{x\in\mathbb{F}_p}\nu_A(x)\left|\sum_{1\leq b\leq B}\chi(x+b)\right| \quad\leq\quad \left(\frac{4.04\,r(d+1)}{d}\right)^{\frac{1}{2r}} H^{1-\frac{1}{r}}p^{\frac{r+1}{4r^2}}(\log p)^{\frac{3}{2r}}\,.$$

Finally, using (5), this gives

$$|S_\chi(H)| \leq \left[\left(\frac{4.04\,r(d+1)}{d}\right)^{\frac{1}{2r}} + \frac{2C}{d^{1-\frac{1}{r}}}\right]E(H) \leq C\,E(H)\,,$$

when $C = 5$. □

5. Explicit results on the least quadratic non-residue

The inequality $n_p \leq p^{\frac{1}{4\sqrt{e}}+\varepsilon}$ for p large enough (depending on ε), is not explicit, in the sense that we haven't mentioned how big p must be depending on the parameter ε. Also, in some cases we want results that work for all p or for all $p \geq p_0$ for some manageable p_0. In this section we'll discuss several explicit results about n_p.

From Pólya–Vinogradov, we can show the explicit estimate $n_p \leq \sqrt{p}\log p$ for all p. As mentioned in Remark 2.3, one can also show $n_p \leq \sqrt{p}+1$ elementarily. But these results are far from the best asymptotic result. One of the first important explicit estimates was proved by Norton [**36**] in 1971. Norton proved

THEOREM 5.1 (Norton, 1971). *Let p be an odd prime number. Let n_p be the least quadratic non-residue modulo p. Then*

$$n_p \leq \begin{cases} 3.9p^{1/4}\log p & \text{if } p \equiv 1 \bmod 4 \\ 4.7p^{1/4}\log p & \text{if } p \equiv 3 \bmod 4. \end{cases}$$

Treviño [**48**], in 2015, improved this to

THEOREM 5.2. *Let p be an odd prime number. Let n_p be the least quadratic non-residue modulo p. Then*

$$n_p \leq \begin{cases} 0.9p^{1/4}\log p & \text{if } p \equiv 1 \bmod 4 \\ 1.1p^{1/4}\log p & \text{if } p \equiv 3 \bmod 4. \end{cases}$$

The proof of the theorem relied on Lemma 4.4 to get a bound when $p \geq 10^{60}$ using $r \approx \log p/4$ and h in terms of r. For the "smaller" p, the idea was to use Lemma 4.4, choosing r and h carefully for different ranges of p.

These estimates are still relatively far from the Burgess estimate of $p^{\frac{1}{4\sqrt{e}}+\varepsilon}$. In that direction, Treviño [**47**], using an explicit Burgess inequality was able to show

THEOREM 5.3. *Let $p \geq 10^{4732}$ be prime. Then*

$$n_p \leq p^{1/6}.$$

The technique of the proof works for any exponent $y > \frac{1}{4\sqrt{e}}$. Indeed, we'll prove the following theorem

THEOREM 5.4. *Let $p \geq 10^{19000}$ be prime. Then*

$$n_p \leq p^{4/25}.$$

To prove it we will need the following explicit version of the Vinogradov trick (Lemma 5.3 in [**47**]):

LEMMA 5.5. *Let $x \geq 286$, and let $y = x^{\frac{1}{\sqrt{e}}+\delta}$ for some $\delta > 0$. Let χ be a non-principal character* $\bmod\, p$ *for some prime p. If $\chi(n) = 1$ for all $n \leq y$, then*

$$\left|\sum_{n\leq x}\chi(n)\right| \geq x\left(2\log\left(\delta\sqrt{e}+1\right) - \frac{1}{\log^2 x} - \frac{1}{\log^2 y} - \frac{1}{x}\right).$$

PROOF OF THEOREM 5.4. Following the proof of Theorem 1.10 in [**47**], let χ be the Legendre symbol modulo p. Then, if $n < p$ and $\chi(n) \neq 1$, n is a quadratic non-residue. Let r be an integer. Let $x \geq 286$ be a real number and let $y =$

$x^{\frac{1}{\sqrt{e}}+\delta} = p^{4/25}$ for some $\delta > 0$. Assume that $\chi(n) = 1$ for all $n \leq y$. Now by Theorem 4.2 and Lemma 5.5 we have

$$5x^{1-\frac{1}{r}}p^{\frac{r+1}{4r^2}}(\log p)^{\frac{3}{2r}} \geq x\left(2\log\left(\delta\sqrt{e}+1\right) - \frac{1}{\log^2 x} - \frac{1}{\log^2 y} - \frac{1}{x}\right).$$

Now, letting $x = p^{\frac{1}{4}+\frac{1}{2r}}$ we get

$$5p^{\frac{3\log\log p}{2r\log p}-\frac{1}{4r^2}} \geq 2\log\left(\delta\sqrt{e}+1\right) - \frac{1}{\log^2 x} - \frac{1}{\log^2 y} - \frac{1}{x}. \tag{6}$$

Picking $r = 40$, one finds that $\delta = 0.002993\ldots$. For $p \geq 10^{19000}$, the right hand side of (6) is bigger than the left hand side, showing that $\chi(n)$ is not always 1 for $n \leq y = p^{4/25}$, and hence the theorem is true. □

We can also consider the case when the modulus is not prime. Suppose q is squarefree and we want to find n_q, the least quadratic non-residue modulo q. Granville, Mollin and Williams [**15**] proved that $n_q \leq \sqrt{q}/2$ for $q > 3705$. The proof used an explicit Pólya–Vinogradov inequality and extensive computations on a special sieving computer called *Manitoba Scalable Sieve Unit* that calculated bounds up to 10^{18}. Using a "smoothed" version of the Pólya–Vinogradov inequality (appeared in [**23**]), Treviño [**45**] improved this result to show that $n_q \leq q^{9/20}$ for $q > 1596$.

One could also consider the n-th least prime quadratic non-residue q_n (the first non-residue is always prime). Explicit bounds for q_2 appear in [**24, 31**]. Ma, McGown, Rhodes, and Wanner [**27**], recently proved that for $p \geq p_0$, there is a constant $C = C(n, p_0)$ depending on n, p_0 such that

$$q_n \leq Cp^{\frac{1}{4}}(\log p)^{\frac{n+1}{2}}.$$

In their paper, they calculate C for different values of n and p_0. For example if $n = 4$ and $p_0 = 10^{20}$, then the constant $C = 2.014$ suffices.

Another related problem is to bound $H(p)$, the maximum number of consecutive integers on which the Legendre symbol $\left(\frac{\cdot}{p}\right)$ is constant, i.e.,

$$H(p) = \max_{a \bmod p}\left\{H : \left(\frac{a+1}{p}\right) = \left(\frac{a+2}{p}\right) = \cdots = \left(\frac{a+H}{p}\right)\right\}.$$

Using work of Burgess [**7**], McGown in [**30**] proved that $H(p) \leq 7.06p^{1/4}\log p$ when $p \geq 5\cdot 10^{18}$, and $H(p) \leq 7p^{1/4}\log p$ if $p \geq 10^{55}$. Treviño ([**46**], [**49**]) improved this to $H(p) \leq 1.55p^{1/4}\log p$ when $p \geq 10^{13}$ and $H(p) \leq 3.38p^{1/4}\log p$ for all odd p.

There are some nice results on a lower bound for $H(p)$. In [**37**], Peralta shows that for any $C < \frac{1}{\log 4}$, there exists a p_0 such that for $p > p_0$, $H(p) > C\log p$. In the interest of getting a clean, explicit lower bound, we prove the following:

THEOREM 5.6. *Let $p > 80000$ be prime. Let $H(p)$ be the maximum number of consecutive integers on which the Legendre symbol $\left(\frac{\cdot}{p}\right)$ is constant. Then*

$$H(p) \geq \frac{1}{2}\log p.$$

PROOF. Let $\chi(a) = \left(\frac{a}{p}\right)$. Note that

$$\frac{\chi(a)+1}{2} = \begin{cases} 1 & \text{if } \chi(a) = 1 \\ 0 & \text{if } \chi(a) = -1 \\ \frac{1}{2} & \text{if } \chi(a) = 0. \end{cases}$$

Suppose $H \geq 3$, then we have

$$\prod_{i=1}^{H} \left(\frac{\chi(a+i)+1}{2} \right) = \begin{cases} 1 & \text{if } \chi(a+1) = \chi(a+2) = \cdots = \chi(a+H) = 1 \\ 0 & \text{otherwise.} \end{cases}$$

Therefore, if

$$S := \sum_{a=0}^{p-1} \prod_{i=1}^{H} \left(\frac{\chi(a+i)+1}{2} \right) > 0,$$

then there exists H consecutive integers that are all quadratic residues. We have

$$S = \frac{1}{2^H} \sum_{a \bmod p} \left(1 + \sum_{j=1}^{2^H} \chi(p_j(a)) \right) = \frac{1}{2^H} \left(p + \sum_{j=1}^{2^H} \sum_{a \bmod p} \chi(p_j(a)) \right),$$

where, for $j \in \{1, 2, \ldots, 2^H\}$, $p_j(a)$ is either a polynomial of degree $\leq H$ or 0. By the Weil estimates (Lemma 4.4),

$$\left| \sum_{a \bmod p} \chi(p_j(a)) \right| \leq (H-1)\sqrt{p}.$$

Therefore

$$|2^H S| \geq p - 2^H (H-1)\sqrt{p} > p - 2^H H \sqrt{p}. \tag{7}$$

When $H = \log p / 2$ we have

$$2^H H \sqrt{p} = \frac{1}{2} p^{\frac{1+\log(2)}{2}} \log p.$$

But the last expression is smaller than p whenever $p > 80000$. Therefore, $S > 0$, which implies that there are at least H consecutive quadratic residues. This means $H(p) \geq \frac{1}{2} \log p$. □

REMARK 5.7. Note that from (7), we can recover Peralta's result, namely that given $C < \frac{1}{\log 4}$, there is a constant p_0 such that for $p > p_0$, $H(p) > C \log p$.

6. The large sieve and Linnik's theorem

In this section we will sketch a proof of the following beautiful result (see [**26**]):

THEOREM 6.1 (Linnik). *Fix $\varepsilon > 0$. Then we have*

$$\#\{p \leq N \mid n_p > p^{\varepsilon}\} = O(N^{\varepsilon}).$$

This result means that the number of exceptions to Vinogradov's conjecture (that the least quadratic non-residue n_p satisfies $n_p \ll p^{\varepsilon}$) is very small; in particular, the theorem implies:

COROLLARY 6.2. *The set of primes where Vinogradov's conjecture fails has density zero.*

Linnik introduced the large sieve (see [**25**]) and it was later studied in detail by Rényi, Roth, Bombieri, Davenport, Halberstam, Gallagher, and other authors. Our discussion here is very brief; more detail, including additional references, can be found in [**9, 10, 18, 34**].

Let $\{a_n\}$ be a sequence of complex numbers and consider the trigonometric polynomial

$$S(\alpha) = \sum_{n=M+1}^{M+N} a_n e(n\alpha) .$$

Let $\alpha_1, \ldots, \alpha_R \in \mathbb{R}$ and suppose that $\|\alpha_r - \alpha_s\| \geq \delta$ for $r \neq s$ where $\|\theta\|$ denotes the distance to the nearest integer. The large sieve refers to an inequality of the form

$$\sum_{r=1}^{R} |S(\alpha_r)|^2 \leq \Delta(N, \delta) \sum_{n=M+1}^{M+N} |a_n|^2.$$

See [**34**] for a short proof that this holds with $\Delta = \delta^{-1} + \pi N$. By taking the α_r to be Farey fractions with denominators up to z, one obtains:

THEOREM 6.3 (The Large Sieve inequality). *For $x, z \in \mathbb{Z}^+$ we have*

$$\sum_{d \leq z} \sum_{\substack{1 \leq a \leq d \\ (a,d)=1}} \left| \sum_{n \leq x} a_n e\left(\frac{na}{d}\right) \right|^2 \leq (z^2 + \pi x) \sum_{n \leq x} |a_n|^2 .$$

One can obtain a sieve out of the previous theorem by the use of so-called Ramanujan sums $\sum_{\substack{a \leq a \leq d \\ (a,d)=1}} e(na/d)$. See [**9**] for the details.

THEOREM 6.4. *Let $\mathcal{A}$ be a positive integers $n \leq x$ and let $\mathcal{P}$ be a set of primes. For each $p \in \mathcal{P}$, suppose we are given a set $\Omega_p = \{w_{1,p}, \ldots, w_{\omega(p),p}\}$ of $\omega(p)$ residue classes modulo p. Let $z > 0$, $P(z)$ the product of the primes $p \in \mathcal{P}$, $p < z$, and set*

$$S(\mathcal{A}, \mathcal{P}, z) := \#\{n \in \mathcal{A} \mid n \not\equiv w_{i,p} \bmod p, \ \forall p | P(z)\} .$$

We have

$$S(\mathcal{A}, \mathcal{P}, z) \leq \frac{z^2 + \pi x}{L(z)},$$

where

$$L(z) := \sum_{d \leq z} \mu^2(d) \prod_{p|d} \frac{\omega(p)}{p - \omega(p)}.$$

REMARK 6.5. In applications, one needs a lower bound on $L(z)$. Often the following estimate suffices:

$$L(z) \geq \sum_{p<z} \frac{\omega(p)}{p - \omega(p)}.$$

The final ingredient we need before proving Linnik's theorem is a lower bound on the numbers of integers up to x with only small prime factors. A number n is y-smooth if all of its prime factors are $\leq y$. The counting function for all y-smooth numbers up to x is defined as

$$\psi(x, y) := \#\{n \leq x \mid \text{all the prime factors of } n \text{ are} \leq y\} .$$

The following is a well-known lower bound on $\psi(x,y)$. The proof we give is suggested in Exercise #12 from Chapter 4 of [**9**].

LEMMA 6.6. *For every $\varepsilon > 0$, there is a constant $c_\varepsilon > 0$ such that for x sufficiently large we have*

$$\psi(x, x^\varepsilon) \geq c_\varepsilon\, x\,.$$

PROOF. Fix $\varepsilon > 0$. We consider numbers of the form $n = m p_1 p_2 \dots p_k$ for $k = [\varepsilon^{-1}]$ where $x^{\varepsilon-\varepsilon^2} < p_j < x^\varepsilon$. We set $\alpha = \varepsilon - \varepsilon^2$ and $\beta = \varepsilon$ and estimate:

$$\sum_{x^\alpha \leq p_1,\dots,p_k \leq x^\beta} \left[\frac{x}{p_1 \dots p_k}\right] \gg \sum_{x^\alpha \leq p_1,\dots,p_k \leq x^\beta} \frac{x}{p_1 \dots p_k} = x\left(\sum_{x^\alpha \leq p \leq x^\beta} \frac{1}{p}\right)^k$$
$$\gg x\left(\log\left(\frac{\beta}{\alpha}\right)\right)^k.$$

□

PROOF OF THEOREM 6.1. Fix $\varepsilon > 0$. Let $\mathcal{A} = \mathbb{Z}^+$, $z = N$, $x = N^2$,

$$\mathcal{P} = \{p \mid n_p > p^\varepsilon\,,\ p \in [N^\varepsilon, N]\}\,,$$

$$\Omega_p = \{\nu \mid (\nu/p) = -1\}\,.$$

Using the Large Sieve, we have:

$$\mathcal{S} = \mathcal{S}(\mathcal{A}, \mathcal{P}, z) \leq \frac{5N^2}{L(z)}.$$

We also have

$$L(z) \geq \sum_{p<z} \frac{\omega(p)}{p - \omega(p)} \geq \frac{1}{3}\#\{p \in \mathcal{P} \mid p \leq N\}.$$

Consider $\mathcal{S}' = \{n \leq N^2 \mid n \text{ has no prime divisors larger than } N^{\varepsilon^2}\}$. We claim that $\mathcal{S}' \subseteq \mathcal{S}$. Let $n \in \mathcal{S}'$. Let $p \in \mathcal{P}$. For any prime q dividing n, we have that $q < N^{\varepsilon^2} < p^\varepsilon$ and therefore $(q/p) = 1$. It follows that $n \in \mathcal{S}$. By Lemma 6.6 we have $\#\mathcal{S}' \geq d_\varepsilon N^2$ for some $d_\varepsilon > 0$. This implies

$$d_\varepsilon N^2 \leq \frac{5N^2}{L(z)} \leq \frac{15N^2}{\#\{p \in \mathcal{P} \mid p \leq N\}},$$

and hence

$$\#\{p \in \mathcal{P} \mid p \leq N\} \leq \frac{15}{d_\varepsilon}.$$

Now we achieve our desired estimate; namely,

$$\{p \leq N \mid n_p > p^\varepsilon\} \leq \pi(N^\varepsilon) + \frac{15}{d_\varepsilon} = O(N^\varepsilon).$$

□

7. Ankeny's Theorem

Recall that Ankeny [**1**] proved $n_p \le C(\log p)^2$ for some constant $C > 0$ under the assumption of the Generalized Riemann Hypothesis (GRH). In particular, this says that the GRH implies Vinogradov's conjecture. Bach [**3**] proved $n_p \le 2(\log p)^2$. The main idea behind Bach's proof appears in [**33**], but to obtain explicit results there are many details to work out; Bach uses a slightly different kernel and introduces a parameter in order to achieve good numerical results. Later, Lamzouri, Li, and Soundararajan [**21**] prove that $n_p \le (\log p)^2$.

Our goal here is prove an explicit version of Ankeny's Theorem without too much effort. We follow Bach's approach in [**3**] as well as some of the exposition given in [**29**]. For ease of exposition, we make the simplifying assumption that $\chi(-1) = 1$. (The same approach still works when $\chi(-1) = -1$ but the functional equation for $L(s, \chi)$ takes a slightly different form.)

Let n_χ denote the least positive integer n for which $\chi(n) \ne 1$. Notice that when $\chi(n) = (n/p)$, the Legendre symbol, one has $n_\chi = n_p$. We prove the following:

THEOREM 7.1. *Let χ be a non-principal Dirichlet character modulo $m \ge 10^9$ with $\chi(-1) = 1$. Assume the RH and the GRH for $L(s, \chi)$. Then $n_\chi < 2(\log m)^2$.*

LEMMA 7.2. *Let χ be a Dirichlet character modulo m. (Here we allow the possibility that χ is the principal character or even that $m = 1$.) For $x > 1$ and $a \in (0, 1)$, we have*

$$-\frac{1}{2\pi i}\int_{2-i\infty}^{2+i\infty} \frac{x^s}{(s+a)^2}\frac{L'(s,\chi)}{L(s,\chi)}\, ds = \sum_{n<x} \chi(n)\Lambda(n)(n/x)^a \log(x/n)\,.$$

PROOF. First, substitute the Dirichlet series

$$\frac{L'(s,\chi)}{L(s,\chi)} = -\sum_{n=1}^{\infty} \chi(n)\Lambda(n)n^{-s}$$

into the left-hand side above and interchange the order of summation and integration. Next, use the fact that for $y > 0$ one has

$$\frac{1}{2\pi i}\int_{2-i\infty}^{2+i\infty} \frac{y^s}{(s+a)^2}\, ds = \begin{cases} y^{-a}\log y & \text{if } y > 1 \\ 0 & \text{otherwise}\,. \end{cases}$$

□

LEMMA 7.3. *Let χ be a non-principal Dirichlet character modulo m with $\chi(-1) = 1$. For $a \in (0, 1)$ and $x > 0$ we have*

$$\begin{aligned} &\frac{x}{(a+1)^2} + \frac{1}{a^2} \\ &= \sum_{\rho \text{ of } \zeta} \frac{x^\rho}{(\rho+a)^2} - \sum_{\rho \text{ of } L_\chi} \frac{x^\rho}{(\rho+a)^2} + \sum_{\substack{n<x \\ \chi(n)\ne 1}} (1-\chi(n))\Lambda(n)(n/x)^a \log(x/n) \\ &+ \frac{\log x}{x^a}\left[\left(\frac{\zeta'}{\zeta}\right)(-a) - \left(\frac{L'_\chi}{L_\chi}\right)(-a)\right] + \frac{1}{x^a}\left[\left(\frac{\zeta'}{\zeta}\right)'(-a) - \left(\frac{L'_\chi}{L_\chi}\right)'(-a)\right]. \end{aligned}$$

PROOF. Evaluate the integral in Lemma 7.2 using the Residue Theorem to obtain:

$$\begin{aligned}\sum_{n<x}\chi(n)\Lambda(n)(n/x)^a\log(x/n) &= -\sum_{\rho\text{ of }L_\chi}\frac{x^\rho}{(\rho+a)^2}-\sum_{n=1}^{\infty}\frac{x^{-2n}}{(a-2n)^2}-\frac{1}{a^2}\\ &\quad -\frac{\log x}{x^a}\left(\frac{L'_\chi}{L_\chi}\right)(-a)-\frac{1}{x^a}\left(\frac{L'_\chi}{L_\chi}\right)'(-a),\end{aligned}$$

$$\begin{aligned}\sum_{n<x}\Lambda(n)(n/x)^a\log(x/n) &= \frac{x}{(a+1)^2}-\sum_{\rho\text{ of }\zeta}\frac{x^\rho}{(\rho+a)^2}-\sum_{n=1}^{\infty}\frac{x^{-2n}}{(a-2n)^2}\\ &\quad -\frac{\log x}{x^a}\left(\frac{\zeta'}{\zeta}\right)(-a)-\frac{1}{x^a}\left(\frac{\zeta'}{\zeta}\right)'(-a).\end{aligned}$$

Subtracting these two equations gives the result. □

Define

$$\psi_{\mathbb{Q}}(s)=\frac{1}{2}\left(\psi\left(\frac{s}{2}\right)-\log\pi\right).$$

In order to expedite the proofs in the rest of this section, we quote some formulae, all of which can be derived from (5.9) of [**20**]. For all $s\in\mathbb{C}$, we have:

$$\frac{\zeta'(s)}{\zeta(s)} = B+\sum_{\rho\text{ of }\zeta}\left(\frac{1}{s-\rho}+\frac{1}{\rho}\right)-\frac{1}{s}-\frac{1}{s-1}-\psi_{\mathbb{Q}}(s). \tag{8}$$

If χ is a non-principal primitive Dirichlet character modulo f, with $\chi(-1)=1$, then for all $s\in\mathbb{C}$ we have:

$$\frac{L'(s,\chi)}{L(s,\chi)} = B_\chi+\sum_{\rho\text{ of }L_\chi}\left(\frac{1}{s-\rho}+\frac{1}{\rho}\right)-\frac{1}{2}\log f-\psi_{\mathbb{Q}}(s). \tag{9}$$

Each sum above is over the non-trivial zeros ρ of the corresponding functions, and is absolutely and uniformly convergent on compact subsets of $\mathbb{C}$. Henceforth we adopt the notation that ρ will always denote a non-trivial zero with $0<\Re(\rho)<1$.

Each of (8), (9) involves a constant B which can be difficult to estimate. Fortunately, in both cases this constant can be eliminated from the equation as follows. Provided the sum is taken in symmetric order[5], one has $B+\sum_{\rho\text{ of }\zeta}\rho^{-1}=0$, and similarly for B_χ. See [**10**] for a simple argument which gives this result for the constant B. The analogous result for B_χ is not obvious; in fact, it was not known until the introduction of the Weil formulas (see [**52, 53**]). Plugging $s=1$ into (9) and comparing against (2.3.1) of [**17**] gives a proof of this result.

LEMMA 7.4. *Let χ be a non-principal primitive Dirichlet character modulo f with $\chi(-1)=1$. Assume the RH and the GRH for $L(s,\chi)$. For $a\in(0,1)$ we have*

$$\sum_{\rho\text{ of }\zeta,L_\chi}\frac{1}{|\rho+a|^2}\le\frac{1}{2a+1}\left(\log f+2\left(\frac{1}{a+1}+\frac{1}{a}\right)+4\psi_{\mathbb{Q}}(a+1)\right).$$

[5]Taking the sum in symmetric order means: $\sum_\rho = \lim_{T\to\infty}\sum_{\substack{\rho=\sigma+it\\ |t|<T}}$

PROOF. Substitute $s = \sigma$ into (9), add the result to its conjugate, and use the fact that the real part of $B_\chi + \sum_\rho \rho^{-1}$ equals 0 to obtain

$$\sum_{\rho \text{ of } L_\chi} \left(\frac{1}{\sigma - \rho} + \frac{1}{\sigma - \overline{\rho}} \right) = \log f + 2\Re \frac{L'(\sigma, \chi)}{L(\sigma, \chi)} + 2\psi_{\mathbb{Q}}(\sigma) .$$

This equation goes back to Landau (see [**22**]). In exactly the same manner, one can show

$$\sum_{\rho \text{ of } \zeta} \left(\frac{1}{\sigma - \rho} + \frac{1}{\sigma - \overline{\rho}} \right) = 2\frac{\zeta'(\sigma)}{\zeta(\sigma)} + 2\left(\frac{1}{\sigma} + \frac{1}{\sigma - 1} \right) + 2\psi_{\mathbb{Q}}(\sigma).$$

Setting $\sigma = a + 1$ and supposing that $\Re(\rho) = 1/2$, we find:

$$\frac{1}{|\rho + a|^2} = \frac{1}{2a+1} \left(\frac{1}{\sigma - \rho} + \frac{1}{\sigma - \overline{\rho}} \right).$$

To complete the proof, we combine everything above and note that

$$\frac{\zeta'(\sigma)}{\zeta(\sigma)} + \Re \frac{L'(\sigma, \chi)}{L(\sigma, \chi)} < 0 ,$$

by considering the Dirichlet series for $(\zeta'/\zeta + L_\chi'/L_\chi)(s)$. □

LEMMA 7.5. *Let χ be a non-principal primitive Dirichlet character modulo f with $\chi(-1) = 1$. For $a \in (0, 1)$ we have*

$$\left| \left(\frac{\zeta'}{\zeta} \right)(-a) - \left(\frac{L_\chi'}{L_\chi} \right)(-a) \right| \le (a+2) \sum_{\rho \text{ of } \zeta, L_\chi} \frac{1}{|(\rho + a)(2 - \rho)|} + 2\left| \frac{\zeta'(2)}{\zeta(2)} \right| + \frac{1}{a} + \frac{1}{a+1} + \frac{3}{2} .$$

PROOF. We begin with the following formulas which hold for all $s \in \mathbb{C}$, provided the sums are taken in symmetric order:

$$\left(\frac{\zeta'}{\zeta} \right)(s) = \sum_{\rho \text{ of } \zeta} \frac{1}{s - \rho} - \frac{1}{s} - \frac{1}{s-1} - \psi_{\mathbb{Q}}(s), \tag{10}$$

$$\left(\frac{L_\chi'}{L_\chi} \right)(s) = \sum_{\rho \text{ of } L_\chi} \frac{1}{s - \rho} - \frac{1}{2} \log f - \psi_{\mathbb{Q}}(s). \tag{11}$$

Formulas (10) and (11) are obtained from (8) and (9) respectively by applying the facts $\sum_{\rho \text{ of } \zeta} \rho^{-1} + B = 0$ and $\sum_{\rho \text{ of } L_\chi} \rho^{-1} + B_\chi = 0$. Plugging $s = 2$ into (10) and subtracting it from itself, and similarly for (11), yields:

$$\left(\frac{\zeta'}{\zeta} \right)(s) = \left(\frac{\zeta'}{\zeta} \right)(2) + \sum_\rho \left(\frac{1}{s - \rho} - \frac{1}{2 - \rho} \right) + \frac{3}{2} - \frac{1}{s} - \frac{1}{s-1} + \psi_{\mathbb{Q}}(2) - \psi_{\mathbb{Q}}(s),$$

$$\left(\frac{L_\chi'}{L_\chi} \right)(s) = \left(\frac{L_\chi'}{L_\chi} \right)(2) + \sum_\rho \left(\frac{1}{s - \rho} - \frac{1}{2 - \rho} \right) + \psi_{\mathbb{Q}}(2) - \psi_{\mathbb{Q}}(s).$$

Using the above, together with the fact

$$\frac{1}{-a - \rho} - \frac{1}{2 - \rho} = -\frac{a+2}{(\rho + a)(2 - \rho)} ,$$

we can write

$$\left(\frac{\zeta'}{\zeta}\right)(-a) - \left(\frac{L'_\chi}{L_\chi}\right)(-a) = (a+2)\left(\sum_{\rho \text{ of } L_\chi} \frac{1}{(\rho+a)(2-\rho)} - \sum_{\rho \text{ of } \zeta} \frac{1}{(\rho+a)(2-\rho)}\right)$$
$$+ \left(\frac{\zeta'}{\zeta}\right)(2) - \left(\frac{L'_\chi}{L_\chi}\right)(2) + \frac{3}{2} + \frac{1}{a} + \frac{1}{a+1}.$$

The result follows upon taking absolute values and using the fact that $\left|\left(L'_\chi/L_\chi\right)(2)\right| \leq \left|(\zeta'/\zeta)(2)\right|$. □

In applying the previous lemma, it will be helpful to note that when $\Re(\rho) = 1/2$, one has

$$|\rho + a|^2 \leq |(\rho+a)(2-\rho)|. \tag{12}$$

LEMMA 7.6. *Let χ be a non-principal primitive Dirichlet character modulo f with $\chi(-1) = 1$. For $a \in (0,1)$ we have*

$$\left|\left(\frac{\zeta'}{\zeta}\right)'(-a) - \left(\frac{L'_\chi}{L_\chi}\right)'(-a)\right| < \sum_{\rho \text{ of } \zeta, L_\chi} \frac{1}{|\rho+a|^2} + \frac{1}{a^2} + \frac{1}{(a+1)^2}.$$

PROOF. We start by differentiating (10) and (11); this gives

$$\left(\frac{\zeta'}{\zeta}\right)'(s) = -\sum_{\rho \text{ of } \zeta} \frac{1}{(s-\rho)^2} + \frac{1}{s^2} + \frac{1}{(s-1)^2} - \psi'_{\mathbb{Q}}(s), \tag{13}$$

$$\left(\frac{L'_\chi}{L_\chi}\right)'(s) = -\sum_{\rho \text{ of } L_\chi} \frac{1}{(s-\rho)^2} - \psi'_{\mathbb{Q}}(s), \tag{14}$$

which allows us to write

$$\left(\frac{\zeta'}{\zeta}\right)'(-a) - \left(\frac{L'_\chi}{L_\chi}\right)'(-a) = \sum_{\rho \text{ of } L_\chi} \frac{1}{(\rho+a)^2} - \sum_{\rho \text{ of } \zeta} \frac{1}{(\rho+a)^2} + \frac{1}{a^2} + \frac{1}{(a+1)^2}.$$

The result follows. □

PROPOSITION 7.7. *Let χ be a non-principal primitive Dirichlet character modulo f with $\chi(-1) = 1$. Assume the RH and the GRH for $L(s,\chi)$. We define*

$$\sum_\rho := \sum_{\rho \text{ of } \zeta, L_\chi} \frac{1}{|\rho+a|^2}.$$

For $x > 0$ we have

$$\frac{x}{(a+1)^2} + \frac{1}{a^2} \leq \sqrt{x}\sum_\rho + 2 \sum_{\substack{n<x \\ \chi(n) \neq 1}} \Lambda(n)(n/x)^a \log(x/n)$$
$$+ \frac{\log x}{x^a}\left((a+2)\sum_\rho + 2\left|\frac{\zeta'(2)}{\zeta(2)}\right| + \frac{1}{a} + \frac{1}{a+1} + \frac{3}{2}\right)$$
$$+ \frac{1}{x^a}\left(\sum_\rho + \frac{1}{a^2} + \frac{1}{(a+1)^2}\right).$$

PROOF. Combine Lemmas 7.3, 7.5, and 7.6 together with (12). □

PROOF OF THEOREM 7.1. The result for a general character follows from the corresponding result for primitive characters and hence we may assume χ is a primitive character modulo f. Define $x := 2(\log f)^2$. Since $f \geq 10^9$, we have $x > 858$. By way of contradiction, suppose that $\chi(n) = 1$ for all $n < x$.

Apply Proposition 7.7 and set $a = 1/2$. We find

$$\frac{x}{9/4} + 4 \leq \sqrt{x} \sum_{\rho} + 2 \sum_{\substack{n < x \\ \chi(n) \neq 1}} \Lambda(n)(n/x)^{1/2} \log(x/n)$$

$$+ \frac{\log x}{\sqrt{x}} \left(\frac{5}{2} \sum_{\rho} + 2 \left| \frac{\zeta'(2)}{\zeta(2)} \right| + \frac{25}{6} \right) + \frac{1}{\sqrt{x}} \left(\sum_{\rho} + \frac{40}{9} \right),$$

where $\zeta'(2)/\zeta(2) \approx -0.56996$. Apply Lemma 7.4, together with the fact $\psi_{\mathbb{Q}}(3/2) \approx -1.1153$, to obtain

$$\sum_{\rho \text{ of } \zeta, L_\chi} \frac{1}{\left|\rho + \frac{1}{2}\right|^2} \leq \frac{1}{2} \log f + 0.437 .$$

This ultimately leads to $\sqrt{x} < 1.2 \log f$ which implies $x < 1.5(\log f)^2$, a contradiction. $\square$

8. Conclusion

We have surveyed some classical results concerning the least quadratic non-residue n_p. Although asking about the size of n_p seems like an innocent question, it has been studied via many different techniques. The proof of the Pólya–Vinogradov inequality in section 3 has a Fourier-analytic flavor, a deep theorem from algebraic geometry is employed to prove the Burgess inequality in section 4, we have the large sieve in section 6 which can be couched in terms of functional analysis, and we have complex analysis and use of the Generalized Riemann Hypothesis in section 7. We have also showcased the techniques that one can use to get some explicit results by showing how to recover constants in most of the sections. Despite how well-studied this problem is, we are still far from proving (unconditional) upper bounds that are anywhere close to the truth. We hope that this survey has given the reader an idea of what is known and not known about the least quadratic non-residue.

Acknowledgments

We would like to thank the referee for the careful reading of our manuscript and the suggestions made that improved the paper.

References

[1] N. C. Ankeny, *The least quadratic non residue*, Ann. of Math. (2) **55** (1952), 65–72, DOI 10.2307/1969420. MR45159

[2] Eric Bach, *Analytic methods in the analysis and design of number-theoretic algorithms*, ACM Distinguished Dissertations, MIT Press, Cambridge, MA, 1985. MR807772

[3] Eric Bach, *Explicit bounds for primality testing and related problems*, Math. Comp. **55** (1990), no. 191, 355–380, DOI 10.2307/2008811. MR1023756

[4] Enrico Bombieri, *Counting points on curves over finite fields (d'après S. A. Stepanov)*, Séminaire Bourbaki, 25ème année (1972/1973), Exp. No. 430, Springer, Berlin, 1974, pp. 234–241. Lecture Notes in Math., Vol. 383. MR0429903

[5] Andrew R. Booker, *Quadratic class numbers and character sums*, Math. Comp. **75** (2006), no. 255, 1481–1492, DOI 10.1090/S0025-5718-06-01850-3. MR2219039

[6] D. A. Burgess, *On character sums and primitive roots*, Proc. London Math. Soc. (3) **12** (1962), 179–192, DOI 10.1112/plms/s3-12.1.179. MR132732

[7] D. A. Burgess, *A note on the distribution of residues and non-residues*, J. London Math. Soc. **38** (1963), 253–256, DOI 10.1112/jlms/s1-38.1.253. MR148628

[8] W. J. Wong, *A characterization of the alternating group of degree* 8, Proc. London Math. Soc. (3) **13** (1963), 359–383, DOI 10.1112/plms/s3-13.1.359. MR147526

[9] Alina Carmen Cojocaru and M. Ram Murty, *An introduction to sieve methods and their applications*, London Mathematical Society Student Texts, vol. 66, Cambridge University Press, Cambridge, 2006. MR2200366

[10] Harold Davenport, *Multiplicative number theory*, 3rd ed., Graduate Texts in Mathematics, vol. 74, Springer-Verlag, New York, 2000. Revised and with a preface by Hugh L. Montgomery. MR1790423

[11] Régis de la Bretèche and Marc Munsch, *Minimizing GCD sums and applications to non-vanishing of theta functions and to Burgess' inequality*, arXiv e-prints (2018), arXiv:1812.03788v4.

[12] V. R. Fridlender, *On the least nth-power non-residue* (Russian), Doklady Akad. Nauk SSSR (N.S.) **66** (1949), 351–352. MR0029940

[13] D. A. Frolenkov and K. Soundararajan, *A generalization of the Pólya-Vinogradov inequality*, Ramanujan J. **31** (2013), no. 3, 271–279, DOI 10.1007/s11139-012-9462-y. MR3081668

[14] S. W. Graham and C. J. Ringrose, *Lower bounds for least quadratic nonresidues*, Analytic number theory (Allerton Park, IL, 1989), Progr. Math., vol. 85, Birkhäuser Boston, Boston, MA, 1990, pp. 269–309. MR1084186

[15] Andrew Granville, R. A. Mollin, and H. C. Williams, *An upper bound on the least inert prime in a real quadratic field*, Canad. J. Math. **52** (2000), no. 2, 369–380, DOI 10.4153/CJM-2000-017-5. MR1755783

[16] Adolf Hildebrand, *On the constant in the Pólya-Vinogradov inequality*, Canad. Math. Bull. **31** (1988), no. 3, 347–352, DOI 10.4153/CMB-1988-050-1. MR956367

[17] Yasutaka Ihara, V. Kumar Murty, and Mahoro Shimura, *On the logarithmic derivatives of Dirichlet L-functions at $s = 1$*, Acta Arith. **137** (2009), no. 3, 253–276, DOI 10.4064/aa137-3-6. MR2496464

[18] Henryk Iwaniec and Emmanuel Kowalski, *Analytic number theory*, American Mathematical Society Colloquium Publications, vol. 53, American Mathematical Society, Providence, RI, 2004, DOI 10.1090/coll/053. MR2061214

[19] Bryce Kerr, Igor E. Shparlinski, and Kam Hung Yau, *A refinement of the Burgess bound for character sums*, Michigan Math. J. **69** (2020), no. 2, 227–240, DOI 10.1307/mmj/1573700737. MR4104371

[20] J. C. Lagarias and A. M. Odlyzko, *Effective versions of the Chebotarev density theorem*, Algebraic number fields: L-functions and Galois properties (Proc. Sympos., Univ. Durham, Durham, 1975), Academic Press, London, 1977, pp. 409–464. MR0447191

[21] Youness Lamzouri, Xiannan Li, and Kannan Soundararajan, *Conditional bounds for the least quadratic non-residue and related problems*, Math. Comp. **84** (2015), no. 295, 2391–2412, DOI 10.1090/S0025-5718-2015-02925-1. MR3356031

[22] Edmund Landau, *Zur Theorie der Heckeschen Zetafunktionen, welche komplexen Charakteren entsprechen* (German), Math. Z. **4** (1919), no. 1-2, 152–162, DOI 10.1007/BF01203394. MR1544358

[23] Mariana Levin, Carl Pomerance, and K. Soundararajan, *Fixed points for discrete logarithms*, Algorithmic number theory, Lecture Notes in Comput. Sci., vol. 6197, Springer, Berlin, 2010, pp. 6–15, DOI 10.1007/978-3-642-14518-6_5. MR2969346

[24] Pierre Lezowski and Kevin J. McGown, *The Euclidean algorithm in quintic and septic cyclic fields*, Math. Comp. **86** (2017), no. 307, 2535–2549, DOI 10.1090/mcom/3169. MR3647971

[25] U. V. Linnik, *"The large sieve."*, C. R. (Doklady) Acad. Sci. URSS (N.S.) **30** (1941), 292–294. MR0004266

[26] U. V. Linnik, *A remark on the least quadratic non-residue*, C. R. (Doklady) Acad. Sci. URSS (N.S.) **36** (1942), 119–120. MR0007758

[27] Shilin Ma, Kevin McGown, Devon Rhodes, and Mathias Wanner, *Explicit bounds for small prime nonresidues*, J. Number Theory **204** (2019), 599–607, DOI 10.1016/j.jnt.2019.04.021. MR3991435

[28] Kevin J. McGown, *Norm-Euclidean cyclic fields of prime degree*, Int. J. Number Theory **8** (2012), no. 1, 227–254, DOI 10.1142/S1793042112500133. MR2887892

[29] Kevin J. McGown, *Norm-Euclidean Galois fields and the generalized Riemann hypothesis* (English, with English and French summaries), J. Théor. Nombres Bordeaux **24** (2012), no. 2, 425–445. MR2950700

[30] Kevin J. McGown, *On the constant in Burgess' bound for the number of consecutive residues or non-residues.* part 2, Funct. Approx. Comment. Math. **46** (2012), no. part 2, 273–284, DOI 10.7169/facm/2012.46.2.10. MR2931671

[31] Kevin J. McGown, *On the second smallest prime non-residue*, J. Number Theory **133** (2013), no. 4, 1289–1299, DOI 10.1016/j.jnt.2012.09.011. MR3004000

[32] Kevin J. McGown and Tim Trudgian, *Explicit upper bounds on the least primitive root*, Proc. Amer. Math. Soc. **148** (2020), no. 3, 1049–1061, DOI 10.1090/proc/14800. MR4055934

[33] Hugh L. Montgomery, *Topics in multiplicative number theory*, Lecture Notes in Mathematics, Vol. 227, Springer-Verlag, Berlin-New York, 1971. MR0337847

[34] Hugh L. Montgomery, *The analytic principle of the large sieve*, Bull. Amer. Math. Soc. **84** (1978), no. 4, 547–567, DOI 10.1090/S0002-9904-1978-14497-8. MR466048

[35] Hugh L. Montgomery and Robert C. Vaughan, *Multiplicative number theory. I. Classical theory*, Cambridge Studies in Advanced Mathematics, vol. 97, Cambridge University Press, Cambridge, 2007. MR2378655

[36] Karl K. Norton, *Numbers with small prime factors, and the least kth power non-residue*, Memoirs of the American Mathematical Society, No. 106, American Mathematical Society, Providence, R.I., 1971. MR0286739

[37] René Peralta, *On the distribution of quadratic residues and nonresidues modulo a prime number*, Math. Comp. **58** (1992), no. 197, 433–440, DOI 10.2307/2153045. MR1106978

[38] Herbert Robbins, *A remark on Stirling's formula*, Amer. Math. Monthly **62** (1955), 26–29, DOI 10.2307/2308012. MR69328

[39] J. Barkley Rosser and Lowell Schoenfeld, *Approximate formulas for some functions of prime numbers*, Illinois J. Math. **6** (1962), 64–94. MR137689

[40] Hans Salié, *Über den kleinsten positiven quadratischen Nichtrest nach einer Primzahl* (German), Math. Nachr. **3** (1949), 7–8, DOI 10.1002/mana.19490030104. MR33844

[41] Hans Salié, *Über die kleinste Primzahl, die eine gegebene Primzahl als kleinsten positiven quadratischen Nichtrest hat* (German), Math. Nachr. **29** (1965), 113–114, DOI 10.1002/mana.19650290109. MR177941

[42] Wolfgang M. Schmidt, *Zur Methode von Stepanov* (German), Acta Arith. **24** (1973), 347–367. (errata insert), DOI 10.4064/aa-24-4-347-367. MR332792

[43] Wolfgang M. Schmidt, *Equations over finite fields. An elementary approach*, Lecture Notes in Mathematics, Vol. 536, Springer-Verlag, Berlin-New York, 1976. MR0429733

[44] S. A. Stepanov, *Elementary method in the theory of congruences for a prime modulus*, Acta Arith. **17** (1970), 231–247, DOI 10.4064/aa-17-3-231-247. MR269657

[45] Enrique Treviño, *The least inert prime in a real quadratic field*, Math. Comp. **81** (2012), no. 279, 1777–1797, DOI 10.1090/S0025-5718-2012-02579-8. MR2904602

[46] Enrique Treviño, *On the maximum number of consecutive integers on which a character is constant*, Mosc. J. Comb. Number Theory **2** (2012), no. 1, 56–72. MR2988388

[47] Enrique Treviño, *The Burgess inequality and the least kth power non-residue*, Int. J. Number Theory **11** (2015), no. 5, 1653–1678, DOI 10.1142/S1793042115400163. MR3376232

[48] Enrique Treviño, *The least k-th power non-residue*, J. Number Theory **149** (2015), 201–224, DOI 10.1016/j.jnt.2014.10.019. MR3296008

[49] Enrique Treviño, *Corrigendum to "On the maximum number of consecutive integers on which a character is constant" [MR2988388]*, Mosc. J. Comb. Number Theory **7** (2017), no. 3, 1–2. MR3749573

[50] Ivan Matveevič Vinogradov, *Selected works*, Springer-Verlag, Berlin, 1985. With a biography by K. K. Mardzhanishvili; Translated from the Russian by Naidu Psv [P. S. V. Naidu]; Translation edited by Yu. A. Bakhturin, DOI 10.1007/978-3-642-15086-9. MR807530

[51] André Weil, *On some exponential sums*, Proc. Nat. Acad. Sci. U.S.A. **34** (1948), 204–207, DOI 10.1073/pnas.34.5.204. MR27006

[52] André Weil, *Sur les "formules explicites" de la théorie des nombres premiers* (French), Comm. Sém. Math. Univ. Lund [Medd. Lunds Univ. Mat. Sem.] **1952** (1952), no. Tome Supplémentaire, 252–265. MR53152

[53] André Weil, *Sur les formules explicites de la théorie des nombres* (French, with Russian summary), Izv. Akad. Nauk SSSR Ser. Mat. **36** (1972), 3–18. MR0379440

California State University, Chico
Email address: kmcgown@csuchico.edu

Lake Forest College
Email address: trevino@lakeforest.edu

Contemporary Mathematics
Volume **775**, 2021
https://doi.org/10.1090/conm/775/15594

Thin homotopy and the holonomy approach to gauge theories

Claudio Meneses

ABSTRACT. We survey several mathematical developments in the holonomy approach to gauge theory. A cornerstone of this approach is the introduction of group structures on spaces of based loops on a smooth manifold, relying on certain homotopy equivalence relations — such as the so-called thin homotopy — and the resulting interpretation of gauge fields as group homomorphisms to a Lie group G satisfying a suitable smoothness condition, encoding the holonomy of a gauge orbit of smooth connections on a principal G-bundle. We also prove several structural results on thin homotopy, and in particular we clarify the difference between thin equivalence and retrace equivalence for piecewise-smooth based loops on a smooth manifold, which are often used interchangeably in the physics literature. We conclude by listing a set of questions on topological and functional analytic aspects of groups of based loops, which we consider to be fundamental to establish a rigorous differential geometric foundation of the holonomy formulation of gauge theory.

CONTENTS

1. Introduction

Since the early developments of gauge theory, the idea of "nonintegrable phase factors" of a gauge field has been recognized as a fundamental notion on which the study of gauge symmetry could be built [**1,65**]. In the same way that gauge fields can be modeled mathematically as orbits in a space of smooth connections on a smooth principal G-bundle $P \to M$ under the action of the infinite dimensional group of bundle automorphisms [**97**], the phase factors associated to a gauge field

2020 *Mathematics Subject Classification.* Primary 53C29, 55P10, 51H25; Secondary 81Q70.
Key words and phrases. Thin homotopy, gauge field, holonomy.
The author was supported by the DFG SPP 2026 priority programme "Geometry at infinity".

can be modeled geometrically in terms of the holonomy (or more precisely, its conjugacy class in G) of any such an orbit representative around any given based loop [**98**]. Although the mathematical idea of reconstructing a gauge orbit of smooth connections on a principal G-bundle in terms of the holonomies of orbit representatives is rather old and seems to have originated in the work of Kobayashi [**54**], considering the latter as a substitute for the former is usually reserved for connections satisfying further integrability conditions such as flatness. An apparent reason that justifies this state of affairs is the following. While the holonomy of a flat connection depends only on the structure of the fundamental group $\pi_1(M,p)$ of the pointed base manifold (M,p), the holonomy of an arbitrary smooth connection would depend on the structure of a considerably more complicated group $\mathscr{L}^{\cdot}(M,p)$ of equivalence classes of based loops, constructed in terms of weaker homotopy equivalences for which holonomy is a well-defined invariant. Thus, in the same way that a gauge orbit of flat connections may be equivalently characterized as a G-conjugacy class of group homomorphisms $\rho : \pi_1(M,p) \to G$, a gauge orbit of smooth connections can be equivalently characterized in terms of a G-conjugacy class of *holonomy homomorphisms* [**14, 18, 54, 58**]

$$H : \mathscr{L}^{\cdot}(M,p) \to G$$

satisfying certain suitable smoothness conditions. However, the topological groups $\mathscr{L}^{\cdot}(M,p)$ are still poorly understood, and do not seem to be widely known to differential geometers.

It is thus not surprising that the study of gauge symmetry in terms of holonomy and parallel transport — the so-called loop approach in the physics literature — has not experienced a similar dramatic development into a mature mathematical discipline when compared to the more standard approaches to mathematical gauge theory. Any serious intent to rectify this unfortunate missed opportunity would necessarily rely on the development of pertinent analytic tools in the study of the geometry of spaces of based loops in a manifold, similar in spirit to the general programme elucidated in the works [**56, 83**].

Two loop homotopy relations that are often used interchangeably in the physics literature are those of *thin homotopy* [**14, 18**] and *retrace homotopy* [**54, 58, 89**]. Both relations lead to equivalence relations that allow to define groups of based loops $\mathscr{L}^{\cdot}(M,p)$. However, while retrace homotopy is a particular case of thin homotopy, it is possible to construct examples of thin homotopies that cannot be factored as a finite concatenation of reparametrizations and retracings. One of the main objectives of this article is to clarify the difference between these relations when defined over the space of piecewise-smooth based loops $\Omega^{\mathrm{ps}}(M,p)$ of a smooth pointed manifold (M,p), which is presented in detail in sections 3 and 4. On the other hand, the groups $\mathscr{L}^{\cdot}(M,p)$ could be topologized as quotients in terms of a suitable function space topology on the set $\Omega^{\mathrm{ps}}(M,p)$. Our results suggest that a topological characterization of thin homotopy is also possible, with retracings as "basic building blocks" through a limiting process, in the spirit of Whitney's approximation theorems [**94**]. We hope to return to these questions in the future.

In section 2 we have tried to compile a sufficiently extensive list of references reflecting the historical development of the holonomy approach to gauge theory in geometry and physics. In this regard, a comprehensive compendium of the physics literature on the subject can be found in the book [**40**]. A rather unfortunate feature of the existing mathematical literature on the subject is its evident scarcity and sparsity. We have attempted to collect what we consider to be a minimal comprehensive set of mathematical bibliography that illustrates our point, to the best of our abilities, while avoiding surveying the standard mathematical approach that originated from Atiyah's school [**10**].[1] Given that the existing mathematical literature on the holonomy approach to gauge theory is the result of the work of many researchers and communities over the last 70 years, it is plausible that we could have made several important omissions. We apologize if that were the case.

Notes added on proof. After the first version of this manuscript was made public, I was notified by Tamer Tlas of his work [**92**]. Among other things, he proves a stronger version of theorem 4. This article has been modified to include his findings. After this article was in final form, I learned of Driver's work [**32**] which precedes most of the proofs of theorem 1, and seems to have been unnoticed by their respective authors.

2. Brief survey

The notions of holonomy and parallel transport in geometry and physics have a long and rich history that flourished through the entire twentieth century. From a physical perspective, the first occurrence of nonintegrable phase factors in the context of field theories with gauge symmetry can be traced back to the seminal work of Dirac [**30**] on the quantization of the electric charge. The fundamental question of the physical significance of electromagnetic potentials was first addressed from both an experimental [**1**] and theoretical [**65**] viewpoints. Inspired by the pioneering ideas of Yang and Mills [**99**] that led to the study of non-abelian gauge theories in physics, Białynicki-Birula [**17**] generalized Mandelstam's work to the case of non-abelian gauge fields, and determined an analog for the Yang–Mills equations in this context. Yet another pioneering idea was proposed by Wilson [**95**], who introduced the notion of *Wilson loops* as a fundamental complete set of gauge field observables in his study of the problem of confinement. It is rather significant that at the same time that the celebrated Wu–Yang dictionary [**97**] was formulated[2] — leading to the standard mathematical axiomatization of gauge field theories — Yang [**98**] also postulated an "integral fomalism" for non-abelian gauge fields based on the idea of non-abelian phase factors, which could be considered as the first general formulation of gauge theories from the holonomy perspective (in this respect, cf. [**53**], [**31**], [**64**], [**33**], [**21**–**23**] for similar considerations), as well as the first attempt at the reconstruction of gauge fields in terms of their non-abelian phase factors (see [**46**] for a more detailed proposal — at the physicists' level of rigor — of the same idea formulated in terms of Wilson loop observables), while leading to non-abelian generalizations of Stokes' theorem [**3**], [**20**]. In yet another interesting direction, Aref'eva and Polyakov [**2, 79**] observed that when an ordinary gauge field A is interpreted as a map $H : \Omega^{\cdot}(M,p) \to G$ in terms of its holonomy

[1]Besides Atiyah's many excellent expositions on the subject, another interesting mathematical survey on the geometry of vector bundles and connections as a very convenient language in physics, and its relation to the theory of moduli spaces, is given in [**93**].

[2]The appearance of the Wu–Yang dictionary was a catalyst for the birth of mathematical gauge theory, identifying the physical notion of a (classical) gauge field with the geometric notion of an orbit of connections on a principal bundle, under the action of the gauge group of bundle automorphisms. However, the first mathematician who studied the relation between gauge fields and connections on principal bundles was Robert Hermann [**50**–**52**]. His insights on the subject have been largely ignored due to the unorthodox nature of his academic career.

(a so-called *chiral field on loop space*) the Yang–Mills equations for A correspond to the zero curvature equation for the "potential" $(\delta H)H^{-1}$.

Lattice gauge theory (LGT) is a physical formalism for quantum gauge theories which originated in the work of Wilson [**95**]. Its basic idea is to replace a gauge field by a collection of "parallel transporters" (elements in G) over a discrete set of paths in the base manifold (usually associated to a lattice in $\mathbb{T}^n$), which play the role of a discretization of the former. Although LGT is sometimes described as a pragmatic mechanism to model theoretical predictions (at the quantum level) that can be tested experimentally [**27**], the insights it offers motivate the very appealing possibility of constructing quantum gauge theories by means of rigorously defined gauge field discretization mechanisms and continuum limits [**74**].

All of the previously described ideas lead to many interesting mathematical questions. The differential geometry of connections on principal bundles [**55**], has ever since been considered as the standard approach towards a solid mathematical foundation of the subject. In this respect, the study of connections on principal bundles in terms of their holonomy seemed to have been motivated by topological questions in the study of fibre bundle theory [**89, 90**], [**70**], [**28, 57**], [**86, 87**], [**29**], while the first purely geometric aspects were considered by Kobayashi [**54**], who posed the question of reconstruction of a connection on a principal bundle in terms of its holonomies for the first time.[3] Yet another fundamental mathematical precedent in the study of the differential geometry of loop spaces was initiated in the work of Chen [**24, 25**] where the idea of *iterated integrals* is introduced.[4]

Regarding topological aspects in the holonomy formulation of gauge theories, the work of Lüscher [**62**] initiated a series of investigations on the reconstruction of topological charges for a principal G-bundle purely in terms of LGT data [**75–78**], which relies on a generic interpolation mechanism in the space of LGT data. A different account for the understanding of these problems is presented in [**68, 69**], which addresses directly the characterization of the topology of principal G-bundles within the holonomy approach.

The holonomy formulation of gauge theory in physics experienced another boost in the 1980s due to several attempts to establish a theory of quantum gravity within the framework of representation theory for groups of based loops.[5] From a more mathematical perspective, and building up on ideas of Gambini and Trias [**37–39, 41–44**], the school of Ashtekar initiated a programme that led to the development of the subject within the context of C^*-algebras [**5, 6**], *generalized connections* [**7, 8**], and *generalized measures* on the spaces $\mathscr{A}_P/\mathscr{G}_P$ of gauge orbits of smooth connections of a principal G-bundle P [**9, 11, 12, 66, 91**]. Generalized connections require to consider a suitable completion of $\mathscr{A}_P/\mathscr{G}_P$ (a rigorous mathematical treatment of generalized connections has also been given in [**34, 35**]). The

[3] This should be contrasted with the work of Narasimhan and Ramanan [**73**], who looked for an alternative reconstruction scheme in terms of a universal connection on the universal bundle of a Lie group G, which was analogous in spirit to the topological trends of that time.

[4] The literature on Chen's iterated integrals is incredibly vast as the subject encountered many applications in both physics and mathematics; we only refer to [**88**], [**49**], and [**84**] for subsequent developments that are directly relevant for the purposes of this work.

[5] In this respect, the survey works [**59–61**] present a short summary of motivations, results and goals of the discipline, while the book [**40**] is a rather comprehensive account of the developments of the subject in those decades.

study of the natural topological and differential geometric aspects of group structures on spaces of based loops on a smooth manifold, and their implications in the study of gauge field theories through holonomy, led to the introduction of the so-called *extended loop groups* [**15, 16**] and their generalized loop representations (cf. [**80**], and the technical problems that this approach seemed to have unearthed).

All of the previous developments provided many insights that paved the way towards a rigorous mathematical formulation of the reconstruction theorem of gauge fields in terms of their holonomy homomorphisms from a group of based loops on a manifold to a Lie group G. Different proofs of the reconstruction theorem appeared in [**13,14**], [**58**], [**48**], [**18**], leading to subsequent mathematical developments of the notion of groups of based loops [**36**], [**45**], [**96**], [**63**], [**26,81**], [**92**]. Somewhat independent, but equally important, are the works of Gross [**47**], who gave an analytic proof of the equivalence between the Yang–Mills and Mandelstam–Białynicki-Birula equations, and Morrison [**72**], who gave a characterization of connections on principal bundles over oriented Riemannian surfaces satisfying the Yang–Mills equations in terms of their corresponding holonomies. Regarding the differential geometry of loop spaces, more recently Stacey [**83, 85**] has studied the problem of defining manifold structures on spaces of piecewise-smooth based loops on a manifold, in the sense of the seminal work of Kriegl and Michor [**56**].[6]

3. Thin homotopy and groups of based loops on a manifold

Let M be a smooth connected manifold, on which a point $p \in M$ is fixed once and for all. A *piecewise-smooth loop based at* p is a continuous map $\gamma : [0,1] \to M$ such that $\gamma(0) = \gamma(1) = p$, and for some finite subset $0 = t_0 < \cdots < t_r = 1$, $\gamma|_{[t_{i-1},t_i]}$ is smooth $\forall\, i = 1, \ldots, r$. Piecewise-analytic based loops are defined analogously.

Let $\Omega^0(M,p)$ denote the set of continuous loops in M based at p. $\Omega^0(M,p)$ admits several natural function space topologies, the most common of which is the compact-open topology. Under such an intrinsic topological space structure, several important subspaces of $\Omega^0(M,p)$ arise if we impose additional smoothness constraints, i.e., the subspaces $\Omega^{\mathrm{ps}}(M,p) \supsetneq \Omega^{\mathrm{pa}}(M,p)$ of piecewise-smooth and piecewise-analytic loops, the subspaces $\Omega^\infty(M,p) \supsetneq \Omega^\omega(M,p)$ of smooth and analytic loops, etc.

$\Omega^\cdot(M,p)$ will denote in general any subspace of $\Omega^0(M,p)$ as in the examples above. A homotopy between two elements $\gamma_0, \gamma_1 \in \Omega^\cdot(M,p)$ will then be understood as a continuous map $\eta : [0,1]^2 \to M$ such that $\eta(s,\cdot) \in \Omega^\cdot(M,p)\ \forall\ s \in [0,1]$, $\eta(i,\cdot) = \gamma_i$ for $i = 0,1$, and that also satisfies the defining property of $\Omega^\cdot(M,p)$, i.e., is smooth, piecewise-smooth on a polygonal paving of $[0,1]^2$, etc. By a slight abuse of notation, we will denote the trivial loop at p simply by p.

The standard concatenation of an ordered pair of based loops (γ, γ') in $\Omega^0(M,p)$, defined as

$$\gamma' \cdot \gamma(t) = \begin{cases} \gamma(2t) & 0 \le t \le 1/2, \\ \gamma'(2t-1) & 1/2 \le t \le 1, \end{cases}$$

[6]It is very tempting to inquire if the latter ideas and analytic techniques could be implemented to build a rigorous "differential calculus" for smooth holonomy homomorphisms, that could serve as an alternative to the standard geometric analysis approach to mathematical gauge theory modeled on the quotient spaces $\mathscr{A}_P/\mathscr{G}_P$ (cf. [**88**]).

determines a binary operation on $\Omega^0(M,p)$, which together with the operation of loop inversion $\gamma^{-1}(t) := \gamma(1-t)$, descends into a group structure in the space of equivalence classes $\pi_1(M,p) := \Omega^0(M,p)/\sim_H$ under homotopy of based loops, giving rise to the fundamental group of M. However, the fundamental group is just one example of a group structure induced on a space of equivalence classes in $\Omega^0(M,p)$ under a suitable homotopy relation, which is also the coarsest in a sense that will be made precise in definition 1. For this purpose, we will call a subspace $\Omega^{\cdot}(M,p) \subset \Omega^0(M,p)$ *concatenable* if concatenation of loops also defines a binary operation on it. $\Omega^{\mathrm{ps}}(M,p)$ and $\Omega^{\mathrm{pa}}(M,p)$ are examples of concatenable subspaces, while $\Omega^{\infty}(M,p)$ and $\Omega^{\omega}(M,p)$ are not. This way, it is also possible to induce a group structure on different orbit spaces in a concatenable subspace with respect to weaker homotopy equivalence relations $\sim_R$.

DEFINITION 1. A *group-like homotopy equivalence relation* in a concatenable subspace $\Omega^{\cdot}(M,p) \subset \Omega^0(M,p)$ is an equivalence relation $\sim_R$ turning loop concatenation and inversion into a group epimorphism

$$\Omega^{\cdot}(M,p)/\sim_R \to \pi_1(M,p) \to e.$$

Several group-like homotopy equivalence relations are relevant to us, and are in fact implicit in the standard construction of the group structure on $\pi_1(M,p)$. The most relevant one for the purposes of this work — thin homotopy — was proposed by Barrett over the space $\Omega^0(M,p)$, although as done by Caetano–Picken, it suffices to define it over $\Omega^{\mathrm{ps}}(M,p)$.

By a *generalized reparametrization* of a based loop $\gamma \in \Omega^{\mathrm{ps}}(M,p)$, we mean a based loop of the form $\gamma \circ \phi$, where $\phi : [0,1] \to [0,1]$ is piecewise-smooth and such that $\phi(0) = 0$, $\phi(1) = 1$, and $0 < \phi(t) < 1$ if $0 < t < 1$. An analogous definition is given over $\gamma \in \Omega^{\mathrm{pa}}(M,p)$. Any such a map leads to a degree 1 map $[\phi] : S^1 \to S^1$ under the identification $S^1 \cong \mathbb{R}/\mathbb{Z}$, but in general the condition $\phi'(t) \neq 0$ is not imposed. By imposing suitable restrictions on ϕ, we can think of generalized reparametrizations as maps $\mathscr{F}_\phi : \Omega^{\cdot}(\Omega,p) \to \Omega^{\cdot}(\Omega,p)$. In particular, it follows that every piecewise-smooth based loop can be reparametrized into a smooth loop.

DEFINITION 2 ([**14, 18**]). Two based loops $\gamma_0, \gamma_1 \in \Omega^0(M,p)$ (resp. $\Omega^{\mathrm{ps}}(M,p)$ or $\Omega^{\mathrm{pa}}(M,p)$) are called *thin homotopic* if there is a homotopy η between them (resp. piecewise-smooth or piecewise-analytic homotopy on a paving of $[0,1]^2$) such that

$$\eta\left([0,1]^2\right) \subseteq \gamma_0([0,1]) \cup \gamma_1([0,1]) \subset M.$$

γ_0, γ_1 are called *thin equivalent* (denoted $\gamma_0 \sim_{R_{\mathrm{thin}}} \gamma_1$) if they are related by a finite composition of thin homotopies. A loop is *thin* if it is thin equivalent to the trivial loop.

Hereafter we will mostly consider thin homotopies on $\Omega^{\mathrm{ps}}(M,p)$, unless stated otherwise. The simplest example of a thin homotopy on a based loop $\gamma \in \Omega^{\mathrm{ps}}(M,p)$ is determined by any piecewise-smooth generalized reparametrization ϕ, given by

$$\eta(s,t) = \gamma(s\phi(t) + (1-s)t)$$

Composition of generalized reparametrizations for the different classes of based loops leads to suitable equivalence relations in the corresponding spaces $\Omega^{\cdot}(M,p)$, which we will call *reparametrization equivalences* (denoted by $\sim_{R_{\mathrm{rep}}}$). Even though reparametrization equivalence turns based loop concatenation into an associative structure on the set of equivalence classes $\Omega^{\text{ps}}(M,p)/\sim_{R_{\text{rep}}}$, it is still not an example of a group-like homotopy equivalence relation in $\Omega^{\text{ps}}(M,p)$.

Yet another example of thin homotopy, which suffices to define a group-like homotopy relation in $\Omega^{\text{ps}}(M,p)$, is that of a *retracing*, which is a homotopy between a based loop of the form $\gamma' = \alpha_2 \cdot \beta^{-1} \cdot \beta \cdot \alpha_1$ and the reduced form $\gamma = \alpha_2 \cdot \alpha_1$, for any piecewise-smooth paths α_1,α_2, and β in M such that $\alpha_1(0) = \alpha_2(1) = p$ and $\alpha_1(1) = \alpha_2(0) = \beta(0)$.[7] Such a homotopy is defined in terms of the standard homotopy between the loop $\beta^{-1} \cdot \beta$ and the constant loop at $\beta(0)$ given by $\eta(s,\cdot) = \beta_s^{-1} \cdot \beta_s$, where $\beta_s(t) = \beta((1-s)t)$.[8]

DEFINITION 3 ([**45, 54, 58, 82, 89**]). Two based loops $\gamma_0, \gamma_1 \in \Omega^{\text{ps}}(M,p)$ are called *retrace equivalent* (denoted $\gamma_0 \sim_{R_{\text{ret}}} \gamma_1$) if they differ by a finite number of retracings under implicit generalized reparametrization.[9] A based loop is *retraceable* if it is retrace equivalent to the trivial loop.

The *holonomy* of a smooth connection A [**55**] on a smooth principal G-bundle $\pi : P \to M$ is a geometric notion associated to any piecewise-smooth loop in M and a choice of base point $b \in \pi^{-1}(p)$. For any $A \in \mathscr{A}_P$ we have an induced map

$$\text{Hol}_A : \Omega^{\text{ps}}(M,p) \to G, \qquad \gamma \mapsto \text{Hol}_A(\gamma).$$

DEFINITION 4 ([**4, 58, 82**]). Given a Lie group G, two based loops $\gamma_0, \gamma_1 \in \Omega^{\text{ps}}(M,p)$ are *G-equivalent* (denoted $\gamma_0 \sim_{R_G} \gamma_1$) if $\text{Hol}_A(\gamma_0) = \text{Hol}_A(\gamma_1)$ for any smooth connection $A \in \mathscr{A}_P$ on any smooth principal G-bundle $P \to M$.

REMARK 1. It is a standard fact that the holonomy of any smooth connection is an invariant under generalized reparametrizations and retracings of any given piecewise-smooth based loop. Hence we have the implication

$$\text{Retrace equivalent} \quad \Rightarrow \quad G\text{-equivalent}$$

This is one of several motivations for considering generalized reparametrizations $\phi : [0,1] \to [0,1]$ that may not be homeomorphisms (see remark 2). Even though in general G-equivalence also depends on the Lie group G by definition, in the case when G is a connected and non-solvable matrix group retrace equivalence leads to strictly smaller classes (see sections 3.1 and 3.2).

Caetano–Picken defined yet another concatenable space of interest. A smooth based loop γ is said to have *sitting instants at p* if there is $0 < \varepsilon \leq 1/2$ such that $\gamma([0,\varepsilon] \cup [1-\varepsilon, 1]) = p$. We will denote by $\Omega_0^\infty(M,p) \subset \Omega^\infty(M,p)$ the subspace of smooth loops with sitting instants at p.

DEFINITION 5 ([**18**]). Two based loops $\gamma_0, \gamma_1 \in \Omega_0^\infty(M,p)$ are *intimate* (denoted $\gamma_0 \sim_{R_{\text{int}}} \gamma_1$) if there is a smooth homotopy $\eta : [0,1]^2 \to M$ such that

(i) $\text{rk}\left(d\eta_{(s,t)}\right) \leq 1 \quad \forall\, (s,t) \in [0,1]^2$ (rank-1 homotopy condition),

[7] The notion of loop retracing is due to Kobayashi [**54**], who used the terms *allongement* and *accourcissement* for the operations transforming γ into γ' and viceversa (cf. [**89**]).

[8] Path concatenation is an associative operation only up to piecewise-smooth reparametrization. Hence the operation of retracing would be an unambiguously defined equivalence relation only over the space of equivalence classes in $\Omega^{\text{ps}}(M,p)$ under generalized reparametrization.

[9] The weakening of reparametrizations is not necessary to define retrace equivalence. The justification for this weakening is discussed in remarks 1 and 2.

(ii) there is $0 < \varepsilon < 1/2$ such that $\forall\, s \in [0,1]$, $\eta(s,\cdot) \in \Omega_0^\infty(M,p)$ with respect to ε, and moreover $\eta(s,\cdot) = \gamma_0$ if $s \in [0,\varepsilon]$, and $\eta(s,\cdot) = \gamma_1$ if $s \in [1-\varepsilon,1]$.

PROPOSITION 1 ([**18**]). *For any given $\gamma \in \Omega^{\mathrm{ps}}(M,p)$ there exists a generalized reparametrization $\phi : [0,1] \to [0,1]$ such that $\gamma \circ \phi \in \Omega_0^\infty(M,p)$.*

REMARK 2. Every thin homotopy η for any given pair $\gamma_0, \gamma_1 \in \Omega^{\mathrm{ps}}(M,p)$ is a particular case of a rank-1 homotopy [**18**]. It follows that for every thin homotopy η, there exists a smooth map $\psi : [0,1]^2 \to [0,1]^2$ such that $\psi(s,\cdot)$ is a generalized reparametrization $\forall\, s \in [0,1]$, $\eta \circ \psi(s,\cdot) \in \Omega_0^\infty(M,p)$ $\forall\, s \in [0,1]$, and condition (ii) in definition 5 is satisfied, i.e., every thin homotopy in $\Omega^{\mathrm{ps}}(M,p)$ can be deformed into an intimate homotopy in $\Omega_0^\infty(M,p)$ through a family of generalized reparametrizations. In particular, it follows that every thin loop in $\Omega_0^\infty(M,p)$ is intimate to the trivial loop. Thus, the equivalence relation $\sim_{R_{\mathrm{int}}}$ can be thought of as a "taming" of thin equivalence in $\Omega^{\mathrm{ps}}(M,p)$ in terms of reparametrization equivalence. Intimacy can also be defined on spaces of based maps $\gamma : [0,1]^k \to M$, leading to generalizations of higher homotopy groups [**19, 67**].

The proof of the following result is relatively straightforward to verify. Its proof is also justified in each of the original articles that introduced each corresponding equivalence relation.

PROPOSITION 2. *The equivalence relations introduced in definitions 2, 3, 4 and 5 are examples of group-like homotopy equivalence relations on their corresponding concatenable subspaces.*

Consequently, there is an induced topological group structure on the quotients

$$\mathscr{L}^{\mathrm{thin}}(M,p) := \Omega^{\mathrm{ps}}(M,p)/\sim_{R_{\mathrm{thin}}}, \qquad \mathscr{L}^{\mathrm{ret}}(M,p) := \Omega^{\mathrm{ps}}(M,p)/\sim_{R_{\mathrm{ret}}},$$

$$\mathscr{L}^{G}(M,p) := \Omega^{\mathrm{ps}}(M,p)/\sim_{R_G}, \qquad \mathscr{L}^{\mathrm{int}}(M,p) := \Omega_0^\infty(M,p)/\sim_{R_{\mathrm{int}}}.$$

When no distinction is necessary, we will refer to any of these groups as $\mathscr{L}^{\cdot}(M,p)$.[10]

Generalized reparametrizations and retracings of piecewise-smooth based loops are particular examples of thin homotopies in $\Omega^{\mathrm{ps}}(M,p)$. It follows from remark 2 that we have two induced group epimorphisms

$$\Phi : \mathscr{L}^{\mathrm{ret}}(M,p) \to \mathscr{L}^{\mathrm{thin}}(M,p), \qquad \Psi : \mathscr{L}^{\mathrm{thin}}(M,p) \to \mathscr{L}^{\mathrm{int}}(M,p)$$

Taking several inclusions and projections into account, these group epimorphisms can be arranged into the following commutative diagram

[10]We will refer to these topological groups generically as *groups of based loops* in M. In the physics literature, the group $\mathscr{L}^G(M,p)$ is commonly referred to as the *group of hoops* for the pair (M,G).

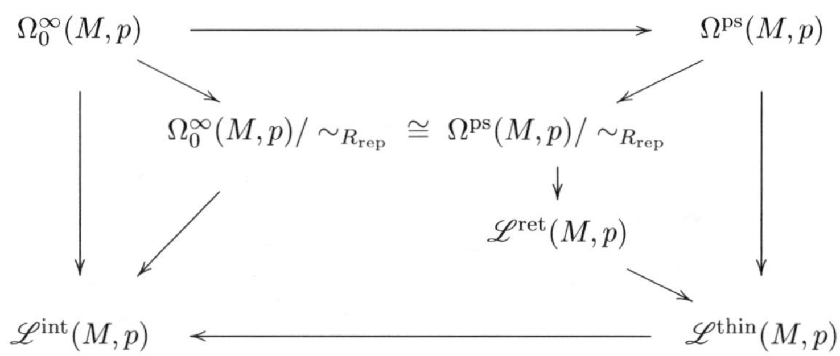

Definition 6. A *G-holonomy relation* in $\Omega^{\cdot}(M,p) \subset \Omega^{\mathrm{ps}}(M,p)$ is any equivalence relation $\sim_R$ for which the holonomy of any smooth connection on any smooth principal G-bundle is an invariant on equivalence classes.

Proposition 3 ([**18**, **54**]). *Retrace equivalence in $\Omega^{\mathrm{ps}}(M,p)$ (definition 3) and intimacy in $\Omega_0^\infty(M,p)$ (definition 5) are examples of G-holonomy relations for any Lie group G.*

The motivation of the notion of a G-holonomy relation is the following. For any such an equivalence relation, the holonomy map of any smooth connection $A \in \mathscr{A}_P$ would descend into a group homomorphism

$$\mathrm{Hol}_A : \Omega^{\cdot}(M,p)/\sim_R \to G.$$

The prototypical example of a G-holonomy relation is G-equivalence itself. One of the main objectives of the works [**14**, **18**, **58**] is to show that each corresponding group-like homotopy equivalence relation is a G-holonomy relation, in such a way that the following reconstruction theorem holds (cf. [**48**]).

Theorem 1 (Reconstruction of gauge fields as holonomy homomorphisms). *For any group homomorphism $H : \mathscr{L} \to G$ satisfying a suitable smoothness condition on smooth families of based loops, where $\mathscr{L} = \mathscr{L}^{\mathrm{thin}}(M,p)$ [**14**], $\mathscr{L}^{\mathrm{ret}}(M,p)$ [**58**], or $\mathscr{L}^{\mathrm{int}}(M,p)$ [**18**] there is a smooth principal G-bundle $\pi : P \to M$ and a smooth connection $A \in \mathscr{A}_P$ such that $\mathrm{Hol}_A = H$ in terms of a choice $b \in \pi^{-1}(p)$.*

Remark 3. As pointed out by Caetano–Picken in [**18**], the definition of thin homotopy in [**14**] makes the realization of thin equivalence as a holonomy relation a nontrivial problem. The introduction of the spaces $\Omega_0^\infty(M,p)$ and the intimacy equivalence relation on them may be thought of as a sort of 'gauge-fixing' condition for smooth based loops, that helps simplifying the proof of theorem 1. It is implicit from the existence of an epimorphism $\Psi : \mathscr{L}^{\mathrm{thin}}(M,p) \to \mathscr{L}^{\mathrm{int}}(M,p)$ for any smooth manifold M, that thin equivalence of piecewise-smooth based loops should be a G-holonomy relation for any Lie group G. In this respect, the results of Tlas ([**92**]; section 3.2) provide a strengthening of this idea since in particular they imply that Ψ is actually an isomorphism.

3.1. A smooth thin loop that is not retraceable. Thin homotopy and retrace homotopy are not equivalent when defined over piecewise-smooth loops, although the latter is always a particular case of the former. This is illustrated by a minimal example of a thin smooth loop γ in $\mathbb{R}^2$ based at the origin that is not retraceable. The example can be constructed explicitly in terms of the bump function $f:[0,1]\to[0,\infty)$ defined as

$$f(t)=\begin{cases} e^{-1/t(1-t)} & \text{if} \quad t\in(0,1), \\ 0 & \text{if} \quad t=0,1. \end{cases} \tag{3.1}$$

whose one-sided nth-derivatives at 0 and 1 are identically zero for any $n\in\mathbb{N}$. Given any pair of sequences $\{r_n\}\subset(0,\infty)$, $\{\theta_n\}\subset[0,2\pi)$ such that

$$\lim_{n\to\infty} r_n\to 0, \qquad \theta_n<\theta_{n+1} \quad \forall\, n\in\mathbb{N},$$

we define the smooth loop $\gamma\in\Omega^\infty\left(\mathbb{R}^2,0\right)$ in a piecewise manner as

$$\gamma|_{[1-2^{-n+1},1-2^{-n}]}:=f\circ\phi_n^{-1}\cdot\left(r_n\cos\left(\theta_n\right),r_n\sin\left(\theta_n\right)\right),$$

for $n=1,2,\ldots$, where

$$\phi_n:[0,1]\to\left[1-2^{-n+1},1-2^{-n}\right], \qquad \phi_n(t):=\left(1-2^{-n}\right)t+\left(1-2^{-n+1}\right)(1-t).$$

By construction, we have that $\gamma^{-1}(0)=\{1-1/2^n \ : \ n\in\mathbb{N}\}\cup\{0,1\}$, and a thin homotopy between γ and the trivial loop is simply given by $\eta(s,\cdot):=(1-s)\gamma$, but clearly γ is not a reparametrization of a finite concatenation of piecewise-smooth retraceable loops (figure 1). Notice that γ would have bounded variation under the standard metric in $\mathbb{R}^2$ if and only if the series $\sum_{n=1}^{\infty} r_n$ is convergent.

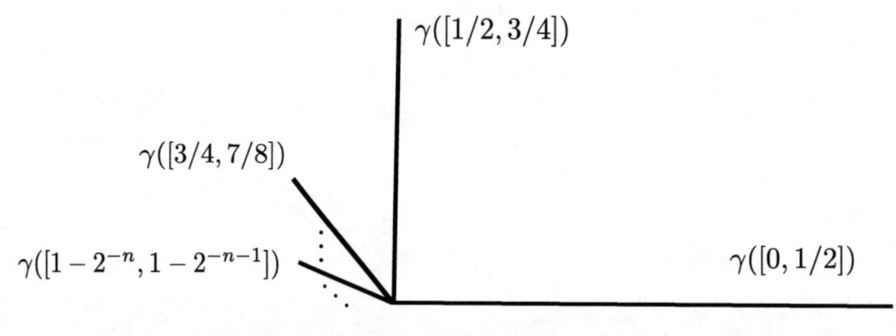

FIGURE 1. The image of a thin smooth loop $\gamma\in\Omega^\infty\left(\mathbb{R}^2,0\right)$

3.2. Results of Tlas. In [**92**], Tlas considered the space $\Omega_0^1(M,p)$ of C^1-based loops in M with vanishing derivative at end points, and studied the homotopy equivalence relation on it induced from compositions of rank-1 homotopies. A crucial ingredient is a factorization theorem (theorem 1 in [**92**]; cf. theorem 4) for any based C^1-loop, which in particular associates a transfinite word to it, and leads to the notion of a *whisker*, which is any element in $\Omega_0^1(M,p)$ whose reduced word is trivial. Similarly, the following result is proved:

THEOREM 2 ([**92**], theorem 3). *The following conditions on a loop $\gamma\in\Omega_0^1(M,p)$ are equivalent:*

(i) *γ is a whisker,*
(ii) *γ is equivalent to the trivial loop via a rank-1 homotopy,*[11]

[11] In [**92**], rank-1 homotopies are called *thin homotopies*. This does not represent a notational inconsistency in view of remark 4.

(iii) *For any semi-simple Lie group G, the holonomy of γ for every smooth connection on every principal G-bundle P is trivial.*[12]

REMARK 4. Thin homotopy is a special case of a rank-1 homotopy [**18**], and after suitable generalized reparametrizations, it follows from Tlas' factorization theorem that every whisker in $\Omega^{\text{ps}}(M,p)$ is a thin loop. Therefore, it also follows from theorem 2 that two loops in $\Omega^{\text{ps}}(M,p)$ are thin homotopic if and only if they are homotopic through a rank-1 homotopy, and we have the isomorphism

$$\mathscr{L}^{\text{thin}}(M,p) \cong \mathscr{L}^{\text{int}}(M,p). \tag{3.2}$$

Moreover, when G is semi-simple, thin homotopy is the largest G-holonomy relation possible in $\Omega^{\text{ps}}(M,p)$, i.e. $\mathscr{L}^{\text{thin}}(M,p) = \mathscr{L}^{G}(M,p)$.

4. Results

We will provide several structural results on the comparison between thin equivalence and retrace equivalence of piecewise-smooth based loops. Ultimately, Tlas' theorem (theorem 2) implies a stronger version of our main result (theorem 4). Nevertheless, we believe that our approach is still interesting, since working with thin homotopies rather that rank-1 homotopies provides a natural mechanism for generalizations to the space $\Omega^0(M,p)$ (remark 6), while the simplicity of the definition of thin homotopy makes many considerations more elementary.

Our strategy consists of studying the structure of the sets $\gamma^{-1}(0)$. This gives a simple proof of the Ashtekar–Lewandowski theorem [**6**] on the equivalence of thin and retrace homotopies for piecewise-analytic loops. First, we provide a general topological result concerning thin homotopy on continuous based loops.

LEMMA 1. *If $\gamma_0, \gamma_1 \in \Omega^0(M,p)$ are thin equivalent, then $\gamma_0([0,1]) \cap \gamma_1([0,1])$ is the image of a based loop $\gamma \in \Omega^0(M,p)$ which is thin equivalent to γ_0 and γ_1.*

PROOF. The images $\gamma_0([0,1])$ and $\gamma_1([0,1])$ are closed since $[0,1]$ is compact and M is Hausdorff. Hence $I_i := \gamma_i^{-1}(\gamma_0([0,1]) \cap \gamma_1([0,1]))$ is closed in $[0,1]$ for $i = 0,1$. The complements $[0,1]\backslash I_i$ are open, and hence equal to an at most countable disjoint union of open intervals in $[0,1]$. Consider the case $i = 0$ and any such a subinterval $(a,b) \subset [0,1] \setminus I_0$. Then we have that $\gamma_0((a,b)) \subset \gamma_0([0,1]) \setminus (\gamma_0([0,1]) \cap \gamma_1([0,1]))$, while $\gamma_0(a), \gamma_0(b) \in \gamma_0([0,1]) \cap \gamma_1([0,1])$. Since γ_0 and γ_1 are thin homotopic, this is only possible if $\gamma_0(a) = \gamma_0(b) = p_0$. Any thin homotopy would then collapse the image $\gamma_0([a,b])$ onto p_0. It follows that the restriction $\gamma_0|_{[a,b]}$ is necessarily a thin loop based at p_0. The map

$$\gamma(t) = \begin{cases} \gamma_0(t) & \text{if} \quad t \in I_0, \\ \gamma_0(a) = \gamma_0(b) & \text{if} \quad t \in (a,b) \subset [0,1] \setminus I_0 \end{cases}$$

is continuous by construction, and such that $\gamma([0,1]) = \gamma_0([0,1]) \cap \gamma_1([0,1])$. It follows that γ_0 is thin homotopic to γ, and consequently γ_1 is too. □

[12] In particular, this indicates that theorem 5 in [**82**], which states the isomorphism

$$\mathscr{L}^{\text{ret}}(M,p) \cong \mathscr{L}^{G}(M,p)$$

when G is connected and non-solvable, is not valid. This is the case since every non-retraceable thin loop (as in section 3.1) will define a nontrivial element in the kernel of the epimorphism $\Phi : \mathscr{L}^{\text{ret}}(M,p) \to \mathscr{L}^{\text{thin}}(M,p)$ following from remark 2 and proposition 2, and consequently $\mathscr{L}^{\text{ret}}(M,p) \ncong \mathscr{L}^{\text{thin}}(M,p)$. Notice that Spallanzani's work suggests that Tlas' theorem is also valid for any connected and non-solvable Lie group G.

Given a based loop $\gamma \in \Omega^{\mathrm{ps}}(M,p)$, we will associate the following closed subsets of $[0,1]$ to it:

$$I_\gamma := \gamma^{-1}(p), \qquad J_\gamma := \{t \in [0,1] \ : \ \gamma \text{ is smooth at } t \text{ and } d\gamma|_t = 0\}.$$

As we have seen in section 3.1, it is easy to provide examples of smooth based loops γ where I_γ and J_γ are not finite subsets. General closed subsets of $[0,1]$ could be rather complicated (e.g. Cantor sets). This complexity is not excluded from the subsets I_γ and J_γ. Namely, we have the following result.

PROPOSITION 4. *For any pair of closed subsets $I, J \subset [0,1]$, there exist smooth loops $\gamma, \gamma' \in \Omega^\infty(M,p)$ such that (i) $I_\gamma = I$; (ii) $I_{\gamma'} = \{0,1\}$ and $J_{\gamma'} = J$.*

PROOF. It is sufficient to verify the claim when $M = \mathbb{R}^2$ and $p = 0$. The proof is modeled on a celebrated theorem of H. Whitney, which states that for any closed subset $I \in \mathbb{R}$, there is a smooth function $f : \mathbb{R} \to \mathbb{R}$ such that $f^{-1}(0) = I$. The basic idea for the proof of Whitney's theorem is to express the open set $[0,1] \setminus I$ as a countable disjoint union of open intervals

$$[0,1] \setminus I_\gamma = \bigsqcup_{n \in \mathscr{I}} (a_n, b_n).$$

Over the closure $[a_n, b_n]$ of any such an interval, we define the restriction of the loop γ to be

$$\gamma|_{[a_n,b_n]} := f \circ \phi_n^{-1} \cdot (r_n \cos(\theta_n), r_n \sin(\theta_n)),$$

where $\phi_n(t) := b_n t + a_n(1-t)$, $r_n \in (0,1]$, $\theta_n \in (0, \pi/2)$ are arbitrary, and f is the bump function (3.1). Then (i) readily follows. As for (ii), the same mechanism yields a smooth loop $\gamma : [0,1] \to \mathbb{R}^2$ such that $I_\gamma = \{0,1\}$ and $\{t \in [0,1] \ : \ d\gamma|_t = 0\} = J$ by consideration of a primitive (whose existence would depend on imposing further constraints on $\{r_n\}$) and scaling it by the bump function f. ☐

REMARK 5. Even though each restriction $\gamma|_{[a_n,b_n]}$ for any connected component $(a_n, b_n) \subset [0,1] \setminus I_\gamma$ of any piecewise smooth loop γ leads to a piecewise-smooth loop γ' for which $I_{\gamma'} = \{0,1\}$, it follows from proposition 4 that the subset $J_{\gamma'}$ could still in principle be an arbitrary closed subset in $[0,1]$. Therefore, based smooth loops on a manifold M could in general traverse the base point p in an arbitrarily wild way, and even over any connected component $(a,b) \subset [0,1] \setminus I_\gamma$, the set $J_\gamma \cap [a,b]$ could be arbitrarily wild as well.

LEMMA 2. *Every based loop $\gamma \in \Omega^{\mathrm{ps}}(M,p)$ is thin homotopic to a loop γ' for which $I_{\gamma'}$ is nowhere dense.*

PROOF. Every connected component in I_γ corresponds to a closed subinterval (which could be a point) where the image of γ is p. Consider the collection of all closed subintervals $I \subset I_\gamma$ which are not a point. We can construct a 1-parameter family of reparametrizations that shrink any such an I into a point. The resulting based loop γ' would be thin homotopic to γ by construction, such that $I_{\gamma'}$ is nowhere dense, and there would be a induced bijection $[0,1] \setminus I_\gamma \leftrightarrow [0,1] \setminus I_{\gamma'}$. ☐

LEMMA 3. *If a based loop $\gamma \in \Omega^{\mathrm{ps}}(M,p)$ is thin, then there exists a thin homotopy η between γ and the trivial loop such that $I_{\gamma^s} = I_\gamma \ \forall \ s \in [0,1)$, where $\gamma^s := \eta(s, \cdot)$. Moreover, η could be chosen to be piecewise-analytic if γ is piecewise-analytic.*

PROOF. Since γ is a thin based loop, there exists a continuous and piecewise-smooth map $\eta : [0,1]^2 \to M$ such that $\gamma^s \in \Omega^{\mathrm{ps}}(M,p)$, $\gamma_0 = \gamma$, $\gamma_1 = p$, and $\eta\left([0,1]^2\right) = \gamma([0,1])$. Then for each fixed $t_0 \in [0,1]$, $\delta_{t_0} := \eta(\cdot, t_0)$ is a path within $\gamma([0,1])$ connecting $\gamma(t_0)$ and p. Assume that $t_0 \notin I_\gamma$. Then $\delta_{t_0}(0) \neq p$. Define

$$s_{t_0} := \min\{\, s \in [0,1] : \delta_{t_0}(s) = p\}.$$

Define a new thin homotopy $\eta' : [0,1]^2 \to M$ by first considering restrictions and reparametrizations along each $[0, s_{t_0}]$ for each $t_0 \in [0,1] \setminus I_\gamma$. By the piecewise-smoothness of η, the extension of η' to each $t_0 \in I_\gamma$ given by $\eta'(\cdot, t_0) = p$ would be continuous and piecewise-smooth, and piecewise-analytic when η is so. By construction η' is a thin homotopy between γ and the trivial loop such that $I_{\gamma'^s} = I_\gamma$ $\forall\, s \in [0,1)$, and which is piecewise-smooth, or piecewise-analytic when η is so. $\square$

The complexity of a thin based loop $\gamma \in \Omega^{\mathrm{ps}}(M,p)$ is encoded in the sets I_γ and J_γ. The general structure of I_γ and J_γ is considerably simpler along the subspace $\Omega^{\mathrm{pa}}(M,p)$ of piecewise-analytic based loops, as illustrated in the subsequent theorem and its corollary (cf. [**6**]). These results illustrate the fundamental difference between analyticity and smoothness of based loops with respect to thin homotopy.

THEOREM 3. *A loop $\gamma \in \Omega^{\mathrm{pa}}(M,p)$ is thin if and only if it is reparametrization equivalent to a finite concatenation of retraceable loops $\gamma_1, \ldots, \gamma_m \in \Omega^{\mathrm{pa}}(M,p)$.*

PROOF. It suffices to verify that if γ is thin, then there exists a piecewise-analytic reparametrization $\gamma' = \gamma \circ \phi$ such that γ' is a finite concatenation of retraceable piecewise-analytic loops. Hence we can assume without any loss of generality that γ is analytic. If γ is the trivial loop, then there is nothing to prove. Otherwise, the analyticity of γ implies that I_γ is finite, i.e., $I_\gamma = \{t_0 = 0 < t_1 < \cdots < t_m = 1\}$. It then follows that $\gamma = \gamma_m \cdot \cdots \cdot \gamma_1$ is a concatenation of m analytic loops $\gamma_1, \ldots, \gamma_m$, such that $I_{\gamma_i} = \{0,1\}$ for $i = 1, \ldots, m$. Each γ_i is explicitly given as

$$\gamma_i = \gamma|_{[t_{i-1}, t_i]} \circ \phi_i, \qquad \phi_i(t) := t_{i-1}(1-t) + t_i t, \qquad i = 1, \ldots, m.$$

We claim that each γ_i is reparametrization equivalent to a based loop in $\Omega^{\mathrm{pa}}(M,p)$ of the form $\beta_i^{-1} \cdot \beta_i$, for some piecewise-analytic analytic path β_i such that $\beta_i(0) = p$.

Consider a piecewise-analytic thin homotopy η between $\eta(0,\cdot) = \gamma$ and the trivial loop $\eta(1,\cdot) = p$ as in lemma 3. Then $\eta(s, t_i) = p$ for each $i = 1, \ldots, m-1$, and $\eta(s,t) \neq p$ if $s \in (0,1)$ and $t \in [0,1] \setminus I_\gamma$. We would also have that $\gamma^s = \gamma_m^s \cdots \gamma_1^s$ $\forall\, s \in [0,1]$. It follows that each γ_i, $i = 1, \ldots, m$, is also thin and analytic, and the sets J_{γ_i} are finite. Since for each $i = 1, \ldots, m$, we have that $\gamma_i^s\left([0,1]\right) \subset \gamma_i\left([0,1]\right)$ $\forall\, s \in [0,1]$, then each $J_{\gamma_i} \setminus \{0,1\}$ is necessarily nonempty. Define

$$t_{\gamma_i} := \min\{t \in J_{\gamma_i} \setminus \{0,1\} \; : \; \gamma_i\left([0,t]\right) = \gamma_i([0,1])\}$$

$$= \min\{t \in (0,1) \; : \; \gamma_i\left([0,t]\right) = \gamma_i([0,1])\},$$

and let

$$\beta_i'(t) := \gamma_i\left(t_{\gamma_i} t\right).$$

Then there exists a generalized analytic reparametrization $\varphi_i : [0,1] \to [0,1]$ such that $\beta_i' = \beta_i \circ \varphi_i$ with $d\beta_i|_t \neq 0$ $\forall\, t \in (0,1)$. If we express $\gamma_i = \delta_i \cdot (\beta_i \circ \varphi_i)$, it follows from the definition of thin homotopy that δ_i and β_i have the same image in M. Similarly, it follows that there is a generalized analytic reparametrization

$\psi_i : [0,1] \to [0,1]$ such that $\delta_i = \beta_i^{-1} \circ \psi_i$. Hence we can express $\gamma_i = \left(\beta_i^{-1} \cdot \beta_i\right) \circ \vartheta_i$ for a piecewise-analytic reparametrization ϑ_i. □

COROLLARY 1. *Two based loops $\gamma_0, \gamma_1 \in \Omega^{\mathrm{pa}}(M,p)$ are thin equivalent if and only if they are retrace equivalent (cf. [**7**]).*

PROOF. It remains to show that thin equivalence implies retrace equivalence. In general, $\gamma_0([0,1]) \cap \gamma_1([0,1])$ is a closed subset of M. It follows from the piecewise-analyticity of γ_0 and γ_1 that the closed subsets $I_i := \gamma_i^{-1}(\gamma_0([0,1]) \cap \gamma_1([0,1]))$, $i = 0,1$, are a finite union of disjoint closed intervals in $[0,1]$, i.e.,

$$I_i = \bigsqcup_{j=0}^{r_i} \left[c_j^i, d_j^i\right], \quad i = 0,1,$$

where $c_j^i \leq d_j^i$ for each $j = 0, \ldots, r_i$. Clearly $c_0^i = 0$ and $d_{r_i}^i = 1$. If γ_0 and γ_1 are thin homotopic, it follows from the general arguments in the proof of lemma 1 that

$$\gamma_i\left(I_i\right) = \gamma_i([0,1]), \quad i = 0,1,$$

and $\gamma_i\left(d_{j-1}^i\right) = \gamma_i\left(c_j^i\right) = p_j^i$ for each $i = 0,1$, $j = 1, \ldots, r_i$. Hence each restriction

$$\gamma_i|_{[d_{j-1}^i, c_j^i]}$$

is a piecewise-analytic loop based at p_j^i, which is thin in the piecewise-analytic sense. By theorem 3, it is reparametrization equivalent to a finite concatenation of retraceable loops. Therefore, γ_0 and γ_1 are retrace equivalent. □

The examples of thin smooth based loops that are not retraceable that we have constructed before are clearly non-analytic. In particular, it follows that there is no analog of theorem 3 for piecewise-smooth based loops. The next result indicates that, nevertheless, the complexity of a piecewise-smooth thin loop γ is entirely encoded in the complexity of the set I_γ. This is so since along the closure of every connected component of its complement $[0,1] \setminus I_\gamma$, the restriction of γ would necessarily reduce to a reparametrization of an elementary retraceable loop.

THEOREM 4. *If a piecewise-smooth based loop $\gamma \in \Omega^{\mathrm{ps}}(M,p)$ is thin, then for every connected component $(a,b) \subset [0,1] \setminus I_\gamma$, the restriction $\gamma|_{[a,b]}$ is reparametrization equivalent to a retraceable loop of the form $\beta^{-1} \cdot \beta$ for some smooth path $\beta : [0,1] \to M$ such that $\beta^{-1}(p) = \{0\}$.*

PROOF. We can assume that γ is smooth after reparametrization. By hypothesis $\gamma(a) = \gamma(b) = p$. Therefore $\gamma' := \gamma|_{[a,b]} \circ \phi$, where $\phi(t) = bt + a(1-t)$, is a smooth based loop, and moreover $I_{\gamma'} = \{0,1\}$. Consider any thin homotopy η between γ and the trivial loop such that $I_{\gamma^s} = I_\gamma$ $\forall s \in [0,1)$ as in lemma 3. The induced restrictions $\gamma'^s := \gamma^s|_{[a,b]} \circ \phi$ also determine a thin homotopy between γ' and the trivial loop. Hence γ' is a thin loop. Similarly to the proof of theorem 3, we consider

$$t_{\gamma'} := \min\{t \in (0,1) \ : \ \gamma'([0,t]) = \gamma'([0,1])\}$$

(notice that in the smooth case, $\min\{t \in J_{\gamma'} \setminus \{0,1\}\}$ may not exist at all). By continuity it follows that, moreover,

$$\gamma'\left([t_{\gamma'}, 1]\right) = \gamma'\left([0, t_{\gamma'}]\right) = \gamma'([0,1]).$$

The smoothness of γ' implies that the set $\gamma'([0, t_{\gamma'}])$ could be parametrized as a path $\beta : [0,1] \to M$ such that $\beta(0) = 0$ and $\beta(1) = \gamma'\left(t_{\gamma'}\right)$, for which $d\beta|_t \neq 0$ if

$t \in (0,1)$. This could be done in terms of arc length rectification for an auxiliary Riemannian metric in M. It follows that the image of the retraceable loop $\beta^{-1}\cdot\beta$ coincides with the image of γ'. It moreover follows that there is a piecewise-smooth generalized reparametrization $\varphi : [0,1] \to [0,1]$ such that $\gamma' = (\beta^{-1}\cdot\beta)\circ\varphi$. □

REMARK 6. It can be verified that the proofs of lemmas 2 and 3, as well as theorem 4 and corollary 2, can be also adapted to hold for thin homotopy defined over the larger spaces $\Omega^0(M,p)$.

We conclude this section with the following corollaries on the structure of thin equivalence for piecewise-smooth based loops. The proof of corollary 2 follows as an immediate consequence of proposition 4 and theorem 4.

COROLLARY 2. *For any $\gamma \in \Omega^{\mathrm{ps}}(M,p)$, there is $\gamma' \in \Omega^{\mathrm{ps}}(M,p)$ such that $\gamma \sim_{\mathrm{thin}} \gamma'$ but $\gamma \nsim_{\mathrm{ret}} \gamma'$. Consequently, for any $\gamma \in \Omega^{\mathrm{ps}}(M,p)$, the equivalence class $[\gamma]_{\mathrm{thin}}$ is strictly larger than the equivalence class $[\gamma]_{\mathrm{ret}}$.*[13]

COROLLARY 3. *Every pair of thin equivalent loops $\gamma \sim_{R_{\mathrm{thin}}} \gamma'$ in $\Omega^{\mathrm{ps}}(M,p)$ is the limit of a sequence of pairs of retrace equivalent loops $\{\gamma_n,\gamma_n'\}$ with respect to the compact-open topology, in the sense that $\{\gamma_n\}\to\gamma$ and $\{\gamma_n'\}\to\gamma'$ as $n\to\infty$, and $\gamma_n \sim_{R_{\mathrm{ret}}} \gamma_n'$ $\forall\, n\in\mathbb{N}$.*

PROOF. As before, it is sufficient to consider the case when $\gamma \in \Omega^{\mathrm{ps}}(M,p)$ is a thin loop. The general case of a pair (γ,γ') of thin homotopic loops in $\Omega^{\mathrm{ps}}(M,p)$ will follow analogously. Consider any given thin loop $\gamma \in \Omega^{\mathrm{ps}}(M,p)$, and let η be a thin homotopy betwen γ and the trivial loop such that $I_{\gamma^s} = I_\gamma$ $\forall\, s\in[0,1)$ as in lemma 3. As before, let

$$[0,1]\setminus I_\gamma = \bigsqcup_{n\in\mathscr{I}} (a_n,b_n).$$

If $\mathscr{I}$ is a finite set then there is nothing to prove. Hence we will assume that $\mathscr{I}=\mathbb{N}$. For any given $n\in\mathbb{N}$, define $\gamma_n^s : [0,1]\to M$ as

$$\gamma_n^s(t) = \begin{cases} \gamma^s(t) & \text{if } \ t\in\bigsqcup_{k\le n}(a_k,b_k),\\ \\ p & \text{otherwise.}\end{cases}$$

and let $\gamma_n := \gamma_n^0$. It readily follows that $\gamma_n^s \in \Omega^{\mathrm{ps}}(M,p)$ $\forall\, n\in\mathbb{N}$ and $s\in[0,1]$, and each γ_n is retraceable in terms of the sequence of thin homotopies η_n defined by the families γ_n^s. By construction, $\gamma_n^s(I_n)=\{p\}$, where $I_n = [0,1]\setminus\bigsqcup_{k\le n}(a_k,b_k)$, while $\gamma_n^s|_{\bigsqcup_{k\le n}[a_k,b_k]} = \gamma^s|_{\bigsqcup_{k\le n}[a_k,b_k]}$ $\forall\, s\in[0,1]$.

A basis element $\mathscr{B}(I,\mathscr{U})$ of the compact-open topology in $\Omega^{\mathrm{ps}}(M,p)$ is determined by a closed set $I\subset[0,1]$ and an open set $\mathscr{U}\subset M$, and consists of all loops γ such that $\gamma(I)\subset\mathscr{U}$. Assume that $\gamma\in\mathscr{B}(I,\mathscr{U})$ for some I and $\mathscr{U}$. Since

$$I = (I\cap I_n)\cup\bigsqcup_{k\le n}(I\cap(a_k,b_k)),$$

there are two possibilities. When $I\cap I_\gamma\neq\emptyset$, it follows that $\gamma_n^s(I)\subset\mathscr{U}$ $\forall\, n\in\mathbb{N}$, i.e. $\gamma_n^s\in\mathscr{B}(I,\mathscr{U})$ $\forall\, n\in\mathbb{N},\ s\in[0,1]$. Similarly, if $I\cap I_\gamma=\emptyset$, then $I\subset\bigsqcup_{k\le n_0}(a_k,b_k)$ for some $n_0\in\mathbb{N}$. Therefore $I\cap I_n=\emptyset$ $\forall\, n>n_0$ and $\gamma_n^s\in\mathscr{B}(I,\mathscr{U})$ $\forall\, n>n_0,\ s\in[0,1]$. In both cases $\{\gamma_n\}\to\gamma$ in the compact-open topology. □

[13] In contrast, for any $\gamma\in\Omega^{\mathrm{pa}}(M,p)$, the equivalence classes of piecewise-analytic loops $[\gamma]_{\mathrm{thin}}$ and $[\gamma]_{\mathrm{ret}}$ coincide (cf. [**7**]).

REMARK 7. We can turn M into a metric space by means of an auxiliary Riemannian metric on it. Under any such a choice, the compact-open topology on $\Omega^{\mathrm{ps}}(M,p)$ is metrizable in terms of

$$d(\gamma,\gamma') := \sup\{d(\gamma(t),\gamma'(t)) \; : \; t \in [0,1]\}.$$

Let $\Omega_b^{\mathrm{ps}}(M,p) \subset \Omega^{\mathrm{ps}}(M,p)$ denote the subspace of loops with bounded variation, which is directly dependent on the choice of Riemannian metric in M. These subspaces play a fundamental role in the construction of smooth structures on spaces of piecewise-smooth based loops [**85**].

COROLLARY 4. *The quotient topology on the groups $\mathscr{L}^{\mathrm{ret}}(M,p)$ induced from the compact-open topology in $\Omega^{\mathrm{ps}}(M,p)$ is never Hausdorff.*

PROOF. it follows from corollary 3 that for every $\gamma \in \Omega^{\mathrm{ps}}(M,p)$, $[\gamma]_{\mathrm{ret}} \subsetneq [\gamma]_{\mathrm{thin}} \subset \overline{[\gamma]_{\mathrm{ret}}}$ and consequently $[\gamma]_{\mathrm{ret}} \neq \overline{[\gamma]_{\mathrm{ret}}}$ for any $\gamma \in \Omega^{\mathrm{ps}}(M,p)$. In particular, the subspace $\Omega^{\mathrm{ps}}(M,p) \sim_{\mathrm{ret}} \Omega^{\mathrm{ps}}(M,p) \subset \Omega^{\mathrm{ps}}(M,p) \times \Omega^{\mathrm{ps}}(M,p)$ of retrace equivalent pairs in $\Omega^{\mathrm{ps}}(M,p)$ is never closed. □

REMARK 8. An alternative proof of the equivalence of (i) and (iii) in theorem 2, concerning the correspondence between thin equivalence and G-equivalence for a semi-simple Lie group G in $\Omega^{\mathrm{ps}}(M,p)$, is also hinted by corollary 3. Since the holonomy of any smooth connection on a smooth principal G-bundle $P \to M$, together with a choice of fiber point $b \in \pi^{-1}(p)$, defines a continuous map $\Omega^{\mathrm{ps}}(M,p) \to G$ satisfying Barrett's smoothness conditions [**14**], it would be sufficient to show that $\overline{[\gamma]_{\mathrm{thin}}} = [\gamma]_{\mathrm{thin}}$ for any given $\gamma \in \Omega^{\mathrm{ps}}(M,p)$.

5. Conclusion and further remarks

One of the motivating purposes of this work was to shed light on the intricacies of thin equivalence in spaces of piecewise smooth loops on a given smooth manifold. Since any satisfactory and mathematically rigorous approach to the study of gauge theories from the holonomy perspective would necessarily rely on functional analytic tools developed on loop spaces, a fundamental step into that direction would be to study these intricacies in careful detail. In this concluding section, we have compiled a short list of remarks and natural questions which arise when following this paradigm. The common feature that unifies all of these questions is the ultimate need to consider natural topologies on the spaces $\Omega^{\mathrm{ps}}(M,p)$ and their induced quotients.

As illustrated by corollary 4, the induced quotient topologies on groups of based loops will depend drastically on the nature of the different equivalence classes of a given based loop $\gamma \in \Omega^{\mathrm{ps}}(M,p)$. It is natural to ask if the subspace of thin equivalence pairs in $\Omega^{\mathrm{ps}}(M,p) \times \Omega^{\mathrm{ps}}(M,p)$ is closed under a suitable choice of function space topology in $\Omega^{\mathrm{ps}}(M,p)$. Otherwise, this would indicate that $\Omega^{\mathrm{ps}}(M,p)$ is "too big", and a suitable subspace of it should be found.

QUESTION 1. Are the subspaces of thin equivalent pairs in $\Omega^{\mathrm{ps}}(M,p)$ closed in $\Omega^{\mathrm{ps}}(M,p) \times \Omega^{\mathrm{ps}}(M,p)$ under the compact-open topology in $\Omega^{\mathrm{ps}}(M,p)$? If not, what is the largest concatenable subspace $\Omega^{\cdot}(M,p) \subset \Omega^{\mathrm{ps}}(M,p)$ leading to a Hausdorff quotient under a suitable function space topology?

QUESTION 2. Is there a Hausdorff subgroup $\mathscr{L}^{\cdot}(M,p) \subset \mathscr{L}^{\mathrm{thin}}(M,p)$ (under a suitable function space topology) for which the reconstruction theorem 1 holds? If so, what is the smallest subgroup with these properties?

Once a Hausdorff topology has been fixed on the group $\mathscr{L}^\bullet(M,p)$, we can consider the induced groups of continuous topological group homomorphisms to G, which we will denote succinctly by $\operatorname{Hom}(\mathscr{L}^\bullet(M,p),G)$. These groups contain a subgroup $\operatorname{Hom}(\mathscr{L}^\bullet(M,p),G)_0$ consisting of those homomorphisms homotopic to the trivial homomorphism. Then we have that

$$\operatorname{Hom}(\mathscr{L}^\bullet(M,p),G)/\operatorname{Hom}(\mathscr{L}^\bullet(M,p),G)_0 \cong \pi_0\left(\operatorname{Hom}(\mathscr{L}^\bullet(M,p),G)\right).$$

Let $\check{H}^1(M,G)$ denote the space of equivalence classes of principal G-bundles over M. Perhaps more interestingly, it is also very reasonable to expect to have a conjectural bijection (cf. [**68, 69**])

$$\operatorname{Hom}(\mathscr{L}^\bullet(M,p),G)/\operatorname{Hom}(\mathscr{L}^\bullet(M,p),G)_0 \cong \check{H}^1(M,G). \tag{5.1}$$

QUESTION 3. Under the previous assumptions on $\Omega^\bullet(M,p)$, do we always have a bijection (5.1)?

Let $\operatorname{Hom}^\infty(\mathscr{L}^\bullet(M,p),G)$ denote the subspace of $\operatorname{Hom}(\mathscr{L}^\bullet(M,p),G)$ satisfying Barrett's smoothness conditions (H3) [**14, 18**]. The reconstruction of gauge fields in terms of holonomy homomorphisms for a given pair (M,G), described in theorem 1, can be expressed succinctly as a statement of the bijectivity of the induced map

$$\bigsqcup_{\{P\}\in\check{H}^1(M,G)} \mathscr{A}_P/\mathscr{G}_P \to \operatorname{Hom}^\infty(\mathscr{L}^\bullet(M,p),G)/G, \tag{5.2}$$

where G acts on $\operatorname{Hom}^\infty(\mathscr{L}^\bullet(M,p),G)$ by conjugation, corresponding to changes of base point $b\in\pi^{-1}(p)\subset P$ under right G-action. While the left-hand side is not an infinite dimensional manifold, it is a stratified space containing an open smooth (infinite dimensional) part [**71**], and in particular, it carries a natural topology. This comparison leads to the following natural and fundamental question.

QUESTION 4. Under the previous assumptions on $\Omega^\bullet(M,p)$, is the bijection (5.2) a homeomorphism?

Following the same train of thought, it would also be desirable to reinterpret Barrett's smoothness conditions (H3) in terms of a suitable "smooth structure" on the induced loop group $\mathscr{L}^\bullet(M,p)$ that is compatible with that of the quotients $\mathscr{A}_P/\mathscr{G}_P$, providing a topological refinement of the correspondence (5.2). It would be interesting to explore this possibility within the context of the work of Stacey [**83, 85**], who introduced a refining condition on spaces of piecewise-smooth loops in such a way that they admit an infinite dimensional manifold structure in the sense of Kriegl-Michor [**56**].

Acknowledgments

The author would like to thank José A. Zapata for many discussions that led to the creation of this work, Roger F. Picken for sharing his opinion and insights on the subject, Tamer Tlas for providing remarks leading to a revision of this work, and the referee for the careful review and the very pertinent suggestions made.

References

[1] Y. Aharonov and D. Bohm, *Significance of electromagnetic potentials in the quantum theory*, Phys. Rev. (2) **115** (1959), 485–491. MR110458

[2] I. Ya. Aref′eva, *The gauge field as chiral field on the path and its integrability*, Lett. Math. Phys. **3** (1979), no. 4, 241–247, DOI 10.1007/BF01821841. MR545399

[3] I. Aref'eva, *Non-abelian Stokes formula*, Theor. Math. Phys. **43** (1980), no. 1, 353–356.

[4] A. Ashtekar, *Lessons from* $(2+1)$*-dimensional quantum gravity*, Strings '90 (College Station, TX, 1990), World Sci. Publ., River Edge, NJ, 1991, pp. 71–88. MR1256495

[5] A. Ashtekar and C. J. Isham, *Representations of the holonomy algebras of gravity and non-abelian gauge theories*, Classical Quantum Gravity **9** (1992), no. 6, 1433–1467. MR1166682

[6] A. Ashtekar and J. Lewandowski, *Completeness of Wilson loop functionals on the moduli space of* $\mathrm{SL}(2,\mathbf{C})$ *and* $\mathrm{SU}(1,1)$ *connections*, Classical Quantum Gravity **10** (1993), no. 6, L69–L74. MR1223399

[7] A. Ashtekar and J. Lewandowski, *Representation theory of analytic holonomy* C^**-algebras*, Knots and quantum gravity (Riverside, CA, 1993), Oxford Lecture Ser. Math. Appl., vol. 1, Oxford Univ. Press, New York, 1994, pp. 21–61. MR1309913

[8] A. Ashtekar and J. Lewandowski, *Differential geometry on the space of connections via graphs and projective limits*, J. Geom. Phys. **17** (1995), no. 3, 191–230, DOI 10.1016/0393-0440(95)00028-G. MR1358736

[9] A. Ashtekar, D. Marolf, and J. Mourão, *Integration on the space of connections modulo gauge transformations*, Proceedings of the Cornelius Lanczos International Centenary Conference (Raleigh, NC, 1993), SIAM, Philadelphia, PA, 1994, pp. 143–160. MR1298229

[10] M. Atiyah, *Collected works. Vol. 5*, Oxford Science Publications, The Clarendon Press, Oxford University Press, New York, 1988. Gauge theories. MR951896

[11] J. C. Baez, *Diffeomorphism-invariant generalized measures on the space of connections modulo gauge transformations*, Proceedings of the Conference on Quantum Topology (Manhattan, KS, 1993), World Sci. Publ., River Edge, NJ, 1994, pp. 21–43. MR1309924

[12] J. C. Baez and S. Sawin, *Functional integration on spaces of connections*, J. Funct. Anal. **150** (1997), no. 1, 1–26, DOI 10.1006/jfan.1997.3108. MR1473623

[13] J. W. Barrett, *Holonomy and gravity*, Gen. Relativity Gravitation **21** (1989), no. 5, 457–466, DOI 10.1007/BF00904497. MR992754

[14] J. W. Barrett, *Holonomy and path structures in general relativity and Yang-Mills theory*, Internat. J. Theoret. Phys. **30** (1991), no. 9, 1171–1215, DOI 10.1007/BF00671007. MR1122025

[15] C. Di Bartolo, R. Gambini, and J. Griego, *The extended loop group: an infinite-dimensional manifold associated with the loop space*, Comm. Math. Phys. **158** (1993), no. 2, 217–240. MR1249593

[16] C. Di Bartolo, R. Gambini, J. Griego, and J. Pullin, *Extended loops: a new arena for nonperturbative quantum gravity*, Phys. Rev. Lett. **72** (1994), no. 23, 3638–3641, DOI 10.1103/PhysRevLett.72.3638. MR1276465

[17] I. Białynicki-Birula, *Gauge-invariant variables in the Yang-Mills theory*, Bull. Acad. Polon. Sci. Sér. Sci. Math Astronom. Phys. **11** (1963), 135–138. MR0149859

[18] A. Caetano and R. F. Picken, *An axiomatic definition of holonomy*, Internat. J. Math. **5** (1994), no. 6, 835–848, DOI 10.1142/S0129167X94000425. MR1298997

[19] A. P. Caetano and R. F. Picken, *On a family of topological invariants similar to homotopy groups*, Rend. Istit. Mat. Univ. Trieste **30** (1998), no. 1-2, 81–90 (1999). MR1704827

[20] A. S. Cattaneo, P. Cotta-Ramusino, and M. Rinaldi, *Loop and path spaces and four-dimensional* BF *theories: connections, holonomies and observables*, Comm. Math. Phys. **204** (1999), no. 3, 493–524, DOI 10.1007/s002200050655. MR1707635

[21] H.-M. Chan, P. Scharbach, and S. T. Tsou, *On loop space formulation of gauge theories*, Ann. Physics **166** (1986), no. 2, 396–421, DOI 10.1016/0003-4916(86)90144-2. MR830518

[22] H.-M. Chan, P. Scharbach, and S. T. Tsou, *Action principle and equations of motion for nonabelian monopoles*, Ann. Physics **167** (1986), no. 2, 454–472, DOI 10.1016/0003-4916(86)90209-5. MR837253

[23] H.-M. Chan and S. T. Tsou, *Gauge theories in loop space*, Acta Phys. Polon. B **17** (1986), no. 3, 259–276. MR844388

[24] K.-t. Chen, *Iterated integrals of differential forms and loop space homology*, Ann. of Math. (2) **97** (1973), 217–246, DOI 10.2307/1970846. MR380859

[25] K. T. Chen, *Iterated path integrals*, Bull. Amer. Math. Soc. **83** (1977), no. 5, 831–879, DOI 10.1090/S0002-9904-1977-14320-6. MR454968

[26] B. Collier, E. Lerman, and S. Wolbert, *Parallel transport on principal bundles over stacks*, J. Geom. Phys. **107** (2016), 187–213, DOI 10.1016/j.geomphys.2016.05.010. MR3521476

[27] M. Creutz, *Quarks, gluons and lattices (vol. 8)*, Cambridge Monographs on Math. Phys., Cambridge University Press, 1983.

[28] M. L. Curtis and R. Lashof, *Homotopy equivalence of fiber bundles*, Proc. Amer. Math. Soc. **9** (1958), 178–182, DOI 10.2307/2033133. MR98378

[29] W. D. Curtis and F. R. Miller, *Gauge groups and classification of bundles with simple structural group*, Pacific J. Math. **68** (1977), no. 2, 331–338. MR482760

[30] P. A. M. Dirac, *Quantised singularities in the electromagnetic field*, Proc. R. Soc. Lond. A **133** (1931), no. 821, 60–72.

[31] L. Dolan, *Hidden symmetry in Yang-Mills theory: a path-dependent gauge transformation*, Phys. Rev. D (3) **22** (1980), no. 12, 3104–3112, DOI 10.1103/PhysRevD.22.3104. MR596942

[32] B. K. Driver, *Classifications of bundle connection pairs by parallel translation and lassos*, J. Funct. Anal. **83** (1989), no. 1, 185–231, DOI 10.1016/0022-1236(89)90035-9. MR993446

[33] B. Durhuus and J. M. Leinaas, *On the loop space formulation of gauge theories*, Phys. Scripta **25** (1982), no. 4, 504–510, DOI 10.1088/0031-8949/25/4/002. MR652896

[34] C. Fleischhack, *Stratification of the generalized gauge orbit space*, Comm. Math. Phys. **214** (2000), no. 3, 607–649, DOI 10.1007/s002200000286. MR1800864

[35] C. Fleischhack, *Regular connections among generalized connections*, J. Geom. Phys. **47** (2003), no. 4, 469–483, DOI 10.1016/S0393-0440(02)00232-2. MR1996337

[36] R. O. Fulp, *Connections on the path bundle of a principal fiber bundle*, Differential Geom. Appl. **4** (1994), no. 3, 201–237, DOI 10.1016/0926-2245(94)00014-X. MR1299396

[37] X. Fustero, R. Gambini, and A. Trias, *Einstein's gravitation as a gauge theory of the Lorentz group*, Phys. Rev. D (3) **31** (1985), no. 12, 3144–3149, DOI 10.1103/PhysRevD.31.3144. MR793155

[38] R. Gambini, *Loop space representation of quantum general relativity and the group of loops*, Phys. Lett. B **255** (1991), no. 2, 180–188, DOI 10.1016/0370-2693(91)90233-G. MR1092675

[39] R. Gambini, L. Leal, and A. Trias, *Loop calculus for lattice gauge theories*, Phys. Rev. D (3) **39** (1989), no. 10, 3127–3135, DOI 10.1103/PhysRevD.39.3127. MR994999

[40] R. Gambini and J. Pullin, *Loops, knots, gauge theories and quantum gravity*, Cambridge Monographs on Mathematical Physics, Cambridge University Press, Cambridge, 1996. With a foreword by Abhay Ashtekar, DOI 10.1017/CBO9780511524431. MR1439964

[41] R. Gambini and A. Trias, *Second quantization of the free electromagnetic field as quantum mechanics in the loop space*, Phys. Rev. D (3) **22** (1980), no. 6, 1380–1384, DOI 10.1103/PhysRevD.22.1380. MR586713

[42] R. Gambini and A. Trias, *Geometrical origin of gauge theories*, Phys. Rev. D (3) **23** (1981), no. 2, 553–555, DOI 10.1103/PhysRevD.23.553. MR599905

[43] R. Gambini and A. Trias, *Chiral formulation of Yang-Mills equations: a geometric approach*, Phys. Rev. D (3) **27** (1983), no. 12, 2935–2939, DOI 10.1103/PhysRevD.27.2935. MR708029

[44] R. Gambini and A. Trias, *Gauge dynamics in the C-representation*, Nuclear Phys. B **278** (1986), no. 2, 436–448, DOI 10.1016/0550-3213(86)90221-X. MR862775

[45] P. Gibilisco, *Bundle-connection pairs and representations of the loop group* (Russian, with Russian summary), Mat. Zametki **61** (1997), no. 4, 503–518, DOI 10.1007/BF02354986; English transl., Math. Notes **61** (1997), no. 3-4, 417–429. MR1619979

[46] R. Giles, *Reconstruction of gauge potentials from Wilson loops*, Phys. Rev. D (3) **24** (1981), no. 8, 2160–2168, DOI 10.1103/PhysRevD.24.2160. MR632522

[47] L. Gross, *A Poincaré lemma for connection forms*, J. Funct. Anal. **63** (1985), no. 1, 1–46, DOI 10.1016/0022-1236(85)90096-5. MR795515

[48] P. M. Hajac, *Axiomatic holonomy maps and generalized Yang-Mills moduli space*, Lett. Math. Phys. **27** (1993), no. 4, 301–309, DOI 10.1007/BF00777377. MR1219503

[49] B. Hambly and T. Lyons, *Uniqueness for the signature of a path of bounded variation and the reduced path group*, Ann. of Math. (2) **171** (2010), no. 1, 109–167, DOI 10.4007/annals.2010.171.109. MR2630037

[50] R. Hermann, *Vector bundles in mathematical physics. Vols. I, II*, W. A. Benjamin, Inc., New York, 1970. MR0266551

[51] R. Hermann, *Gauge fields and Cartan-Ehresmann connections. Part A*, Interdisciplinary Mathematics, Vol. X, Math Sci Press, Brookline, Mass., 1975. MR0493849

[52] R. Hermann, *Yang-Mills, Kaluza-Klein, and the Einstein program*, Interdisciplinary Mathematics, XIX, Math Sci Press, Brookline, Mass., 1978. With contributions by Frank Estabrook and Hugo Wahlquist. MR502252

[53] R. Jackiw, *Gauge-covariant conformal transformations*, Phys. Rev. Lett. **41** (1978), no. 24, 1635–1638.

[54] S. Kobayashi, *La connexion des varietés fibrées I, II*, Comptes Rendus Acad. Sci. **238** (1954), no. 3–4, 318–319,443–444.
[55] S. Kobayashi and K. Nomizu, *Foundations of differential geometry. Vol I*, Interscience Publishers, a division of John Wiley & Sons, New York-London, 1963. MR0152974
[56] A. Kriegl and P. W. Michor, *The convenient setting of global analysis*, Mathematical Surveys and Monographs, vol. 53, American Mathematical Society, Providence, RI, 1997, DOI 10.1090/surv/053. MR1471480
[57] R. Lashof, *Classification of fibre bundles by the loop space of the base*, Ann. of Math. (2) **64** (1956), 436–446, DOI 10.2307/1969594. MR82099
[58] J. Lewandowski, *Group of loops, holonomy maps, path bundle and path connection*, Classical Quantum Gravity **10** (1993), no. 5, 879–904. MR1219436
[59] R. Loll, *Loop approaches to gauge field theories* (English, with English and Russian summaries), Teoret. Mat. Fiz. **93** (1992), no. 3, 481–505, DOI 10.1007/BF01016398; English transl., Theoret. and Math. Phys. **93** (1992), no. 3, 1415–1432 (1993). MR1233555
[60] R. Loll, *Gauge theory and gravity in the loop formulation*, Canonical gravity: from classical to quantum (Bad Honnef, 1993), Lecture Notes in Phys., vol. 434, Springer, Berlin, 1994, pp. 254–288, DOI 10.1007/3-540-58339-4_21. MR1297337
[61] R. Loll, *The loop formulation of gauge theory and gravity*, Knots and quantum gravity (Riverside, CA, 1993), Oxford Lecture Ser. Math. Appl., vol. 1, Oxford Univ. Press, New York, 1994, pp. 1–19. MR1309912
[62] M. Lüscher, *Topology of lattice gauge fields*, Comm. Math. Phys. **85** (1982), no. 1, 39–48. MR667766
[63] M. Mackaay and R. Picken, *Holonomy and parallel transport for abelian gerbes*, Adv. Math. **170** (2002), no. 2, 287–339, DOI 10.1006/aima.2002.2085. MR1932333
[64] Yu. M. Makeenko and A. A. Migdal, *Quantum chromodynamics as dynamics of loops*, Nuclear Phys. B **188** (1981), no. 2, 269–316, DOI 10.1016/0550-3213(81)90258-3. MR629499
[65] S. Mandelstam, *Quantum electrocynamics without potentials*, Ann. Physics **19** (1962), 1–24, DOI 10.1016/0003-4916(62)90232-4. MR142329
[66] D. Marolf and J. M. Mourão, *On the support of the Ashtekar-Lewandowski measure*, Comm. Math. Phys. **170** (1995), no. 3, 583–605. MR1337134
[67] J. F. Martins and R. Picken, *On two-dimensional holonomy*, Trans. Amer. Math. Soc. **362** (2010), no. 11, 5657–5695, DOI 10.1090/S0002-9947-2010-04857-3. MR2661492
[68] C. Meneses and J. A. Zapata, *Homotopy classes of gauge fields and the lattice*, Adv. Theor. Math. Phys. **23** (2019), no. 8, 2207–2254, DOI 10.4310/ATMP.2019.v23.n8.a7. MR4101663
[69] C. Meneses and J. A Zapata, *Macroscopic observables from the comparison of local reference systems*, Classical Quantum Gravity **36** (2019), no. 23, 235011, 28, DOI 10.1088/1361-6382/ab49a7. MR4062720
[70] J. W. Milnor, *Construction of universal bundles I-II*, Ann. Math. **63** (1956), no. 2, 272–284,430–436.
[71] P. K. Mitter and C.-M. Viallet, *On the bundle of connections and the gauge orbit manifold in Yang-Mills theory*, Comm. Math. Phys. **79** (1981), no. 4, 457–472. MR623962
[72] K. Morrison, *Yang-Mills connections on surfaces and representations of the path group*, Proc. Amer. Math. Soc. **112** (1991), no. 4, 1101–1106, DOI 10.2307/2048660. MR1069292
[73] M. S. Narasimhan and S. Ramanan, *Existence of universal connections*, Amer. J. Math. **83** (1961), 563–572, DOI 10.2307/2372896. MR133772
[74] K. Osterwalder and E. Seiler, *Gauge field theories on a lattice*, Ann. Physics **110** (1978), no. 2, 440–471, DOI 10.1016/0003-4916(78)90039-8. MR459367
[75] A. Phillips, *Characteristic numbers of* U_1*-valued lattice gauge fields*, Ann. Physics **161** (1985), no. 2, 399–422, DOI 10.1016/0003-4916(85)90086-7. MR793820
[76] A. Phillips and D. Stone, *Lattice gauge fields, principal bundles and the calculation of topological charge*, Comm. Math. Phys. **103** (1986), no. 4, 599–636. MR832541
[77] A. V. Phillips and D. A. Stone, *The computation of characteristic classes of lattice gauge fields*, Comm. Math. Phys. **131** (1990), no. 2, 255–282. MR1065672
[78] A. V. Phillips and D. A. Stone, *A topological Chern-Weil theory*, Mem. Amer. Math. Soc. **105** (1993), no. 504, vi+79, DOI 10.1090/memo/0504. MR1169229
[79] A. M. Polyakov, *Gauge fields as rings of glue*, Nuclear Phys. B **164** (1980), no. 1, 171–188, DOI 10.1016/0550-3213(80)90507-6. MR561638

[80] T. A. Schilling, *Non-covariance of the generalized holonomies: examples*, J. Math. Phys. **37** (1996), no. 8, 4041–4052, DOI 10.1063/1.531615. MR1400832

[81] U. Schreiber and K. Waldorf, *Parallel transport and functors*, J. Homotopy Relat. Struct. **4** (2009), no. 1, 187–244. MR2520993

[82] P. Spallanzani, *Groups of loops and hoops*, Comm. Math. Phys. **216** (2001), no. 2, 243–253, DOI 10.1007/s002200000284. MR1814846

[83] A. Stacey, *Constructing smooth manifolds of loop spaces*, Proc. Lond. Math. Soc. (3) **99** (2009), no. 1, 195–216, DOI 10.1112/plms/pdn058. MR2520355

[84] A. Stacey, *Comparative smootheology*, Theory Appl. Categ. **25** (2011), No. 4, 64–117. MR2805746

[85] A. Stacey, *The smooth structure of the space of piecewise-smooth loops*, Glasg. Math. J. **59** (2017), no. 1, 27–59, DOI 10.1017/S0017089516000033. MR3576326

[86] J. D. Stasheff, *Parallel transport and classification of fibrations*, Algebraic and geometrical methods in topology (Conf. Topological Methods in Algebraic Topology, State Univ. New York, Binghamton, N.Y., 1973), Springer, Berlin, 1974, pp. 1–17. Lecture Notes in Math., Vol. 428. MR0375305

[87] J. Stasheff, *"Parallel" transport—revisited*, Mat. Contemp. **41** (2012), 191–204. MR3087578

[88] J. N. Tavares, *Chen integrals, generalized loops and loop calculus*, Internat. J. Modern Phys. A **9** (1994), no. 26, 4511–4548, DOI 10.1142/S0217751X94001795. MR1295759

[89] C. Teleman, *Généralisation du groupe fondamental* (French), Ann. Sci. École Norm. Sup. (3) **77** (1960), 195–234. MR0124068

[90] C. Teleman, *Sur les connexions infinitésimales qu'on peut définir dans les structures fibrées différentiables de base donnée* (French), Ann. Mat. Pura Appl. (4) **62** (1963), 379–412, DOI 10.1007/BF02410657. MR164300

[91] T. Tlas, *On the Ashtekar-Lewandowski measure as a restriction of the product one*, J. Math. Phys. **55** (2014), no. 12, 122302, 11, DOI 10.1063/1.4902931. MR3390544

[92] T. Tlas, *On the holonomic equivalence of two curves*, Internat. J. Math. **27** (2016), no. 6, 1650055, 22, DOI 10.1142/S0129167X16500555. MR3516982

[93] V. S. Varadarajan, *Vector bundles and connections in physics and mathematics: some historical remarks*, A tribute to C. S. Seshadri (Chennai, 2002), Trends Math., Birkhäuser, Basel, 2003, pp. 502–541. MR2017600

[94] H. Whitney, *Analytic extensions of differentiable functions defined in closed sets*, Trans. Amer. Math. Soc. **36** (1934), no. 1, 63–89, DOI 10.2307/1989708. MR1501735

[95] K. G. Wilson, *Confinement of quarks*, Phys. Rev. D **10** (1974), no. 8, 2445–2459.

[96] E. E. Wood, *Reconstruction theorem for groupoids and principal fiber bundles*, Internat. J. Theoret. Phys. **36** (1997), no. 5, 1253–1267, DOI 10.1007/BF02435815. MR1452246

[97] T. T. Wu and C. N. Yang, *Concept of nonintegrable phase factors and global formulation of gauge fields*, Phys. Rev. D (3) **12** (1975), no. 12, 3845–3857, DOI 10.1103/PhysRevD.12.3845. MR426712

[98] C. N. Yang, *Integral formalism for gauge fields*, Phys. Rev. Lett. **33** (1974), 445–447, DOI 10.1103/PhysRevLett.33.445. MR401005

[99] C. N. Yang and R. L. Mills, *Conservation of isotopic spin and isotopic gauge invariance*, Phys. Rev. (2) **96** (1954), 191–195. MR65437

Mathematisches Seminar, Christian-Albrechts-Universität zu Kiel, Heinrich-Hecht-Platz 6, 24118 Kiel, Germany

Email address: meneses@math.uni-kiel.de

Contemporary Mathematics
Volume **775**, 2021
https://doi.org/10.1090/conm/775/15595

On fractional higher-order Dirichlet boundary value problems: Between the Laplacian and the bilaplacian

Alberto Saldaña

ABSTRACT. The solutions of boundary value problems for the Laplacian and the bilaplacian exhibit very different qualitative behaviors. Particularly, the failure of general maximum principles for the bilaplacian implies that solutions of higher-order problems are less rigid and more complex. One way to better understand this transition is to study the intermediate Dirichlet problem in terms of fractional Laplacians. This survey aims to be an introduction to this type of problems; in particular, the different pointwise notions for these operators is introduced considering a suitable natural extension of the Dirichlet boundary conditions for the fractional setting. Solutions are obtained variationally and, in the case of the ball, via explicit kernels. The validity of maximum principles for these intermediate problems is also discussed as well as the limiting behavior of solutions when approaching the Laplacian or the bilaplacian case.

1. Introduction

Let $U \subset \mathbb{R}^N$ ($N \in \mathbb{N}$) be an open bounded set with smooth boundary and $\beta \in (0,1)$. The (inhomogeneous) Dirichlet boundary value problem for the *Laplacian* $\Delta u = \sum_{i=1}^N \partial_{ii} u$ is to find the unique solution $u \in C^{2+\beta}(U) \cap C(\overline{U})$ of

$$-\Delta u = f \quad \text{in } U, \quad u = g \quad \text{on } \partial U, \tag{1.1}$$

where $f \in C^\beta(U)$ and $g \in C(\partial B)$ are given data. Analogously, the Dirichlet boundary value problem for the *bilaplacian* is to find the unique solution $u \in C^{4+\beta}(U) \cap C^1(\overline{U})$ of

$$\Delta^2 u = f \quad \text{in } U, \quad u = g \quad \text{on } \partial U, \quad -\partial_\nu u = h \quad \text{on } \partial U, \tag{1.2}$$

where ν denotes the unit outward-pointing normal vector on ∂U, $f \in C^\beta(U)$, $g \in C^1(\partial B)$, and $h \in C(\partial B)$ are given data. The problems (1.1) and (1.2) are at the core of the linear theory for elliptic operators, and there is an extensive study of their solutions, primarily focusing on existence, uniqueness, regularity, qualitative properties (positivity, symmetry), and representation formulas, see [**18, 19**].

Regarding existence and uniqueness of solutions to (1.1) and (1.2), a very powerful approach is given by *variational methods*, where solutions can be found as (unique) minimizers of some energy functional on a suitable Sobolev space. Once

2020 *Mathematics Subject Classification.* Primary 35C05, 35S15, 35B50.

the existence of a solution is established, one of the most important questions regarding its qualitative properties is the following:

Given *nonnegative* data, is the corresponding solution *nonnegative*?

This is sometimes called a *positivity preserving property* or a maximum principle. To be more precise, for the Laplacian this amounts to the following question: if $f \geq 0$ and $g \geq 0$, is it true that the solution u of (1.1) is nonnegative?, similarly for the bilaplacian: if $f \geq 0$, $g \geq 0$, and $h \geq 0$, is it true that the solution u of (1.2) is nonnegative?

For the Laplacian, this positivity preserving property *always* holds, and this is a formidable tool in the study of linear and nonlinear elliptic equations (and systems). Maximum principles are at the heart of a priori bounds, symmetry characterizations, existence, nonexistence, uniqueness, multiplicity, and regularity results, among others; we refer to [**11**, **19**, **30**–**34**] for a glimpse of the diversity of problems and methods in which maximum principles play an essential role.

For the bilaplacian the situation is more delicate: maximum principles do *not* hold in general and this is linked to the fact that higher-order problems are "less rigid", which produces a larger and more complex set of solutions, see for example [**4**, **8**, **12**, **18**, **28**] and the references therein. There are, however, some partial results regarding positivity preserving properties in the higher-order setting; for example, if U is a ball (or a small perturbation of the ball), $f \geq 0$, $h \geq 0$, and $g \equiv 0$, then $u \geq 0$ [**18**]. One of the most important (and long-standing) open problems in the theory of higher-order linear elliptic problems is to understand which domains U allow a positivity preserving property for (1.2) with $f \geq 0$ and $g \equiv h \equiv 0$ [**18**].

This paper focuses on the intermediate problem between (1.1) and (1.2) by considering fractional powers of the Laplacian $(-\Delta)^s$ with $s \in (1,2)$. This fractional higher-order Dirichlet problem has a very rich structure resulting from the mixture between the Laplacian, the bilaplacian, and new purely nonlocal phenomena. The understanding of this complex structure is interesting in its own right, but another motivation to study these problems comes from the fact that the fractional setting offers a bridge between the two Dirichlet problems (1.1) and (1.2), one which is very well-behaved and rigid, whereas the other is laxer and less constrained by the properties of the given data. Such a bridge has the potential to offer a new perspective and novel insights on the many open problems regarding linear and nonlinear higher-order equations.

Boundary value problems for higher-order fractional powers of the Laplacian have not been studied much in the literature so far. Some known results are the following. General regularity results have been proved in [**21**] (see also the survey [**23**] and the references therein), a Pohožaev identity and an integration by parts formula is given in [**29**], a comparison between different notions of higher-order fractional operators is done in [**27**], and spectral results are obtained in [**22**]. A discussion on the pointwise definition of $(-\Delta)^s$ can be found in [**7**], explicit integral representations of solutions in [**3**–**6**], and a study of positivity preserving properties in [**4**, **8**].

Our discussion below is based on the results from [**5**–**8**] and is guided by the following natural questions. For $s \in (1,2)$:

(Q1) What would be a suitable pointwise evaluation for the operator $(-\Delta)^s$?

(Q2) Is it possible to study higher-order fractional Dirichlet problems variationally?
(Q3) What can be said about positivity preserving properties for $(-\Delta)^s$?
(Q4) What is the natural extension of (inhomogeneous) Dirichlet boundary conditions for $(-\Delta)^s$ and how can we find solutions?
(Q5) What happens with the solutions of $(-\Delta)^s u = f$ in U as $s \to 2$ or as $s \to 1$? Do we recover solutions of (1.1) and (1.2)?

We answer the first two questions in Sections 2 and 3 in a rather general setting. Section 4 is devoted to answer (Q3) and for (Q4) and (Q5) we study in detail the case of the ball in Sections 5 and 6. Finally, in Appendix A we include a brief discussion on the composition of Green functions.

To close this introduction we remark that this survey is focused only in the case $s \in (1,2)$ for simplicity and to fix ideas; however, most of the definitions and results discussed below are available for any $s > 1$, see [**3**, **5**–**8**] and Remark 5.9 for some comments on the case $s > 2$. We also mention that a more thorough study of the convergence of solutions is possible, see [**24**].

1.1. Notations. In the reminder of the paper we use the following standing notation. We fix $B := B_1(0)$ and $B_r := B_r(0)$ for $r > 0$. For $m \in \mathbb{N}_0$ and U open we write $C^{m,0}(U)$ to denote the space of m-times continuously differentiable functions in U and, for $\sigma \in (0,1]$ and $s = m + \sigma$, we write $C^s(U) := C^{m,\sigma}(U)$ to denote the space of functions in $C^{m,0}(U)$ whose derivatives of order m are (locally) σ-Hölder continuous in U or (locally) Lipschitz continuous in U if $\sigma = 1$. We denote by $C^s(\overline{U})$ the set of functions $u \in C^s(U)$ such that

$$\|u\|_{C^s(U)} := \sum_{|\alpha|\leq m} \|\partial^\alpha u\|_{L^\infty(U)} + \sum_{|\alpha|=m} \sup_{\substack{x,y\in U\\ x\neq y}} \frac{|\partial^\alpha u(x) - \partial^\alpha u(y)|}{|x-y|^\sigma} < \infty. \tag{1.3}$$

Moreover, for $s \in (0, \infty]$,

$$C^s_c(U) := \{u \in C^s(\mathbb{R}^N) : \operatorname{supp} u \subset\subset U\},$$
$$C^s_0(U) := \{u \in C^s(\mathbb{R}^N) : u = 0 \text{ on } \mathbb{R}^N \setminus U\},$$

where $\operatorname{supp} u := \overline{\{x \in U \,:\, u(x) \neq 0\}}$ is the support of u. We also write $A \subset\subset B$ to denote that A is *compactly contained* in B, that is, that $\overline{A}$ is a compact set and $\overline{A} \subset B$.

We use $u^+ := u_+ := \max\{u, 0\}$ to denote the positive part of u. For $\beta \in \mathbb{R}$ we set

$$\delta(x)^\beta := \begin{cases} (1-|x|^2)^\beta, & \text{if } 1-|x|^2 > 0, \\ 0, & \text{if } 1-|x|^2 \leq 0. \end{cases}$$

If $\beta = 1$ we simply write $\delta(x)$. The fractional Sobolev space $H^s(\mathbb{R}^N)$ is given by

$$H^s(\mathbb{R}^N) := \left\{u \in L^2(\mathbb{R}^N) \,:\, (1+|\xi|^2)^{\frac{s}{2}}\, \mathcal{F}(u) \in L^2(\mathbb{R}^N)\right\},$$

where $\mathcal{F}$ denotes the Fourier transform and, for $U \subset \mathbb{R}^N$ open, the homogeneous Dirichlet fractional Sobolev space is

$$\mathcal{H}^s_0(U) := \{u \in H^s(\mathbb{R}^N) \,:\, u \equiv 0 \text{ on } \mathbb{R}^N \setminus U\}.$$

Furthermore, $H^s(U) := \{u\chi_U \,:\, u \in H^s(\mathbb{R}^N)\}$, where χ_U is the characteristic function of U, namely, $\chi_U(x) = 1$ if $x \in U$ and $\chi_U(x) = 0$ if $x \notin U$. We frequently use the following normalization constants:

$$\omega_N := 2\pi^{\frac{N}{2}}\,\Gamma\left(\frac{N}{2}\right)^{-1}, \quad k_{N,s} := \frac{2^{1-2s}}{\omega_N \Gamma(s)^2}, \quad \gamma_{N,\sigma} := \frac{2}{\Gamma(\sigma)\,\Gamma(1-\sigma)\omega_N}, \tag{1.4}$$

where Γ denotes the usual *Gamma function.* Finally, we recall that, in dimension one ($N = 1$), the boundary integral is meant in the sense $\int_{\partial B} f(\theta)\, d\theta = f(-1) + f(1)$.

2. Pointwise evaluations

The pointwise definition of the higher-order fractional Laplacian $(-\Delta)^s$ for $s \in (1,2)$ can be a delicate issue and some of its aspects may seem a bit counter-intuitive at first glance. Here we present three ways to understand this operator pointwisely and discuss some of their advantages and disadvantages. The first one is a classical definition via the Fourier transform, the second one is based on a composition of operators (similarly as in the definition of bilaplacian), and finally the third one is based on higher-order finite differences. This last pointwise notion is the most general and is the one we use in the rest of the paper. For smooth functions ($C_c^\infty(\mathbb{R}^N)$, for example), all these evaluations agree; but, as soon as one considers less regular elements, differences—which are crucial to study boundary value problems—appear.

We also emphasize that, in the fractional setting, the pointwise definition of the operator is *closely linked* to the type of boundary conditions that is being studied. In this survey we concentrate only on *Dirichlet* boundary conditions. To see how different boundary conditions may require a change in the pointwise notion of $(-\Delta)^s$, we refer to [**1, 26**] and the references therein, where Neumann and Robin-type boundary conditions are considered for powers $s \in (0,1)$.

2.1. Via Fourier transform.

Fractional Laplacians can be seen as a *pseudo-differential* operator, that is, they can be defined via the Fourier transform $\mathcal{F}$ prescribing the *symbol* of the operator, namely, for $s > 0$,

$$(-\Delta)^s \varphi(x) = \mathcal{F}^{-1}(|\cdot|^{2s}\mathcal{F}(\varphi))(x) \quad \text{for all } \varphi \in C_c^\infty(\mathbb{R}^N). \tag{2.1}$$

This notion has the advantage of relating the structure and properties of the Fourier transform with the higher-order fractional Laplacian but it is a rather indirect pointwise definition, which makes it difficult to perform some explicit pointwise calculations.

2.2. Via a composition of operators.

The bilaplacian operator Δ^2 can be simply defined by iterating the Laplacian, that is,

$$\Delta^2 u(x) = (-\Delta)(-\Delta)u(x) \quad \text{for all } u \in C^4(\mathbb{R}^N).$$

Analogously, for $s \in (1,2)$, one can define the higher-order Laplacian $(-\Delta)^s$ as a composition of $(-\Delta)$ and $(-\Delta)^{s-1}$, where $(-\Delta)^{s-1}$ is given by

$$(-\Delta)^{s-1}u(x) := e_{N,s} \int_{\mathbb{R}^N} \frac{2u(x) - u(x+y) - u(x-y)}{|y|^{N+2(s-1)}}\, dy \tag{2.2}$$

with

$$e_{N,s} = -\frac{4^{s-1}\Gamma(\frac{N}{2} + s - 1)}{\pi^{\frac{N}{2}}\Gamma(1-s)}. \tag{2.3}$$

Here $e_{N,s}$ is a suitable normalization constant such that (2.1) holds and u is such that the integral (2.2) is finite. The right-hand side of (2.2) is sometimes called a *hypersingular integral*, because the singularity of the kernel $|y|^{N+2(s-1)}$ at zero is not integrable and requires some local smoothness of u to guarantee integrability, for instance, that u is of class $C^{2(s-1)+\alpha}$ at x for some $\alpha > 0$. Moreover, to ensure integrability at infinity, one must impose some growth restrictions; this is usually done by requiring that u belongs to the space $\mathcal{L}^1_{s-1}$, where

$$\mathcal{L}^1_t := \left\{ u \in L^1_{loc}(\mathbb{R}^N) \ : \ \int_{\mathbb{R}^N} \frac{|u(x)|}{1+|x|^{N+2t}}\, dx < \infty \right\} \quad \text{for any } t > 0. \tag{2.4}$$

Let U be an open set in $\mathbb{R}^N$, then the integral (2.2) is finite for $x \in U$ if $u \in C^{2(s-1)+\alpha}(U) \cap \mathcal{L}^1_{s-1}$. Furthermore, one can show that $(-\Delta)^{s-1}u \in C^{2+\alpha}(U)$ if $u \in C^{2s+\alpha}(U) \cap \mathcal{L}^1_{s-1}$, see [**35**, Proposition 2.7]. Therefore, we can define, for $u \in C^{2s+\alpha}(U) \cap \mathcal{L}^1_s$ and $x \in U$,

$$(-\Delta)^s u(x) = (-\Delta)(-\Delta)^{s-1}u(x). \tag{2.5}$$

This pointwise evaluation is very helpful for explicit calculations, since the operator $(-\Delta)^{s-1}$ can be computed in some cases (see Appendix A or [**15,16**]). However, the evaluation (2.5) has the following disadvantage: to compute $(-\Delta)^{s-1}u$ one requires the growth restriction $u \in \mathcal{L}^1_{s-1}$, which is not optimal for $(-\Delta)^s$ (see Theorem 5.5 below).

We emphasize that the order of the operators in (2.5) is very important and it *cannot* be freely interchanged in general, namely, it is *not* true that $(-\Delta)$ $(-\Delta)^{s-1}u(x)$ equals $(-\Delta)^{s-1}(-\Delta)u(x)$; the equality holds only for smooth enough functions, which is not the case in general for solutions of Dirichlet boundary value problems, see Appendix A for an explicit computation in this regard. Finally, we mention that other compositions such as $(-\Delta)^r(-\Delta)^t u$ with $r, t \in (0,1)$ and $r+t = s$ are not well suited for the study of boundary value problems; the reason—similarly as in the case in Appendix A—is that, although for $u \in C^\infty_c(\mathbb{R}^N)$ all these pointwise notions are equivalent, solutions of boundary value problem are not regular enough to guarantee that these compositions are always well defined (note that $(-\Delta)^{s-1}(-\Delta)u(x)$ requires that u is twice weakly differentiable in $\mathbb{R}^N$, because $(-\Delta)^{s-1}$ is a nonlocal operator).

2.3. Via finite differences. We now introduce the most general pointwise evaluation of the higher-order fractional Laplacian, which, similarly as in (2.2), is in terms of hypersingular integrals but involves higher-order finite differences. For $s \in (1,2)$, $U \subset \mathbb{R}^N$ open, $\beta \in (0,1)$, $u \in \mathcal{L}^1_s \cap C^{2s+\beta}(U)$, and $x \in U$, let

(2.6)
$$(-\Delta)^s u(x) := c_{N,s} \int_{\mathbb{R}^N} \frac{u(x+2y) - 4u(x+y) + 6u(x) - 4u(x-y) + u(x-2y)}{|y|^{N+2s}}\, dy,$$

where

$$c_{N,s} = \frac{\Gamma(\frac{N}{2}+s)}{\pi^{\frac{N}{2}}\Gamma(-s)(1-4^{1-s})} \tag{2.7}$$

is a normalization constant such that (2.1) holds (see [**7**, Theorem 1.9] for the details). In the following, whenever we write $(-\Delta)^s u(x)$ as a pointwise evaluation, we always mean it in the sense of (2.6).

Explicit pointwise calculations using (2.6) are slightly more involved than those for (2.5), and typically require some combinatorial identities, see [**7**]. We remark that, if $u \in \mathcal{L}^1_{s-1}$, then (2.6) is equivalent to (2.5) (note that $\mathcal{L}^1_{s-1} \subset \mathcal{L}^1_s$), we state this result next.

LEMMA 2.1 (Particular case of Corollary 1.4 in [**7**]). *Let* $\beta \in (0,1)$, $s \in (1,2)$, $U \subset \mathbb{R}^N$ *be smooth open domain, and* $u \in C^{2s+\beta}(U) \cap \mathcal{L}^1_{s-1}$, *then, for* $x \in U$,

$$\begin{aligned}&(-\Delta)(-\Delta)^{s-1}u(x)\\ &\quad = e_{N,s}(-\Delta)\int_{\mathbb{R}^N} \frac{2u(x)-u(x+y)-u(x-y)}{|y|^{N+2s}}\,dy\\ &\quad = c_{N,s}\int_{\mathbb{R}^N} \frac{u(x+2y)-4u(x+y)+6u(x)-4u(x-y)+u(x-2y)}{|y|^{N+2s}}\,dy;\end{aligned}$$

in particular, the pointwise evaluations (2.6) *and* (2.5) *are equivalent for* $u \in C^{2s+\beta}(U) \cap \mathcal{L}^1_{s-1}$.

3. Variational framework

The variational study of the higher-order fractional Laplacian with homogeneous Dirichlet boundary conditions can be framed in suitable fractional Sobolev spaces. To be precise, let $s \in (1,2)$ and recall the definition of the usual fractional Sobolev space

$$H^s(\mathbb{R}^N) := \left\{u \in L^2(\mathbb{R}^N)\ :\ (1+|\xi|^2)^{\frac{s}{2}}\ \mathcal{F}(u) \in L^2(\mathbb{R}^N)\right\},$$

where $\mathcal{F}$ denotes the Fourier transform and, for $U \subset \mathbb{R}^N$ open, define the homogeneous Dirichlet fractional Sobolev space

$$\mathcal{H}^s_0(U) := \{u \in H^s(\mathbb{R}^N)\ :\ u \equiv 0 \text{ on } \mathbb{R}^N \setminus U\} \tag{3.1}$$

equipped with the norm $\|u\|_{\mathcal{H}^s_0(U)} := (\|u\|^2_{L^2(U)} + \sum_{i=1}^N \|\partial_i u\|^2_{L^2(U)} + \mathcal{E}_s(u,u))^{\frac{1}{2}}$, where $\mathcal{E}_s$ is a suitable scalar product in $\mathcal{H}^s_0(U)$. Similarly as in the previous section, we can have three formulas for this scalar product, each one naturally associated to each pointwise evaluation, however, since $\mathcal{H}^s_0(U) \subset L^2(\mathbb{R}^N) \subset \mathcal{L}^1_{s-1} \subset \mathcal{L}^1_s$, these three expressions are *equivalent* for functions in $\mathcal{H}^s_0(U)$. Nevertheless, sometimes one expression can be better suited than the other, depending on the object of study. The three formulas—via Fourier transform, composition of operators, and finite differences respectively—are the following: For $u, v \in \mathcal{H}^s_0(U)$ and $s \in (1,2)$, let

(3.2)
$$\mathcal{E}_s(u,v) = \int_{\mathbb{R}^N} |\xi|^{2s}\mathcal{F}u(\xi)\mathcal{F}v(\xi)\ d\xi,$$
(3.3)
$$= \frac{e_{N,s}}{2}\int_{\mathbb{R}^N}\int_{\mathbb{R}^N} \frac{(\nabla u(x)-\nabla u(y))\cdot(\nabla v(x)-\nabla v(y))}{|x-y|^{N+2(s-1)}}\,dx\,dy,$$
(3.4)
$$= \frac{c_{N,s}}{2}\int_{\mathbb{R}^N}\int_{\mathbb{R}^N} \frac{(2u(x)-u(x+y)-u(x-y))(2v(x)-v(x+y)-v(x-y))}{|y|^{N+2s}}\,dxdy,$$

where the normalization constants $e_{N,s}$ and $c_{N,s}$ are given in (2.3) and (2.7). The proof of the equivalence between these expressions can be found in [**7**, Theorem 1.8].

Now, let $f \in L^2(\Omega)$, we say that a function $u \in \mathcal{H}_0^s(\Omega)$ is a *weak solution* of

$$(-\Delta)^s u = f \quad \text{in } \Omega, \quad u = 0 \quad \text{on } \mathbb{R}^N \setminus \Omega, \tag{3.5}$$

if

$$\mathcal{E}_s(u, \varphi) = \int_\Omega f(x)\varphi(x)\, dx \quad \text{for all } \varphi \in \mathcal{H}_0^s(\Omega). \tag{3.6}$$

In this setting we can use Riesz theorem to yield the following existence result.

THEOREM 3.1 (Corollary 3.6 in [**8**]). *Let $U \subset \mathbb{R}^N$ be an open bounded set. Then for any $f \in L^2(U)$ there is a unique weak solution $u \in \mathcal{H}_0^s(U)$ of $(-\Delta)^s u = f$ in U.*

The scalar product $\mathcal{E}_s$ also satisfies the following integration-by-parts-type formula (see [**6**, Lemma 2.4] and [**7**, Theorem 1.8]).

LEMMA 3.2. *Let $U \subset \mathbb{R}^N$ be an open bounded set with Lipschitz boundary, $\alpha \in (0,1)$, $s \in (1,2)$, $u \in C^{2s+\alpha}(U) \cap \mathcal{L}_{s-1}^1 \cap \mathcal{H}_0^s(U)$. Then*

$$\mathcal{E}_s(u, \varphi) = \int_{\mathbb{R}^N} u\,(-\Delta)^s \varphi\, dx = \int_{\mathbb{R}^N} \varphi\,(-\Delta)^s u\, dx \quad \textit{for all } \varphi \in C_c^\infty(U).$$

About the regularity of weak solutions, the following is known.

LEMMA 3.3 (Theorem 2.2 in [**20**]). *Let $U \subset \mathbb{R}^N$ be a bounded smooth domain and $\beta \in (0,1)$ such that $2s + \beta \notin \mathbb{N}$. If $f \in C^\beta(\overline{U})$ and $u \in \mathcal{H}_0^s(U)$ is a weak solution of $(-\Delta)^s u = f$ in U, then $u \in C_0^s(U) \cap C^{2s+\beta}(U)$.*

4. Positivity preserving properties

As mentioned in the introduction, the Laplacian possesses the following well-known general maximum principle. Let U be a bounded domain[1] with smooth boundary in $\mathbb{R}^N$ and let $H_0^1(U)$ denote the usual Sobolev space of weakly differentiable functions with zero trace at ∂U.

LEMMA 4.1. *If $f \in L^2(U)$ is nonnegative in U and $u \in H_0^1(U)$ is a weak solution of*

$$(-\Delta)u = f \quad \textit{in } U, \quad u = 0 \quad \textit{on } \partial U,$$

that is,

$$\int_U \nabla u \nabla \varphi\, dx = \int_U \varphi f\, dx \quad \textit{for all } \varphi \in H_0^1(U), \tag{4.1}$$

then $u \geq 0$ a.e. in U.

PROOF. Since $u^- := \min\{0, u\} \in H_0^1(U)$, then (4.1) implies that

$$0 \leq \|u^-\|_{H_0^1(U)}^2 = \int_U |\nabla u^-|^2\, dx = \int_U \nabla u \nabla u^-\, dx = \int_U u^- f\, dx \leq 0,$$

that is, $\|u^-\|_{H_0^1(U)} = 0$ and therefore $u^- \equiv 0$ a.e. in U. □

[1] A domain is an open connected set.

A very similar proof can be done to show the validity of maximum principles for the fractional Laplacian $(-\Delta)^s$ for $s \in (0,1)$, see [**8**]. Observe that there are two important ingredients in the proof of Lemma 4.1: the variational characterization of the solution (4.1) and the belonging of the negative part u^- to the test space $H_0^1(U)$. Since the gradient ∇u^- has a jump discontinuity at the level set $\{x \in U : u(x) = 0\}$, we have that u^- is not twice weakly differentiable in general, and therefore $u^- \notin H^2(U)$, which prevents that a similar proof can be performed for the bilaplacian[2].

Interestingly, in [**13**, Théorème 1] it is shown that

$$u^- \in H^s(U) \quad \text{if } u \in H^s(U) \text{ and } s \in \Big(0\,,\frac{3}{2}\Big),$$

and, as explained in the previous section, the problem $(-\Delta)^s u = f$ has a variational structure. Since these are the main ingredients in the proof of maximum principles for $s = 1$—which uses u^- as a test function—it was conjectured that maximum principles would hold for the higher-order fractional Laplacian if $s \in (0, \frac{3}{2})$. However, our next result reveals that the positivity preserving property fails to hold in general for $s \in (1,2)$, therefore it is *not* the belonging of u^- to the space of test functions the reason why maximum principles hold for $s = 1$.

THEOREM 4.2. *[Particular case of Theorem 1.1 in* [**8**]*] Let $N \in \mathbb{N}$, $s \in (1,2)$, $U \subset \mathbb{R}^N$ be an open bounded smooth domain, let B be an open ball compactly contained in $\mathbb{R}^N \setminus U$, and let $\Omega := U \cup B$. There are $f \in C^\infty(\overline{\Omega})$ and a sign-changing $u \in C^s(\mathbb{R}^N) \cap C^\infty(\Omega) \cap L^\infty(\mathbb{R}^N) \cap \mathcal{H}_0^s(\Omega)$ such that*

$$(-\Delta)^s u = f > 0 \text{ in } \Omega, \quad u = 0 \text{ on } \mathbb{R}^N \backslash \Omega, \quad u \lneq 0 \text{ in } U, \text{ and } u > 0 \text{ in } B.$$

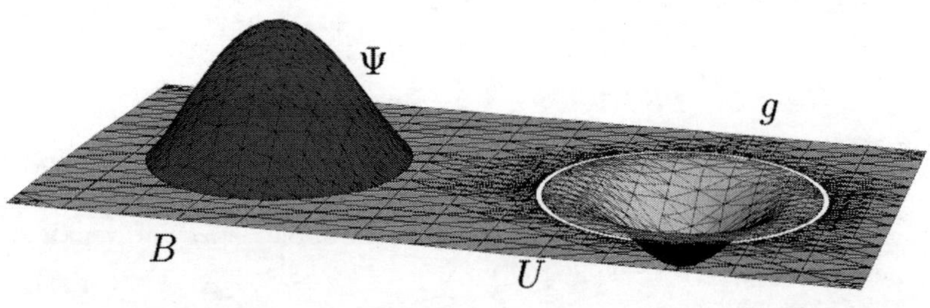

FIGURE 1. Theorem 4.2 is shown via an explicit counterexample $u = \psi + g$ with the above shape.

In fact, one can show that the Green function G_s^Ω associated to two disjoint balls is positive if $s \in (0,1)$ but sign changing if $s \in (1,2)$.

THEOREM 4.3 (Particular case of Theorem 1.10 in [**6**]). *Let $N \in \mathbb{N}$, $e_1 = (1,0,\ldots,0) \in \mathbb{R}^N$, $B = B_1(0)$, $V = B_1(3e_1)$, and $\Omega = B \cup V$. Then*

$$\begin{aligned} G_s^\Omega &> 0 \quad \text{in} \quad \{(x,y) \in (B \times B) \cup (V \times V) : x \neq y\}, \\ G_s^\Omega &> 0 \quad \text{in} \quad (B \times V) \cup (V \times B), \quad \text{if} \quad s \in (0,1), \\ G_s^\Omega &< 0 \quad \text{in} \quad (B \times V) \cup (V \times B), \quad \text{if} \quad s \in (1,2). \end{aligned}$$

[2]Here $H^2(U)$ denotes the Sobolev space of functions which are twice weakly differentiable in U and we say that $u \in H^2(U) \cap H_0^1(U)$ is a weak solutions of $(\Delta)^2 u = f$ in U and $\partial_\nu u = u = 0$ on ∂U if $\int_U \Delta u \Delta \varphi \, dx = \int_U \varphi f \, dx$ for all $\varphi \in H^2(U) \cap H_0^1(U)$.

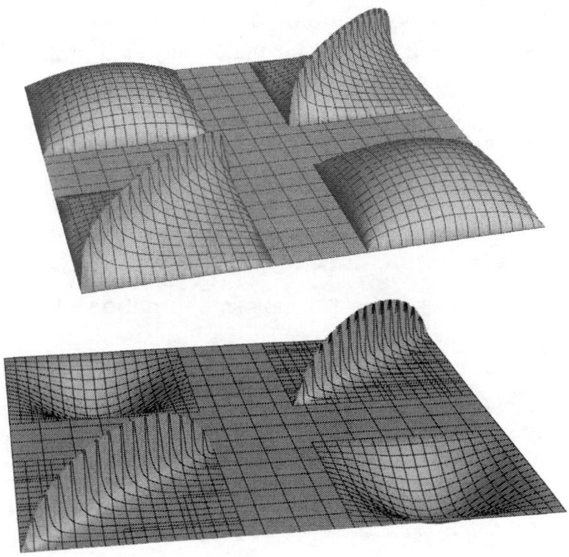

FIGURE 2. The Green function G_s^Ω for $(-\Delta)^s$ in $\Omega = (-1,1) \cup (2,4)$, on the left for $s \in (0,1)$ and on the right for $s \in (1,2)$.

The domain considered in Theorem 4.2 is disconnected, however one can use a perturbation argument to join the domains with a thin tube and find a sign-changing solution in a *connected domain*[3].

THEOREM 4.4 (Theorem 1.11 in [**6**]). *Let* $N \geq 2$, $s \in (1,2)$, $\Omega = B_1(0) \cup B_1(3e_1)$, $L := \{te_1 : 0 < t < 3\}$, *and*

$$\Omega_n = \Omega \cup \{ \, x \in \mathbb{R}^N : \operatorname{dist}(x \; , \; L) < \frac{1}{n} \, \} \quad \text{for } n \in \mathbb{N}. \tag{4.2}$$

There is $n \in \mathbb{N}$, *a nonnegative function* $f_n \in L^\infty(\Omega_n)$, *and a weak solution* $u_n \in \mathcal{H}_0^s(\Omega_n)$ *of* $(-\Delta)^s u_n = f_n \geq 0$ *in* Ω_n, $u_n = 0$ *on* $\mathbb{R}^N \backslash \overline{\Omega_n}$, *such that* $\operatorname{essinf}_{\Omega_n} u_n < 0$ *and* $\operatorname{esssup}_{\Omega_n} u_n > 0$.

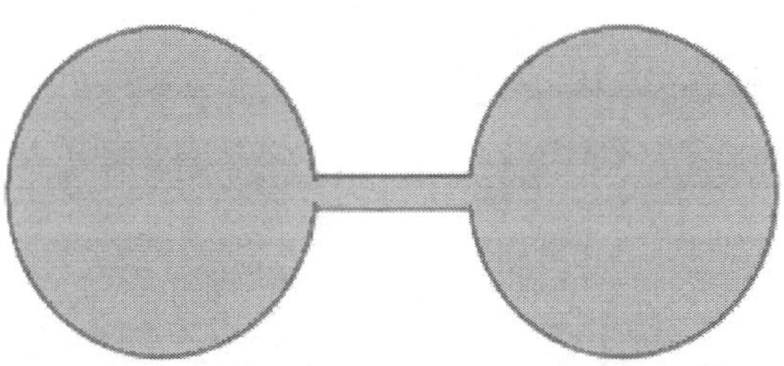

FIGURE 3. Example of the domain Ω_n given in (4.2) for $N = 2$.

[3]A slight variation of this argument can be used to find sign-changing solutions also in connected *smooth* domains, see also [**4**] for a sign-changing solution in an ellipse.

Although maximum principles do not hold in general domains in the higher-order fractional case, we shall see in the next section that positivity preserving properties *do* hold whenever the domain is a ball.

5. Explicit formulas for the inhomogeneous Dirichlet problem in balls

In the reminder of the paper we focus our attention on the case of a unitary ball

$$B := \{x \in \mathbb{R}^N : |x| < 1\} \quad \text{in } \mathbb{R}^N,\ N \in \mathbb{N}.$$

The purpose of this section is to explore the notion of the inhomogeneous Dirichlet problem for the higher-order fractional Laplacian, namely, we study what is the natural extension of Dirichlet boundary conditions in order to have well-posed problems for $(-\Delta)^s$ and such that one recovers (1.1) and (1.2) in the limit as $s \to 1^+$ or $s \to 2^-$.

A first remark is that the operator $(-\Delta)^s$ is *nonlocal*, since the pointwise computation of $(-\Delta)^s u(x)$ involves all the values of u in $\mathbb{R}^N$, see (2.6). Therefore, a natural "boundary condition" is to prescribe values in the complement of the domain, that is,

$$u = v \quad \text{on } \mathbb{R}^N \backslash \overline{B},$$

for some suitable $v \in \mathcal{L}^1_s$. We shall see below that the behavior of v close to the boundary ∂B is particularly important when trying to find an explicit representation formula for the solution.

Furthermore, one can also prescribe data at the boundary ∂B using appropriate *traces*. To be precise, for $\beta \in \mathbb{R}$ let

$$\delta(x)^\beta := \begin{cases} (1-|x|^2)^\beta, & \text{if } 1-|x|^2 > 0, \\ 0, & \text{if } 1-|x|^2 \le 0. \end{cases}$$

Then, for suitable u and for all $z \in \partial B$, let

$$(5.1) \qquad \begin{aligned} D^{s-2}u(z) &:= \lim_{\substack{x\to z\\ x\in B}} \delta(x)^{2-s}\Big(u(x) - \lim_{\substack{y\to z\\ y\in\mathbb{R}^N\backslash B}} u(y)\Big), \\ D^{s-1}u(z) &:= -\lim_{\substack{x\to z\\ x\in B}} \frac{\partial}{\partial |x|}\Big[\delta(x)^{2-s}\Big(u(x) - \lim_{\substack{y\to z\\ y\in\mathbb{R}^N\backslash B}} u(y)\Big)\Big], \end{aligned}$$

where all the limits are always meant in the normal direction, that is, with $\frac{x}{|x|} = \frac{y}{|y|} = z$. At first glance, the traces (5.1) might look strange and cumbersome, but let us analyze closely these definitions. A first remark is that, if $u = 0$ in $\mathbb{R}^N \backslash B$, then the traces reduce to

$$(5.2) \qquad D^{s-2}u(z) = \lim_{\substack{x\to z\\ x\in B}} \delta(x)^{2-s}u(x), \quad D^{s-1}u(z) = -\lim_{\substack{x\to z\\ x\in B}} \frac{\partial}{\partial |x|}[\delta(x)^{2-s}u(x)].$$

For $s = 2$ the weight δ^{2-s} disappears and (5.2) are exactly the Dirichlet boundary conditions for the bilaplacian ($D^0 u = u$ and $D^1 u = -\partial_\nu u$ on ∂B). For $s \in (1,2)$ note that $2 - s \in (0,1)$ and therefore, if $u \in C(B)$ and $D^{s-2}u(z) = \lim_{x\to z} \delta(x)^{2-s}u(x) \neq 0$ for some $z \in \partial B$, then u must be singular at z (see Figure 5 below). Solutions satisfying these kind of boundary conditions are sometimes called *very large solutions* and they have been studied in **[2]** for the case $s \in (0,1)$ using a similar trace operator. We also mention that trace operators combining weights and derivatives as in (5.2) were also used in [**21**, Theorem 6.1] (see also [**20**]), to study solvability of pseudodifferential operators in a more general setting.

REMARK 5.1. The traces in (5.1) are different from the ones used in [**6**], which are given by

$$\begin{aligned} \widetilde{D}^{s-2}u(z) &= \lim_{\substack{x\to z\\ x\in B}}[\delta(x)^{2-s}u(x)],\\ \widetilde{D}^{s-1}u(z) &= -\lim_{\substack{x\to z\\ x\in B}}\frac{\partial}{\partial|x|^2}[\delta(x)^{2-s}u(x)]. \end{aligned} \tag{5.3}$$

The two main differences between (5.1) and (5.3) are the use of the differential operator $\frac{\partial}{\partial|x|}$ instead of $\frac{\partial}{\partial|x|^2}$ and the limit $\lim\limits_{\substack{y\to z\\ y\in\mathbb{R}^N\setminus B}} u(y)$, which does not appear in (5.3). The reason for the first change is that the operator $\frac{\partial}{\partial|x|}$ is simply the normal derivative ∂_ν at ∂B, which is more common in the study of boundary value problems and substituting $\frac{\partial}{\partial|x|}$ instead of $\frac{\partial}{\partial|x|^2}$ does not imply many changes for $s\in(1,2)$, in fact,

$$\frac{\partial}{\partial|x|^2}f(x)=\frac{1}{2}\frac{\partial}{\partial|x|}f(x)\quad\text{at }\partial B.$$

The second change, that is, the limit from outside the ball $\mathbb{R}^N\setminus B$, is necessary to consider more general data in $\mathbb{R}^N\setminus B$ which may not vanish close to ∂B, see Theorem 5.6 below.

In the next subsections we show how to construct solutions to the fractional higher-order inhomogeneous Dirichlet boundary value problem via explicit kernels.

5.1. The Green function. For $x,y\in\mathbb{R}^N$, $x\neq y$ let $\rho(x,y)=\delta(x)\delta(y)|x-y|^{-2}$ and define

$$G_s(x,y) \;=\; k_{N,s}|x-y|^{2s-N}\int_0^{\rho(x,y)}\frac{t^{s-1}}{(t+1)^{\frac{N}{2}}}\,dt,\quad s>0,\quad x,y\in\mathbb{R}^N,\quad x\neq y, \tag{5.4}$$

where $k_{N,s}$ is a positive normalization constant given in (1.4). The kernel (5.4) is known as Boggio's formula (see [**5, 9, 10, 14, 18**]) or Green function for $(-\Delta)^s$ in B. Using this kernel we state the following existence and uniqueness result. Observe also that, since G_s is a positive kernel, a maximum principle is automatically satisfied.

THEOREM 5.2 (Theorem 1.1 in [**5**] and Theorem 1.4 in [**6**]). *Let $s\in(1,2)$, $N\in\mathbb{N}$, $f\in C^\alpha(\overline{B})$ for some $\alpha\in(0,1)$ such that $2s+\alpha\notin\mathbb{N}$, and*

$$u:\mathbb{R}^N\to\mathbb{R}\quad\text{be given by}\quad u(x):=\int_B G_s(x,y)\,f(y)\,dy, \tag{5.5}$$

then $u\in C^{2s+\alpha}(B)\cap C_0^s(B)$ is the unique pointwise solution (in $\mathcal{H}_0^s(B)$) of

$$(-\Delta)^s u=f\quad\text{in }B,\quad u\equiv 0\quad\text{on }\mathbb{R}^N\setminus B,\quad D^{s-2}u=D^{s-1}u=0\quad\text{on }\partial B, \tag{5.6}$$

and there is $C>0$ such that $\|\operatorname{dist}(\cdot,\partial B)^{-s}u\|_{L^\infty(B)}<C\|f\|_{L^\infty(B)}$. Furthermore, since G_s is a positive kernel, if $f\geq 0$ in U and $f\not\equiv 0$, then $u>0$ in U.

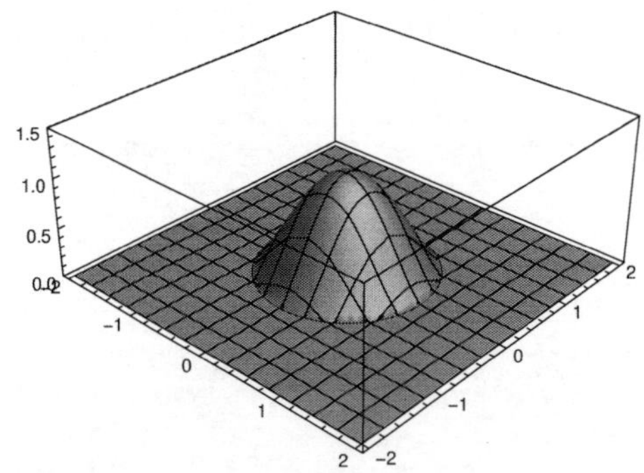

FIGURE 4. Solution of (5.6)with $f \equiv 1$.

This is a good moment to make an interesting remark. If $g = h = 0$ on ∂B, then the function u given by (5.5) is the solution of (1.2) if $s = 2$ and of (1.1) if $s = 1$. We can also consider the function arising from using twice the kernel G_1, namely,

$$\widetilde{v}(x) := \int_B G_1(x,z) \int_B G_1(z,y) f(y)\; dy\; dz.$$

Then $\widetilde{v}$ is a solution of $\Delta^2 \widetilde{v} = f$ in B and $\widetilde{v}$ satisfies *Navier boundary conditions*, that is, $\widetilde{v} = \Delta\widetilde{v} = 0$ on ∂B. Observe that $\widetilde{v}$ is not a solution of (1.2) (for example, if $f > 0$ in B then $-\partial_\nu \widetilde{v} > 0$ on ∂B, by Hopf Lemma). In the fractional case $s \in (1, 2)$, one can also consider the function u given by (5.5) with $f \equiv 1$ in B and compare it with $v, w : \mathbb{R}^N \to \mathbb{R}$ given by

$$v(x) := \int_B G_{s-1}(x,z) \int_B G_1(z,y)\; dy\, dz, \quad w(x) := \int_B G_1(x,z) \int_B G_{s-1}(z,y)\; dy\, dz.$$

We show in Appendix A that u, v, and w solve different boundary conditions. Furthermore, we also show the following surprising fact:

$$(-\Delta)^s u = (-\Delta)^s v = 1 \text{ in } B, \text{ but } (-\Delta)^s w \neq 1 \text{ in } B.$$

There are several reasons for this somewhat unexpected behavior. A particularly important factor is the regularity of these functions at the boundary ∂B. We refer to the explicit calculations in Appendix A for the details.

5.2. The boundary Poisson kernels. For $x \in \mathbb{R}^N$ and $z \in \partial B$ with $x \neq z$ let

$$\begin{aligned} E_{s-1}(x,z) &:= \frac{1}{2\omega_N} \frac{\delta(x)^s}{|x-z|^N}, \\ E_{s-2}(x,z) &:= \frac{1}{4\omega_N} \frac{\delta(x)^s}{|x-z|^{N+2}} (N\delta(x) - (N-4)|x-z|^2), \end{aligned} \tag{5.7}$$

where ω_N is a normalization constant given in (1.4). The kernels (5.7) are called the *boundary Poisson kernels* for $(-\Delta)^s$ in B or *Edenhofer kernels* in honor of Johann Edenhofer who first state their formula in the case $s \in \mathbb{N}$, see **[17]**. Using these kernels we can now solve the following boundary value problems.

THEOREM 5.3 (Theorem 1.4 in [**6**]). *Let $g \in C^{1,0}(\partial B)$ and $u : \mathbb{R}^N \to \mathbb{R}$ be given by*

$$u(x) = \int_{\partial B} E_{s-2}(x,\theta)\ g(\theta)\ d\theta \quad \textit{for } x \in \mathbb{R}^N.$$

Then, $u \in C^\infty(B)$, $\delta^{2-s}u \in C^{1,0}(\overline{B})$, and

(5.8)
$$(-\Delta)^s u = 0 \textit{ in } B, \quad u = 0 \textit{ on } \mathbb{R}^N \backslash \overline{B}, \quad D^{s-2}u = g \textit{ on } \partial B, \quad D^{s-1}u = 0 \textit{ on } \partial B.$$

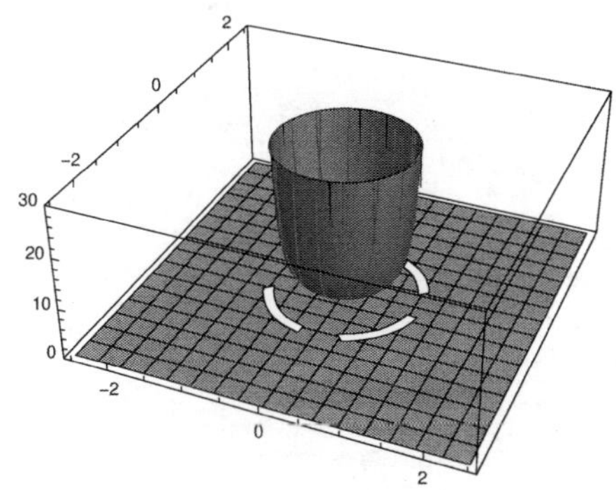

FIGURE 5. Example of a solution of (5.8) with $g \equiv 1$.

THEOREM 5.4 (Theorem 1.4 in [**6**]). *Let $g \in C(\partial B)$ and $u : \mathbb{R}^N \to \mathbb{R}$ be given by*

$$u(x) = \int_{\partial B} E_{s-1}(x,\theta)\ g(\theta)\ d\theta \quad \textit{for } x \in \mathbb{R}^N.$$

Then, $u \in C^\infty(B)$, $\delta^{1-s}u \in C(\overline{B})$, and

(5.9)
$$(-\Delta)^s u = 0 \textit{ in } B, \quad u = 0 \textit{ on } \mathbb{R}^N \backslash \overline{B}, \quad D^{s-2}u = 0 \textit{ on } \partial B, \quad D^{s-1}u = g \textit{ on } \partial B.$$

As mentioned at the beginning of the section, we see in Figure 5 that solutions of (5.8) are *singular* whenever $g \neq 0$ on ∂B. In this sense, the trace $D^{s-2}u$ can also be seen as a *singular trace* or *singular profile* of u. Figures 4–6 show the very different boundary behaviors that come into play when considering higher-order fractional boundary value problems. We also note that there is a relationship between the Green function G_s and the boundary Poisson kernels, namely,

$$E_{s-1}(x,z) = c_1 \lim_{y \to z} \frac{\partial^2}{\partial(|y|^2)^2}[\delta(y)^{2-s}G_s(x,y)],$$
$$E_{s-2}(x,z) = c_2 \lim_{y \to z} \frac{\partial^3}{\partial(|y|^2)^3}[\delta(y)^{2-s}G_s(x,y)]$$

for $z \in \partial B$ and $x \in \mathbb{R}^N$, where c_1 and c_2 are suitable constants, see [**6**, Lemma 1.8].

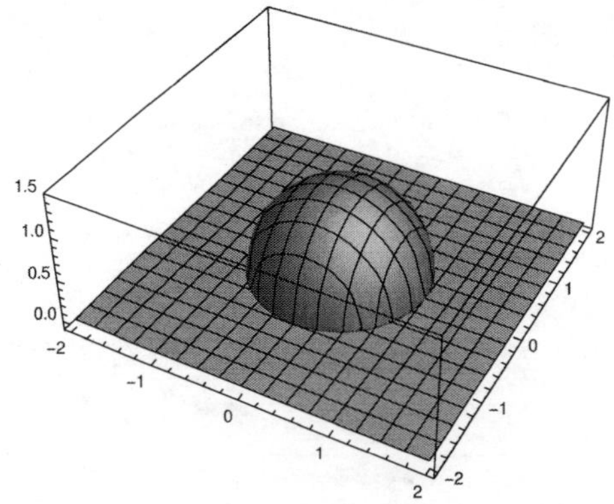

FIGURE 6. Example of a solution of (5.9) with $g \equiv 1$.

5.3. The nonlocal Poisson kernel. Let $m \in \{0,1\}$, $\sigma \in (0,1)$, and $s = m+\sigma \in (0,2)$. For $x \in \mathbb{R}^N$ and $y \in \mathbb{R}^N \backslash \overline{B}$ let

$$\Gamma_s(x,y) \;:=\; (-1)^m \frac{\gamma_{N,s-1}}{|x-y|^N} \frac{(1-|x|^2)_+^s}{(|y|^2-1)^s}, \tag{5.10}$$

where $\gamma_{N,s-1}$ is a positive normalization constant given in (1.4). The kernel (5.10) is called the *nonlocal Poisson kernel* for $(-\Delta)^s$ in B, and it can be used to construct s-harmonic solutions with prescribed values in $\mathbb{R}^N \backslash \overline{B}$.

THEOREM 5.5 (Theorem 1.1 in [**6**] and Theorem 1.6 in [**7**]). *Let $s \in (1,2)$, G_s as in (5.4), and Γ_s as in (5.10), then*

$$\Gamma_s(x,y) = \; -(-\Delta)_y^s G_s(x,y) \quad \textit{for } x \in \mathbb{R}^N,\ y \in \mathbb{R}^N \setminus \overline{B} \tag{5.11}$$

and, if $\psi \in \mathcal{L}_s^1$ with $\psi = 0$ in B_r for some $r > 1$, and $u : \mathbb{R}^N \to \mathbb{R}$ is given by

$$u(x) \;=\; \int_{\mathbb{R}^N \backslash \overline{B}} \Gamma_s(x,y)\psi(y)\, dy \;+\; \psi(x), \tag{5.12}$$

then $u \in C^\infty(B) \cap C^s(B_r) \cap H^s(B_\rho)$ for any $\rho \in (1,r)$ and u is the unique pointwise solution in the space $C^s(\overline{B}) \cap H^s(B)$ of

$$(-\Delta)^s u = 0 \quad \textit{in } B, \quad u = \psi \quad \textit{on } \mathbb{R}^N \backslash \overline{B}, \quad D^{s-2}u = D^{s-1}u = 0 \quad \textit{on } \partial B. \tag{5.13}$$

Observe that in Theorem 5.5 the data ψ is assumed to be zero close to the boundary ∂B, which implies that the traces (5.1) reduce to (5.2). If ψ is not zero close to ∂B then (5.12) might not be integrable and in this case one needs a different kernel. In fact, this case requires to use a lower-order nonlocal Poisson kernel together with a suitable correction using the boundary Poisson kernel E_{s-1}. Recall that χ_U denotes the characteristic function of U, that is, $\chi_U(y) = 1$ if $y \in U$ and $\chi_U(y) = 0$ if $y \notin U$. The next result is new.

THEOREM 5.6. *Let $\psi \in \mathcal{L}_s^1 \cap C^{s-1+\alpha}(\overline{B_r})$, $\psi_1 := \psi\chi_{B_r}$, $\psi_2 := \psi\chi_{\mathbb{R}^N \backslash B_r}$ for some $r > 1$, $\alpha \in (0,1)$, and assume that*

$$z \mapsto \int_{\mathbb{R}^N \backslash \overline{B}} \frac{\psi_1(y) - \psi_1(z)}{|z-y|^N(|y|^2-1)^{s-1}}\, dy \quad \textit{belongs to } C(\partial B). \tag{5.14}$$

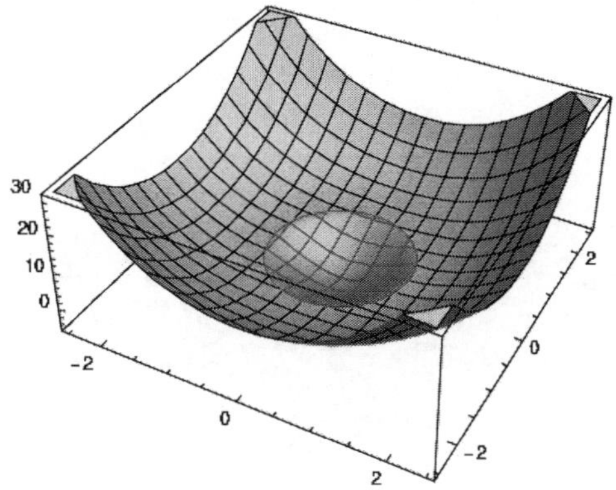

FIGURE 7. Example of a solution of (5.13). Observe that although the data ψ is positive in $\mathbb{R}^N \backslash B$, the solution u is negative in B. This is a higher-order phenomenon, which does not happen for $s \in (0,1)$, [**6**].

Let $u : \mathbb{R}^N \to \mathbb{R}$ *be given by*

$$u(x) = \int_{\mathbb{R}^N\backslash\overline{B}} \Gamma_s(x,y)\psi_2(y)\,dy + \int_{\mathbb{R}^N\backslash\overline{B}} \Gamma_{s-1}(x,y)\psi_1(y)\,dy$$
$$- 2\gamma_{N,s-1}\int_{\partial B} E_{s-1}(x,z)\int_{\mathbb{R}^N\backslash\overline{B}} \frac{\psi_1(y)-\psi_1(z)}{|z-y|^N(|y|^2-1)^{s-1}}\,dy\,dz \;+\; \psi(x)\chi_{\mathbb{R}^N\backslash B},$$

then $u \in C^\infty(B)\cap C^{s-1}(\overline{B})\cap \mathcal{L}^1_s$ *is a pointwise solution of*

$$(-\Delta)^s u = 0 \quad \text{in } B, \quad u = \psi \quad \text{on } \mathbb{R}^N\backslash\overline{B}, \quad D^{s-2}u = D^{s-1}u = 0 \quad \text{on } \partial B.$$

We remark that, if $\psi \in \mathcal{L}^1_{s-1}$, then the claim in Theorem 5.6 holds with $\psi_1 \equiv \psi$ and $\psi_2 \equiv 0$. Property (5.14) is needed to use Theorem 5.4 and its verification usually involves long computations but it can be easily verified in some simple situations, for example, if $\psi = c$ in B_r for some $r > 1$ and $c \in \mathbb{R}$, since in this case

$$\int_{\mathbb{R}^N\backslash\overline{B}} \frac{\psi_1(y)-\psi_1(z)}{|z-y|^N(|y|^2-1)^{s-1}}\,dy = \int_{\mathbb{R}^N\backslash\overline{B_r}} \frac{\psi_1(y)-\psi_1(z)}{|z-y|^N(|y|^2-1)^{s-1}}\,dy$$

and $|z-y| > r$ for all $z \in \partial B$ and $y \in \mathbb{R}^N\backslash\overline{B_r}$. Observe also that, if $N=1$, then $\partial B = \{-1,1\}$ and (5.14) holds provided the integral is finite, which can be shown for any

$$\psi \in \mathcal{L}^1_s \cap C^{s-1+\alpha,0}((-r,r)) \quad \text{for some } r>1 \text{ and } \alpha \in (0,1).$$

PROOF OF THEOREM 5.6. Let ψ, ψ_1, ψ_2, r, and u be as in the statement. By [**6**, Theorem 1.1] we know that $v : \mathbb{R}^N \to \mathbb{R}$ given by

$$v(x) = \int_{\mathbb{R}^N\backslash\overline{B}} \Gamma_{s-1}(x,y)\psi_1(y)\,dy + \psi_1(x)\chi_{\mathbb{R}^N\backslash B}$$

belongs to $C^\infty(B)\cap\mathcal{L}^1_{s-1}\cap C^{s-1}(B_r)$ and solves pointwisely $(-\Delta)^{s-1}v(x) = 0$ in B. Then, by Lemma 2.1, we also have that

$$(-\Delta)^s v(x) = (-\Delta)(-\Delta)^{s-1}v(x) = 0 \text{ in } B. \tag{5.15}$$

For $z\in\partial B$ let

$$\varphi(z):=\gamma_{N,s-1}\int_{\mathbb{R}^N\setminus\overline{B}}\frac{\psi_1(y)-\psi_1(z)}{(|y|^2-1)^{s-1}|z-y|^N}\,dy. \tag{5.16}$$

By (5.14) we have that $\varphi\in C(\partial B)$. By Theorems 5.4, 5.5 (see also [**6**, Theorem 1.1]), and by (5.15) it follows that $u\in C^\infty(B)\cap C^{s-1}(\overline{B})\cap\mathcal{L}^1_s$ is a pointwise solution of $(-\Delta)^s u=0$ in B. It remains to verify the boundary conditions. By definition, $u=\psi$ in $\mathbb{R}^N\setminus B$. Moreover, since u is bounded in B_r for some $r>1$ we have that $D^{s-2}u=0$ on ∂B. On the other hand, fix $z\in\partial B$ and recall that

$$D^{s-1}u(z):=-\lim_{\substack{x\to z\\ x\in B}}\frac{\partial}{\partial|x|}\Big[\delta(x)^{2-s}\Big(u(x)-\lim_{\substack{y\to z\\ y\in\mathbb{R}^N\setminus B}}u(y)\Big)\Big].$$

Using that the nonlocal Poisson kernel is normalized ($\int_{\mathbb{R}^N\setminus\overline{B}}\Gamma_{s-1}(\cdot,y)\,dy=1$ in B) we have that

(5.17)

$$\psi_1(z)=\psi_1(z)\int_{\mathbb{R}^N\setminus\overline{B}}\Gamma_{s-1}(x,y)\,dy=\gamma_{N,s-1}\int_{\mathbb{R}^N\setminus\overline{B}}\frac{\delta(x)^{s-1}\psi_1(z)}{(|y|^2-1)^{s-1}|z-y|^N}\,dy \quad \text{for } x\in B,\ z\in\partial B.$$

Moreover, by [**6**, Lemma 2.1] we know that, if a function f satisfies that $f=0$ in $\mathbb{R}^N\setminus\overline{B}$ and $D^{s-2}f=0$ on ∂B, then

$$D^{s-1}f(z)=2\lim_{x\to z}\frac{f(x)}{\delta(x)^{s-1}}\qquad\text{for } z\in\partial B. \tag{5.18}$$

For $x\in\mathbb{R}^N$, let $F(x):=\int_{\mathbb{R}^N\setminus\overline{B}}\Gamma_{s-1}(x,y)\psi_1(y)\,dy+\psi_1(x)\chi_{\mathbb{R}^N\setminus B}(x)$. Observe that

$$\lim_{\substack{y\to z\\ y\in\mathbb{R}^N\setminus B}}F(y)=\lim_{\substack{y\to z\\ y\in\mathbb{R}^N\setminus B}}\psi_1(y)\chi_{\mathbb{R}^N\setminus B}(y)=\psi_1(z)\quad\text{for } z\in\partial B$$

and, by (5.17)

$$F(x)-\lim_{\substack{w\to z\\ w\in\mathbb{R}^N\setminus B}}F(w)=\gamma_{N,s-1}\int_{\mathbb{R}^N\setminus\overline{B}}\frac{\delta(x)^{s-1}(\psi_1(y)-\psi_1(z))}{(|y|^2-1)^{s-1}|x-y|^N}\,dy \quad \text{for } x\in B,\ z\in\partial B. \tag{5.19}$$

Then, using (5.14), (5.18), and (5.19),

$$D^{s-1}F(z)=2\gamma_{N,s-1}\int_{\mathbb{R}^N\setminus\overline{B}}\frac{\psi_1(y)-\psi_1(z)}{(|y|^2-1)^{s-1}|z-y|^N}\,dy=2\varphi(z)\quad\text{for } z\in\partial B.$$

But then, since

$$D^{s-1}\left(\int_{\mathbb{R}^N\setminus\overline{B}}\Gamma_s(\cdot,y)\psi_2(y)\,dy+\psi_2\right)=0\quad\text{on }\partial B,$$

$$D^{s-1}\left(-\int_{\partial B}E_{s-1}(\cdot,w)\varphi(w)\,dw\right)=-\varphi\quad\text{on }\partial B,$$

by Theorems 5.4 and 5.5, we obtain that

$$D^{s-1}u = D^{s-1}\left(\int_{\mathbb{R}^N\setminus\overline{B}} \Gamma_s(\cdot,y)\psi_2(y)\,dy + F - 2\int_{\partial B} E_{s-1}(\cdot,w)\varphi(w)\,dw\right) = 0 \quad \text{on } \partial B,$$

as claimed. □

5.4. The Dirichlet boundary value problem in balls and representation formulas. We can now put together all the previous results to obtain the following.

THEOREM 5.7 (Theorem 1.4 in [**6**] and Theorem 1.6 in [**7**]). *Let $\alpha \in (0,1]$ such that $2s+\alpha \notin \mathbb{N}$, $g_k \in C^{1-k,0}(\partial B)$ for $k=0,1$, $f \in C^\alpha(\overline{B})$, $h \in \mathcal{L}^1_s$ such that $h=0$ in B_r, $r>1$, and $u:\mathbb{R}^N \to \mathbb{R}$ be given by $u(x)=h(x)$ for $x \in \mathbb{R}^N\setminus\overline{B}$ and*

$$u(x) = \int_B G_s(x,y)f(y)\,dy + \int_{\mathbb{R}^N\setminus\overline{B}} \Gamma_s(x,y)h(y)\,dy$$
$$+ \int_{\partial B} E_{s-2}(x,\theta)\,g_0(\theta)\,d\theta + \int_{\partial B} E_{s-1}(x,\theta)\,g_1(\theta)\,d\theta \quad \text{for } x \in B.$$

Then, $u \in C^{2s+\alpha}(B)$, $\delta^{2-s}u \in C^{1,0}(\overline{B})$ and

$$(-\Delta)^s u = f \text{ in } B, \quad u = h \text{ on } \mathbb{R}^N\setminus\overline{B}, \quad D^{s-2}u = g_0 \text{ on } \partial B, \quad D^{s-1}u = g_1 \text{ on } \partial B.$$

This solution given by Theorem 5.7 is, in fact, unique. As a consequence, we have the following representation formula.

THEOREM 5.8 (Theorem 1.5 in [**6**]). *Let $\alpha \in (0,1)$ such that $2s+\alpha \notin \mathbb{N}$, $r>1$, $u \in \mathcal{L}^1_s \cap C^{2s+\alpha}(B)$ be such that*

$$\delta^{2-s}u \in C^{1+\alpha}(\overline{B}), \quad (-\Delta)^s u \in C^\alpha(\overline{B}), \quad \text{and} \quad u = 0 \quad \text{in } B_r\setminus\overline{B}.$$

Then, for $x \in B$,

$$u(x) = \int_B G_s(x,y)(-\Delta)^s u(y)\,dy + \int_{\mathbb{R}^N\setminus\overline{B}} \Gamma_s(x,y)u(y)\,dy$$
$$+ \int_{\partial B} E_{s-2}(x,\theta)D^{s-2}u(\theta)\,d\theta + \int_{\partial B} E_{s-1}(x,\theta)D^{s-1}u(\theta)\,d\theta.$$

For a uniqueness statement in the case where u is nonzero in $B_r\setminus\overline{B}$, we refer to [**6**, Theorem 1.6].

REMARK 5.9. As mentioned in the introduction, similar results as those presented in this survey hold for $s>2$; however, one can also find surprising differences. To mention a few, let $s=m+\sigma>2$ with $m \in \mathbb{N}$ and $\sigma \in (0,1)$.

(1) A pointwise evaluation of $(-\Delta)^s u(x)$ can be calculated in terms of finite differences of order $2(m+1)$, namely,

$$(-\Delta)^s u(x) := \frac{c_{N,m+1,s}}{2}\int_{\mathbb{R}^N} \frac{\delta_{m+1}u(x,y)}{|y|^{N+2s}}\,dy,$$

where $\delta_{m+1}u(x,y) := \sum_{k=-m+1}^{m+1}(-1)^k\binom{2(m+1)}{m+1-k}u(x+ky)$ and $c_{N,m+1,s}>0$ is a suitable normalization constant (see [**3**, Equation (1.2)]). In particular, direct calculations for the case $s>2$ are harder and they often require fine combinatorial identities.

(2) The nonlocal Poisson kernel Γ_s given in (5.10) has the same formula for $s = m + \sigma > 2$, namely,

$$\Gamma_s(x,y) \ := \ (-1)^m \frac{\gamma_{N,\sigma}}{|x-y|^N} \frac{(1-|x|^2)_+^s}{(|y|^2-1)^s}.$$

Observe that the Poisson kernel is *positive* if m is even and *negative* if m is odd. As a consequence, the counterexample from Theorem 4.2 can be adjusted to show that $(-\Delta)^s$ does not have a general positivity preserving property if $s \in (m, m+1)$ with m odd. Remarkably, if $s \in (m, m+1)$ with m even, this counterexample does not work anymore, and in fact, one can show that the Green's function for two disjoint balls is positive! (see [**6**, Theorem 1.10]). A counterexample for $s \in (2,3)$ can be found in [**4**] in an ellipsoidal domain. Although maximum principles are not expected to hold for any $s > 1$, a counterexample for $s \in (m, m+1)$ with $m \geq 4$ even, is still missing.

(3) Problems with a nonlocal boundary condition $u = \psi$ in $\mathbb{R}^N \backslash B$ with $\psi \neq 0$ close to the boundary ∂B are more intricate to characterize if $s > 2$, and a result such as Theorem 5.6 is not yet available if $s > 2$.

(4) Similarly as in Section 5.2, for $s = m + \sigma > 2$ one has $m+1$ boundary Poisson kernels. Together with the nonlocal boundary condition (b.c.), we have that linear problems involving $(-\Delta)^s$ require $m+2$ b.c. to be well-posed. Note that $(-\Delta)^m$ and $(-\Delta)^{m+1}$ need only m and $m+1$ b.c. respectively to be well-posed. The "extra" boundary condition is due to the nonlocality of $(-\Delta)^s$.

6. Asymptotic behavior of solutions

The convergence of the kernels G_s, E_{s-2}, E_{s-1}, and the corresponding solutions as $s \to 2^-$ is well behaved, in the sense that the (pointwise) limits exist and the resulting function is also a solution. This can be easily verified (for suitable data) in virtue of Theorem 5.7 (see also [**6**, Theorem 1.4]) and the dominated convergence theorem.

Nevertheless, the limit as $s \to 1^+$ is more delicate and may not always yield a meaningful solution. We show this with a simple example: let $N = 1$, $\sigma \in (0,1)$, and $s = 1 + \sigma$; then

$$u_s(x) := \int_{\partial B} E_{s-2}(x,y)\, dy = E_{s-2}(x,-1) + E_{s-2}(x,1) = (1-x^2)^{s-2} = (1-x^2)^{\sigma-1}$$

is a solution (by Theorem 5.7) of $(-\Delta)^s u_s = 0$ in B satisfying $D^{\sigma-1}u_s(z) = 1$ and $D^\sigma u_s(z) = 0$ for $z \in \partial B$. If $\sigma \to 0$ (*i.e.*, if $s \to 1^+$), then $u_s(x) \to (1-x^2)^{-1}$, which is *not* harmonic in B. Note that $u_1 \notin L^1(B)$ and that the extra boundary condition ($D^{\sigma-1}u_s(z) = 1$) required in the higher-order case ($s \in (1,2)$) is *incompatible in the limit* with problems of lower order ($s = 1$). In conclusion, there are sequences of s-harmonic functions which converge pointwisely to a function which is not harmonic; in other words, s-harmonicity is not a property that is always preserved in the limit without additional assumptions.

On the other hand, note that if $\sigma \to 1^-$ (*i.e.*, if $s \to 2^-$), then $u_s(x) \to 1$ pointwisely for $x \in B$, which is, in fact, a solution of

$$(-\Delta)^2 u_2 = 0 \quad \text{in } B, \quad D^0 u_2(z) = u_2(z) = 1, \quad D^1 u_2(z) = \frac{1}{2}\partial_\nu 1 = 0 \quad \text{for } z \in \partial B.$$

Regarding the kernel Γ_s, observe that the solution u given by (5.12) goes uniformly to 0 as s approaches 1 or 2, due to the constant $\gamma_{N,s-1}$ given in (1.4) and the assumption that $\psi = 0$ in B_r with $r > 1$. Finally, for the convergence of the Poisson kernel Γ_σ to the Poisson kernel for the Laplacian as $\sigma \to 1^-$ see [**25**, footnote on page 121].

Appendix A. On the composition of Green functions

Let G_t be given by Boggio's formula (5.4). In this appendix we show that the functions $u, v, w : \mathbb{R}^N \to \mathbb{R}$ given by

$$\begin{aligned} u(x) &:= \int_B G_s(x,y)\,dy, \\ v(x) &:= \int_B G_{s-1}(x,z) \int_B G_1(z,y)\,dy\,dz, \\ w(x) &:= \int_B G_1(x,z) \int_B G_{s-1}(z,y)\,dy\,dz. \end{aligned} \tag{A.1}$$

solve each a different problem. In particular, this also illustrates (see (A.4) below) the fact that the operators $(-\Delta)$ and $(-\Delta)^{s-1}$ can only be interchanged whenever the involved functions are well defined, and this can be a very subtle issue.

To fix ideas, we focus on the case $s = \frac{3}{2}$ and $N = 1$. A first important difference between u, v, and w comes from their optimal regularity. Indeed, by the regularity properties of Boggio's formula (see Theorem 5.2 or [**6**]), we know that

$$u \in C^\infty(B) \cap C_0^s(\overline{B}), \quad v \in C^\infty(B) \cap C_0^{\frac{1}{2}}(\overline{B}), \quad \text{and} \quad w \in C^\infty(B) \cap C_0^1(B).$$

In particular, we see that $(-\Delta)^s$ can always be computed in B. We now argue that u, v, and w solve in fact very different problems. First, we note that, by Theorem 5.2, u is the unique solution of (5.6) with $f = 1$ in B. Next, by construction, we see that $(-\Delta)^s v = (-\Delta)(-\Delta)^{\frac{1}{2}} v = 1$ in $(-1,1)$ satisfying the boundary conditions

$$v = 0 \quad \text{in } \mathbb{R}\backslash\overline{B} \quad \text{and} \quad \lim_{\substack{x\to z\\ x\in B}} (-\Delta)^{\frac{1}{2}} v(x) = 0 \quad \text{for } z \in \partial B.$$

Using a (fractional) Hopf Lemma (see [**6**, Corollary 1.9]), one can show that $D^{s-1}v > 0$ on ∂B and therefore v does not satisfy the boundary conditions in (5.6). We now turn our attention to the function w given by (A.1). By [**15**, Table 1 and Table 2], we know that

$$(-\Delta)^{\frac{1}{2}} \delta(x)^{\frac{1}{2}} = 1 \quad \text{in } B \quad \text{and} \quad (-\Delta)^{\frac{1}{2}} (x\delta(x)^{\frac{1}{2}}) = 2x \quad \text{in } B. \tag{A.2}$$

Then,

$$w_1 : \mathbb{R} \to \mathbb{R} \quad \text{given by} \quad w_1(x) = \int_B G_{\frac{1}{2}}(x,y)\,dy$$

is a solution of

$$(-\Delta)^{\frac{1}{2}} w_1 = 1 \quad \text{in } B, \quad w_1 = 0 \quad \text{in } \mathbb{R}\backslash B$$

and, by uniqueness (Theorem 3.1) and (A.2),

$$w_1(x) = \int_B G_{\frac{1}{2}}(x,y)\,dy = \delta(x)^{\frac{1}{2}} \quad \text{for } x \in \mathbb{R}. \tag{A.3}$$

Since G_1 is the Dirichlet Green function of $(-\Delta)$ in $B = (-1,1)$, we have that w given by (A.1) is a solution of

$$(-\Delta)w = w_1 = \delta^{\frac{1}{2}} \quad \text{in } B, \quad w = 0 \quad \text{in } \mathbb{R}\backslash B$$

and w satisfies

$$w = 0 \quad \text{in } \mathbb{R}\backslash \overline{B}, \quad \lim_{x\to z}(-\Delta)w(x) = 0 \quad \text{for } z \in \partial B.$$

Moreover, by Lemma 2.1, we have that $(-\Delta)^{\frac{3}{2}}w(x) = -\Delta(-\Delta)^{\frac{1}{2}}w(x)$ for $x \in B$. If we naïvely interchange the Laplacians, we would have that

$$(-\Delta)^{\frac{3}{2}}w(x) = -\Delta(-\Delta)^{\frac{1}{2}}w(x) = (-\Delta)^{\frac{1}{2}}(-\Delta)w(x) = (-\Delta)^{\frac{1}{2}}w_1(x) = 1 \text{ for } x \in B. \tag{A.4}$$

However, such an interchange is not possible, because Δw is not well defined on ∂B.[4] In fact, we can show the following result.

LEMMA A.1. *If w is given by* (A.1), *then* $(-\Delta)^{\frac{3}{2}}w \neq 1$ *in* $B = (-1,1)$.

PROOF. Let $\sin^{-1}$ denote the arcsine function and d and $'$ denote derivatives. By direct computation, we have that

$$W(y) = \frac{1}{12}\left(-2\sqrt{1-y^2}\,(y^2+2) - 6y\sin^{-1}(y) + 3\pi\right)\chi_{(-1,1)}(y)$$

is the unique solution of

$$-d^2W = -W'' = \delta^{\frac{1}{2}} \quad \text{in } (0,1), \quad W(1) = 0 = W'(0), \tag{A.5}$$

and therefore, by (A.3),

$$W(x) = w(x) = \int_B G(x,z)\delta^{\frac{1}{2}}(z)\,dz \quad \text{for } x \in B.$$

Then

$$w'(y) = \frac{1}{2}\left(-\sqrt{1-y^2}y - \sin^{-1}(y)\right)\chi_{(-1,1)}(y) \quad \text{in } \mathbb{R}\backslash\{\pm 1\}$$

and w' has a jump discontinuity[5] at ± 1. By [**7**, Corollary 1.4] and [**5**, Proposition B.2] we have that

$$(-\Delta)^{1+\frac{1}{2}}w = -d^2(-\Delta)^{\frac{1}{2}}w = -d(-\Delta)^{\frac{1}{2}}dw = -d(-\Delta)^{\frac{1}{2}}w' \quad \text{in } (-1,1).$$

By linearity and using (A.2),

$$(-\Delta)^{\frac{1}{2}}\left[\frac{1}{2}\left(-\delta(y)^{\frac{1}{2}}y - \sin^{-1}(y)\right)\chi_{(-1,1)}\right] = \frac{1}{2}\left(-2y - (-\Delta)^{\frac{1}{2}}[\sin^{-1}(y)\chi_{(-1,1)}(y)]\right),$$

therefore,

$$(-\Delta)^{\frac{3}{2}}w(y) = -d(-\Delta)^{\frac{1}{2}}dw(y) = 1 + d(-\Delta)^{\frac{1}{2}}[\sin^{-1}(y)\chi_{(-1,1)}(y)] \quad \text{for } y \in (-1,1).$$

---

[4]As can be seen in the proof of Lemma A.1, w is not twice weakly differentiable on ∂B because its derivative w' has a jump discontinuity at ∂B. Since $(-\Delta)^{\frac{1}{2}}$ is a nonlocal operator, the function Δw must be well defined in $\mathbb{R}^N$.

[5]Thus w' is not weakly differentiable in $\mathbb{R}$ and one cannot compute $(-\Delta)^{\frac{1}{2}}(-\Delta)w = (-\Delta)^{\frac{1}{2}}(-d^2)w$.

Let $\zeta(y) = \sin^{-1}(y)\chi_{(-1,1)}(y)$. To finish the proof, we show that $d(-\Delta)^{\frac{1}{2}}\zeta \neq 0$ in $(-1,1)$. Assume by contradiction that $d(-\Delta)^{\frac{1}{2}}\zeta = 0$ in B, then, $(-\Delta)^{\frac{1}{2}}\zeta = c$ in B for some $c \in \mathbb{R}$. Now, we reach a contradiction using a representation result [6]. Since $\delta^{\frac{1}{2}}\zeta \in C^{\frac{1}{2}}([-1,1])$, $\zeta \in L^\infty([-1,1])$, and $(-\Delta)^{\frac{1}{2}}\zeta \in C^{\frac{1}{2}}([-1,1])$, (actually, $(-\Delta)^{\frac{1}{2}}\zeta = c \in C^\infty([-1,1])$), we can use the representation of solutions [**6**, Theorem 1.5] and obtain that

$$\sin^{-1}(y)\chi_{(-1,1)}(y) = \zeta(y) = \int_B G_{\frac{1}{2}}(x,y)c\,dy = c(1-|y|^2)_+^{\frac{1}{2}} \quad \text{in } \mathbb{R},$$

a contradiction. Therefore $d(-\Delta)^{\frac{1}{2}}[\sin^{-1}(y)\chi_{(-1,1)}] \neq 0$ in B and $(-\Delta)^s w \neq 1$ in B. □

Acknowledgments

The author thanks N. Abatangelo, S, Jarohs, and F. M. López García for their comments and suggestions to improve this survey.

References

[1] Nicola Abatangelo, *A remark on nonlocal Neumann conditions for the fractional Laplacian*, Arch. Math. (Basel) **114** (2020), no. 6, 699–708, DOI 10.1007/s00013-020-01440-9. MR4102340

[2] Nicola Abatangelo, *Large S-harmonic functions and boundary blow-up solutions for the fractional Laplacian*, Discrete Contin. Dyn. Syst. **35** (2015), no. 12, 5555–5607, DOI 10.3934/dcds.2015.35.5555. MR3393247

[3] Nicola Abatangelo, Serena Dipierro, Mouhamed Moustapha Fall, Sven Jarohs, and Alberto Saldaña, *Positive powers of the Laplacian in the half-space under Dirichlet boundary conditions*, Discrete Contin. Dyn. Syst. **39** (2019), no. 3, 1205–1235, DOI 10.3934/dcds.2019052. MR3918215

[4] Nicola Abatangelo, Sven Jarohs, and Alberto Saldaña, *Fractional Laplacians on ellipsoids*, Math. Eng. **3** (2021), no. 5, Paper No. 038, 34, DOI 10.3934/mine.2021038. MR4181195

[5] Nicola Abatangelo, Sven Jarohs, and Alberto Saldaña, *Green function and Martin kernel for higher-order fractional Laplacians in balls*, Nonlinear Anal. **175** (2018), 173–190, DOI 10.1016/j.na.2018.05.019. MR3830727

[6] Nicola Abatangelo, Sven Jarohs, and Alberto Saldaña, *Integral representation of solutions to higher-order fractional Dirichlet problems on balls*, Commun. Contemp. Math. **20** (2018), no. 8, 1850002, 36, DOI 10.1142/S0219199718500025. MR3883404

[7] Nicola Abatangelo, Sven Jarohs, and Alberto Saldaña, *Positive powers of the Laplacian: from hypersingular integrals to boundary value problems*, Commun. Pure Appl. Anal. **17** (2018), no. 3, 899–922, DOI 10.3934/cpaa.2018045. MR3809107

[8] Nicola Abatangelo, Sven Jarohs, and Alberto Saldaña, *On the loss of maximum principles for higher-order fractional Laplacians*, Proc. Amer. Math. Soc. **146** (2018), no. 11, 4823–4835, DOI 10.1090/proc/14165. MR3856149

[9] R. M. Blumenthal, R. K. Getoor, and D. B. Ray, *On the distribution of first hits for the symmetric stable processes*, Trans. Amer. Math. Soc. **99** (1961), 540–554, DOI 10.2307/1993561. MR126885

[10] T. Boggio, *Sulle funzioni di Green d'ordine m*, Rend. Circ. Mat. Palermo, **20** (1905) 97–135.

[6] More direct approaches are also possible, but they require lengthy computations. For example, since ζ has a jump discontinuity at ± 1, one can show that $\lim_{x\to 1^-}(-\Delta)^{\frac{1}{2}}\zeta(x) = \infty$. Or, using integration by parts and the mean value theorem (see [**5**, Lemma B.1]), one can even compute explicitly the value of $(-\Delta)^{\frac{1}{2}}\zeta(x)$ at any $x \in (-1,1)$, but to keep this survey short we do not include the details of these calculations.

[11] Denis Bonheure, Juraj Földes, Ederson Moreira dos Santos, Alberto Saldaña, and Hugo Tavares, *Paths to uniqueness of critical points and applications to partial differential equations*, Trans. Amer. Math. Soc. **370** (2018), no. 10, 7081–7127, DOI 10.1090/tran/7231. MR3841843

[12] Denis Bonheure, Juraj Földes, and Alberto Saldaña, *Qualitative properties of solutions to mixed-diffusion bistable equations*, Calc. Var. Partial Differential Equations **55** (2016), no. 3, Art. 67, 23, DOI 10.1007/s00526-016-0987-6. MR3506201

[13] Gérard Bourdaud and Yves Meyer, *Fonctions qui opèrent sur les espaces de Sobolev* (French), J. Funct. Anal. **97** (1991), no. 2, 351–360, DOI 10.1016/0022-1236(91)90006-Q. MR1111186

[14] Serena Dipierro and Hans-Christoph Grunau, *Boggio's formula for fractional polyharmonic Dirichlet problems*, Ann. Mat. Pura Appl. (4) **196** (2017), no. 4, 1327–1344, DOI 10.1007/s10231-016-0618-z. MR3673669

[15] Bartłomiej Dyda, *Fractional calculus for power functions and eigenvalues of the fractional Laplacian*, Fract. Calc. Appl. Anal. **15** (2012), no. 4, 536–555, DOI 10.2478/s13540-012-0038-8. MR2974318

[16] Bartłomiej Dyda, Alexey Kuznetsov, and Mateusz Kwaśnicki, *Fractional Laplace operator and Meijer G-function*, Constr. Approx. **45** (2017), no. 3, 427–448, DOI 10.1007/s00365-016-9336-4. MR3640641

[17] J. Edenhofer, *Eine Integraldarstellung der Lösung der Dirichletschen Aufgabe bei der Polypotentialgleichung im Falle einer Hyperkugel* (German), Math. Nachr. **69** (1975), 149–162, DOI 10.1002/mana.19750690115. MR393524

[18] Filippo Gazzola, Hans-Christoph Grunau, and Guido Sweers, *Polyharmonic boundary value problems*, Lecture Notes in Mathematics, vol. 1991, Springer-Verlag, Berlin, 2010. Positivity preserving and nonlinear higher order elliptic equations in bounded domains, DOI 10.1007/978-3-642-12245-3. MR2667016

[19] David Gilbarg and Neil S. Trudinger, *Elliptic partial differential equations of second order*, Classics in Mathematics, Springer-Verlag, Berlin, 2001. Reprint of the 1998 edition. MR1814364

[20] Gerd Grubb, *Local and nonlocal boundary conditions for μ-transmission and fractional elliptic pseudodifferential operators*, Anal. PDE **7** (2014), no. 7, 1649–1682, DOI 10.2140/apde.2014.7.1649. MR3293447

[21] Gerd Grubb, *Fractional Laplacians on domains, a development of Hörmander's theory of μ-transmission pseudodifferential operators*, Adv. Math. **268** (2015), 478–528, DOI 10.1016/j.aim.2014.09.018. MR3276603

[22] Gerd Grubb, *Spectral results for mixed problems and fractional elliptic operators*, J. Math. Anal. Appl. **421** (2015), no. 2, 1616–1634, DOI 10.1016/j.jmaa.2014.07.081. MR3258341

[23] Gerd Grubb, *Fractional-order operators: boundary problems, heat equations*, Mathematical analysis and applications—plenary lectures, Springer Proc. Math. Stat., vol. 262, Springer, Cham, 2018, pp. 51–81, DOI 10.1007/978-3-030-00874-1_2. MR3880448

[24] Sven Jarohs, Alberto Saldaña, and Tobias Weth, *A new look at the fractional Poisson problem via the logarithmic Laplacian*, J. Funct. Anal. **279** (2020), no. 11, 108732, 50, DOI 10.1016/j.jfa.2020.108732. MR4141492

[25] N. S. Landkof, *Foundations of modern potential theory*, Springer-Verlag, New York-Heidelberg, 1972. Translated from the Russian by A. P. Doohovskoy; Die Grundlehren der mathematischen Wissenschaften, Band 180. MR0350027

[26] Roberta Musina and Alexander I. Nazarov, *Sobolev inequalities for fractional Neumann Laplacians on half spaces*, Adv. Calc. Var. **14** (2021), no. 1, 127–145, DOI 10.1515/acv-2018-0020. MR4193437

[27] Roberta Musina and Alexander I. Nazarov, *On fractional Laplacians—2*, Ann. Inst. H. Poincaré Anal. Non Linéaire **33** (2016), no. 6, 1667–1673, DOI 10.1016/j.anihpc.2015.08.001. MR3569246

[28] L. A. Peletier and W. C. Troy, *Spatial patterns*, Progress in Nonlinear Differential Equations and their Applications, vol. 45, Birkhäuser Boston, Inc., Boston, MA, 2001. Higher order models in physics and mechanics, DOI 10.1007/978-1-4612-0135-9. MR1839555

[29] Xavier Ros-Oton and Joaquim Serra, *Local integration by parts and Pohozaev identities for higher order fractional Laplacians*, Discrete Contin. Dyn. Syst. **35** (2015), no. 5, 2131–2150, DOI 10.3934/dcds.2015.35.2131. MR3294242

[30] A. Saldaña, *Partial symmetry of solutions to parabolic problems via reflection methods*, Rend. Semin. Mat. Univ. Politec. Torino **74** (2016), no. 2, 105–112. MR3772604

[31] Alberto Saldaña, *Qualitative properties of coexistence and semi-trivial limit profiles of nonautonomous nonlinear parabolic Dirichlet systems*, Nonlinear Anal. **130** (2016), 31–46, DOI 10.1016/j.na.2015.09.020. MR3424607

[32] Alberto Saldaña and Tobias Weth, *Asymptotic axial symmetry of solutions of parabolic equations in bounded radial domains*, J. Evol. Equ. **12** (2012), no. 3, 697–712, DOI 10.1007/s00028-012-0150-6. MR2960850

[33] Alberto Saldaña and Tobias Weth, *On the asymptotic shape of solutions to Neumann problems for non-cooperative parabolic systems*, J. Dynam. Differential Equations **27** (2015), no. 2, 307–332, DOI 10.1007/s10884-014-9395-9. MR3368155

[34] Alberto Saldaña and Hugo Tavares, *Least energy nodal solutions of Hamiltonian elliptic systems with Neumann boundary conditions*, J. Differential Equations **265** (2018), no. 12, 6127–6165, DOI 10.1016/j.jde.2018.07.013. MR3865107

[35] Luis Enrique Silvestre, *Regularity of the obstacle problem for a fractional power of the Laplace operator*, ProQuest LLC, Ann Arbor, MI, 2005. Thesis (Ph.D.)–The University of Texas at Austin. MR2707618

Instituto de Matemáticas, Universidad Nacional Autónoma de México, Circuito Exterior, Ciudad Universitaria, 04510 Coyoacán, CDMX, Mexico

Email address: `alberto.saldana@im.unam.mx`

Contemporary Mathematics
Volume **775**, 2021
https://doi.org/10.1090/conm/775/15596

H^1 solutions on curve integrable spacetimes

Yafet Sanchez Sanchez

ABSTRACT. In this work we use a framework based on scalar linear hyperbolic equations that characterises gravitational singularities as the obstruction to well-posedness. In particular, we show existence of H^1 solutions on a certain class of curve integrable spacetimes. This implies that these spacetimes are not singular in the sense of obstructions to well-posedness even if their regularity is below C^2.

1. Introduction

An important requirement of any classical physical theory is that given suitable initial data one can determine the evolution of the system. Within the theory of general relativity the condition of global hyperbolicity [**1**] plays a key role, because this condition is a sufficient condition for the well-posedness of hyperbolic equations for several physically important fields (see e.g.[**2**] for details). Mathematically a spacetime region $\mathcal{N}$ is said to be globally hyperbolic if the causality condition is satisfied, and for any two points $p, q \in \mathcal{N}$ the causal diamond $J^+(p) \cap J^-(q)$ is compact and contained in $\mathcal{N}$ [**3**].

One way of interpreting the compactness of $J^+(p) \cap J^-(q)$ is that this set "does not contain any points on the edge of spacetime, i.e. at infinity or at a singularity" [**2**, §6.6]. In this context one has to choose a definition of singularity and therefore choose the regularity for the metric in the region $\mathcal{N}$. If one considers the existence and uniqueness of geodesics, then a sufficient condition to ensure this is that the metric is $C^{1,1}$ [1]. This seems to be the natural regularity for the singularity theorems [**4, 5**].

In this paper, we would like to explore a further option. We will consider a singularity on a Lorentzian manifold as an obstruction to the evolution of a test-field u with dynamics governed by the the hyperbolic equation

$$\Box_g u := g^{ab}\nabla_a \nabla_b u = f \tag{1.1}$$

where g^{ab} is an inverse metric tensor, ∇_a is the covariant derivative and f is a source.

This point of view was called generalised hyperbolicity by Clarke [**6**]. It involves regarding certain traditional singularities as interior points in a spacetime with low

---

2020 *Mathematics Subject Classification.* Primary 83C75.

This work was supported by a Conacyt Graduate Fellowship.

[1] Lipschitz functions with bounded first derivatives

regularity and then proving local and global well-posedness of the wave equation in the rough extension.

Clarke's definition of $\Box_g$*-globally hyperbolicity* was motivated by two things. Firstly, there are a number of spacetimes in which points are removed due to the presence of weak singularities and therefore are not globally hyperbolic but still of physical interest. These include spacetimes with thin shells of matter [**7**], impulsive gravitational waves [**8**] and shell-crossing singularities [**9**]. The second motivating factor was that of using test fields (given by solutions of the wave equation) rather than test particles (given by solutions of the geodesic equation) to probe the structure of singularities. An advantage of using test fields is that the behaviour of the naturally defined energy-momentum tensor T^{ab} of the test-field u given by:

$$T^{ab}[u] = \left(g^{ac}g^{bd} - \frac{1}{2}g^{ab}g^{cd}\right)\frac{\partial u}{\partial x^c}\frac{\partial u}{\partial x^d} \tag{1.2}$$

gives a direct measurement of the physical effect of the singularity which is not easily obtained when considering families of test particles.

The structure of this paper is as follows. In Section 2 we give some preliminaries and establish the notation. In Section 3 we introduce the class of curve integrable spacetimes considered by Clarke [**6**], give a general overview on how to show well-posedness and prove the existence of solutions with an integrable energy momentum tensor on a certain class of curve integrable spacetimes.

2. Preliminaries and notation

A function f defined on a nonempty open set $\Omega \subset \mathbb{R}^k$ is called a *test function* if $f \in C^\infty(\Omega)$ and there is a compact set $K \subset \Omega$ such that the support of f lies in K. The set of all test functions is denoted by $C_0^\infty(\Omega)$.

Let $\tilde{L}^p(\Omega,\mu)$ be the set of all continuous functions $g : \bar{\Omega} \to \mathbb{R}$ for which

$$\|g\|_p = \left(\int_\Omega |g|^p\right)^{\frac{1}{p}} d\mu < \infty \tag{2.1}$$

The completion of $\tilde{L}^p(\Omega,\mu)$ with respect to $\|\cdot\|_p$ is the space called $L^p(\Omega,\mu)$. We say f is in $L^p_{loc}(\Omega,\mu)$ if f is in $L^p(K,\mu)$ for all compact sets K properly contained in Ω . The space $L^\infty(\Omega,\mu)$ is defined as $\{f \text{ s.t } |f(x)| \leq C\ \mu\text{-almost everywhere}\}$ i.e. bounded functions almost everywhere.

Let u in $L^1_{loc}(\Omega,\mu)$. The weak derivative $v = \frac{\partial^\alpha u}{\partial x^\alpha}$ of u, if it exists, is a v in $L^1_{loc}(\Omega,\mu)$ that satisfies:

$$\int_\Omega u\frac{\partial^\alpha \phi}{\partial x^\alpha} d\mu = (-1)^{|\alpha|}\int_\Omega v\phi d\mu$$

for all $\phi \in C^\infty(\Omega)$.

Let k be a non- negative integer and let $1 \leq p \leq \infty$. Then we define the *Sobolev space* $W^{k,p}(\Omega,\mu)$ to be the set of all distributions $u \in L^p(\Omega,\mu)$ such that the weak derivatives $\frac{\partial^\alpha u}{\partial x^\alpha} \in L^p(\Omega,\mu)$ for $|\alpha| \leq k$.

When $p = 2$ we can define an inner product:

$$(u,v)_k = \Sigma_{|\alpha|\leq k}\int_\Omega \frac{\partial^\alpha u}{\partial x^\alpha}\frac{\partial^\alpha v}{\partial x^\alpha} d\mu \tag{2.2}$$

that defines a norm

$$(u,u)_k = (\|u\|^k_\Omega)^2$$

and we denote this Sobolev space as $H^k(\Omega, \mu)$.

We define also the *local Sobolev spaces* $W^{k,p}_{loc}(\Omega, \mu)$. We say f is in $W^{k,p}_{loc}(\Omega, \mu)$ if given any compact set K properly contained in Ω, f is in $W^{k,p}(K, \mu)$.

When the measure is the one associated to the volume form dx^{k-1}, (or dx^k when the domain is the whole spacetime), we will omit the measure and just write $L^2(\Omega)$. We will also use this convention when considering the details of the function spaces $L^2(\Omega, \nu_h)$. However, this will be mentioned explicitly in the text when it occurs.

We also might want to approximate functions in a Sobolev space through C^∞ functions. The way to do this is by smoothing with *strict delta nets*. A net $\{(\rho_n)\} \in [1, \infty)$ of smooth functions on $\mathbb{R}^k$ is called a strict delta net, if

- $supp(\rho_n) = \{p \in \mathbb{R}^k | \rho_n \neq 0\} \to 0$ for $n \to \infty$.
- $\int \rho_n d\mu \to 1$ for $n \to \infty$.
- ρ_n is uniformly bounded in $L^1(\Omega, \mu)$.

For any strict delta net $\{(\rho_n)\}$ we denote by $d\rho_n$ the *diameter of the support* of ρ_n , i.e. $d\rho_n := sup\{|x| : x \in supp(\rho_n)\}$. For any $f \in L^1_{loc}(\Omega, \mu)$ we call the *convolution* f_n of f with a strict delta net $\{(\rho_n)\}$ a smoothing of f , given by

$$f_n(x) := \rho_n * f(x) = \int_{B(x, d\rho_n)} f(x-y)\rho_n(y)dy \tag{2.3}$$

where $B(x, r)$ denotes the open ball of radius r around x.

The following theorem states how the convolution f_n of f can be used to approximate f. The proof can be found in [**10**].

THEOREM 2.1. *The smoothing of any $f \in L^1_{loc}(\Omega)$ has the following properties:*

- *$f_n \in C^\infty(\Omega_{d\rho_n})$ and $f_n \to f$ almost everywhere. The convergence here is pointwise.*
- *If f is continuous the convergence is actually uniform on compact subsets of Ω.*
- *If $f \in W^{m,p}(\Omega) for 1 \leq p < \infty$ then $f_n \to f \in W^{m,p}(\Omega)$. The convergence here is in the norm given by $W^{m,p}(\Omega)$.*
- *The derivative operator commute with smoothing, i.e. $\partial_x f_n(x) = \rho_n * \partial_x f$*

This result implies that the C^∞ functions are *dense* in any Sobolev space, i.e. we can always find sequences of C^∞ functions that converge to any desired function in a given Sobolev space. In fact, in the case of L^p spaces the functions $C^\infty_0$are dense.

3. Curve integrable spacetimes

In this section, we provide the formal definition that Clarke's work presented in [**6**] on, what he calls, curve integrable spacetimes. Roughly speaking, a curve integrable spacetime is one in which the integrals of both the connection and curvature along a curve are bounded (see condition (4) of the main theorem for a precise description). From a physical point of view a spacetime is a curve integrable spacetime if there is a set C that defines a range of timelike directions which are transverse to any shock or caustic that may be present.

The general structure of of this section is as follows. First we introduce in 3.1 the general geometric setting and give a precise definition of curve integrable spacetimes. Then in 3.2 we give the formal definitions of weak solutions and give an overview how to prove well posedness. Finally in 3.3 we show the existence of H^1 solutions.

3.1. The general setting. The general geometric background we use to define a generalised solution is the existence of a *lens-shaped* domain $\Sigma_{[T',T]}$ properly contained in an open bounded subset of $\mathbb{R}^4$.

This means that there is a smooth map $\Theta : \Sigma \times (-a, a) \to M$ where Σ is a compact, C^1 co-dimension one sub-manifold with boundary that has the following properties:

- $\Theta(\cdot, 0) : \Sigma \to \Sigma_0$ is the identity map.
- $\Theta(x, T) = \Theta(x, s)$ for any $x \in \partial\Sigma$, $T, s \in (-a, a)$.
- For any fixed $s \in (-a, a)$, $\Theta(\Sigma, s)$ is an 3-dimensional spacelike hypersurface.
- Away from $\partial\Sigma \times (-a, a)$, Θ is a diffeomorphism.

We denote the region from 0 to $T \geq 0$ by $\Sigma_{[0,T]}$ and from $T' \leq 0$ to 0 by $\Sigma_{[T',0]}$. Notice that given coordinates x^i on Σ, Θ provides coordinates (t, x^i) for the region $\Sigma_{[0,T]}$ away from the image of $\partial\Sigma$. We will therefore always choose charts such that the time coordinate coincides with the time coordinate given by Θ when working in coordinates.

We now state the main conditions that a curve integrable spacetime has to satisfy and therefore the precise regularity of the spacetime we are considering.

CONDITIONS 3.1. **(Curve integrable spacetimes)** *Let* $(\Sigma_{[0,T]}, g)$ *be a lens-shaped domain and* p *a point in the domain satisfying:*

1. *The components* g_{ab} *and* g^{ab} *are* C^0 ;
2. $g_{ab,c}$ *exist and are* $L^p_{loc}(\Sigma_{(0,T)})$ *with* $p \geq 3$;
3. *The components* g_{ab} *are* C^2 *in* $\Sigma_{(0,T)} \backslash J^+(p)$;
4. *There is a non-empty set,* $C \subset \mathbb{R}^4$, *and positive functions* $M, N : \mathbb{R}^+ \to \mathbb{R}^+$ *such that, if* γ *is a curve with* $d\gamma/ds \in C$ *for all* s *then*
   - γ *is future timelike;*
   - *the integrals*

$$I_\gamma(a) := \int_0^a |\Gamma^a_{bc}(\gamma(s))|^2 ds \tag{3.1}$$

and

$$J_\gamma(a) := \int_0^a |R^a_{bcf}(\gamma(s))| ds \tag{3.2}$$

(where Γ *is defined using the weak derivatives of* g *and* R *is understood in the sense of distributions) are finite for all* $a < T$, *with*

$$I_\gamma(a) < M(a), J_\gamma(a) < N(a) \tag{3.3}$$

and $M(a), N(a) \to 0$, *as* $a \to 0$.

Then $\Sigma_{[0,T]}$ *is a curve integrable spacetime.*

The first and second condition are the minimal requirements needed to recover the curvature tensor as a distribution defined in terms of g^{ab} and $g_{ab,c}$ [**11**]. Moreover, this regularity is needed for Stokes's theorem [**12**] to be valid. The third and fourth conditions formalise our statement that a curve integrable spacetime admits a set C that defines a range of timelike directions which are transverse to any shock or caustic that may be present. Moreover the integrals of the curvature and the square of the connection along the integral curves of this timelike directions are bounded.

Notice that our definition of curve integrable spacetimes differ from [**6**] and [**13**]. The reason for this is that in [**6**] Stokes's theorem is used , however the geometric conditions established there do not guarantee its applicability. In the case of [**13**] there is a mistake in the appendix and therefore to guarantee the existence of a timelike vector field with bounded covariant derivative one has to used the square integrable condition on the connection that appears on [**6**] and is explicitly equation (3.1) in this paper.

3.2. Weak solutions and well-posedness. In this subsection we define an appropriate definition of a weak solution suitable for the low regularity geometric setting.

Our main motivation will be the fact that $\Box_g$ is formally self-adjoint in $\Sigma_{(0,T)}$. That is for $v, w \in C_0^\infty(\Sigma_{(0,T)})$, we have

$$(v, \Box_g w)_{L^2(\Sigma_{(0,T)},\nu_g)} = (\Box_g v, w)_{L^2(\Sigma_{(0,T)},\nu_g)} \tag{3.4}$$

This leads us to define a weak solution of

$$\Box_g u = f \tag{3.5}$$

in the region $\Sigma_{(0,T)}$ where g_{ab} is a curve integrable spacetime, $f \in L^2(\Sigma_{(0,T)})$ and $f|_{\Sigma_{t<0}} = 0$ satisfying

$$u|_{\Sigma_{t<0}} = 0 \tag{3.6}$$

$$\frac{\partial u}{\partial t}|_{\Sigma_{t<0}} = 0 \tag{3.7}$$

as follows

DEFINITION 3.2. **(Weak solution)** *We say a function:*

$$u \in H^1(\Sigma_{(0,T)}, \nu_g)$$

is a local weak solution of the initial value problem (3.5), (3.6) *and* (3.7) *provided that:*

For all $v \in C_0^\infty(\Sigma_{(0,T)})$

$$\int_{\Sigma_{(0,T)}} u \Box_g v \nu_g = \int_{\Sigma_{(0,T)}} f v \nu_g \tag{3.8}$$

where $f \in L^2(\Sigma_{(0,T)})$ *and* $f|_{\Sigma_{t<0}} = 0$ *and* u *satisfies*

$$u|_{\Sigma_{t<0}} = 0 \tag{3.9}$$

$$\frac{\partial u}{\partial t}|_{\Sigma_{t<0}} = 0 \tag{3.10}$$

It is a remarkable characteristic of inhomogeneous wave equations that the problem (3.5),(3.6) and (3.7) admits an a priori estimate

$$(\tilde{\|}\phi\|^1_{\Sigma_t})^2 \leq C\left(\|f\|_{L^2(M,\nu_g)}\right)^2 \tag{3.11}$$

for some constant C (see [**6**, **13**] for a derivation) and where we have introduced the norm

$$\tilde{\|}\phi\|^{1}_{\Sigma_t} = \left[\int_{\Sigma_t}\left(\left(\frac{\partial\phi}{\partial t}\right)^2 + \sum_{i=1}^{3}\left(\frac{\partial\phi}{\partial x^i}\right)^2 + \phi^2\right)\nu_h\right]^{\frac{1}{2}} \tag{3.12}$$

As we can see the structure of the energy estimates works naturally where the function and the derivatives satisfy L^2 conditions. In fact, the reason for using Hilbert spaces, such as H^k spaces or $L^2((0,T),H^k)$ rather than C^k spaces in the analysis of first order and second order hyperbolic PDEs is because of the existence of this kind of estimate. We also notice that although there is a natural norm in the C^k spaces given by

$$||\phi||_{C^k(U)} = \sum_{|\alpha|\le k}\sup_{x\in U}\left|\frac{\partial^\alpha f}{\partial x^\alpha}(x)\right|,$$

where we have used multi-index notation, the use of this norm misses the point as there is no simple relation between the C^k norm of the function at the initial time and the C^k norm at future time steps.

Moreover, subject only to the regularity of the source function and the coefficients of $\Box_g$, one can obtain "higher energy estimates" of the form

$$(\tilde{\|}\phi\|^{k}_{\Sigma_t})^2 \le C(\|f\|^{k-1}_{\Sigma_{[0,T]}})^2 \tag{3.13}$$

which correspond to norms that include derivatives of k-th order [**14**]. For big enough k, one can guarantee the continuity of the function in terms of the continuity of f using Sobolev embedding theorems.

To show uniqueness and stability in [**6**, **13**], one assumes that the weak solutions satisfy an energy estimate. This naturally leads to the following definition.

DEFINITION 3.3. **(Regular wave solution)** *We say a weak solution $u \in H^1(\Sigma_{(0,T)},\nu_g)$ is* **regular** *if u satisfies the energy estimate*

$$(\tilde{\|}u\|^{1}_{\Sigma_t})^2 \le C\|f\|^2_{L^2(\Sigma_{(0,T)},\nu_g)} \tag{3.14}$$

where

$$\tilde{\|}u\|^{1}_{\Sigma_t} = \left[\int_{\Sigma_t}\left(\left(\frac{\partial u}{\partial t}\right)^2 + \sum_{i=1}^{3}\left(\frac{\partial u}{\partial x^i}\right)^2\right)\nu_h\right]^{\frac{1}{2}} \tag{3.15}$$

for all $0\le t\le T$ and ν_h is the volume form associated to the induced metric on Σ_t

We now state the main result of linear scalar wave equations in curve integrable spacetimes.

THEOREM 3.4. *Given the wave equation*

$$\Box_g u = f \tag{3.16}$$

in the region $\Sigma_{(0,T)}$ where g_{ab} is a curve integrable spacetime, $f\in L^2(\Sigma_{(0,T)})$ and $f|_{\Sigma_{t<0}} = 0$ satisfying

$$u|_{\Sigma_{t<0}} = 0 \tag{3.17}$$

$$\frac{\partial u}{\partial t}|_{\Sigma_{t<0}} = 0 \tag{3.18}$$

then there exists a unique regular weak solution $u \in H^1(\Sigma_{(0,T)}, \nu_g)$. *Furthermore this solution is continuous with respect the source function.*

We now give an overview of the proof this result. Details can be found in [**6**, **13**]. We would like to remark that in both papers the H^1 regularity of the weak solution is never shown. However, it is used to prove uniqueness and stability. The last part of this paper gives a proof that the solutions are indeed of H^1 regularity.

The proof of existence uses the Hahn-Banach Theorem and the Riesz Representation Theorem.

We define the functional

$$k_f(\Box_g \omega) = (f, \omega)_{L^2(\Sigma_{(0,T)}, \nu_g)} \tag{3.19}$$

$$k_f : \Box C_0^\infty(\Sigma_{(0,T)}) \to \mathbb{R} \tag{3.20}$$

for $\omega \in C_0^\infty(\Sigma_{(0,T)})$

Now using the energy inequality given by equation (3.11) and Cauchy-Schwartz we obtain:

$$k(\Box_g \omega) = (f, \omega)_{L^2(\Sigma_{(0,T)}, \nu_g)} \tag{3.21}$$

$$\leq (f, f)_{L^2(\Sigma_{(0,T)}, \nu_g)} (\omega, \omega)_{L^2(\Sigma_{(0,T)}, \nu_g)} \tag{3.22}$$

$$\leq c(\|f\|_{L^2(\Sigma_{(0,T)}, \nu_g)})^2 (\|\Box_g \omega\|_{L^2(\Sigma_{(0,T)}, \nu_g)})^2 \tag{3.23}$$

The Hahn-Banach theorem allows us to extend the functional to the whole $L^2(\Sigma_{(0,T)}, \nu_g)$ without increasing the norm. Moreover, the estimate implies that k_f is bounded in $L^2(\Sigma_{(0,T)}, \nu_g)$.

Then using the Riesz's Representation Theorem there is a $u \in L^2(\Sigma_{(0,T)}, \nu_g)$ such that $k_f(\Box_g \omega) = (u, \Box_g \omega)_{L^2(\Sigma_{(0,T)}, \nu_g)}$ which implies

$$\int_{\Sigma_{(0,T)}} u \Box_g \omega \nu_g = \int_{\Sigma_{(0,T)}} f \omega \nu_g$$

for all $\omega \in C_0^\infty(\Sigma_{(0,T)})$.

Notice that we can extend u to an L^2 function on $(-\infty, T) \times \Sigma$ by setting $u|_{t<0} = 0$ and we can extend f in the same manner. This shows that u satisfies the requirements to be a weak solution.

If we assume that our solutions are regular then we have uniqueness and stability. This can be seen by taking the difference of two weak solutions, $u = u_1 - u_2$, and noticing that it satisfies the homogeneous problem and the energy estimate given by equation (3.11). The stability of the solution can be shown by using the fact that the norm of the difference between two solutions is bounded from above by the norm of the source functions.

3.3. H^1 regularity. In [**6**, **13**] it is assumed that the solution $u \in L^2(\Sigma_{(0,T)}, \nu_g)$ constructed in fact satisfies the regularity condition $u \in H^1(\Sigma_{(0,T)}, \nu_g)$. In this section we prove that assumption, however we will need an extra condition on the regularity of the metric which we state below

CONDITIONS 3.5. *Let* $(\Sigma_{(0,T)}, g)$ *be a lens-shaped domain satisfying:*

(1) *The components* g_{ab} *and* g^{ab} *are* $W^{1,\infty}(\Sigma_{(0,T)})$.
(2) *The inverse of the metric tensor satisfies*

$$g^{ab}(\mathbf{x}) - g^{ab}(\mathbf{x}') \leq C(d_E(\mathbf{x}, \mathbf{x}'))^2$$

where C is a constant and $(d_E(\mathbf{x},\mathbf{x}'))^2 = \left((t-t')^2 + \sum_{i=1}^{3}|x^i - x^{i'}|^2\right)$ *with* $\mathbf{x} = (t, x^1, x^2, x^3)$ *and* $\mathbf{x}' = (t', x^{1'}, x^{2'}, x^{3'})$

This additional regularity condition is strong enough to guarantee the existence of a timelike vector field with a well-defined bounded covariant derivative. Nevertheless, we did not assume this before because to show the existence of L^2 solution it is not needed. The boundness of the covariant derivative under the weaker conditions is precisely the content of Clarke's paper [**6**]. However, we will make this extra assumption in order to show that the solution has H^1-regularity.

The main idea will be to regularise the solution u using a strict delta net $\{\rho_n\}$ and then using the additional conditions 3.5 show that there are good estimates for the commutator $[\rho_n, \Box_g]u$.

The strict delta net we will use is such that the mollification $\rho_n * u = u_n$ is a smooth function with compact support vanishing at $t \leq 0$ and therefore the energy estimate given by equation (3.11) gives

$$(\tilde{\|}u_n\|^1_{\Sigma_t})^2 \leq C(\|f_n\|_{L^2(\Sigma_{(0,T)},\nu_g)})^2 \tag{3.24}$$

for all $0 \leq t \leq T$.

Now notice that

$$f_n = \Box_g u_n = \rho_n * f + [\Box_g, \rho_n *]u \tag{3.25}$$

and using the properties of the strict delta net we have that $(\|\rho_n * f\|_{L^2(\Sigma_{(0,T)},\nu_g)})^2$ converge to $(\|f\|_{L^2(\Sigma_{(0,T)},\nu_g)})^2$. Therefore, the key point is to show that $[\Box_g, \rho_n *]u$ is bounded in $L^2(\Sigma_{(0,T)}, \nu_g)$. We do that in two steps. First we show in the following lemma good estimates for $L^2(\Sigma_{(0,T)})$. Then using the equivalency between the norms $L^2(\Sigma_{(0,T)}, \nu_g)$ and $L^2(\Sigma_{(0,T)})$ (see [**2**]) we obtain the result for $L^2(\Sigma_{(0,T)}, \nu_g)$.

LEMMA 3.6. *If $u \in L^2(\Sigma_{(0,T)}, \nu_g)$, then the commutator $[\Box_g, \rho_n *]u$ is bounded in $L^2(\Sigma_{(0,T)})$.*

Proof.

We start by doing the following direct calculation in which for the moment we use $\mathbf{x} := (t, x^i)$ and $\mathbf{y} := (t, y^i)$ for convenience. For the moment, we assume $u \in C_0^\infty(\Sigma_{(0,T)})$ then

$$\begin{aligned}
[\Box_g, \rho_n *]u &= (g^{ab}\rho_n * u_{ab} - \Gamma_a^{ab}\rho_n * u_b) - \rho_n * (g^{ab}u_{ab} - \Gamma_a^{ab}u_b) \\
&= \left(g^{ab}\rho_n * u_{ab} - \rho_n * g^{ab}u_{ab}\right) - \left(\Gamma_a^{ab}\rho_n * u_b - \rho_n * \Gamma_a^{ab}u_b\right)
\end{aligned}$$

Now we will analyse both terms.

$$\begin{aligned}
&\left(g^{ab}\rho_n * u_{ab} - \rho_n * g^{ab}u_{ab}\right) \\
&\qquad = \int_{\mathbb{R}^4} \rho_n(\mathbf{x}-\mathbf{y})\left[g^{ab}(\mathbf{x}) - g^{ab}(\mathbf{y})\right]\partial_a\partial_b u(\mathbf{x})dx^4 \\
&\qquad = \int_{\mathbb{R}^4} \partial_a\partial_b\rho_n(\mathbf{x}-\mathbf{y})\left[g^{ab}(\mathbf{x}) - g^{ab}(\mathbf{y})\right]u(\mathbf{x})dx^4 \\
&\qquad + 2\int_{\mathbb{R}^4} \partial_a\rho_n(\mathbf{x}-\mathbf{y})\partial_b\left[g^{ab}(\mathbf{x}) - g^{ab}(\mathbf{y})\right]u(\mathbf{x})dx^4 \\
&\qquad + \int_{\mathbb{R}^4} \rho_n(\mathbf{x}-\mathbf{y})\partial_a\partial_b\left[g^{ab}(\mathbf{x}) - g^{ab}(\mathbf{y})\right]u(\mathbf{x})dx^4
\end{aligned}$$

Now the first term gives

$$\left\| \int_{\mathbb{R}^4} \partial_a \partial_b \rho_n(\mathbf{x}-\mathbf{y}) \left[g^{ab}(\mathbf{x}) - g^{ab}(\mathbf{y})\right] u(\mathbf{x}) dx^4 \right\|_{L^2(\Sigma_{(0,T)})} \tag{3.26}$$

$$\leq K \left\| \left[|\mathbf{x}|^2 \partial_a \partial_b \rho_n(\mathbf{x}) * u \right] \right\|_{L^2(\Sigma_{(0,T)})} \tag{3.27}$$

$$\leq K' \left\| |\mathbf{x}|^2 \partial_a \partial_b \rho_n(\mathbf{x}) \right\|_{L^1(\Sigma_{(0,T)})} \|u\|_{L^2(\Sigma_{(0,T)})} \tag{3.28}$$

where we have used Young's inequality and the first condition given by the Geometric Conditions 3.5.

Notice that

$$\begin{aligned} \left\| |\mathbf{x}|^2 \partial_a \partial_b \rho_n(\mathbf{x}) \right\|_{L^1(\Sigma_{(0,T)})} &= \frac{1}{\epsilon^6} \int_{\Sigma_{(0,T)}} |\mathbf{x}|^2 \partial_a \partial_b \rho(\frac{\mathbf{x}}{\epsilon}) dx^4 \\ &= \int_{\Sigma_{(0,T)}} |\mathbf{y}|^2 \partial_a \partial_b \rho(\mathbf{y}) dx^4 \end{aligned}$$

which shows that $\left\| |\mathbf{x}|^2 \partial_a \partial_b \rho_n(\mathbf{x}) \right\|_{L^1(\Sigma_{(0,T)})}$ is independent of ϵ.

This gives

$$\left\| \int_{\mathbb{R}^4} \partial_a \partial_b \rho_n(\mathbf{x}-\mathbf{y}) \left[g^{ab}(\mathbf{x}) - g^{ab}(\mathbf{y})\right] u(\mathbf{x}) dx^4 \right\|_{L^2(\Sigma_{(0,T)})} \tag{3.29}$$

$$\leq K \|u\|_{L^2(\Sigma_{(0,T)})} \tag{3.30}$$

We also have that

$$\left\| \int_{\mathbb{R}^4} \partial_a \rho_n(\mathbf{x}-\mathbf{y}) \partial_b \left[g^{ab}(\mathbf{x}) - g^{ab}(\mathbf{y})\right] u(\mathbf{x}) dx^4 \right\|_{L^2(\Sigma_{(0,T)})} \tag{3.31}$$

$$\leq K \left\| \left[|\mathbf{x}| \partial_a \partial_b \rho_n(\mathbf{x}) * u \right] \right\|_{L^2(\Sigma_{(0,T)})} \tag{3.32}$$

$$\leq K' \left\| |\mathbf{x}| \partial_a \partial_b \rho_n(\mathbf{x}) \right\|_{L^1(\Sigma_{(0,T)})} \|u\|_{L^2(\Sigma_{(0,T)})} \tag{3.33}$$

where we have used Young's inequality and the Lipschitz condition given in the Geometric Conditions 3.5.

Notice that

$$\begin{aligned} \left\| |\mathbf{x}| \partial_a \partial_b \rho_n(\mathbf{x}) \right\|_{L^1(\Sigma_{(0,T)})} &= \frac{1}{\epsilon^5} \int_{\Sigma_{(0,T)}} |\mathbf{x}| \partial_a \partial_b \rho(\frac{\mathbf{x}}{\epsilon}) dx^4 \\ &= \int_{\Sigma_{(0,T)}} |\mathbf{y}| \partial_a \partial_b \rho(\mathbf{y}) dx^4 \end{aligned}$$

which shows that $\left\| |\mathbf{x}| \partial_a \partial_b \rho_n(\mathbf{x}) \right\|_{L^1(\Sigma_{(0,T)})}$ is independent of ϵ.

Therefore, we can conclude

$$\left\| \int_{\mathbb{R}^4} \partial_a \rho_n(\mathbf{x}-\mathbf{y}) \partial_b \left[g^{ab}(\mathbf{x}) - g^{ab}(\mathbf{y})\right] u(\mathbf{x}) dx^4 \right\|_{L^2(\Sigma_{(0,T)})} \tag{3.34}$$

$$\leq K'' \|u\|_{L^2(\Sigma_{(0,T)})} \tag{3.35}$$

The last term gives

$$\int_{\mathbb{R}^4} \rho_n(\mathbf{x}-\mathbf{y}) \partial_a \partial_b \left[g^{ab}(\mathbf{x}) - g^{ab}(\mathbf{y})\right] u(\mathbf{x}) dx^4 \tag{3.36}$$

$$\leq \quad K \int_{\mathbb{R}^4} \rho_n(\mathbf{x}-\mathbf{y}) u(\mathbf{x}) dx^4 \tag{3.37}$$

$$\leq \quad K''' \|u\|_{L^2(\Sigma_{(0,T)})} \tag{3.38}$$

where we have used Young's inequality and the Lipschitz condition given in the Geometric Conditions 3.5.

Finally, using (3.29), (3.34) and (3.36) we have that

$$\| \left(g^{ab}\rho_n * u_{ab} - \rho_n * g^{ab}u_{ab}\right) \|_{L^2(\Sigma_{(0,T)})} \leq K \|u\|_{L^2(\Sigma_{(0,T)})} \tag{3.39}$$

Now we analyse the second term

$$\begin{aligned}
\left(\Gamma^{ab}_{a}\rho_n * u_b - \rho_n * \Gamma^{ab}_{a}u_b\right) & \\
&= \int_{\mathbb{R}^4} \rho_n(\mathbf{x}-\mathbf{y}) \left[\Gamma^{ab}_{a}(\mathbf{y}) - \Gamma^{ab}_{a}(\mathbf{x})\right] \partial_b u(\mathbf{x}) dx^4 \\
&= \int_{\mathbb{R}^4} \partial_b\rho_n(\mathbf{x}-\mathbf{y}) \left[\Gamma^{ab}_{a}(\mathbf{y}) - \Gamma^{ab}_{a}(\mathbf{x})\right] u(\mathbf{x}) dx^4 \\
&+ \int_{\mathbb{R}^4} \rho_n(\mathbf{x}-\mathbf{y})\partial_b \left[\Gamma^{ab}_{a}(\mathbf{y}) - \Gamma^{ab}_{a}(\mathbf{x})\right] u(\mathbf{x}) dx^4
\end{aligned}$$

The first term can be bounded in an analogous manner as in (3.34) while the second term can be bounded as in (3.36). Putting together these results give

$$\| \left(\Gamma^{ab}_{a}\rho_n * u_b - \rho_n * \Gamma^{ab}_{a}u_b\right) \|_{L^2(\Sigma_{(0,T)})} \leq K' \|u\|_{L^2(\Sigma_{(0,T)})} \tag{3.40}$$

We can conclude then that

$$\|[\Box_g, \rho_n]u\|_{L^2(\Sigma_{(0,T)})} \leq C \|u\|_{L^2(\Sigma_{(0,T)})} \tag{3.41}$$

for all $u \in C_0^\infty(\Sigma_{(0,T)})$.

Therefore the linear functional $F(u) = [\Box_g, \rho_n]u$ is bounded and densely defined in $L^2(\Sigma_{(0,T)})$ with a unique extension to $L^2(\Sigma_{(0,T)})$.

This concludes the proof $\boxdot$.

Therefore using Lemma 3.6 we can prove now the regularity of the solution as in definition 3.3.

We begin by noticing that

$$\begin{aligned}
(\|u_n\|_{H^1(\Sigma_{(0,T)},\nu_g)})^2 &\leq (\tilde{\|}u_n\|^1_{\Sigma_t})^2 \\
&\leq (\|f_n\|_{L^2(\Sigma_{(0,T)},\nu_g)})^2 \\
&= (\|\rho_n * f + [\Box_g, \rho_n]u\|_{L^2(\Sigma_{(0,T)},\nu_g)})^2 \\
&\leq \tilde{C}\left(\left(\|f\|_{L^2(\Sigma_{(0,T)},\nu_g)}\right)^2 + \left(\|u\|_{L^2(\Sigma_{(0,T)},\nu_g)}\right)^2\right)
\end{aligned}$$

for all $0 \leq t \leq T$

Now using Banach-Alaoglu, we know there is a subsequence that converge weakly to an element $\tilde{u} \in H^1(\Sigma_{(0,T)}, \nu_g)$ and therefore also converge weakly in $L^2(\Sigma_{(0,T)}, \nu_g)$. However we know u_n converges strongly to $u \in L^2(\Sigma_{(0,T)}, \nu_g)$ and therefore weakly. By uniqueness of limits we have that $\tilde{u} = u$. This proves $u \in H^1(\Sigma_{(0,T)}, \nu_g)$

Moreover using that the norm is weakly semi continuous we have that

$$\begin{aligned}
(\|u\|_{H^1(\Sigma_{(0,T)},\nu_g)})^2 &\leq \liminf_{n\to\infty} (\|u_n\|_{H^1(\Sigma_{(0,T)},\nu_g)})^2 \\
&\leq (\tilde{\|}u_n\|^1_{\Sigma_t})^2 \\
&\leq (\|f_n\|_{L^2(\Sigma_{(0,T)},\nu_g)})^2 \\
&= (\|\rho_n * f + [\Box_g, \rho_n]u\|_{L^2(\Sigma_{(0,T)},\nu_g)})^2 \\
&\leq \tilde{C}\left(\left(\|f\|_{L^2(\Sigma_{(0,T)},\nu_g)}\right)^2 + \left(\|u\|_{L^2(\Sigma_{(0,T)},\nu_g)}\right)^2\right)
\end{aligned}$$

Acknowledgments

The author would like to thank J. A. Vickers and the referee for their comments on an early draft.

References

[1] Jean Leray, *Hyperbolic differential equations*, The Institute for Advanced Study, Princeton, N. J., 1953. MR0063548

[2] S. W. Hawking and G. F. R. Ellis, *The large scale structure of space-time*, Cambridge University Press, London-New York, 1973. Cambridge Monographs on Mathematical Physics, No. 1. MR0424186

[3] Antonio N. Bernal and Miguel Sánchez, *Globally hyperbolic spacetimes can be defined as 'causal' instead of 'strongly causal'*, Classical Quantum Gravity **24** (2007), no. 3, 745–749, DOI 10.1088/0264-9381/24/3/N01. MR2294243

[4] Michael Kunzinger, Roland Steinbauer, Milena Stojković, and James A. Vickers, *Hawking's singularity theorem for $C^{1,1}$-metrics*, Classical Quantum Gravity **32** (2015), no. 7, 075012, 19, DOI 10.1088/0264-9381/32/7/075012. MR3322151

[5] Michael Kunzinger, Roland Steinbauer, Milena Stojković, and James A. Vickers, *A regularisation approach to causality theory for $C^{1,1}$-Lorentzian metrics*, Gen. Relativity Gravitation **46** (2014), no. 8, Art. 1738, 18, DOI 10.1007/s10714-014-1738-7. MR3245957

[6] C. J. S. Clarke, *Generalized hyperbolicity in singular spacetimes*, Classical Quantum Gravity **15** (1998), no. 4, 975–984, DOI 10.1088/0264-9381/15/4/018. MR1620233

[7] W. Israel, *Il Nuovo Cimento*, B Series 10 11 Luglio , **44**, 1, 1-14 (1966)

[8] Roger Penrose, *The geometry of impulsive gravitational waves*, General relativity (papers in honour of J. L. Synge), Clarendon Press, Oxford, 1972, pp. 101–115. MR0503490

[9] C. J. S. Clarke and N. O'Donnell, *Dynamical extension through a space-time singularity*, Rend. Sem. Mat. Univ. Politec. Torino **50** (1992), no. 1, 39–60 (1993). Singularities in curved space-times. MR1218679

[10] Robert A. Adams and John J. F. Fournier, *Sobolev spaces*, 2nd ed., Pure and Applied Mathematics (Amsterdam), vol. 140, Elsevier/Academic Press, Amsterdam, 2003. MR2424078

[11] R. Steinbauer and J. A. Vickers, *Class. Quantum Grav.*, **26** 6 (2008)

[12] Piotr T. Chruściel, *On maximal globally hyperbolic vacuum space-times*, J. Fixed Point Theory Appl. **14** (2013), no. 1, 325–353, DOI 10.1007/s11784-013-0141-4. MR3202035

[13] Yafet Sanchez Sanchez, *Regularity of curve integrable spacetimes*, Gen. Relativity Gravitation **47** (2015), no. 7, Art. 80, 25, DOI 10.1007/s10714-015-1922-4. MR3359079

[14] Lawrence C. Evans, *Partial differential equations*, Graduate Studies in Mathematics, vol. 19, American Mathematical Society, Providence, RI, 1998. MR1625845

Mathematical Sciences and STAG Research Centre, University of Southampton, Southampton, SO17 1BJ, United Kingdom

Current address: Institut für Analysis, Gottfried Wilhelm Leibniz Universität Hannover Welfengarten 1 30167, Germany

Email address: yess@math.uni-hannover.de

Contemporary Mathematics
Volume **775**, 2021
https://doi.org/10.1090/conm/775/15597

On the pseudospectra of Schrödinger operators on Zoll manifolds

David Sher, Alejandro Uribe, and Carlos Villegas-Blas

ABSTRACT. We consider non-self-adjoint Schrödinger operators $\Delta + V$ where Δ is the Laplace-Beltrami operator on a Zoll manifold X and $V \in C^\infty(X, \mathbb{C})$. We obtain asymptotic results on the pseudo-spectrum and numerical range of such operators.

1. Introduction and statement of results

Let Δ denote the Laplace-Beltrami operator on a compact Zoll manifold X and let $V \in C^\infty(X, \mathbb{C})$. We will study asymptotic properties of the pseudospectrum and numerical range of the Schrödinger operator $\Delta + M_V$, where M_V is the operator of multiplication by V. Here we are taking Δ as a self-adjoint operator with a suitable domain in $L^2(X)$.

Recall that, given an operator P densely defined on some Hilbert space $\mathcal{H}$, and $\epsilon > 0$, the ϵ-pseudospectrum of P is the set of complex numbers λ such that

$$\|(P - \lambda I)^{-1}\| \geq \frac{1}{\epsilon},$$

where, by definition, the norm on the left-hand side is infinite if λ is an eigenvalue of P. Equivalently, λ is in the ϵ-pseudospectrum of P iff there exists $\psi \in \mathcal{H} \setminus \{0\}$ such that

$$\frac{\|(P - \lambda I)\psi\|}{\|\psi\|} \leq \epsilon.$$

This motivates the following definition:

DEFINITION 1.1. We say that a sequence of complex numbers $\{\lambda_k\}$ is in the *asymptotic pseudospectrum* of $\Delta + M_V$ if there exists a sequence $\{\psi_k\}$ of functions in $L^2(X)$ such that as $k \to \infty$,

$$\frac{\|(\Delta + M_V - \lambda_k I)\psi_k\|}{\|\psi_k\|} = O(k^{-\infty}) \quad \text{and} \quad \lim_{k\to\infty} |\lambda_k| = +\infty. \tag{1.1}$$

The sequence $\{\psi_k\}$ is referred to as an associated *pseudo mode.*

2020 *Mathematics Subject Classification.* Primary 47A10; Secondary 35J10, 35S05.

The first author was supported in part by NSF grant EMSW21-RTG 1045119. The third author was partially supported by projects PAPIIT-UNAM IN105718 and CONACYT CB-000000000283531.

Note that (1.1) implies that, $\forall \epsilon > 0$, λ_k will be in the ϵ-pseudospectrum of $\Delta + M_V$ for all sufficiently large k. The previous definition has an analogue in the semi-classical setting, see for example [**18**] §12.5 (and references therein). In the present case the role of Planck's constant is played by $\hbar = 1/k$.

It is well-known that the spectrum of a non-self-adjoint operator is very unstable. The pseudospectrum, however, is much more stable, which makes it a natural object to study. It is also known that points in the pseudospectrum do not have to be close to the spectrum, even in the semi-classical limit.

In this paper we consider the case where X is a closed, connected Zoll manifold. More specifically, we make the assumption there is a $T > 0$ such that each geodesic on X has minimal period T (we rule out the existence of exceptional short geodesics). Such manifolds are often called C_l-manifolds [**1**]. The simplest examples are the spheres with the round metric. It is known that the spectrum of the Laplacian Δ on a Zoll manifold consists of clusters of eigenvalues of uniformly bounded width centered at the points

$$\Lambda_k = \frac{4\pi^2}{T^2}\left(k + \frac{\beta}{4}\right)^2, \quad k = 1, 2, \ldots,$$

where β is the Morse index of one periodic geodesic (hence all periodic geodesics). Specifically

$$\exists\, C > 0 \text{ such that} \qquad \sigma(\Delta) \subset \bigcup_{k\in\mathbb{N}} [\Lambda_k - C\,,\, \Lambda_k + C], \tag{1.2}$$

where $\sigma(\Delta)$ denotes the spectrum of Δ. Note that the distance between the clusters Λ_k and Λ_{k+1} is linear in k, which allows us to separate the eigenvalues of the Laplacian unambiguously into clusters, at least for large k.

More precisely, as proved in [**4**], there exist self-adjoint, commuting pseudo-differential operators A of order one and Q_0 of order zero such that the spectrum of A consists of the eigenvalues $k = 1, 2, \ldots$ and

$$\Delta = \frac{4\pi^2}{T^2}\left(A + \frac{\beta}{4}\right)^2 + Q_0. \tag{1.3}$$

In particular, we can take C equal to the operator norm $\|Q_0\|$ with $Q_0 \in \mathcal{B}(L^2(X))$. Moreover, if for each k we let

$$E_k := \text{eigenspace of } A \text{ corresponding to the eigenvalue } k, \tag{1.4}$$

the eigenvalues in the k-th cluster consist of the eigenvalues of the restriction $Q_0|_{E_k}$, shifted by Λ_k.

The operator $(A + \frac{\beta}{4})^2$ has spectrum contained in $(\mathbb{N} + \frac{\beta}{4})^2$, and has the same principal and sub principal symbols as Δ up to the factor $\frac{4\pi^2}{T^2}$. For example, in case $X = S^n$, the unit n-dimensional sphere, $T = 2\pi$ and the eigenvalues of the Laplacian are $k(k + n - 1) = (k + \frac{n-1}{2})^2 - \frac{(n-1)^2}{4}$. Therefore in this case

$$\beta = 2(n-1) \quad \text{and} \quad Q_0 = -\frac{(n-1)^2}{4} I.$$

For the round sphere the principal symbol of Q_0 is constant. In the general Zoll case the symbol of Q_0 is not constant (see [**17**], Theorem 3, for an expression for it), but it is always constant along geodesics since $[Q_0, \Delta] = 0$. The fact that the symbol of Q_0 is not constant corresponds to the non-trivial clustering phenomenon of the Laplace eigenvalues seen in Zoll metrics, as the asymptotic behavior of the eigenvalue clusters is given to first order by the principal symbol of Q_0 via a Szegö limit theorem.

In conclusion, our Schrödinger operator $\Delta + M_V$ has the form

$$\Delta + M_V = \frac{4\pi^2}{T^2}\left(A + \frac{\beta}{4}\right)^2 + Q_0 + M_V. \tag{1.5}$$

Thus we can think of M_V as a non-self-adjoint perturbation added to Q_0.

1.1. Statements of our results. We can now state the main results of this paper.

1.1.1. *On the spectrum.* First we will show that the spectrum of $\Delta + M_V$ also has a cluster structure (at least if we are sufficiently far from the origin) generalizing (1.2) (see also [**5**]). Let us denote by D the closed disk around the origin with radius $||Q_0 + M_V||$.

THEOREM 1.2. *Let $V : X \to \mathbb{C}$ be a bounded measurable function. Then the spectrum of $H := \Delta + M_V$ is contained in the union*

$$\bigcup_{k=0}^{\infty} D_k, \quad D_k := \Lambda_k + D. \tag{1.6}$$

The spectrum of H consists entirely of isolated eigenvalues with finite algebraic multiplicities and, for k sufficiently large, the spectrum of H in the the disk D_k consists of eigenvalues with total algebraic multiplicity equal to the dimension of E_k. Moreover, the generalized eigenspaces of $\Delta + M_V$ span $L^2(X)$.

Note that the sets D_k are pairwise disjoint for large k.

1.1.2. *Results on the asymptotic pseudospectrum.* We now turn to the statement of our results on the asymptotic pseudospectrum of $\Delta + M_V$.

Let $H_0(x,\xi) = \|\xi\|$, $H_0 \in C^\infty(T^*X\backslash\{0\})$. This Hamiltonian generates geodesic flow re-parametrized by arc length. In what follows we shall think of oriented geodesics as trajectories of the Hamiltonian H_0 on the unit tangent bundle of X. Let $\mathcal{O}$ be the manifold of oriented geodesics on X. This is the symplectic manifold obtained by symplectic reduction of $T^*X \setminus \{0\}$ under the circle action generated by H_0. Given any smooth function V on X, one can define its Radon transform to be the function $\widetilde{V}$ on $\mathcal{O}$ defined by

$$\widetilde{V}(\gamma) = \frac{1}{T}\int_\gamma V\, ds,$$

where s denotes arc length along the oriented geodesic $\gamma \in \mathcal{O}$, and we are abusing the notation and continue to denote by V the pull-back of the potential to the unit tangent bundle. Note that $\mathcal{O}$ has an involution corresponding to reversing the orientation of the geodesics. This involution is anti-symplectic, and $\widetilde{V}$ is even with respect to it.

Now recall that Q_0 commutes with Δ (and with $(A+\frac{\beta}{4})$). Therefore its principal symbol q_0 is invariant along geodesics and may thus be viewed as a function on $\mathcal{O}$, also denoted q_0. We reviewed above that, in case X is the unit sphere, $q_0 = -\frac{(n-1)^2}{4}$ (constant). With this notation, our result is as follows:

THEOREM 1.3. *(1) Let $\mu = (\widetilde{V} + q_0)(\gamma)$, where $\gamma \in \mathcal{O}$ is such that*

$$\{\Re\widetilde{V} + q_0\,,\, \Im\widetilde{V}\}(\gamma) < 0.$$

Then $\{\Lambda_k + \mu\}$ is in the asymptotic pseudospectrum of $\Delta + M_V$. In fact, for each $m \in \gamma$, there exists a pseudo mode $\{\psi_k\}$ such that the sum $\sum_{k=1}^{\infty} \psi_k$ is a distribution whose wave-front set is the ray in $T^*X \setminus \{0\}$ through m.

(2) If $\{\Lambda_k + \mu\}$ is in the asymptotic pseudospectrum with $\mu = (\widetilde{V} + q_0)(\gamma)$, and

$$\{\Re\widetilde{V} + q_0\,,\, \Im\widetilde{V}\}(\gamma) > 0,$$

then for any associated pseudo mode $\{\psi_k\}$, the wave-front set of $\sum_k \psi_k$ is disjoint from the conic set in $T^*X \setminus \{0\}$ generated by γ.

Note that although the potential $\widetilde{V}$ is even under the involution of reversing the orientation of γ, the Poisson bracket $\{\Re\widetilde{V} + q_0\,,\, \Im\widetilde{V}\}(\gamma)$ is odd under the same involution. Together with part (1) of the theorem, this implies that the set of sequences of the form

$$\left\{\lambda_k + \widetilde{V}(\gamma)\ ;\ \{\Re\widetilde{V} + q_0\,,\, \Im\widetilde{V}\}(\gamma) \neq 0\right\}$$

is contained in the asymptotic pseudospectrum. However, in general this set does not equal the asymptotic pseudospectrum. For example, in the case of spheres, if V is odd (with respect to the antipodal map) then $\widetilde{V}$ is identically zero. In that case, however, there is a substitute function for $\widetilde{V}$, namely the function

$$\widetilde{\widetilde{V}} = \frac{1}{4}\widetilde{V^2} - \frac{1}{8\pi}\int_0^{2\pi} dt \int_0^t \{\phi_t^* V\,,\, \phi_s^* V\}\, ds \tag{1.7}$$

where ϕ_t is geodesic flow, see [**14**] (Theorem 3.2). More subtly we expect that one can have that $\{\Re\widetilde{V}+q_0\,,\, \Im\widetilde{V}\}(\gamma) = 0$ and have pseudo modes concentrated at γ with pseudo eigenvalue $\widetilde{V}(\gamma)$, in the spirit of Theorem 2.1 in [**13**], where non-vanishing of higher-order derivatives is assumed.

Theorem 1.3 is analogous to Theorem 12.8 in [**18**] (originally due to Davies), where a similar Poisson bracket condition appears. It is also analogous in spirit to the main result in [**2**].

1.1.3. *Results on the numerical range.* Recall that the numerical range of an operator $\hat{H}$ on $L^2(X)$ with domain $\mathcal{D}(\hat{H})$ is the subset of the complex plane defined by:

$$\mathcal{R}(\hat{H}) = \left\{\frac{\langle \hat{H}(\psi), \psi\rangle}{\langle \psi, \psi\rangle}\ ;\ \psi \in \mathcal{D}(\hat{H})\right\}.$$

We are interested in the numerical range of $\hat{H} = \Delta + M_V$. We will only consider the asymptotics of the sets of values

$$\left\{\frac{\langle \hat{H}(\psi), \psi\rangle}{\langle \psi, \psi\rangle}\ ;\ \psi \in E_k\right\}.$$

To state our result we shift these sets by Λ_k:

THEOREM 1.4. *In case $X = S^n$ or X a Zoll surface, the limit $\mathcal{R}_\infty$ of the sets*

$$\mathcal{R}_k = \left\{\frac{\langle (\Delta + M_V)\psi, \psi\rangle}{\langle \psi, \psi\rangle} - \Lambda_k\ ;\ \psi \in E_k \setminus \{0\}\right\}$$

as k tends to infinity is the convex hull of the image of $\widetilde{V} + q_0 : \mathcal{O} \to \mathbb{C}$.

For a similar result for Berezin-Toeplitz operators see [**2**].

2. Proof of Theorem 1.2

Let $\widetilde{A}$ be the operator $\frac{4\pi^2}{T^2}\left(A+\frac{\beta}{4}\right)^2$. Let z be in the resolvent set $\rho(\widetilde{A}^2)$ of the operator $\widetilde{A}^2$, with the distance $d(z,\sigma(\widetilde{A}^2))$ from z to the spectrum $\sigma(\widetilde{A}^2)$ of $\widetilde{A}^2$ greater than the norm $||Q_0+M_V||$. Since $\widetilde{A}^2$ is self adjoint then $||\left(\widetilde{A}^2-z\right)^{-1}||=1/d(z,\sigma(\widetilde{A}^2))$ which implies $||\,(Q_0+M_V)\left(\widetilde{A}^2-z\right)^{-1}||<1$. Therefore the operator $\left(I+(Q_0+M_V)\left(\widetilde{A}^2-z\right)^{-1}\right)^{-1}$ is a bounded operator which in turn implies that $(\Delta+M_V-z)^{-1}$ exists and it is a bounded operator, i.e. z is in the resolvent set $\rho(H)$ of H. Thus the spectrum $\sigma(H)$ of H must be contained in the union of the closed disks D_k as indicated in (1.6).

Consider $z\in\mathbb{C}$ such that $d(z,\sigma(\Delta))>2\|M_V\|$. Since Δ has compact resolvent then writing $H-z=\left(I+M_V(\Delta-z)^{-1}\right)(\Delta-z)$ we see that $(H-z)^{-1}$ exists and it must be compact (note that since Δ is self adjoint then $\|(\Delta-z)^{-1}\|=1/d(z,\sigma(\Delta))$). Thus we have that the spectrum of H consists entirely of eigenvalues with finite algebraic multiplicities, see reference [**12**], theorem III, section 6.8.

Let $g(k)=\Lambda_k-\Lambda_{k-1}=O(k)$ and let $\mathcal{C}_k$ be the circle of radius $g(k)/2$ and center Λ_k. Note that, for k sufficiently large, both the interval of radius $||Q_0||$ around Λ_k and the disk D_k are inside $\mathcal{C}_k$. Moreover,

$$\forall z\in\mathcal{C}_k,\quad d(z,\sigma(\Delta))\geq g(k)/3 \text{ and } d(z,\sigma(\hat{H}))\geq g(k)/3.$$

In particular, $\mathcal{C}_k$ is contained in both $\rho(\Delta)$ and $\rho(H)$. Thus the following two projections are well defined:

$$P_k=\frac{1}{2\pi\imath}\oint_{\mathcal{C}_k}(z-H)^{-1}\,dz \tag{2.1}$$

and

$$\Pi_k=\frac{1}{2\pi\imath}\oint_{\mathcal{C}_k}(z-\Delta)^{-1}\,dz. \tag{2.2}$$

Since Δ is self adjoint then $\|(z-\Delta)^{-1}\|=1/d(z,\sigma(\Delta))=O(k^{-1})$ uniformly in $\mathcal{C}_k$. This implies for k sufficiently large that $\|M_V(z-\Delta)^{-1}\|\leq 1/2$ which in turn implies, using a Neumann series, that

$$\begin{aligned} \|\,(z-H)^{-1}\,\| &\leq \|(z-\Delta)^{-1}\|\;\|\left(I+M_V(z-\Delta)^{-1}\right)^{-1}\| && (2.3)\\ &\leq \frac{1}{d(z,\sigma(\Delta))}\,\frac{1}{1-\|M_V(z-\Delta)^{-1}\|}=O(k^{-1}) && (2.4)\end{aligned}$$

uniformly in $\mathcal{C}_k$.

Thus we obtain

$$\begin{aligned} \|P_k-\Pi_k\| &= \left\|\frac{1}{2\pi\imath}\oint_{\mathcal{C}_k}(z-H)^{-1}M_V\,(z-\Delta)^{-1}\,dz\right\| \\ &\leq \frac{1}{2\pi}\oint_{\mathcal{C}_k}\|\,(z-H)^{-1}\,\|\;\|M_V\|\;\|\,(z-\Delta)^{-1}\,\||dz| \\ &\leq \frac{1}{2\pi}(2\pi g(k))O(k^{-1})\;||M_V||\;O(k^{-1})=O(k^{-1}). && (2.5)\end{aligned}$$

Therefore, for k sufficiently large, the norm $\|P_k - \Pi_k\|$ is less than one which in turn implies that the dimension of the range of P_k, $\dim(\mathrm{Ran}(P_k))$, and the dimension of the range of Π_k, $\dim(\mathrm{Ran}(\Pi_k))$, must be the same, see reference [**12**], chapter I, section 4.6. Moreover, $\dim(\mathrm{Ran}(P_k))$ must be finite which implies that the spectrum of H inside $\mathcal{C}_k$ consists only of eigenvalues, see reference [**12**], chapter III, section 6.5. We conclude that the total algebraic multiplicity of the eigenvalues of H inside D_k must be the same as the initial geometric multiplicity of the cluster of eigenvalues of Δ in the interval $[\Lambda_k - \|Q_0\|, \Lambda_k + \|Q_0\|]$, namely the dimension of E_k.

Finally, for $1 < p < \infty$, let us denote by $\mathcal{S}_p$ the space of all completely continuous operators S for which $\sum_{j=1}^{\infty} s_j^p(S) < \infty$ where $s_j(S)$ are the eigenvalues of $|S| \equiv \sqrt{S^*S}$. The fact that the generalized eigenspaces of $\Delta + M_V$ span $L^2(X)$ is a consequence of theorem 10.1 in reference [**5**] noting that Δ has discrete spectrum and for all λ in the resolvent set of Δ, $(\Delta-\lambda)^{-1}V(\Delta-\lambda)^{-1}$ belongs to a Schatten class $\mathcal{S}_p$ for some $1 < p < \infty$. This last fact is a consequence of (i) the property that $\mathcal{S}_p$ is an ideal in $\mathcal{B}(L^2(X))$, (ii) $V(\Delta-\lambda)^{-1}$ is a bounded operator and (iii) $(\Delta-\lambda)^{-1}$ belongs to $\mathcal{S}_p$ for some $1 < p < \infty$ (the multiplicity of the eigenvalues of Δ in the interval $[\Lambda_k - C\,,\,\Lambda_k + C]$ is given by a polynomial in k for k sufficiently large, see reference [**4**]).

3. Proof of Theorem 1.3

3.1. The averaging method. We begin by recalling the "averaging method" ([**8, 16**]):

PROPOSITION 3.1. *If $V : X \to \mathbb{C}$ is smooth, there exists a pseudo-differential operator of order (-1), F, such that*

$$e^F(\Delta + M_V)e^{-F} = e^F\big(\frac{4\pi^2}{T^2}(A+\frac{\beta}{4})^2 + Q_0 + M_V\big)e^{-F} = \frac{4\pi^2}{T^2}(A+\frac{\beta}{4})^2 + B + S,$$

where B is a pseudo-differential operator of order zero such that $[A,B]=0$, and S is a smoothing operator. Moreover, the symbol of B is $\widetilde{V} + q_0$, regarded as a function on $T^*X^n \setminus \{0\}$.

PROOF. We'll sketch the proof, for completeness. Let

$$\widetilde{A} = \frac{2\pi}{T}(A + \frac{\beta}{4});$$

then $\widetilde{A}^2 + Q_0 = \Delta$, and $\widetilde{A}$ takes the value $\sqrt{\Lambda_k}$ on the space E_k. We proceed inductively, with F_1 an operator or order (-1) solving the equation

$$[F_1, \widetilde{A}^2] = B_0 - (M_V + Q_0) + \text{order } (-1),$$

where B_0 is the operator

$$B_0 = \frac{1}{T}\int_0^T e^{it\widetilde{A}}(M_V + Q_0)e^{-it\widetilde{A}}\,dt.$$

It is easy to see that B_0 commutes with $\widetilde{A}$ and hence with A.

For F_1 we take an operator with principal symbol

$$f_1 := -\frac{H_0^{-1}}{2T}\int_0^T dt \int_0^t \phi_s^*(V + q_0)\,ds$$

where $H_0(x,\xi) = \|\xi\|$ and ϕ_s is its Hamilton flow. Note that the principal symbol of $\widetilde{A}^2$ is the same as the principal symbol of Δ, i.e. H_0^2. The symbol of $[F_1, \widetilde{A}^2]$ is

$$\{f_1\,,\, H_0^2\} = 2H_0\{f_1\,,\, H_0\} = -\frac{1}{T}\int_0^T \big(\phi_t^*(V+q_0) - (V+q_0)\big)\,dt = V - V^{\rm ave},$$

where $V^{\rm ave} + q_0 = \frac{1}{T}\int_0^T \phi_t^*(V)\,dt + q_0$ is the principal symbol of B_0. Note that the factors of q_0 cancel since $\phi_t^*(q_0) = q_0$ (i.e. q_0 is already averaged over the flow). It follows that

$$e^{F_1}\big(\widetilde{A}^2 + Q_0 + M_V\big)e^{-F_1} = \widetilde{A}^2 + B_0 + R_{-1},$$

where R_{-1} is an operator of order (-1).

Using the same constructions, one can modify F_1 by an operator G_1 of order (-2) so that, if $F_2 = F_1 + G_1$,

$$e^{F_2}\big(\widetilde{A}^2 + Q_0 + M_V\big)e^{-F_2} = \widetilde{A}^2 + B_0 + B_{-1} + R_{-2}$$

where R_{-2} is now an operator of order (-2) and both B_0 and B_{-1} commute with A. This process can be continued indefinitely, and after a Borel summation we can find an operator F_∞, with the same principal symbol as F_1, such that

$$e^{F_\infty}\big(\widetilde{A}^2 + Q_0 + M_V\big)e^{-F_\infty} = \widetilde{A}^2 + B + S,$$

where $[A, B] = 0$ and S is a smoothing operator, and where the principal symbol of B is the same as that of B_0. □

3.2. Constructing the pseudo mode. We now proceed to construct a pseudo mode for $\Delta + M_V$, under the hypotheses of part (1) of Theorem 1.3. We will do this by first constructing a pseudo mode for $\widetilde{A}^2 + B$, then showing that it is also a pseudo mode for $\tilde{A}^2 + B + S$, and finally applying the averaging lemma.

If S^*X denotes the unit cosphere bundle of X, one has a diagram

(3.1)
$$\begin{array}{ccc} \iota : S^*X & \hookrightarrow & T^*X\setminus\{0\} \\ \pi\downarrow & & \\ \mathcal{O} & & \end{array}$$

and $\pi^*\widetilde{V}$ is the symbol of B restricted to S^*M. Let us assume the hypotheses of part (1) of the theorem, taking without loss of generality $\sigma_B(m) = \mu = 0$. Thus, if $m \in \gamma$

$$\{\Re\widetilde{V}\,,\,\Im\widetilde{V}\}(m) < 0. \tag{3.2}$$

Let $\mathcal{R} \subset T^*X\setminus\{0\}$ be the ray through m. $\mathcal{R}$ is a conic isotropic submanifold of T^*X. Associated to any such manifold are spaces of distributions $J^k(X,\mathcal{R})$, the Hermite distributions of Boutet de Monvel and Guillemin [**3**].

LEMMA 3.2. *For every ℓ there exists $\psi \in J^\ell(X,\mathcal{R})$ (with non-zero principal symbol) such that $B(\psi) \in C^\infty(X)$.*

PROOF. The construction is symbolic. The symbol of an element $\psi \in J^\ell(X,\mathcal{R})$ at $m \in \mathcal{R}$ is an object in the space

$$\sigma_m \in \bigwedge^{1/2}\mathcal{R}_m \otimes \mathcal{S}(\Sigma_m),$$

where:

(1) $\mathcal{R}_m = T_m\mathcal{R}$ and $\wedge^{1/2}$ stands for half forms on it,

(2) $\Sigma_m = \mathcal{R}_m^\circ/\mathcal{R}_m$ is the symplectic normal space to $\mathcal{R}_m$ (here $\mathcal{R}_m^\circ$ is the symplectic orthogonal to $\mathcal{R}_m$ inside $T_m S^*X$).
(3) $\mathcal{S}(\Sigma_m)$ is the space of smooth vectors for the representation of the Heisenberg group of Σ_m in a quantization of Σ_m.

Given that $\mathcal{R}$ is one-dimensional and the symbol is homogeneous, the half form portion of the symbol is trivial: It is $dr^{\ell/2}$. The interesting part of the symbol is the vector in $\mathcal{S}(\Sigma_m)$.

We first note that Σ_m is naturally isomorphic with $T_\gamma\mathcal{O}$. The isomorphism arises from the fact that $\omega(\partial_r, \Xi) = 1$, where Ξ is the Hamilton field of $\|\xi\|$ and ∂_r is the radial field on T^*X. Therefore the intersection $\mathcal{R}_m^\circ \cap T_m(S^*X)$ is both naturally isomorphic to Σ_m and to $T_\gamma\mathcal{O}$.

The argument is now identical to that in [**2**], §3. To find the symbol of ψ, we solve the equation $B(\psi) = 0 \mod C^\infty$ using the symbol calculus of Hermite distributions. For any $\psi \in J^\ell(X, \mathcal{R})$ the symbol of $B(\psi)$ as an element in $J^\ell(X, \mathcal{R})$ is zero, because $\sigma_B(m) = 0$. Therefore $B(\psi) \in J^{\ell-1/2}(X, \mathcal{R})$. To find the symbol of $B(\psi)$ as an element of this space we need to use the first transport equation for Hermite distributions (Theorem 10.2 in [**3**]). The symbol of $B(\psi)$ is $\mathcal{L}(\sigma_m)$, where $\mathcal{L} : \mathcal{S}(\Sigma_m) \to \mathcal{S}(\Sigma_m)$ is the operator in the infinitesimal representation of the Heisenberg group of Σ_m associated to the Hamilton field of σ_B at m (which is tangent to $\mathcal{R}$). By Lemma 3.1 in [**2**], the condition (3.2) on the Poisson bracket implies that $\mathcal{L}$ is surjective with non-trivial kernel. The symbol of ψ is chosen in the kernel of $\mathcal{L}$. If $\psi^{(1)} \in J^\ell(X, \mathcal{R})$ has such a symbol, then $B(\psi^{(1)}) \in J^{\ell-1}(X, \mathcal{R})$. We then modify $\psi^{(1)}$ to $\psi^{(2)} = \psi^{(1)} + \chi^{(1)}$ where $\chi^{(1)} \in J^{\ell-1/2}(X, \mathcal{R})$ is chosen so that $B(\psi^{(2)}) \in J^{\ell-3/2}(X, \mathcal{R})$. This will be the case if the symbol, τ, of $\chi^{(1)}$ is such that $\mathcal{L}(\tau)$ cancels the symbol of $B(\psi^{(1)})$. Such a symbol τ exists because the operator $\mathcal{L}$ is surjective. Continuing in this fashion indefinitely (and doing a Borel-type summation) gives the result. □

Let ψ be as in the lemma and $\psi = \sum_{k=1}^\infty \psi_k$ be its decomposition into sums of eigenfunctions corresponding to the eigenvalue clusters. Since $[\Delta, B] = 0$, it follows that

$$B(\psi) = \sum_{k=1}^\infty B(\psi_k)$$

is the cluster decomposition of $B(\psi)$, and therefore

$$\|B(\psi_k)\| = O(k^{-\infty})$$

since $B(\psi)$ is smooth. Since $\psi \in J^\ell(X, \mathcal{R})$, proceeding similarly as in [**2**], one can prove that the sequence of norms $\{\psi_k\}$ has an asymptotic expansion as $k \to \infty$ in decreasing powers of k with non-trivial leading term. It follows that $\{\psi_k\}$ is a pseudo mode for $\Delta + B$ with pseudo-eigenvalues $\lambda_k = \Lambda_k$.

At this point, we should have a pseudo mode for $\widetilde{A}^2 + B$ with ψ_k an eigenfunction of $\widetilde{A}^2$ for each k, with eigenvalue precisely Λ_k.

Of course, the averaging lemma related $\Delta + M_V$ to $\widetilde{A}^2 + B + S$, not to $\widetilde{A}^2 + B$. However:

LEMMA 3.3. $\frac{||S\psi_k||}{||\psi_k||} = \mathcal{O}(k^{-\infty})$.

PROOF. For any $N \in \mathbb{N}$, the operator $S\widetilde{A}^N$ is smoothing (since S is) and hence is bounded on L^2 by some constant C_N. Then $||S\widetilde{A}^N\psi_k|| \leq C_N||\psi_k||$. On the other hand, $||S\widetilde{A}^N\psi_k|| = ||S\Lambda_k^{N/2}\psi_k|| = \Lambda_k^{N/2}||S\psi_k||$. We conclude that $||S\psi_k|| \leq C_N\Lambda_k^{-N/2}||\psi_k||$; however, Λ_k is a quadratic polynomial in k with positive leading coefficient. Since N is arbitrary, this completes the proof. □

COROLLARY 3.4. *The sequence* $\{\psi_k\}$ *is also a pseudo mode for* $\widetilde{A}^2 + B + S$.

PROOF. We may as well assume ψ_k are L^2-normalized; the proof then follows immediately from the previous lemma. □

We now finish the proof of the first part of Theorem 1.3. Let us define $\varphi_k := e^{-F}\psi_k$ where $\{\psi_k\}$ is a pseudo mode as in the previous corollary such that, without loss of generality, $\|\psi_k\| \equiv 1$. Then

$$\|(\Delta + M_V - \Lambda_k I)\varphi_k\| = \|e^{-F}(\widetilde{A}^2 + B + S - \Lambda_k I)(\psi_k)\|;$$

since e^{-F} is a bounded operator and and $\{\psi_k\}$ is a pseudomode for $\tilde{A}^2 + B + S$, this is $\mathcal{O}(k^{-\infty})$. On the other hand, since F is of order (-1),

$$(e^{-F})^* e^{-F} = I + T$$

where T is an operator of order (-1). Therefore

$$\|\varphi_k\|^2 = \|\psi_k\|^2 + \langle T\psi_k, \psi_k\rangle.$$

Since the second term on the right-hand side tends to zero as $k \to \infty$, $\{\varphi_k\}$ is a pseudo mode for $\Delta + M_V$ with pseudo-eigenvalues $\lambda_k = \Lambda_k$. (Recall that we assumed without loss of generality that $\mu = 0$.) This proves the first part of Theorem 1.3.

3.3. Proof of part 2. Let us assume that $\{\Lambda_k + \mu\}$, where $\mu = (\widetilde{V} + q_0)(\gamma)$, is in the asymptotic pseudospectrum, By definition, there exists a pseudo-mode $\{\psi_k \in E_k\}$, a sequence that satisfies $\|\psi_k\| = 1$ for all k and

$$\|(M_V^{\text{ave}} + Q_0 - \mu)(\psi_k)\| = O(k^{-\infty}).$$

Let ψ be the distribution $\psi = \sum_{k=1}^{\infty} \psi_k$. This is clearly non-smooth, while

$$(M_V^{\text{ave}} + Q_0 - \mu I)(\psi) \in C^\infty(X) \tag{3.3}$$

by the Sobolev embedding theorem.

Assume now that

$$\{\Re\widetilde{V} + q_0\,,\, \Im V\}(\gamma) > 0.$$

By Theorem 27.1.11 of [**11**], the operator $M_V^{\text{ave}} + Q_0 - \mu I$ is microlocally subelliptic on the cone over the geodesic $\gamma \subset S^*X$, with loss of 1/2 derivatives. Given (3.3), it follows that the wave-front set of ψ must be disjoint from this cone.

4. Proof of Theorem 1.4

Recall our notation: X is a Zoll manifold, and let A be the operator appearing in (1.5). A is a first-order pseudo-differential operator with symbol $\sigma_A(x, \xi) = \|\xi\|_x$ and spectrum the eigenvalues $k = 0, 1, \ldots$. Let $L^2(X) = \oplus_{k=0}^{\infty} E_k$ be the decomposition of $L^2(X)$ into eigenspaces of A.

Referring to (1.5) note first that, for each k,

$$\mathcal{R}_k = \left\{ \frac{\langle (Q_0 + M_V)\psi, \psi\rangle}{\langle \psi, \psi\rangle}\ ;\ \psi \in E_k \setminus \{0\}; \right\}$$

that is, $\mathcal{R}_k$ is the numerical range of the operator $Q_0 + M_V + R_{-1}$ restricted to the finite-dimensional subspace E_k. It is clear that $\mathcal{R}_k$ is closed for each k and, by the Toeplitz-Hausdorff theorem, $\mathcal{R}_k$ is convex. An elementary argument shows that the limit $\mathcal{R}_\infty$ is also closed and convex.

4.1. Existence of modes with microsupport on geodesics. The purpose of this section is to prove:

PROPOSITION 4.1. *Let X be either a standard S^n or a Zoll surface, and let $\gamma \subset S^*X$ be a geodesic. Then there exist sequences $\{u_k\}$ of functions such that:*

(1) *$\forall k$ $u_k \in E_k$ and $\|u_k\| = 1$,*
(2) *the semi-classical wave front set of the sequence is equal to γ, and*
(3) *For all pseudo-differential operators Q of order zero on X one has*

$$\langle Q(u_k), u_k\rangle = \frac{1}{2\pi}\int_\gamma \sigma_Q\, ds + O(1/\sqrt{k}) \tag{4.1}$$

where σ_Q is the principal symbol of Q.

PROOF. First suppose X is a sphere. Since the construction is $\mathrm{SO}(n+1)$ equivariant, without loss of generality γ corresponds to the intersection γ_0 of $S^n \subset \mathbb{R}^{n+1}$ with the x_1x_2 plane, where $(x_1, \ldots, x_{n+1})$ are the standard coordinates in $\mathbb{R}^{n+1}$. Then one can take

$$u_k(x) = a_k\,(x_1 + ix_2)^k \tag{4.2}$$

where $a_k \in \mathbb{R}$ is chosen so that the L^2 norm of u_k is equal to one. It is known that these functions have the required properties (see Proposition 7.6 in [**7**]).

Now assume that X is a Zoll surface, which we normalize so that its geodesics have length 2π. Then it is known that there exists an invertible Fourier integral operator $U : L^2(X) \to L^2(S^2)$ such that

$$U\,\Delta\,U^{-1} = \Delta_0 + R, \tag{4.3}$$

where Δ_0 is the standard Laplacian on S^2 and R is a pseudo-differential operator of order zero. (See [**15**].) The FIO U is associated to a homogeneous canonical transformation $T : T^*S^2 \setminus \{0\} \to T^*X \setminus \{0\}$ that intertwines the geodesic flows. Applying once again the averaging method (more precisely Lemma 1 in [**8**]) to $\Delta_0 + R$, one can assume without loss of generality that $[\Delta_0, R] = 0$. (The result cited says that any operator of the form $\Delta_0 + R$ with R self-adjoint pseudo-differential of order zero can be conjugated, this time by a unitary pseudo-differential operator, to an operator of the same form but now such that $[\Delta_0, R] = 0$.) For each k, the operator U maps the space of spherical harmonics of degree k onto E_k.

Pre-composing U with a rotation, we can also assume that T maps the geodesic γ_0 to γ. We then define u_k to be the result of applying the operator U to the right-hand side of (4.2). Since the desired properties are equivariant with respect to actions of unitary FIOs, we are done. □

This existence result immediately implies:

COROLLARY 4.2. *The image of $\widetilde{V} + q_0$, and its convex hull, are contained in $\mathcal{R}_\infty$.*

PROOF. Given $\gamma \in \mathcal{O}$, let $\{u_k\}$ be as in the previous Proposition. By (4.1), the limit of the matrix coefficients

$$\langle (Q_0 + M_V)u_k, u_k\rangle \in \mathcal{R}_k$$

as $k \to \infty$ is precisely $(q_0 + \widetilde{V})(\gamma)$. □

4.2. The converse. We now prove that $\mathcal{R}_\infty$ is contained in the convex hull of the image of $\widetilde{V}+q_0$. (The proof is modeled on the proof of [**2**, Prop. 5.2].) To show this, we will show that for all lines in $\mathbb{C}$ such that one of the half planes cut out by the line contains the range of $\widetilde{V}+q_0$, every element of $\mathcal{R}_\infty$ is contained in the same half plane. This will show that $\mathcal{R}_\infty$ is contained in, hence equal to, the convex hull in question.

We will use the following Lemma (closely related to the sharp Gårding inequality):

LEMMA 4.3. *Let Q be a zeroth order, self-adjoint pseudo-differential operator on S^n. Assume that its principal symbol $q \in C^\infty(T^*S^n \setminus \{0\})$ is non-negative: $q \geq 0$. Let $\{\psi_k \in E_k\}$ be a sequence of spherical harmonics such that $\|\psi_k\| = 1$ for all k. Assume furthermore that*

$$\lim_{k\to\infty} \langle Q(\psi_k), \psi_k\rangle = \ell \in \mathbb{R}.$$

Then $\ell \geq 0$.

PROOF. By the sharp Gårding inequality (see [**10**], Exercise 4.9), there exists a pseudo-differential operator B of order -1 such that $\forall u \in L^2(S^n)$

$$\langle Q(u), u\rangle \geq -\langle B(u), u\rangle.$$

As before, let A be the operator which is equal to multiplication by k when restricted to E_k. Recall that A is first-order pseudo-differential, so that BA is of order zero and therefore bounded in L^2. Therefore $\exists C > 0$ such that, with ψ_k as in the hypotheses of the Lemma, $\forall k$

$$k\,|\langle B(\psi_k), \psi_k\rangle| = |\langle BA(\psi_k), \psi_k\rangle| \leq C.$$

Therefore $\forall k$

$$\langle Q(\psi_k), \psi_k\rangle \geq -\frac{C}{k}.$$

Taking limits as $k \to \infty$ yields the desired result. □

To finish the proof of the converse, consider a line

$$\{u + iv \in \mathbb{C}\ ;\ au + bv = c,\ a, b, c \in \mathbb{R}\}$$

in $\mathbb{C}$, and assume that the range of $\widetilde{V}+q_0$ is contained in the region $au + bv \geq c$. Let $u_0 + iv_0 \in \mathcal{R}_\infty$. We need to show that $au_0 + bv_0 \geq c$. By definition of $\mathcal{R}_\infty$, there exists a sequence $\{\psi_k \in E_k\}$ such that $\forall k$ $\|\psi_k\| = 1$ and such that the sequence of complex numbers

$$u_k + iv_k := \langle (Q_0 + M_V^{\mathrm{ave}})\psi_k, \psi_k\rangle$$

converges to $u_0 + iv_0$ as k goes to infinity. Consider now the operator

$$Q = a\,(Q_0 + M_{\Re V}^{\mathrm{ave}}) + bM_{\Im V}^{\mathrm{ave}} - cI. \tag{4.4}$$

This is a self-adjoint pseudo-differential operator of order zero with non-negative symbol, and

$$\langle Q(\psi_k), \psi_k\rangle = au_k + bv_k - c \xrightarrow[k\to\infty]{} au_0 + bv_0 - c. \tag{4.5}$$

By the Lemma this limit is non-negative, and the proof is complete.

5. Examples on S^2

5.1. Preliminaries on the averaging operator. We present here preliminary results on the operator $V \mapsto \widetilde{V}$ on the two sphere $S^2 \subset \mathbb{R}^3$ (see also the appendix in [**6**]). The space, $\mathcal{O}$, of oriented geodesics on S^2 can be identified with a copy of S^2. Indeed an oriented great circle on S^2 is the intersection of S^2 with an oriented plane through the origin:

$$\gamma = S^2 \cap \pi_\gamma.$$

The identification $\mathcal{O} \cong S^2$ is via the map $\gamma \mapsto$ the unit normal vector to π_γ defining the orientation. Clearly this identification is equivariant with respect to the action of SO(3). We can therefore regard the averaging operator as an operator on $C^\infty(S^2)$:

$$\begin{array}{ccc} S^2 & \xrightarrow{ave} & S^2 \\ V & \mapsto & \widetilde{V} \end{array}, \tag{5.1}$$

where $\widetilde{V}(\gamma) = \frac{1}{2\pi}\int_\gamma V\, ds$. This operator is obviously SO(3) equivariant, and therefore it maps each space of spherical harmonics, E_k, into itself. Furthermore, by Schur's lemma, it is a constant c_k times the identity on E_k.

To proceed further, let (x, y, z) denote the ambient ($\mathbb{R}^3$) coordinates and let $\zeta = x + iy$. Note that, for each k,

$$\zeta^k \in E_k \tag{5.2}$$

(we will abuse the notation and denote the restrictions of x, y, z to the sphere by the same letters). Indeed $(x+iy)^k$ is a homogeneous polynomial of degree k and it is harmonic, by the Cauchy-Riemann equations. Following the identification $\mathcal{O} \cong S^2$ one finds that

$$c_k = \widetilde{\zeta^k}(1,0,0) = \frac{1}{2\pi}\int_0^{2\pi} (i\cos(t))^k\, dt = \begin{cases} 0 & \text{if } k \text{ is odd} \\ (-1)^l \frac{(2l-1)(2l-3)\cdots 3\cdot 1}{(2l)(2l-2)\cdots 4\cdot 2} & \text{if } k = 2l \end{cases}$$

Using Wallis' formula for π it follows that $c_{2l} \sim (-1)^l (l\pi)^{-1/2}$.

5.2. Analytic functions. Our first example is actually a class of examples, namely potentials of the form

$$V(x,y,z) = f(\zeta), \quad f(\zeta) = \sum_{l=1}^{\infty} a_l \zeta^{2l}$$

where the series is assumed to have a radius of convergence > 1. In §5 of [**9**] it was shown that the eigenvalues of $\Delta + M_V$ are exactly the same as those of Δ (note that there is no constant term in the series), the intuition being that the average of the operator M_V, restricted to E_k, is nilpotent for each k.

To see what Theorem 1.3 says about the asymptotic pseudospectrum in this example, we need to investigate the Poisson bracket condition on $\widetilde{V}$. Note that, by the previous discussion,

$$\widetilde{V} = \widetilde{f}(\zeta) = \sum_{l=1}^{\infty} a_l\, c_{2l} \zeta^{2l}$$

which shows that $\widetilde{V}$ is of the same form as V.

LEMMA 5.1. *Let $\widetilde{V} = F + iG$ with F and G real-valued. Then*

$$\{F\,,\,G\} = z\left(F_x^2 + F_y^2\right). \tag{5.3}$$

PROOF. Denote by (θ, z) toric coordinates on $\mathcal{O} \cong S^2$, so that the symplectic form is $dz \wedge d\theta$. Then

$$\{F\,,\,G\} = F_z G_\theta - F_\theta G_z =$$

$$= (x_z F_x + y_z F_y)(x_\theta G_x + y_\theta G_y) - (x_\theta F_x + y_\theta F_y)(x_z G_x + y_z G_y).$$

Using the Cauchy-Riemann equations and simplifying one obtains

$$\{F\,,\,G\} = (x_z F_x + y_z F_y)(-x_\theta F_y + y_\theta F_x) - (x_\theta F_x + y_\theta F_y)(-x_z F_y + y_z F_x)$$

$$= (x_z y_\theta - x_\theta y_z)\,(F_x^2 + F_y^2).$$

But $\{x\,,\,y\} = z$. $\square$

As the pre-image under $\widetilde{V}$ of any complex number contains points whose z coordinates differ by a sign, we get:

COROLLARY 5.2. *For each $\mu \in \{\tilde{f}(\zeta)\,;\,|\zeta| < 1$ and $\tilde{f}'(\zeta) \neq 0\}$ the sequence $\{\Lambda_k + \mu\}$ is in the asymptotic pseudospectrum of $\Delta + M_V$.*

5.3. Quadratic examples. In this section we take V of the form $V = (ax + iy)^2$, where a is a real constant not equal to either zero or one. If $\widetilde{\Delta}$ denotes the (negative) Laplacian on $\mathbb{R}^3$, then $\widetilde{\Delta} V = 2(a^2 - 1)$. It follows that

$$h = (ax + iy)^2 + \frac{1}{3}(1 - a^2)(x^2 + y^2 + z^2)$$

is a harmonic, homogeneous polynomial on $\mathbb{R}^3$. Therefore, the decomposition of V into spherical harmonics is

$$V|_{S^2} = h|_{S^2} - \frac{1}{3}(1 - a^2).$$

This allows us to compute $\widetilde{V}$, which is, up to constants, basically V itself:

$$\widetilde{V} = -\frac{1}{2}h - \frac{1}{3}(1 - a^2) \quad \Rightarrow \quad \widetilde{V} = -\frac{1}{2}V - \frac{1}{2}(1 - a^2).$$

Therefore

$$\{\Re\widetilde{V}\,,\,\Im\widetilde{V}\} = \frac{1}{4}\{a^2x^2 - y^2\,,\,2axy\} = \frac{a}{2}\left(\{a^2x^2\,,\,xy\} - \{y^2\,,\,xy\}\right) =$$

$$= \frac{a}{2}\left(2a^2x^2z - 2y^2(-z)\right) = az\left(a^2x^2 + y^2\right).$$

COROLLARY 5.3. *For each $\mu \in \{\widetilde{V}(\zeta)\,;\,0 < |\zeta| < 1\}$ the sequence $\{\Lambda_k + \mu\}$ is in the asymptotic pseudospectrum of $\Delta + M_{(ax+iy)^2}$.*

References

[1] Arthur L. Besse, *Manifolds all of whose geodesics are closed*, Ergebnisse der Mathematik und ihrer Grenzgebiete [Results in Mathematics and Related Areas], vol. 93, Springer-Verlag, Berlin-New York, 1978. With appendices by D. B. A. Epstein, J.-P. Bourguignon, L. Bérard-Bergery, M. Berger and J. L. Kazdan. MR496885

[2] D. Borthwick and A. Uribe, *On the pseudospectra of Berezin-Toeplitz operators*, Methods Appl. Anal. **10** (2003), no. 1, 31–65, DOI 10.4310/MAA.2003.v10.n1.a3. MR2014161

[3] L. Boutet de Monvel and V. Guillemin, *The spectral theory of Toeplitz operators*, Annals of Mathematics Studies, vol. 99, Princeton University Press, Princeton, NJ; University of Tokyo Press, Tokyo, 1981, DOI 10.1515/9781400881444. MR620794

[4] Yves Colin de Verdière, *Sur le spectre des opérateurs elliptiques à bicaractéristiques toutes périodiques* (French), Comment. Math. Helv. **54** (1979), no. 3, 508–522, DOI 10.1007/BF02566290. MR543346

[5] I. C. Gohberg and M. G. Kreĭn, *Introduction to the theory of linear nonselfadjoint operators*, Translations of Mathematical Monographs, Vol. 18, American Mathematical Society, Providence, R.I., 1969. Translated from the Russian by A. Feinstein. MR0246142

[6] Victor Guillemin, *The Radon transform on Zoll surfaces*, Advances in Math. **22** (1976), no. 1, 85–119, DOI 10.1016/0001-8708(76)90139-0. MR426063

[7] Victor Guillemin, *Some microlocal aspects of analysis on compact symmetric spaces*, Seminar on Microlocal Analysis, Ann. of Math. Stud., vol. 93, Princeton Univ. Press, Princeton, N.J., 1979, pp. 79–111. MR560313

[8] Victor Guillemin, *Band asymptotics in two dimensions*, Adv. in Math. **42** (1981), no. 3, 248–282, DOI 10.1016/0001-8708(81)90042-6. MR642393

[9] V. Guillemin and A. Uribe, *Spectral properties of a certain class of complex potentials*, Trans. Amer. Math. Soc. **279** (1983), no. 2, 759–771, DOI 10.2307/1999566. MR709582

[10] Alain Grigis and Johannes Sjöstrand, *Microlocal analysis for differential operators*, London Mathematical Society Lecture Note Series, vol. 196, Cambridge University Press, Cambridge, 1994. An introduction, DOI 10.1017/CBO9780511721441. MR1269107

[11] Lars Hörmander, *The analysis of linear partial differential operators. II*, Classics in Mathematics, Springer-Verlag, Berlin, 2005. Differential operators with constant coefficients; Reprint of the 1983 original, DOI 10.1007/b138375. MR2108588

[12] Tosio Kato, *Perturbation theory for linear operators*, Classics in Mathematics, Springer-Verlag, Berlin, 1995. Reprint of the 1980 edition. MR1335452

[13] Karel Pravda-Starov, *A general result about the pseudo-spectrum of Schrödinger operators*, Proc. R. Soc. Lond. Ser. A Math. Phys. Eng. Sci. **460** (2004), no. 2042, 471–477, DOI 10.1098/rspa.2003.1194. MR2034648

[14] Alejandro Uribe, *Band invariants and closed trajectories on S^n*, Adv. in Math. **58** (1985), no. 3, 285–299, DOI 10.1016/0001-8708(85)90120-3. MR815359

[15] Alan Weinstein, *Fourier integral operators, quantization, and the spectra of Riemannian manifolds* (English, with French summary), Géométrie symplectique et physique mathématique (Colloq. Internat. CNRS, No. 237, Aix-en-Provence, 1974), Éditions Centre Nat. Recherche Sci., Paris, 1975, pp. 289–298. With questions by W. Klingenberg and K. Bleuler and replies by the author. MR0650990

[16] Alan Weinstein, *Asymptotics of eigenvalue clusters for the Laplacian plus a potential*, Duke Math. J. **44** (1977), no. 4, 883–892. MR482878

[17] Steve Zelditch, *Fine structure of Zoll spectra*, J. Funct. Anal. **143** (1997), no. 2, 415–460, DOI 10.1006/jfan.1996.2981. MR1428823

[18] Maciej Zworski, *Semiclassical analysis*, Graduate Studies in Mathematics, vol. 138, American Mathematical Society, Providence, RI, 2012, DOI 10.1090/gsm/138. MR2952218

Mathematics Department, DePaul University, Chicago, Illinois 60604
Email address: `dsher@depaul.edu`

Mathematics Department, University of Michigan, Ann Arbor, Michigan 48109
Email address: `uribe@umich.edu`

Instituto de Matemáticas, UNAM, Unidad Cuernavaca
Email address: `villegas@matcuer.unam.mx`

Contemporary Mathematics
Volume **775**, 2021
https://doi.org/10.1090/conm/775/15598

On a conjecture of Naito-Sagaki: Littelmann paths and Littlewood-Richardson Sundaram tableaux

Jacinta Torres

ABSTRACT. In recent work with Schumann we have proven a conjecture of Naito-Sagaki giving a branching rule for the decomposition of the restriction of an irreducible representation of the special linear Lie algebra to the symplectic Lie algebra, therein embedded as the fixed-point set of the involution obtained by the folding of the corresponding Dyinkin diagram. It provides a new approach to branching rules for non-Levi subalgebras in terms of Littelmann paths. In this paper we motivate this result, provide examples, and give an overview of the combinatorics involved in its proof.

Introduction

Let $\mathfrak{g}$ be a semi-simple Lie algebra over the field of complex numbers and $\sigma : \mathfrak{g} \to \mathfrak{g}$ an automorphism of finite order. In [**8**], Kac has shown that σ is conjugate to the automorphism induced by a Dynkin diagram automorphism, and that the set of fixed points $\mathfrak{g}^{\sigma}$ is again a semi-simple Lie algebra, of one of six possible types. The basic problem of decomposing the restriction of an irreducible representation of $\mathfrak{g}$ to $\mathfrak{g}^{\sigma}$ (known as a "branching rule") is, in general, open.

If, however, the automorphism σ is semi-simple and has infinite order, then $\mathfrak{g}^{\sigma}$ is a Levi subalgebra [**20**]. In this case, there exist natural and beautiful branching rules in terms of Littelmann paths [**10**]. Littelmann paths are generalisations of Young tableaux to all Kac-Moody Lie algebras, and were used [**10**] to give a remarkable generalisation of the Littelwood-Richardson rule for the decomposition of the tensor product of two irreducible finite-dimensional representations.

In [**12**] a new approach to the branching problem was suggested by Naito-Sagaki in terms of Littelmann paths for $\mathfrak{g} = \mathfrak{sl}(2n, \mathbb{C})$ and $\mathfrak{g}^{\sigma} = \mathfrak{sp}(2n, \mathbb{C})$ in the form of a conjecture. In this paper we reformulate the conjecture in terms of symplectic Littelwood-Richardson tableaux, which were previously studied by Sundaram in [**16**]. We call them Littlewood-Richardson Sundaram tableaux. We give a proof of the conjecture in some cases using this reformulation, which was used in [**15**] to give a general proof. Our methods imply a bijection between lattice points of polytopes, in the spirit of Berenstein-Zelevinsky and Knutson-Tao [**2**] [**9**].

2020 *Mathematics Subject Classification.* Primary 17B10; Secondary 20C30.
Key words and phrases. Representation theory, algebraic combinatorics.

1. Notation for the Lie algebras

Let $\mathfrak{h} \subset \mathfrak{sl}(2n,\mathbb{C})$ be the Cartan sub-algebra of diagonal matrices. Let ε_i be the linear map $\mathfrak{h} \to \mathbb{C}$ defined by $\mathrm{diag}(a_1, \cdots, a_{2n}) \mapsto a_i$. We write $\mathfrak{sl}(2n,\mathbb{C}) = \langle x_i, y_i, h_i \rangle_{i\in\{1,\cdots,2n-1\}}$ where $h_i = \mathrm{E}_{ii} - \mathrm{E}_{i+1,i+1}$ and where x_i and y_i are the Chevalley generators corresponding to the simple root $\alpha_i := \varepsilon_i - \varepsilon_{i+1}$. The automorphism σ is given by

$$\sigma(x_i) = x_{2n-i},$$
$$\sigma(y_i) = y_{2n-i}, \text{ and}$$
$$\sigma(h_i) = h_{2n-i}.$$

The fixed point set $\mathfrak{g}^\sigma$ is generated as a Lie algebra by $\{\hat{x}_i, \hat{y}_i, \hat{h}_i | \ i \in \{1,\cdots,n\}\}$ (see Proposition 7.9 in [**8**]), where

$$\hat{x}_i = \begin{cases} x_i + x_{2n-i} & \text{if } i \in [1,n) \\ x_n & \text{if } i = n \end{cases}$$

$$\hat{y}_i = \begin{cases} y_i + y_{2n-i} & \text{if } i \in [1,n) \\ y_n & \text{if } i = n \end{cases}$$

$$\hat{h}_i = \begin{cases} h_i + h_{\alpha_{2n-i}} & \text{if } i \in [1,n) \\ h_n & \text{if } i = n. \end{cases}$$

This Lie algebra is isomorphic to $\mathfrak{sp}(2n,\mathbb{C})$ (see Proposition 7.9 in [**8**]) and

$$\mathfrak{h}^\sigma = \bigoplus_{i=1}^{n} \hat{h}_i = \mathfrak{h} \cap \mathfrak{g}^\sigma \subset \mathfrak{h}$$

is a Cartan subalgebra. Let $\mathrm{P}_{C_n} \subset (\mathfrak{h}^\sigma)^*$ be the set of integral weights of $\mathfrak{g}^\sigma$ with respect to $\mathfrak{h}^\sigma$, and let $\mathrm{P}_{A_{2n-1}} \subset \mathfrak{h}^*$ be the set of integral weights of $\mathfrak{g}$ with respect to $\mathfrak{h}$.

To avoid confusion we will denote elements of $\mathrm{P}_{A_{2n-1}}$ by λ and elements of P_{C_n} by $\hat{\lambda}$; in particular, the fundamental weights in $\mathrm{P}_{A_{2n-1}}$ will be denoted by $\omega_1, \cdots, \omega_{2n-1}$, and the fundamental weights in P_{C_n} by $\hat{\omega}_1, \cdots, \hat{\omega}_n$. We can then write

$$\mathrm{P}_{A_{2n-1}} = \bigoplus_{i=1}^{2n-1} \mathbb{Z}\omega_i$$

and

$$\mathrm{P}_{C_{2n}} = \bigoplus_{i=1}^{n} \mathbb{Z}\hat{\omega}_i,$$

where the direct sums are as $\mathbb{Z}$-modules. Also, for $\lambda = a_1\omega_1 + \cdots + a_n\omega_n$ we write $\hat{\lambda} = a_1\hat{\omega}_1 + \cdots + a_n\hat{\omega}_n$ for the corresponding element in P_{C_n}. Weights with non-negative coefficients are called **dominant**. We will denote the corresponding sets by $\mathrm{P}^+_{A_{2n-1}}$ and $\mathrm{P}^+_{C_n}$. We will consider the real vector spaces $\mathfrak{h}_\mathbb{R}$ and $\mathfrak{h}^\sigma_\mathbb{R}$ spanned by the fundamental weights $\omega_1, \cdots, \omega_{2n-1}$ respectively $\hat{\omega}_1, \cdots, \hat{\omega}_n$. The dominant Weyl chamber is the convex hull of the dominant weights in $\mathfrak{h}^*_\mathbb{R}$ (respectively $(\mathfrak{h}^\sigma)^*_\mathbb{R}$).

2. Tableaux, words and their paths

A **shape** is a finite sequence of non-negative integers $\underline{d} = (d_1, \cdots, d_k)$. The Young diagram of shape $\underline{d}$ consists of $d_1 + \cdots + d_k$ top-left aligned columns such that the first d_1 columns (from right to left) have one box, and the first d_s boxes after the $(d_1 + \cdots + d_{s-1})$-th column have d_s boxes. To a dominant integral weight $\lambda = a_1\omega_1 + \cdots + a_{2n-1}\omega_{2n-1} \in \mathrm{P}^+_{A_{2n-1}}$ we assign the shape $\underline{d}_\lambda = (a_1, \cdots, a_{2n-1})$.

EXAMPLE 2.1. For $\lambda = 3\omega_1 + \omega_2$ we have $\underline{d}_\lambda = (3,1)$. To it is associated the following Young diagram:

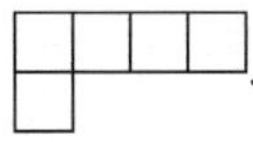.

A **semi-standard tableau** of shape $\underline{d}$ is a filling of the Young diagram of shape $\underline{d}$ with letters from either one of the ordered alphabets

$$\mathscr{A}_{2n} = \{1 < \cdots < 2n\}$$

or

$$\mathscr{C}_n = \{1 < \cdots < n < \overline{n} < \cdots < \overline{1}\}$$

such that entries are strictly increasing downwards along each column of boxes and weakly increasing along each row from left to right. If the entries of a given semi-standard tableau belong to $\mathscr{A}_{2n}$ we will call it a **semi-standard Young tableau**, and if they belong to $\mathscr{C}_n$, we will call it a **symplectic semi-standard tableau**. For a dominant integral weight $\lambda \in \mathrm{P}^+_{A_{2n-1}}$, we will denote the set of semi-standard Young tableaux of shape $\underline{d}_\lambda$ by $\mathrm{SSYT}_{A_{2n}}(\lambda)$

EXAMPLE 2.2. A semi-standard Young tableau of shape $(3,1)$:

$$\begin{array}{|c|c|c|c|}\hline 1 & 1 & 2 & 5 \\ \hline 2 & \multicolumn{3}{c}{} \\ \cline{1-1}\end{array}.$$

Let $\mathscr{W}(\mathscr{A}_{2n})$ denote the word monoid on $\mathscr{A}_{2n}$ and $\mathscr{W}(\mathscr{C}_n)$ be the word monoid on $\mathscr{C}_n$. The **word** $w(\mathscr{T})$ of a semi-standard tableau $\mathscr{T}$ is obtained by reading its rows, from right to left and top to bottom.

EXAMPLE 2.3. The symplectic semi-standard tableau $\mathscr{T} = \begin{array}{|c|c|c|c|}\hline 1 & 1 & 2 & 5 \\ \hline \overline{3} & \multicolumn{3}{c}{} \\ \cline{1-1}\end{array}$ has word $w(\mathscr{T}) = 5211\overline{3}$.

3. Restriction of paths that come from words

3.1. Paths and their restrictions. We will consider paths to be continuous maps $\pi : [0,1] \to \mathfrak{h}^*_{\mathbb{R}}, \pi' : [0,1] \to (\mathfrak{h}^\sigma_{\mathbb{R}})^*$ starting at the origin and ending at an integral weight:

$$\pi(0) = 0 = \pi'(0), \quad \pi(1) \in \mathrm{P}_{A_{2n-1}}, \quad \pi'(1) \in \mathrm{P}_{C_n}.$$

The map $\mathfrak{h}^* \to (\mathfrak{h}^\sigma)^*, \varphi \mapsto \varphi|_{\mathfrak{h}^\sigma}$ induces a map $\mathrm{res}' : \mathrm{P}_{A_{2n-1}} \to \mathrm{P}_{C_n}$. Given a path $\pi : [0,1] \to \mathfrak{h}^*_{\mathbb{R}}$, we define a restricted path $\mathrm{res}(\pi)$ by $\mathrm{res}(\pi) : [0,1] \to (\mathfrak{h}^\sigma_{\mathbb{R}})^*, t \mapsto \mathrm{res}'(\pi(t))$.

We will also consider the concatenation $\pi_2 * \pi_1$ of two paths π_1 and π_2 with the same codomain: it is the path obtained by translating π_2 to the endpoint $\pi_1(1)$ of π_1. A path is **dominant** if its image is contained in the dominant Weyl chamber.

3.2. Paths that come from words. In the first part of this section we follow Section 2 of [**12**]. Let $w = w_1 \cdots w_k$ be a word, either in $\mathscr{W}(\mathscr{A}_{2n})$ or in $\mathscr{W}(\mathscr{C}_n)$. To it we assign the path:

$$\pi_w = \pi_{w_k} * \cdots * \pi_{w_1},$$

where, for $w_i \in \mathscr{A}_{2n}$ (respectively $w_i \in \mathscr{C}_n$), the path $\pi_{w_i} : [0,1] \to \mathfrak{h}_{\mathbb{R}}^*$ (respectively $\pi_{w_i} : [0,1] \to (\mathfrak{h}_{\mathbb{R}}^{\sigma})^*$) is given by $t \mapsto t\varepsilon_{w_i}$, where we define $\varepsilon_{\overline{l}} := -\varepsilon_l$ for $1 \le l \le n$. Also, in general, for paths $\pi_1, \cdots, \pi_k : [0,1] \to \mathfrak{h}_{\mathbb{R}}^*$, we have

$$\operatorname{res}(\pi_1 * \cdots * \pi_k) = \operatorname{res}(\pi_1) * \cdots * \operatorname{res}(\pi_k). \tag{3.1}$$

Set $\hat{\varepsilon}_i = \operatorname{res}'(\varepsilon_i)$ for $i \in \{1, \cdots, 2n\}$. Then, for $i \in \{1, \cdots, 2n\}$ and $j \in \{1, \cdots, n\}$ we have

$$\hat{\varepsilon}_i(\hat{h}_j) = \begin{cases} 1 & \text{if } i \in \{j, 2n-j\} \\ -1 & \text{if } i \in \{j+1, 2n-j+1\} \\ 0 & \text{otherwise.} \end{cases} \tag{3.2}$$

Therefore $\hat{\varepsilon}_i = -\hat{\varepsilon}_{2n-i+1}$, which means we can describe $\operatorname{res}(\pi_w)$ in the following simple way for $w \in \mathscr{W}(\mathscr{A}_{2n})$: First obtain from w a word $\operatorname{res}(w)$ in the alphabet $\mathscr{C}_n$ by replacing a letter w_i in w with $\overline{2n - w_i + 1}$ if $n < w_i \le 2n$. All other letters stay the same. The corresponding path $\pi_{\operatorname{res}(w)}$ coincides with $\operatorname{res}(\pi_w)$ by (3.2) and (3.1).

EXAMPLE 3.1. Let $n = 2$ and $w = 121223341$. Then $\operatorname{res}(w) = 12122\overline{2}\,\overline{2}\,\overline{1}1$.

4. The Naito-Sagaki conjecture

Let $\lambda \in \mathrm{P}^+_{A_{2n-1}}$ be dominant and let $\mathrm{L}(\lambda)$ be the associated simple module for $\mathfrak{sl}(2n, \mathbb{C})$. Recall the set $\mathrm{SSYT}_{\mathcal{A}_{2n}}(\lambda)$ of semi-standard Young tableaux of shape $\underline{d}_\lambda$. Let

$$\operatorname{domres}(\lambda) = \{\eta = \operatorname{res}(\pi_{w(\mathrm{T})}) \text{ dominant} : \mathrm{T} \in \mathrm{SSYT}_{\mathcal{A}_{2n}}(\lambda)\}$$

be the set of restricted paths associated to words of elements of $\mathrm{SSYT}_{\mathcal{A}_{2n}}(\lambda)$ that are dominant, and for $\nu \in \mathrm{P}^+_{\mathrm{C}_{2n}}$, let

$$\operatorname{domres}(\lambda, \nu) = \{\delta \in \operatorname{domres}(\lambda) : \delta(1) = \nu\}.$$

EXAMPLE 4.1. Let $n = 2$ and $\lambda = \omega_1 + \omega_2$. Then

$$\operatorname{domres}(\lambda) = \left\{\operatorname{res}(\pi_{w(\mathrm{T})}) | \mathrm{T} \in \left\{\begin{array}{|c|c|}\hline 1 & 1 \\ \hline 2 \\ \cline{1-1}\end{array}, \begin{array}{|c|c|}\hline 1 & 1 \\ \hline 3 \\ \cline{1-1}\end{array}\right\}\right\}$$

$$\operatorname{domres}(\lambda, \lambda) = \left\{\operatorname{res}(\pi_{w(\mathrm{T})}) | \mathrm{T} \in \left\{\begin{array}{|c|c|}\hline 1 & 1 \\ \hline 2 \\ \cline{1-1}\end{array}\right\}\right\}$$

$$\operatorname{domres}(\lambda, \omega_1) = \left\{\operatorname{res}(\pi_{w(\mathrm{T})}) | \mathrm{T} \in \left\{\begin{array}{|c|c|}\hline 1 & 1 \\ \hline 3 \\ \cline{1-1}\end{array}\right\}\right\}.$$

THEOREM 4.2. *(S-T., [**15**]) Let $\lambda \in \mathrm{P}^+_{A_{2n-1}}$ be dominant, and let $\mathrm{L}(\lambda)$ be the associated simple module for $\mathfrak{sl}(2n, \mathbb{C})$. Then*

$$\operatorname{res}^{\mathfrak{g}}_{\mathfrak{g}^\sigma}(\mathrm{L}(\lambda)) = \bigoplus_{\delta \in \operatorname{domres}(\lambda)} \mathrm{L}(\delta(1)).$$

EXAMPLE 4.3. Let $n = 2$ and $\lambda = \omega_1 + \omega_2$ as in Example 4.1. Then

$$\operatorname{res}^{\mathfrak{g}}_{\mathfrak{g}^\sigma}(\mathrm{L}(\lambda)) = \mathrm{L}(\hat{\omega}_1) \oplus \mathrm{L}(\hat{\omega}_1 + \hat{\omega}_2).$$

REMARK 4.4. Theorem 4.2 was conjectured by Naito-Sagaki in [**12**] in 2005. In [**12**] it is stated for $\mathrm{L}(\lambda)$ a representation of $\mathfrak{gl}(2n,\mathbb{C})$ for λ non-negative and dominant. However, the representation of $\mathfrak{gl}(2n,\mathbb{C})$ induced by an irreducible representation of $\mathfrak{sl}(2n,\mathbb{C})$ has the same highest weight and restricts back to itself. See §15.3 in [**5**].

5. Littlewood-Richardson tableaux and n-symplectic Sundaram tableaux: branching

DEFINITION 5.1. Let $\lambda,\nu \in \mathrm{P}_{A_{2n-1}}$ be two dominant weights such that $\underline{d}_\nu \subset \underline{d}_\lambda$ (this means that one shape is contained in the other when aligned with respect to their top left corners. In Example 5.5 below we see that $\underline{d}_\nu \subset \underline{d}_\lambda$). A tableau $\mathscr{T}$ of skew shape λ/ν is a filling of the Young diagram of shape λ leaving the boxes that belong to $\nu \subset \lambda$ blank, with the others having entries in the alphabet $\mathscr{A}_{2n}$, and such that these entries are strictly increasing along the columns and weakly increasing along the rows. The word $w(\mathscr{T})$ of $\mathscr{T}$ is obtained just as for semi-standard Young tableaux, reading from right to left and from top to bottom, ignoring the blank boxes.

A shape $\underline{d} = (d_1,\cdots,d_k)$ is **even** if $d_i = 0$ unless i is an even number, for $i \in \{1,\cdots,k\}$. Also, for a shape $\underline{d}$ define $l(\underline{d})$ to be the length of the longest column of the associated Young diagram.

DEFINITION 5.2. Let $\lambda,\nu,\eta \in \mathrm{P}^+_{A_{2n-1}}$ be dominant weights such that the shapes $\underline{d}_\nu$ and $\underline{d}_\eta$ are contained in the shape $\underline{d}_\lambda$ of λ. A **Littlewood-Richardson** (respectively **Littlewood-Richardson Sundaram** or **n-symplectic Sundaram**) tableau of skew shape λ/ν and weight η is a tableau of skew shape λ/ν that is semi-standard, and has a dominant word of weight η (respectively $\underline{d}_\eta$ is even and $2i+1$ does not appear strictly below row $n+i$ for $i \in \{0,1,\cdots,\frac{1}{2}l(\underline{d}_\eta)\}$). Here a word $w \in \mathscr{W}(\mathscr{A}_{2n})$ is dominant if the path π_w is dominant. We will denote these skew tableaux by $\mathrm{LR}(\lambda/\nu,\eta)$ (respectively $\mathrm{LRS}(\lambda/\nu,\eta)$). Littlewood-Richardson Sundaram tableaux were introduced by Sundaram in [**16**], as **n-symplectic** tableaux.

REMARK 5.3. Note that if $l(\lambda) \le n$ (such weights are called **stable**) then $\mathrm{LRS}(\lambda/\nu,\eta) = \mathrm{LR}(\lambda/\nu,\eta)$.

REMARK 5.4. If λ is stable and $\mathscr{T}$ is a Littlewood-Richardson tableau of skew shape λ/ν then its entries belong to the set $\{1,\cdots,n\}$. This is because if, say, k appears in row l_k of $\mathscr{T}$, then, since the word of $\mathscr{T}$ is dominant, a $k-1$ must appear either directly above k in the same column, or in a column to the right, and since $\mathscr{T}$ is semi-standard, it appears in at most row $l_k - 1$.

EXAMPLE 5.5. The tableau $\mathscr{L} = \begin{array}{|c|c|c|}\hline & 1 & 1 \\ \hline & 2 \\ \cline{1-2} 2 \\ \cline{1-1}\end{array}$ is a Littlewood-Richardson tableau of skew shape λ/ν and weight η for $\lambda = \omega_1+\omega_2+\omega_3, \nu = \omega_2$, and $\eta = 2\omega_2$ and the tableau $\mathscr{T} = \begin{array}{|c|}\hline \ \\ \hline \ \\ \hline 1 \\ \hline\end{array}$ is a Littlewood-Richardson tableau of skew shape λ'/ν' and weight η' for $\lambda' = \omega_3, \nu' = \omega_2$, and $\eta' = \omega_1$. Notice that $\mathscr{L}$ is 2-symplectic Sundaram while $\mathscr{T}$ is not.

DEFINITION 5.6. The Littlewood-Richardson coefficient is defined as the number $c^{\lambda}_{\nu,\eta} \in \mathbb{Z}_{\geq 0}$ such that

$$\mathrm{L}(\nu) \otimes \mathrm{L}(\eta) = \bigoplus_{\underline{d}_\nu \subset \underline{d}_\lambda} c^{\lambda}_{\nu,\eta} \mathrm{L}(\lambda)$$

where $\mathrm{L}(\lambda), \mathrm{L}(\nu)$, and $\mathrm{L}(\eta)$ are all representations of $\mathfrak{sl}(2n, \mathbb{C})$.

Theorem 5.7 below is known as the **Littlewood-Richardson rule**. It was first stated in 1943 by Littlewood and Richardson.

THEOREM 5.7. [**4**] *The Littlewood-Richardson coefficients are obtained by counting Littlewood-Richardson tableaux:*

$$c^{\lambda}_{\nu,\eta} = |\mathrm{LR}(\lambda/\nu, \eta)|.$$

REMARK 5.8. Theorem 5.7 implies that $c^{\lambda}_{\nu,\eta} = c^{\lambda}_{\eta,\nu}$.

We will use the notation $c^{\lambda}_{\nu,\eta}(\mathrm{S}) = |\mathrm{LRS}(\lambda/\nu, \eta)|$. The following theorem was proven by Sundaram in Chapter IV of her PhD thesis [**16**]. See also Corollary 3.2 of [**17**]. For stable weights it was proven by Littlewood in [**11**] and is known as the Littlewood branching rule.

THEOREM 5.9. [**16**] *Let $\lambda \in \mathrm{P}_{A_{2n-1}}$ be dominant. Then*

$$\mathrm{res}^{\mathfrak{g}}_{\mathfrak{g}^\sigma}(\mathrm{L}(\lambda)) = \bigoplus_{\substack{\underline{d}_\nu \subset \underline{d}_\lambda \\ l(\underline{d}_\nu) \leq n}} \mathrm{N}_{\lambda,\nu} \mathrm{L}(\nu)$$

where $\mathrm{N}_{\lambda,\nu} = \sum_{\underline{d}_\eta \text{ even}} c^{\lambda}_{\nu,\eta}(\mathrm{S})$.

6. On the proof of the Naito-Sagaki conjecture

To prove the conjecture, in [**15**] we construct a bijection

$$\mathrm{domres}(\lambda, \nu) \overset{1:1}{\longleftrightarrow} \bigcup_{\substack{\underline{d}_\eta \subseteq \underline{d}_\lambda; \\ \underline{d}_\eta \text{ even}}} \mathrm{LRS}(\lambda/\eta, \nu). \tag{6.1}$$

To do so we have used combinatorics of "up-down tableaux", and a result by Sundaram and Berele which is in spirit of a symplectic Robinson-Schensted Knuth correspondence. In the last section of this manuscript we provide an overview of these techniques. In what follows we present a direct bijection in the special cases of $n = 2$ and $\lambda = a_1\omega_1 + a_2\omega_2 + a_3\omega_3$, for all n. The advantage of this proof is highlighted in Remark 6.5. The following construction should provide some insight. Given a tableau $\mathscr{T} \in \mathrm{domres}(\lambda)$ we will construct a weight $\eta_{\mathscr{T}}$ with even shape $\underline{d}_{\eta_{\mathscr{T}}}$. To do this, first replace, in $\mathscr{T}$, all letters w such that $n < w \leq 2n$ by $\overline{2n - w + 1}$, and denote the resulting symplectic semi-standard tableau by $\mathrm{res}(\mathscr{T})$. Its word $w(\mathrm{res}(\mathscr{T}))$ is equal to the restricted word $\mathrm{res}(w(\mathscr{T}))$. Now, in each column of $\mathrm{res}(\mathscr{T})$, replace an entry w by a blank square if $\overline{w}$ appears to its left in the word $\mathrm{res}(w(\mathscr{T})) = w(\mathrm{res}(\mathscr{T}))$. In that case, replace the entry $\overline{w}$ by a blank square as well. Count the number of blank squares in each column, and order these squares to obtain a Young diagram of shape $\underline{d}_{\eta_{\mathscr{T}}}$.

EXAMPLE 6.1. Let $n = 2$ and

$$\mathscr{T} = \begin{array}{|c|c|c|}\hline 1 & 1 & 1 \\ \hline 2 & 2 & \\ \cline{1-2} 3 & & \\ \cline{1-1}\end{array} .$$

Then

$$\mathrm{res}(\mathscr{T}) = \begin{array}{|c|c|c|}\hline 1 & 1 & 1 \\ \hline 2 & 2 & \\ \cline{1-2} \bar{2} & & \\ \cline{1-1}\end{array}, \qquad \eta_{\mathscr{T}} = \omega_2, \text{ and } \underline{d}_{\eta_{\mathscr{T}}} = (0,1)$$

where the Young diagram $\begin{array}{|c|}\hline \ \\ \hline \ \\ \hline\end{array}$ of shape $\underline{d}_{\eta_{\mathscr{T}}}$ is obtained by replacing $\begin{array}{|c|}\hline 2 \\ \hline \bar{2} \\ \hline\end{array}$ in $\mathrm{res}(\mathscr{T})$ by blank squares.

LEMMA 6.2. *Let $\mathscr{T}$ be as above. Then the shape $\underline{d}_{\eta_{\mathscr{T}}}$ is even.*

PROOF. We will call a column standard if its entries are consecutive integers, starting with 1. The proof is by induction on the number of right-most aligned consecutive standard columns in $\mathscr{T}$, counted from right to left. For example: the tableau $\begin{array}{|c|c|c|}\hline 1 & 1 & 1 \\ \hline 2 & 2 & \\ \cline{1-2} 4 & & \\ \cline{1-1}\end{array}$ has two right-most aligned consecutive standard columns, the tableau $\begin{array}{|c|c|c|}\hline 1 & 1 & 1 \\ \hline 2 & 4 & \\ \cline{1-2} 3 & & \\ \cline{1-1}\end{array}$ has one, and the tableau $\begin{array}{|c|c|}\hline 1 & 1 \\ \hline 2 & 4 \\ \hline 3 & \\ \cline{1-1}\end{array}$ has none. Let $\mathscr{D}$ be the first column in $\mathrm{res}(\mathscr{T})$ (counted from right to left) that is not standard. Then there exists $s > 0$ such that the first s boxes of $\mathscr{D}$ are filled in with the numbers i such that $i \leq s$, and its $s+1$-th box is filled in with $\bar{l}$ for some $l \leq n$. Since $\mathrm{res}(\mathscr{T})$ has a dominant word, it must even hold that $l \leq s$. The same is true for the rest of the entries in $\mathscr{D}$, which are barred since entries are strictly increasing. The boxes in $\mathscr{D}$ with barred entries together with the boxes in $\mathscr{D}$ that have as entries their non-barred versions (they all exist, since the word of $\mathscr{T}$ is dominant) make up one of the columns of the Young diagram of shape $\underline{d}_{\eta}$. This column has an even number of boxes. Let us now ignore these entries. Let $\mathscr{C}$ be the closest column to the left of $\mathscr{D}$ that is not standard. For the induction step, we construct a new semi-standard Young tableau in which $\mathscr{C}$ is the first column to not be standard. Let $\mathscr{L}$ be the semi-standard Young tableau made up of the first right-most aligned standard columns of $\mathscr{T}$. Since the word of $\mathrm{res}(\mathscr{T})$ is dominant, we may use the non-ignored entries in $\mathscr{D}$ to construct a new tableau $\mathscr{L}'$ from $\mathscr{L}$. We do this by adding one of these boxes either at the end of $\mathscr{L}$ or at the end of a column, in such a way that the resulting arrangement is still a semi-standard Young tableau with standard columns only. For example, if $n = 3$ and $\mathscr{T} = \begin{array}{|c|c|c|c|c|}\hline 1 & 1 & 1 & 1 & 1 \\ \hline 2 & 2 & 2 & 2 & \\ \cline{1-4} 3 & 3 & 3 & & \\ \cline{1-3} 4 & 5 & & & \\ \cline{1-2} 6 & & & & \\ \cline{1-1}\end{array}$, then $\mathrm{res}(\mathscr{T}) = \begin{array}{|c|c|c|c|c|}\hline 1 & 1 & 1 & 1 & 1 \\ \hline 2 & 2 & 2 & 2 & \\ \cline{1-4} 3 & 3 & 3 & & \\ \cline{1-3} \bar{3} & \bar{2} & & & \\ \cline{1-2} \bar{1} & & & & \\ \cline{1-1}\end{array}$ has a dominant word.

Also, $\mathscr{L} = \begin{array}{|c|c|c|}\hline v1 & 1 & 1 \\ \hline 2 & 2 & \\ \cline{1-2} 3 & & \\ \cline{1-1}\end{array}$, $\mathscr{D} = \begin{array}{|c|}\hline 1 \\ \hline 2 \\ \hline 3 \\ \hline \bar{2} \\ \hline\end{array}$, $\mathscr{C} = \begin{array}{|c|}\hline 1 \\ \hline 2 \\ \hline 3 \\ \hline 4 \\ \hline 6 \\ \hline\end{array}$, $\mathscr{L}' = \begin{array}{|c|c|c|c|}\hline 1 & 1 & 1 & 1 \\ \hline 2 & 2 & & \\ \cline{1-2} 3 & 3 & & \\ \cline{1-2}\end{array}$, and the ignored boxes in $\mathscr{D}$ are $\begin{array}{|c|}\hline 2 \\ \hline\end{array}$ and $\begin{array}{|c|}\hline \bar{2} \\ \hline\end{array}$. Define a new tableau $\mathrm{res}(\mathscr{T})'$ by concatenating all the columns in

$\mathrm{res}(\mathscr{T})$, from left to right and up to $\mathscr{D}$, with $\mathscr{L}'$. It follows from the construction that $\mathrm{res}(\mathscr{T})'$ has a dominant word and is semi-standard. In the previous example, this new tableau is $\mathrm{res}(\mathscr{T})' = \begin{array}{|c|c|c|c|c|}\hline 1&1&1&1&1\\ \hline 2&2&2& \multicolumn{2}{c}{} \\ \cline{1-3} 3&3&3& \multicolumn{2}{c}{} \\ \cline{1-3} \bar{3} & \multicolumn{4}{c}{} \\ \cline{1-1} \bar{1} & \multicolumn{4}{c}{} \\ \cline{1-1} \end{array}$. We may then apply the induction hypothesis. This concludes the proof. □

REMARK 6.3. Lemma 6.2 is only true for $\mathscr{T}$ a semi-standard Young tableau. Consider for example $n = 2$ and the key (as in [**7**], [**19**]) $\mathscr{T} = \boxed{1}\boxed{4}\boxed{1}$. Then $\mathrm{res}(\mathscr{T}) = \boxed{1}\boxed{\bar{1}}\boxed{1}$ is dominant, however, the shape $\eta_{\mathscr{T}} = (1,0)$, is not even.

LEMMA 6.4. *If $\lambda = a_1\omega_1 + a_2\omega_2 + a_3\omega_3$ and ν and η are dominant weights in* $\mathrm{P}_{A_{2n-1}}$ *such that $\underline{d}_\eta$ is even and both $\underline{d}_\eta$ and $\underline{d}_\nu$ are contained in $\underline{d}_\lambda$, then*

$$c^\lambda_{\nu,\eta}(\mathrm{S}) = c^\lambda_{\nu,\eta}.$$

PROOF. Assume that $\mathscr{T}$ is a Littlewood-Richardson tableau of skew shape λ/ν and weight η that is not Sundaram. This means that there is at least a "1" in the third row. Since $\mathscr{T}$ is semi-standard, all the "1"'s in the third row must appear left-most and all next to one another. But since $\underline{d}_\eta$ is even, for each of these 1's there must exist a "2" that appears before it, in the word reading order. But this means, since the word is dominant, that there must have appeared a "1" before this "2". This contradicts the evenness of $\underline{d}_\eta$! □

PROOF OF (6.1) FOR $n = 2$ OR $\lambda = a_1\omega_1 + a_2\omega_2 + a_3\omega_3$. Let $\mathscr{T} \in \mathrm{domres}(\lambda,\nu)$, and set

$$m_1 = \#\text{ columns in } \mathscr{T} \text{ of the form } \begin{array}{|c|}\hline 1\\ \hline 2\\ \hline \bar{1}\\ \hline\end{array},$$

$$m_2 = \#\text{ columns in } \mathscr{T} \text{ of the form } \begin{array}{|c|}\hline 1\\ \hline 2\\ \hline \bar{2}\\ \hline\end{array},$$

$$m_3 = \#\text{ columns in } \mathscr{T} \text{ of the form } \begin{array}{|c|}\hline 1\\ \hline \bar{1}\\ \hline\end{array} \text{ and}$$

$$m_4 = \#\text{ columns in } \mathscr{T} \text{ of the form } \begin{array}{|c|}\hline 1\\ \hline 2\\ \hline 3\\ \hline\end{array}.$$

If $n = 2$ then $m_2 = m_4$. It follows from semi-standardness and dominance of $\mathrm{res}(w(\mathscr{T}))$ that these are the only possible columns aside from columns of the form $\begin{array}{|c|}\hline 1\\ \hline 2\\ \hline\end{array}$ and the single box columns $\boxed{1}$. Note that since $\mathrm{res}(w(\mathscr{T}))$ is dominant, the following condition holds

$$m_1 \leq \lambda_1 - \lambda_2. \tag{6.2}$$

Actually (6.2) is equivalent to the dominance of $\mathrm{res}(w(\mathscr{T}))$, once the λ_i's are set. We assign to $\mathscr{T}$ a Littlewood-Richardson tableau $\varphi(\mathscr{T}) \in \mathrm{LR}(\lambda/\eta_{\mathscr{T}},\nu)$. By Lemma 6.2, $\eta_{\mathscr{T}}$ is even.

Write $\lambda = \lambda_1\varepsilon_1 + \lambda_2\varepsilon_2 + \lambda_3\varepsilon_3$. Note that $\eta_{\mathscr{T}}$ has $m = m_1 + m_2 + m_3$ columns, all of length 2. Fill in the first $\lambda_1 - m$ right-most boxes in the first row with a "1" , and the first $\lambda_2 - b$ right-most boxes in the second row with a "2". If $n \neq 2$ fill in the first m_4 right-most entries of the third row with a "3". Then fill in the next right-most m_1 entries in the third row with a "2", and the remaining entries with a "1". The resulting tableau $\varphi(\mathscr{T})$ is a Littlewood-Richardson tableau by construction. Now we will show that any element in $\bigcup_{\substack{\underline{d}_\eta \subset \underline{d}_\lambda; \\ \eta \text{ even}}} \mathrm{LR}(\lambda/\eta, \nu)$ can be obtained in this way. Let $\eta \in \mathrm{P}^+_{A_{2n-1}}$ have an even shape $\underline{d}_\eta \subset \underline{d}_\lambda$ (this means $\underline{d}_\eta$ consists of size 2 columns) and let $\mathscr{L} \in \mathrm{LR}(\lambda/\eta, \nu)$. Set

$$l_1 = \# \text{ of 1's in } \mathscr{L}$$
$$l_2 = \# \text{ of 2's in } \mathscr{L}, \text{ and}$$
$$b = \# \text{ of columns of } \eta.$$

Note that this information determines $\mathscr{L}$ together with λ. In view of the previous construction, we would like to find non-negative integers m_1, m_2, m_3 and m_4 such that

$$l_1 = \lambda_1 - m_1 - m_3 \tag{6.3}$$
$$l_2 = \lambda_2 - m_3 - m_2 \tag{6.4}$$
$$m = m_1 + m_2 + m_3 \tag{6.5}$$
$$m_4 = \# \text{ of 3's in } \mathscr{L}. \tag{6.6}$$

Since $\mathscr{L}$ has a dominant word, we have

$$l_2 - (\lambda_2 - b) \leq \lambda_1 - \lambda_2. \tag{6.7}$$

Substituting (6.4) and (6.5) in (6.7) we get precisely (6.2), so if we find solutions m_1, m_2, m_3, the resulting tableau will automatically belong to domres(λ, ν). Claim 1 below assures that this is the case. □

CLAIM 1. The system determined by (6.3), (6.4), and (6.5), has integer solutions (possibly zero) b_1, b_2 and b_3 if and only if

$$m \geq \lambda_1 - l_1 \tag{6.8}$$
$$m \geq \lambda_2 - l_2 \tag{6.9}$$
$$\lambda_1 + \lambda_2 \geq m + l_1 + l_2. \tag{6.10}$$

IDEA OF PROOF. The proof is a straightforward solution of the system of equations determined by (6.3) - (6.5). □

It follows from the definitions that these conditions are satisfied by all elements of

$$\bigcup_{\substack{\underline{d}_\eta \subset \underline{d}_\lambda; \\ \underline{d}_\eta \text{ even}}} \mathrm{LR}(\lambda/\eta, \nu). \tag{6.11}$$

REMARK 6.5. A first direct consequence of our proof is a description of the left and right hand sides of 6.1 as lattice points of two convex polytopes. In fact, the inequalities and variables used to describe (6.11) figure in Section 3 of [**13**]. In [**15**] we exhibit this as a general phenomenon. The approach taken in [**15**] to prove Theorem 4.2 is to construct a combinatorial bijection (6.1) as is explained in Section 6. It would be interesting to know if there is a bijection, for all n, (as in Section 6) that comes from a unimodular equivalence.

7. Symplectic combinatorics of King, Berele and Sundaram

In this section, which is somewhat independent from the rest, we provide an overview of the combinatorics involved in the proof of (6.1).

7.1. King tableaux. Let n be a positive integer and λ a Young diagram of at most $2n$ rows. In [**6**] King used the following alphabet:

$$\mathcal{K}_n = \{1 < \bar{1} < \cdots < n < \bar{n}\}.$$

We denote the set of all semi-standard Young tableaux in this alphabet by $\mathrm{SSYT}_{\mathcal{K}_n}(\lambda)$. Recall that this means that the entries are weakly increasing along the rows and weakly increasing along the columns.

DEFINITION 7.1. A semistandard Young tableau in the alphabet $\mathcal{K}_n$ of shape a Young diagram of at most n rows, is a **King tableau**, if, in row i, all appearing entries are greater than or equal to i. We will denote the set of all King tableaux of shape ν by $\mathrm{King}(\nu)$. For $\mathrm{T} \in \mathrm{SSYT}_{\mathcal{K}_n}(\lambda)$ we say that all entries k in row j smaller than j are **symplectic violations**. For a semi-standard tableau $\mathrm{T} \in \mathrm{SSYT}_{\mathcal{K}_n}(\lambda)$ of shape of at most $2n$ rows, we will use the same terminology, and by convention will call symplectic violations any entries in a row $j > n$ larger than n.

REMARK 7.2. Let V be a complex vector space of even dimension $2n$ with a skew-symmetric bilinear form $\langle -, - \rangle$ and let

$$\mathrm{Sp}(2n, \mathbb{C}) = \{\mathrm{A} \in \mathrm{GL}(\mathrm{V}) : \langle \mathrm{A}\, v, \mathrm{A}\, v\rangle = \langle v, v\rangle\}$$

be the symplectic group of transformations invariant under the form $\langle -, - \rangle$. Under an appropriate choice of basis, the set

$$\mathrm{H} = \mathrm{diag}(x_1, x_1^{-1}, \ldots, x_n, x_n^{-1})$$

of diagonal matrices is a maximal torus in $\mathrm{Sp}(2n, \mathbb{C})$. Finite dimensional irreducible representations of $\mathrm{Sp}(2n, \mathbb{C})$ are parametrised by shapes of at most n parts/rows. King tableaux are used to express the characters of these representations: if $\rho : \mathrm{Sp}(2n, \mathbb{C}) \to \mathrm{L}(\lambda)$ is the irreducible representation of highest weight the shape λ, then its character, the trace $\mathrm{sp}_\lambda = \mathrm{tr}(\rho)$ can be expressed as a polynomial function on H as

$$\mathrm{sp}_\lambda = \sum_{\mathrm{T} \in \mathrm{King}(\lambda)} \mathrm{wt}(\mathrm{T}) \tag{7.1}$$

where, if $w(\mathrm{T}) = w_1 \cdots w_r$ denotes the word of T, read from right to left and top to bottom,

$$\mathrm{wt}(\mathrm{T}) = \mathrm{sgn}(w_1) \cdots \mathrm{sgn}(w_r)$$

with $\mathrm{sgn}(i) = x_i$ and $\mathrm{sgn}(\bar{i}) = x_i^{-1}$ for $i = 1, \ldots, n$. It is in this sense that King tableaux are analogues of usual semi-standard Young tableaux, which are used in the representation theory of $\mathrm{SL}(n, \mathbb{C})$. The formula (7.1) is the analogue of the original formula for the usual Schur functions/characters of representations of $\mathrm{SL}(n, \mathbb{C})$, where $\mathrm{SSYT}_{\mathcal{K}_n}(\lambda)$ is replaced by the usual set of semi-standard Young tableaux in the alphabet $1 < \ldots < n$.

An **up-down sequence of shapes** is a sequence $(\varnothing = \mu_0, \mu_1, \cdots, \mu_k)$ where each μ_i is a Young diagram and such that for every $1 \le i \le n$, the Young diagram μ_{i+1} is obtained from μ_i by adding or removing one box, and such that each μ_i has, at most, n rows.

EXAMPLE 7.3. An up-down sequence of shapes of length 10:

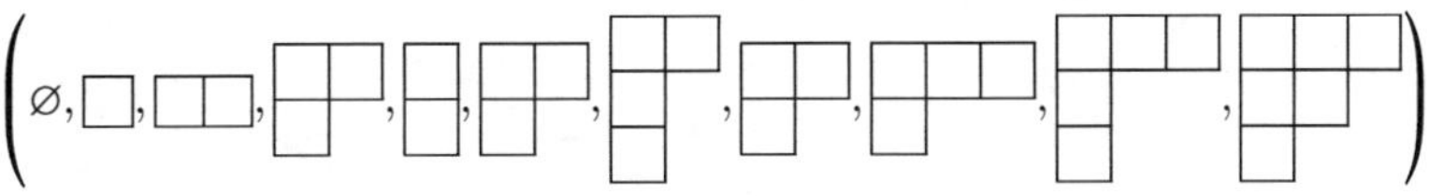

7.2. Berele's algorithm. Berele's theorem reads as follows.

THEOREM 7.4 (Berele). *There is a bijection between the set of words of length k in the alphabet $\mathcal{K}_n$ and pairs* $(\mathrm{P}_\mu, \mathrm{S})$*, where* P_μ *is a King tableau of shape* μ *and* S *is an up-down sequence of length* k *and final shape* μ*.*

This bijection, given by what we refer to as the Berele algorithm, assigns, to each word in the alphabet $\mathcal{K}_n$, a King tableau and an up-down sequence of shapes in the following way. The idea is to follow the usual Robinson-Schensted-Knuth algorithm - as long as there are no symplectic violations. Set $i = 0$, $\mu^0 = \varnothing$ and $\mathrm{P}^0 = \varnothing$. Repeat the following procedure until $i = r$.

(1) Set $i \to i+1$ and $\mathrm{P} = \mathrm{P}^{i-1} \leftarrow w_i$ (the row-bumping of w_i into P^{i-1}).
(2) If P is a King tableau, then set $\mathrm{P}^i = \mathrm{P}$, $\mu^i = \mathrm{sh(P)}$ and go to step (1). Otherwise, there must be a symplectic violation. It is not complicated to realise that the first time such a violation is introduced in the tableau will happen precisely when a letter i bumps an $\bar{i}$ out of row i and into row $i+1$. If this is the case, bump the entries that you can without producing a symplectic violation, and then, when you reach the first point at which an i is about to bump a $\bar{i}$ out of row i and into row $i+1$, stop the bumping, that is, erase this i and, instead, do the following:
   (a) Replace the $\bar{i}$ by an i.
   (b) Replace the first i in that column by a black dot.
   (c) Apply jeu-de taquin to the dot, sliding it east and south, out of the tableau.

   Set $\mathrm{P}^i = \mathrm{P}$, $\mu^i = \mathrm{sh(P)}$. If $i < k$, go back to Step (1). Else stop the procedure.

For a proof we refer the reader to Sundaram's thesis [**16**] or to the original paper by Berele [**1**].

EXAMPLE 7.5. Let $w = \bar{1}321\bar{2}21211$. If we apply Berele's algorithm to w, the resulting sequence of King tableaux and up-down sequence of shapes are as follows:

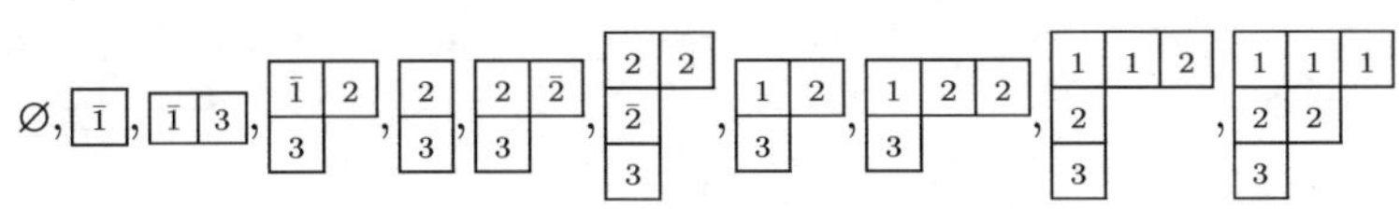

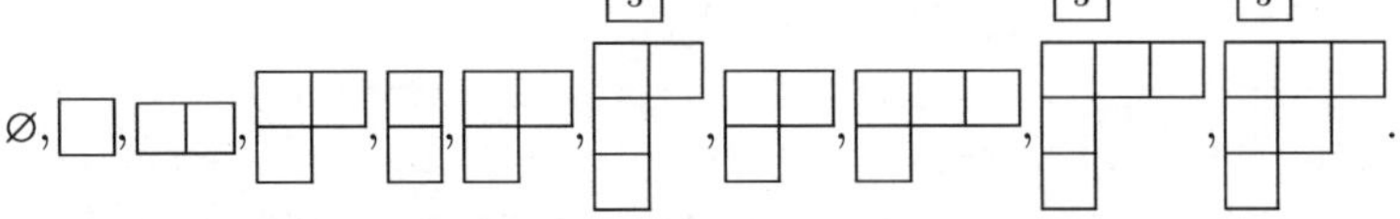

In her thesis, Sundaram gave a very useful refinement of Berele's bijection. The idea is as follows. To each up-down tableau, one may associate two things: a standard Young tableau and a Littlewood-Richardson Sundaram tableau of the same shape. The contents of this section are in their majority already exposed in [**15**] but we provide a quick description of the comfort of the reader.

7.3. The Burge correspondence. We recall some facts about the combinatorics of two-line arrangements which we need in this next section.

DEFINITION 7.6. A *special two-line array* is a two-line array of distinct positive integers

$$\begin{bmatrix} j_1 & \cdots & j_r \\ i_1 & \cdots & i_r \end{bmatrix}$$

such that

1. $j_1 < \cdots < j_r$
2. $j_s > i_s, s \in \{1, \cdots, r\}$.

Consider a special two-line array as above and let $s_1, \cdots, s_r$ be the re-ordering of the index set $\{1, \cdots, r\}$ such that $i_{s_1} < \cdots < i_{s_r}$. We can obtain a strict Young tableau from our array by column bumping its entries:

$$j_{s_1} \to \cdots \to j_{s_r} \to i_1 \to \cdots \to i_r \to \varnothing,$$

where by strict we mean that it is semi-standard and that all its entries are distinct.

EXAMPLE 7.7. Consider the two-line arrangement

$$\begin{bmatrix} 6 & 10 \\ 3 & 1 \end{bmatrix}$$

We have

$$1 \to \varnothing = \begin{array}{|c|}\hline 1 \\ \hline\end{array}, \quad 3 \to \begin{array}{|c|}\hline 1 \\ \hline\end{array} = \begin{array}{|c|}\hline 1 \\ \hline 3 \\ \hline\end{array}, \quad 6 \to \begin{array}{|c|}\hline 1 \\ \hline 3 \\ \hline\end{array} = \begin{array}{|c|}\hline 1 \\ \hline 3 \\ \hline 6 \\ \hline\end{array}, \quad 10 \to \begin{array}{|c|}\hline 1 \\ \hline 3 \\ \hline 6 \\ \hline\end{array} = \begin{array}{|c|}\hline 1 \\ \hline 3 \\ \hline 6 \\ \hline 10 \\ \hline\end{array}$$

which is the same as

$$10 \to 6 \to 3 \to 1 \to \varnothing = \begin{array}{|c|}\hline 1 \\ \hline 3 \\ \hline 6 \\ \hline 10 \\ \hline\end{array}.$$

The following theorem is known as the Burge correspondence [**3**] (see also Theorem 3.31 in [**16**] and Lemma 10.7 in Sundaram's thesis [**16**]).

THEOREM 7.8. *The assignment above defines a bijection between special two-line arrangements and standard Young tableaux of even shape.*

7.4. Sundaram's refinement. We may associate a standard Young tableau to every shape in a given up-down sequence, ending up with a standard Young tableau of shape μ and a standard Young tableau of even shape ν. We do this in the spirit of the Robinson-Schensted-Knuth correspondence: At each step j of the sequence of shapes, write a "j" inside the new box if a new box was added at this step, and, if a box was removed, column-bump the entry corresponding to that box out of the standard Young tableau that you had at step $j-1$. This procedure defines a standard Young tableau of shape μ which will be obtained at the end of the sequence. We will denote it by Q^p and call it the partial Q-symbol. To obtain the even permutation we do the following. Save the step j together with the bumped-out entry r_j in a special two-line array as $\begin{bmatrix} j \\ r_j \end{bmatrix}$ and concatenate the two-line arrays obtained every time there is a box-removal, with the first one always left-most. In the end we get a special two-line array. We use the method described in Section 7.3 to produce from it a standard Young tableau of even shape and denote it by E. We have:

$$\mathrm{E} = j_{s_1} \to \cdots \to j_{s_r} \to i_1 \to \cdots \to i_r \to \varnothing.$$

Now we produce one last standard Young tableau, the *final* Q-symbol:

$$\mathrm{Q} = j_{s_1} \to \cdots \to j_{s_r} \to i_1 \to \cdots \to i_r \to \mathrm{Q}^p\,.$$

The first step towards the refinement is the following result.

PROPOSITION 7.9 (Sundaram [**16**]). *Let* $\mathrm{T} \in \mathrm{SSYT}_{\mathcal{K}_n}(\lambda)$. *Then the final* Q-*symbol for the up-down sequence associated to the row word* $w_r(\mathrm{T})$ *of* T *has shape* λ. *In fact, this final* Q-*symbol only depends on* λ.

Recall that Littlewood-Richardson Sundaram tableaux are certain skew tableaux, filled with numbers from 1 to n, and such that its word is a dominant path, with weight, a shape consisting only of even columns. Such a shape is said to be **even**. We will denote the set of Littlewood-Richardson Sundaram tableaux of skew shape λ/μ and weight ν by $\mathrm{LRS}(\lambda/\mu,\nu)$. When ν needs not be specified we denote the larger set by $\mathrm{LRS}(\lambda/\mu)$. We also say that elements in this set have shape λ, when μ is not immediately relevant, and denote the corresponding set by $\mathrm{LRS}(\lambda)$. The first step towards the refinement is the following theorem.

THEOREM 7.10 (Sundaram). *There is a one-to-one correspondence between the set of up-down sequences of shapes* $\underline{\mu}$ *with final shape* μ, *and the union* $\bigcup_{\lambda\supset\mu} \mathrm{LRS}(\lambda/\mu)$.

To assign a Littlewood-Richardson tableau to a final Q-symbol, one " fills up" the skew shape λ/μ as follows: for each entry j in $\mathrm{E_T}$, let r_j be the row to which it belongs (in $\mathrm{E_T}$). Write this number in the skew shape λ/μ in the row of $\mathrm{Q_T}$ where j lies. In the proof of Theorem 8.11 (and Theorem 9.4) in [**16**], it is shown that this process yields a Littlewood Richardson Sundaram tableau $\Phi(\mathrm{T}) \in \mathrm{LRS}(\lambda/\mu,\eta)$ In fact, the following theorem holds (it is stated and proven in Theorems 8.14 and 9.4 of [**16**]).

THEOREM 7.11 (Sundaram). *Berele's algorithm gives a bijection between the set* $\mathrm{SSYT}_{\mathcal{K}_n}(\lambda)$ *of semi-standard Young tableaux of shape* λ *in the alphabet* $\{1 < \bar{1} < \cdots < n < \bar{n}\}$ *and the set* $\bigcup_{\substack{\mu\subset\lambda\\ \nu\ even}} \mathrm{King}_n(\mu) \times \mathrm{LRS}(\lambda/\mu,\nu)$.

The details of the proof can be found in [**16**]. The most important step in the proof of (6.1) is to assign to each element in $\mathrm{domres}(\lambda,\mu)$ an up-down sequence of shapes just as in this section, and to use Theorem 7.11.

Acknowledgments

The author would like to thank Stéphane Gaussent for introducing her to this problem, also Nicolas Ressayre and Bea Schumann for some very interesting conversations and a fruitful collaboration, respectively. Thanks to Daisuke Sagaki for the nice correspondence and to Peter Littelmann for his support and guiding remarks. Thank you as well to Michael Ehrig, Stéphane Gaussent and Noah White for their comments on a preliminary version of this paper. This manuscript is the full version of the extended abstract [**18**] which was presented at FPSAC'17 as a poster.

The last section is mostly expository. One of the first talks that the author gave on the results exposed in this paper was at the second reunion of MMM (Matemáticos Mexicanos en el Mundo).

References

[1] Allan Berele, *A Schensted-type correspondence for the symplectic group*, J. Combin. Theory Ser. A **43** (1986), no. 2, 320–328, DOI 10.1016/0097-3165(86)90070-1. MR867655

[2] A. D. Berenstein and A. V. Zelevinsky, *Tensor product multiplicities and convex polytopes in partition space*, J. Geom. Phys. **5** (1988), no. 3, 453–472, DOI 10.1016/0393-0440(88)90033-2. MR1048510

[3] William H. Burge, *Four correspondences between graphs and generalized Young tableaux*, J. Combinatorial Theory Ser. A **17** (1974), 12–30, DOI 10.1016/0097-3165(74)90024-7. MR351884

[4] William Fulton, *Young tableaux*, London Mathematical Society Student Texts, vol. 35, Cambridge University Press, Cambridge, 1997. With applications to representation theory and geometry. MR1464693

[5] William Fulton and Joe Harris, *Representation theory*, Graduate Texts in Mathematics, vol. 129, Springer-Verlag, New York, 1991. A first course; Readings in Mathematics, DOI 10.1007/978-1-4612-0979-9. MR1153249

[6] R. C. King, *Weight multiplicities for the classical groups*, Group theoretical methods in physics (Fourth Internat. Colloq., Nijmegen, 1975), Springer, Berlin, 1976, pp. 490–499. Lecture Notes in Phys., Vol. 50. MR0480895

[7] Stéphane Gaussent, Peter Littelmann, and An Hoa Nguyen, *Knuth relations, tableaux and MV-cycles*, J. Ramanujan Math. Soc. **28A** (2013), 191–219. MR3115193

[8] Victor G. Kac, *Infinite-dimensional Lie algebras*, 3rd ed., Cambridge University Press, Cambridge, 1990, DOI 10.1017/CBO9780511626234. MR1104219

[9] Allen Knutson and Terence Tao, *The honeycomb model of* $\mathrm{GL}_n(\mathbf{C})$ *tensor products. I. Proof of the saturation conjecture*, J. Amer. Math. Soc. **12** (1999), no. 4, 1055–1090, DOI 10.1090/S0894-0347-99-00299-4. MR1671451

[10] Peter Littelmann, *Paths and root operators in representation theory*, Ann. of Math. (2) **142** (1995), no. 3, 499–525, DOI 10.2307/2118553. MR1356780

[11] Dudley E. Littlewood, *The Theory of Group Characters and Matrix Representations of Groups*, Oxford University Press, New York, 1940. MR0002127

[12] Satoshi Naito and Daisuke Sagaki, *An approach to the branching rule from* $\mathfrak{sl}_{2n}(\mathbb{C})$ *to* $\mathfrak{sp}_{2n}(\mathbb{C})$ *via Littelmann's path model*, J. Algebra **286** (2005), no. 1, 187–212, DOI 10.1016/j.jalgebra.2004.12.014. MR2124814

[13] Igor Pak and Ernesto Vallejo, *Combinatorics and geometry of Littlewood-Richardson cones*, European J. Combin. **26** (2005), no. 6, 995–1008, DOI 10.1016/j.ejc.2004.06.008. MR2143205

[14] B. Pasquier and N. Ressayre, *The saturation property for branching rules—examples*, Exp. Math. **22** (2013), no. 3, 299–312, DOI 10.1080/10586458.2013.813746. MR3171094

[15] Bea Schumann and Jacinta Torres, *A non-Levi branching rule in terms of Littelmann paths*, Proc. Lond. Math. Soc. (3) **117** (2018), no. 5, 1077–1100, DOI 10.1112/plms.12175. MR3877772

[16] Sheila Sundaram, *ON THE COMBINATORICS OF REPRESENTATIONS OF THE SYMPLECTIC GROUP*, ProQuest LLC, Ann Arbor, MI, 1986. Thesis (Ph.D.)–Massachusetts Institute of Technology. MR2941115

[17] Sheila Sundaram, *Tableaux in the representation theory of the classical Lie groups*, Invariant theory and tableaux (Minneapolis, MN, 1988), IMA Vol. Math. Appl., vol. 19, Springer, New York, 1990, pp. 191–225. MR1035496

[18] Jacinta Torres, *On a conjecture of Naito-Sagaki: Littelmann paths and Littlewood-Richardson Sundaram tableaux* (English, with English and French summaries), Sém. Lothar. Combin. **78B** (2017), Art. 53, 11. MR3678635

[19] Jacinta Torres, *Word reading is a crystal morphism*, Transform. Groups **21** (2016), no. 2, 577–591, DOI 10.1007/s00031-016-9364-y. MR3492049

[20] È. B. Vinberg, *The Weyl group of a graded Lie algebra* (Russian), Izv. Akad. Nauk SSSR Ser. Mat. **40** (1976), no. 3, 488–526, 709. MR0430168

Institute of Mathematics of the Jagiellonian University
Email address: jacinta.torres@uj.edu.pl

SELECTED PUBLISHED TITLES IN THIS SUBSERIES

775 **Fernando Galaz-García, Cecilia González-Tokman, and Juan Carlos Pardo Millán, Editors,** Mexican Mathematicians in the World, 2021

774 **Randall J. Swift, Alan Krinik, Jennifer M. Switkes, and Jason H. Park, Editors,** Stochastic Processes and Functional Analysis, 2021

773 **Nicholas R. Baeth, Thiago H. Freitas, Graham J. Leuschke, and Victor H. Jorge Pérez, Editors,** Communicative Algebra, 2021

772 **Anatoly M. Vershik, Victor M. Buchstaber, and Andrey V. Malyutin, Editors,** Topology, Geometry, and Dynamics, 2021

771 **Nicolás Andruskiewitsch, Gongxiang Liu, Susan Montgomery, and Yinhuo Zhang, Editors,** Hopf Algebras, Tensor Categories and Related Topics, 2021

770 **Stéphane Ballet, Gaetan Bisson, and Irene Bouw, Editors,** Arithmetic, Geometry, Cryptography and Coding Theory, 2021

769 **Kiyoshi Igusa, Alex Martsinkovsky, and Gordana Todorov, Editors,** Representations of Algebras, Geometry and Physics, 2021

768 **Dražen Adamović, Andrej Dujella, Antun Milas, and Pavle Pandžić, Editors,** Lie Groups, Number Theory, and Vertex Algebras, 2021

767 **Moshe Jarden and Tony Shaska, Editors,** Abelian Varieties and Number Theory, 2021

766 **Paola Comparin, Eduardo Esteves, Herbert Lange, Sebastián Reyes-Carocca, and Rubí E. Rodríguez, Editors,** Geometry at the Frontier, 2021

765 **Michael Aschbacher,** Quaternion Fusion Packets, 2021

764 **Gabriel Cunningham, Mark Mixer, and Egon Schulte, Editors,** Polytopes and Discrete Geometry, 2021

763 **Tyler J. Jarvis and Nathan Priddis, Editors,** Singularities, Mirror Symmetry, and the Gauged Linear Sigma Model, 2021

762 **Atsushi Ichino and Kartik Prasanna,** Periods of Quaternionic Shimura Varieties. I., 2021

761 **Ibrahim Assem, Christof Geiß, and Sonia Trepode, Editors,** Advances in Representation Theory of Algebras, 2021

760 **Olivier Collin, Stefan Friedl, Cameron Gordon, Stephan Tillmann, and Liam Watson, Editors,** Characters in Low-Dimensional Topology, 2020

759 **Omayra Ortega, Emille Davie Lawrence, and Edray Herber Goins, Editors,** The Golden Anniversary Celebration of the National Association of Mathematicians, 2020

758 **Jan Šťovíček and Jan Trlifaj, Editors,** Representation Theory and Beyond, 2020

757 **Kaïs Ammari and Stéphane Gerbi, Editors,** Identification and Control: Some New Challenges, 2020

756 **Joeri Van der Veken, Alfonso Carriazo, Ivko Dimitrić, Yun Myung Oh, Bogdan D. Suceavă, and Luc Vrancken, Editors,** Geometry of Submanifolds, 2020

755 **Marion Scheepers and Ondřej Zindulka, Editors,** Centenary of the Borel Conjecture, 2020

754 **Susanne C. Brenner, Igor Shparlinski, Chi-Wang Shu, and Daniel B. Szyld, Editors,** 75 Years of Mathematics of Computation, 2020

753 **Matthew Krauel, Michael Tuite, and Gaywalee Yamskulna, Editors,** Vertex Operator Algebras, Number Theory and Related Topics, 2020

752 **Samuel Coskey and Grigor Sargsyan, Editors,** Trends in Set Theory, 2020

751 **Ashish K. Srivastava, André Leroy, Ivo Herzog, and Pedro A. Guil Asensio, Editors,** Categorical, Homological and Combinatorial Methods in Algebra, 2020

For a complete list of titles in this series, visit the
AMS Bookstore at **www.ams.org/bookstore/conmseries/**.